Association for Women in Mathematics Series

Volume 39

Series Editor

Kristin Lauter, Meta AI, Seattle, USA

Focusing on the groundbreaking work of women in mathematics past, present, and future, Springer's Association for Women in Mathematics Series presents the latest research and proceedings of conferences worldwide organized by the Association for Women in Mathematics (AWM). All works are peer-reviewed to meet the highest standards of scientific literature, while presenting topics at the cutting edge of pure and applied mathematics, as well as in the areas of mathematical education and history. Since its inception in 1971, The Association for Women in Mathematics has been a non-profit organization designed to help encourage women and girls to study and pursue active careers in mathematics and the mathematical sciences and to promote equal opportunity and equal treatment of women and girls in the mathematical sciences. Currently, the organization represents more than 3000 members and 200 institutions constituting a broad spectrum of the mathematical community in the United States and around the world.

Titles from this series are indexed by Scopus.

Shabnam Akhtari · Yu-Ru Liu · Rachel Newton ·
Jennifer Park
Editors

Research Directions in Number Theory

Women in Numbers VI

ASSOCIATION FOR WOMEN IN MATHEMATICS

Springer

Editors
Shabnam Akhtari
Department of Mathematics
Pennsylvania State University
State College, PA, USA

Yu-Ru Liu
Department of Pure Mathematics
University of Waterloo
Waterloo, ON, Canada

Rachel Newton
Department of Mathematics
King's College London
London, UK

Jennifer Park
Department of Mathematics
The Ohio State University
Columbus, OH, USA

ISSN 2364-5733 ISSN 2364-5741 (electronic)
Association for Women in Mathematics Series
ISBN 978-3-032-11182-1 ISBN 978-3-032-11183-8 (eBook)
https://doi.org/10.1007/978-3-032-11183-8

This work was supported by UK Research and Innovation.

This Springer imprint is published by the registered company Springer Nature Switzerland AG
The registered company address is: Gewerbestrasse 11, 6330 Cham, Switzerland

Preface

"The Women in Numbers (WIN) network was created in 2008 for the purpose of increasing the number of active women researchers in number theory. For this purpose, WIN sponsors regular conferences, taking place approximately every three years, where women scholars gather to collaborate on cutting-edge research in the field and produce publishable scientific results. The WIN workshops provide an ongoing forum for involving each new generation of junior faculty and graduate students in state-of-the-art research in number theory. The WIN6 workshop was organized in partnership with the Clay Mathematics Institute and held at the Banff International Research Station, in Alberta, Canada, from March 26 to March 31, 2023."

State College, USA — Shabnam Akhtari
Waterloo, Canada — Yu-Ru Liu
London, UK — Rachel Newton
Columbus, USA — Jennifer Park

Acknowledgements Besides the Banff International Research Station (BIRS) for providing excellent research facilities for participants during the Women in Numbers workshop in 2023, we are grateful to the following organizations for their support: The National Science Foundation (NSF), the Number Theory Foundation, the Pacific Institute for the Mathematical Sciences (PIMS), Journal of Number Theory, and the Clay Mathematics Institute.

Contents

On Rational Points on Classifying Stacks and Malle's Conjecture

Shabnam Akhtari, Jennifer Park, Marta Pieropan, and Soumya Sankar

Abstract In this expository article, we compare Malle's conjecture on counting number fields of bounded discriminant with recent conjectures of Ellenberg–Satriano–Zureick-Brown and Darda–Yasuda on counting points of bounded height on classifying stacks. We illustrate the comparisons via the classifying stacks $B(\mathbb{Z}/n\mathbb{Z})$ and $B\mu_n$.

1 Introduction

In 2004, Malle [Mal02, Mal04] put forward a conjecture for the asymptotic growth of the count of number field extensions of bounded discriminant and fixed Galois group. Field extensions of fixed Galois group correspond to certain points on suitable classifying stacks. Recently, Ellenberg et al. [ESZB23] proposed a framework for heights on stacks and a conjecture for the order of growth of the number of rational points of bounded height on stacks that is compatible with Manin's conjecture [FMT89, BM90] for varieties. They interpret Malle's conjecture as a special case of their conjecture in the case of classifying stacks of finite constant group schemes. Later, Darda and Yasuda [DY23, DY22a] proposed a different framework for heights and an asymptotic formula that extends Malle's conjecture to classifying stacks of not necessarily constant finite group schemes.

S. Akhtari (✉)
Department of Mathematics, Pennsylvania State University, University Park, PA, USA
e-mail: akhtari@psu.edu

J. Park
Department of Mathematics, The Ohio State University, Columbus, OH, USA
e-mail: park.2720@osu.edu

M. Pieropan · S. Sankar
Utrecht University, Mathematical Institute, Utrecht, The Netherlands
e-mail: m.pieropan@uu.nl

S. Sankar
e-mail: s.sankar@uu.nl

S. Akhtari et al. (eds.), *Research Directions in Number Theory*, Association for Women in Mathematics Series 39, https://doi.org/10.1007/978-3-032-11183-8_1

The goal of this paper is to clarify the relation between these three conjectures over number fields focusing on the special cases of $B(\mathbb{Z}/n\mathbb{Z})$ and $B\mu_n$. While several of the results and conjectures mentioned may be generalized to global function fields, we will not consider that case in this article. A working knowledge of algebraic number theory and familiarity with the basic language of schemes is required, but the reader is not expected to have any background on stacks.

The structure of the paper is as follows. In Sect. 2, we start by recalling the theory of classifying stacks of finite group schemes. We then describe in detail the sets of rational points on $B\mu_n$ in Sect. 2.3 and on $B(\mathbb{Z}/n\mathbb{Z})$ in Sect. 2.4. In Sect. 3, we summarize the literature of heights on classifying stacks of finite groups, with particular focus on the [ESZB23] framework. Finally, in Sect. 4, we overview the literature on Malle's conjecture and highlight the connections to the two conjectures for stacks and to various results in the same spirit. Along the whole paper, the concepts that are introduced, though often more general, are illustrated using the examples of $B(\mathbb{Z}/n\mathbb{Z})$ and $B\mu_n$.

1.1 Notation

Throughout the paper, we will use K to denote a number field and $\mathcal{O}_K$ to denote the ring of integers of K. We will use M_K to denote the set of places of K. For a finite $v \in M_K$, q_v will denote the size of the residue field at v. If $v \in M_K$ is infinite, q_v will denote the Euler number e. We will use S to denote a scheme, and in most cases, we will only be interested in S of the form Spec $\mathbb{Z}$, Spec $\mathcal{O}_K$, or Spec K.

A group scheme over S is a group object in the category of schemes. Throughout this article, we will assume that our group schemes are finite and smooth. In particular, they are of the form Spec A for some ring A. If a group scheme G_S (or just G when the base scheme is clear from context) corresponds to a constant sheaf of abelian groups, it is called a constant group scheme. A finite constant group scheme thus comes from some finite group G and we will denote it by $\underline{G}$.

Throughout this article, we will fix a positive integer n. Let $\mathbb{Z}/n\mathbb{Z}$ denote the additive cyclic group of order n, and μ_n the multiplicative group scheme of order n. Their group scheme structures are explained in Sect. 2. For every positive integer m, we will use ζ_m to denote a primitive m-th root of unity.

For a number field K we let $\Gamma_K = \mathrm{Gal}(\overline{K}/K)$ be the Galois group of a fixed algebraic closure $\overline{K}$ of K. For a finite group G, an étale G-algebra L/K is an étale algebra of degree $\#G$ where $G \hookrightarrow \mathrm{Aut}_K(L)$ and G acts transitively on the idempotents of L (this is the notion of a G structured algebra as in [Woo10]). When L is a field, it is Galois with Galois group G. For an extension of étale algebras $L/K/\mathbb{Q}$, $\Delta(L/K)$ will denote the relative discriminant, while $|\Delta(L/K)|$ will denote the absolute discriminant, i.e., the absolute value of the norm of $\Delta(L/K)$ from K to $\mathbb{Q}$ of $\Delta(L/K)$.

If $L = K(\alpha)$ is an extension of degree n, then its Galois group G acts transitively on the n conjugates of α and hence gives a map $G \hookrightarrow S_n$. In particular, if L is

Galois over K, we get a map $G \to S_{\#G}$. The regular representation of G is the map $G \to S_{\#G}$ induced by the action of G acting on itself by left multiplication.

Given a group scheme G over S, we will write BG_S for the classifying stack of G over S, defined in Sect. 2.2. If S is clear from context, we will suppress the subscript. Our assumptions on G imply that BG_S is an algebraic stack, and in the case when G is étale over S, also Deligne–Mumford. See [Ols16, E.g. 8.1.12, Definition 8.3.1]. We will not introduce any technical definitions of stacks unless absolutely necessary, but we refer to the reader to [Fan01, Voi04, Ols16, Alp24] for further reading.

Given a scheme X or a stack $\mathcal{X}$ over a field K and a height function Ht on $\mathcal{X}(\overline{K})$, we say the function Ht has the Northcott property if for any real $B > 0$ and étale algebra L/K,

$$\#\{x \in \mathcal{X}(L) \mid \mathrm{Ht}(x) < B\} < \infty.$$

In general, if Ht denotes some height function, we will use lowercase ht to denote the logarithmic height $\log \mathrm{Ht}$, and vice versa.

Given two functions $f, g : \mathbb{R} \to \mathbb{R}$, we say that $f \ll g$ if there is a constant c such that for x sufficiently large, $f(x) \leq cg(x)$. We say that $f \sim g$ if

$$\lim_{x \to \infty} \frac{f(x)}{g(x)} = 1.$$

2 Rational Points on BG

In this section, G is a finite group scheme over $\operatorname{Spec} \mathbb{Z}$. In particular, it is affine. In many cases in this article, G will be induced from a finite group (also denoted G via abuse of notation) as the constant sheaf $\underline{G}$, which is representable over $\operatorname{Spec} \mathbb{Z}$ by the scheme

$$G_{\mathbb{Z}} = \bigsqcup_{g \in G} \operatorname{Spec} \mathbb{Z},$$

see [Sta23, Tag 047F] and [Mil80, Above Corollary II.1.7]. Notice that $G_{\mathbb{Z}}$ is a finite disjoint union of $\#G$ $\mathbb{Z}$-points in $\mathbb{A}^1_{\mathbb{Z}}$. It is a smooth affine scheme. For every ring A we write $G_A := G_{\mathbb{Z}} \times_{\operatorname{Spec} \mathbb{Z}} \operatorname{Spec} A$.

We will define the classifying stack BG, which parametrizes G-torsors. We begin by recalling the definition of G-torsors in Sect. 2.1, the categorical definition of BG in Sect. 2.2, and we give explicit descriptions of points on this stack in the cases where G is μ_n and $\mathbb{Z}/n\mathbb{Z}$ in Sects. 2.3 and 2.4.

In this section, we will state everything for the base scheme $\operatorname{Spec} \mathbb{Z}$. Several definitions work for a more general base scheme, but this makes our exposition easier.

2.1 G-Torsors

Definition 1 Let T be a scheme (over $\operatorname{Spec}\mathbb{Z}$). A G-torsor over T (which, in this case of affine group schemes is the same as a principal G-bundle) is a map of schemes $\pi\colon P \to T$ with a left G-action on P such that on some (étale or fppf) cover $\{U_i \to T\}$, $P \times_T U_i$ is the trivial torsor $G|_{U_i} \times U_i$. Equivalently, we can require that the map $G \times_T P \to P \times_T P$ given by $(g, p) \mapsto (p, gp)$ is an isomorphism by [Ols16, Proposition 4.5.6].

We refer the reader to [Ols16, Sect. 4.5], [Mil80, Sect. III.4], and [Sko01] for further reading on torsors and their properties.

Remark 1 If $T = \operatorname{Spec} K$ for a field K, and $G = \operatorname{Spec} A$ is finite, then a G-torsor over T is a morphism $P \to \operatorname{Spec} K$ with an action of G on P and such that under suitable field extensions L_i/K, $P_{L_i} \cong G_{L_i}$ compatible with the G_{L_i}-action by left multiplication on G_{L_i}.

We now look at two special cases.

Example 1 (*G constant*) In the case where the group scheme is a constant scheme given by a finite group, the set of G-torsors over $T = \operatorname{Spec} K$ is in bijection with the set of étale G-algebras over K (in the sense of Sect. 1.1). Furthermore, the set of connected G-torsors over $T = \operatorname{Spec} K$ is in bijection with the set of Galois extensions of K with Galois group G.

Indeed, let $P = \operatorname{Spec} L$ be a G-torsor over K. Since the map $G \times P \to P \times P$ is an isomorphism, we have $L \otimes_K L \cong L^{\#G}$ as L-vector spaces, and hence, $[L : K] = \#G$. Since the map $G \times P \to P \times P$ is an isomorphism, G acts on P by automorphisms of P as a K-scheme, or equivalently of L as a K-algebra. Hence, there is a group homomorphism $G \to \operatorname{Aut}_K(L)$. Let H be its image. Then trivializing P over L and taking pullback induce a sequence of group homomorphisms

$$G \to H \to \operatorname{Aut}_{G\text{-torsors}/L}(P_L) \cong \operatorname{Aut}_{G\text{-torsors}/L}(G_L) = G$$

whose composition is the identity on G. Hence, $[L : K] = \#G \leq \#\operatorname{Aut}_K(L)$, and L/K is a Galois extension.

Example 2 (*G étale-locally constant*) Assume that over a separable closure $\overline{K}$ of K, $G_{\overline{K}}$ is a constant group scheme. Write $P = \operatorname{Spec} L$. Then there exist suitable field extensions L'/K such that $L' \otimes_K L \cong L'^{\oplus \#G}$.

For example, if $G = \mu_3$, $K = \mathbb{Q}$ and $L = \mathbb{Q}(\sqrt[3]{2})$. Let $L' = L(\zeta_3)$. Then $P_{L'} = \operatorname{Spec}(L \otimes_K L')$, and

$$\begin{aligned} L \otimes_K L' = \mathbb{Q}(\sqrt[3]{2}) \otimes_{\mathbb{Q}} \mathbb{Q}(\sqrt[3]{2}, \zeta_3) &\cong \mathbb{Q}[x]/(x^3 - 2) \otimes_{\mathbb{Q}} \mathbb{Q}(\sqrt[3]{2}, \zeta_3) \\ &\cong \mathbb{Q}(\sqrt[3]{2}, \zeta_3)[x]/(x^3 - 2) \cong \mathbb{Q}(\sqrt[3]{2}, \zeta_3)^{\oplus 3}, \end{aligned}$$

where the last isomorphism holds because L' contains all the cube roots of 2. We will see in Sect. 2.5 that μ_3 becomes constant after base change to $\mathbb{Q}(\zeta_3)$.

2.2 The Classifying Stack BG

Let G be a group scheme that is finite over a base scheme S and acts trivially on S. The classifying stack BG over S is the quotient stack $[S/G]$. In this section, we briefly recall the definition of a quotient stack, and give interpretations of rational points of BG in some special cases. As stated before, we work with $S = \operatorname{Spec}\mathbb{Z}$ for ease of exposition. Since BG_S is the base change of $BG_{\operatorname{Spec}\mathbb{Z}}$ to S for a general scheme S, this reduction is harmless for this article. We refer the reader to [Ols16, Sect. 8.1] for a more detailed exposition.

2.2.1 Quotient Stacks

Let X be a noetherian scheme. Let G be an affine group scheme with a left action on X given by $\rho : G \times X \to X$.

Definition 2 The quotient stack $[X/G]$ is the (pseudo)functor from the category of schemes to the category of groupoids,

$$[X/G] : (Sch) \to (Gpds)$$

such that

(1) On the level of objects, a scheme T is sent to the category $[X/G](T)$, whose objects are the G-torsors $\pi : \mathcal{E} \to T$ together with a G-equivariant morphisms $\alpha : \mathcal{E} \to X$; and the morphisms of this category are isomorphisms of G-torsors commuting with the G-equivariant morphisms.
(2) The morphisms are naturally induced by pullbacks along the morphism $f : T' \to T$ in (Sch).

By setting $X = \operatorname{Spec}\mathbb{Z}$ with the trivial G-action, we obtain the classifying stack $BG_{\operatorname{Spec}\mathbb{Z}} := [X/G]$ [Ols16, Definition 8.1.14].

Notation 1 Although BG is a stack valued in groupoids, by abuse of notation, we will use $BG(T)$ to denote the (pointed) set of isomorphism classes of equivariant G-torsors over T.

When working with $BG_{\operatorname{Spec}\mathbb{Z}}$, the condition of α being G-equivariant becomes automatic. Indeed, α must be the unique map $\mathcal{E} \to \operatorname{Spec}\mathbb{Z}$, which must in turn agree with the composition of the two G-equivariant maps $\pi : \mathcal{E} \to T$ and the structure map $T \to \operatorname{Spec}\mathbb{Z}$. In particular, this implies that $BG_{\operatorname{Spec}\mathbb{Z}}(T)$ is the set of isomorphism classes of G-torsors over T. Further, it is well known that the latter set is classified by $H^1_{fppf}(T, G)$. See for instance, [Sko01, Sect. 2.2.] or [Ols16, Chapter 12]. In what follows, we will omit the subscript $\operatorname{Spec}\mathbb{Z}$ from $BG_{\operatorname{Spec}\mathbb{Z}}$.

Example 3 (*Twisting by Galois descent, see* [Sko01, Sect. 2.1] *or* [Ser02, Sect. III.1]) Let $T = \operatorname{Spec} K$ and G a smooth finite group scheme over K. By the observation above and using the fact that G is smooth, we may classify the G-torsors over T by $H^1_{fppf}(\operatorname{Spec} K, G) = H^1_{\acute{e}t}(\operatorname{Spec} K, G)$, which in turn is equal to the Galois cohomology group $H^1(\Gamma_K, G(\overline{K}))$. Given a G-torsor P, and a continuous cocycle $\varphi : \Gamma_K \to G(\overline{K})$, a standard way to obtain the twist of P by φ is by taking invariants of the twisted Galois action,

$$\Gamma_K \times P_{\overline{K}} \to P_{\overline{K}}, \qquad (\sigma, x) \mapsto \varphi(\sigma) \cdot (\sigma(x)).$$

Conversely, any other G torsor over K can be obtained in this way. This construction will be useful in later sections.

2.3 The Stack $B\mu_n$

2.3.1 The Group Scheme μ_n

Let n be a positive integer, then $\mu_n := \operatorname{Spec} \mathbb{Z}[t]/(t^n - 1)$ is a group scheme with the group structure given by

$$\operatorname{Spec} \mathbb{Z}[t]/(t^n - 1) \times_{\operatorname{Spec} \mathbb{Z}} \operatorname{Spec} \mathbb{Z}[t]/(t^n - 1) \to \operatorname{Spec} \mathbb{Z}[t]/(t^n - 1), \qquad (a, b) \mapsto ab,$$

which is induced by the ring homomorphism

$$\varphi : \mathbb{Z}[t]/(t^n - 1) \to \mathbb{Z}[u]/(u^n - 1) \otimes_{\mathbb{Z}} \mathbb{Z}[v]/(v^n - 1), \qquad t \mapsto u \otimes v.$$

Indeed, for every field K, given two points with coordinates $a, b \in \mu_n(K)$, the ideal of the point $(a, b) \in \mu_{n,K} \times_{\operatorname{Spec} K} \mu_{n,K}$ is

$$I_{(a,b)} = ((u - a) \otimes 1, 1 \otimes (v - b)) \subseteq K[u]/(u^n - 1) \otimes_K K[v]/(v^n - 1).$$

We observe that

$$\varphi(t - ab) = u \otimes v - ab \otimes 1 = ((u - a) \otimes 1)(1 \otimes (v - b)) + (u - a) \otimes b + a \otimes (v - b) \in I_{(a,b)}.$$

Thus $t - ab \in \varphi^{-1}(I_{(a,b)})$ and since $(t - ab)$ is a maximal ideal in $K[t]/(t^n - a)$ we have $\varphi^{-1}(I_{(a,b)}) = (t - ab)$. This shows that φ is the ring homomorphism that defines the group structure on μ_n.

2.3.2 Rational Points of $B\mu_n$

In this section, for any non-zero element a in a field K, we will denote by $a^{1/n} \in \overline{K}^{\times}$ a choice of an n-th root of a. The following theorem describes the isomorphism classes of K-points on $B\mu_n$.

Theorem 1 *Let K be a field with a fixed algebraic closure $\overline{K}$. Assume that n is invertible in K, and let ζ_n be a primitive n-th root of 1 in $\overline{K}$. Then there are bijections between the following objects:*

(1) the set $B\mu_n(K)$, and
(2) the set $K^\times/(K^\times)^n = H^1_{\acute{e}t}(K, \mu_{n,K})$.

The bijection is given by

$$K^\times/(K^\times)^n \to (B\mu_n)(K), \qquad [a] \mapsto [\operatorname{Spec} K_{a^{-1}}],$$

where $K_a := K[t]/(t^n - a)$ for all $a \in K^\times$.

Remark 2 Note that K_a is a field if and only if the polynomial $t^n - a$ is irreducible. Otherwise, K_a is just an étale K-algebra of dimension n, as $\operatorname{char}(K) \nmid n$. By [Lan02, Theorem VI.9.1] irreducibility holds if and only if $a \notin K^p$ for any prime number $p \mid n$ and $a \notin -4K^4$ whenever $4 \mid n$.

Proof As in Sect. 2.2 and Notation 1, $B\mu_n(K)$ denotes the set of isomorphism classes of torsors under μ_n over K. By [Ols16, Proposition 4.5.6] and [Mil80, Corollary III.4.7] (see also [Sko01, (2.11)]), such isomorphism classes are classified by the flat cohomology group $H^1_{fl}(K, \mu_{n,K})$. Since n is coprime to the characteristic of K, the group scheme $\mu_{n,K}$ is smooth over K, and flat cohomology can be replaced by étale cohomology by [Mil80, Theorem III.3.9]. Hence, there is a bijection between $(B\mu_n)(K)$ and $H^1_{\acute{e}t}(K, \mu_{n,K})$.

We prove that $H^1_{\acute{e}t}(K, \mu_{n,K}) \cong K^\times/(K^\times)^n$ via étale cohomology on the Kummer exact sequence

$$1 \to \mu_n(\overline{K}) \to \overline{K}^\times \overset{x \mapsto x^n}{\to} \overline{K}^\times \to 1. \tag{1}$$

Since K is a field, $H^1_{\acute{e}t}(K, \mu_{n,K})$ can be computed as the Galois cohomology group $H^1(\Gamma_K, \mu_n(\overline{K}))$, where Γ_K is the absolute Galois group of K. Taking Galois cohomology of (1) gives the long exact sequence

$$1 \to H^0(\Gamma_K, \mu_n(\overline{K})) \to H^0(\Gamma_K, \overline{K}^\times) \to H^0(\Gamma_K, \overline{K}^\times) \overset{\delta}{\to} H^1(\Gamma_K, \mu_n(\overline{K})) \to H^1(\Gamma_K, \overline{K}^\times) \to \cdots$$

Since $H^1(\Gamma_K, \overline{K}^\times) = 0$ by Hilbert 90 [Mil80, Proposition III.4.9], we get the exact sequence

$$1 \to \mu_n(\overline{K}) \cap K \to K^\times \overset{x \mapsto x^n}{\to} K^\times \overset{\delta}{\to} H^1(\Gamma_K, \mu_n(\overline{K})) \to 0,$$

which shows that $K^\times/(K^\times)^n \cong H^1(\Gamma_K, \mu_n(\overline{K}))$, via the boundary map, δ.

Consider the function

$$K^\times/(K^\times)^n \to \{K_a : a \in K^\times\}/\sim, \qquad [a] \mapsto [K_{a^{-1}}],$$

where $\sim$ indicates isomorphisms of K-algebras. We will show that for any $a \in K^\times$, $K_{a^{-1}}$ is a $\mu_{n,K}$-torsor over K corresponding to the image $\delta([a]) \in H^1_{ét}(K, \mu_{n,K})$, thus showing that the diagram

$$\begin{array}{ccccc} & & B\mu_n(K) = \left\{\begin{matrix}\text{isomorphism classes}\\ \text{of } \mu_n \text{ torsors over } K\end{matrix}\right\} & & \\ & \overset{[a]\mapsto \operatorname{Spec} K_{a^{-1}}}{\nearrow} & & \overset{\sim}{\nwarrow} & \\ K^\times/(K^\times)^n & & \xrightarrow[\sim]{\delta} & & H^1_{ét}(K, \mu_{n,K}) = \left\{\begin{matrix}\text{Galois cohomology}\\ \text{classes}\end{matrix}\right\} \end{array}$$

commutes and that the map $a \mapsto K_{a^{-1}}$ is indeed a bijection.

Fix $\alpha \in \overline{K}$ such that $\alpha^n = a$. For every $r \in \{0, 1, \ldots, n-1\}$, there is a cocycle

$$\delta_r : \Gamma_K \to \mu_n(\overline{K}), \qquad \sigma \mapsto \sigma(\zeta_n^r \alpha)(\zeta_n^r \alpha)^{-1},$$

whose image in $H^1(\Gamma_K, \mu_{n,K})$ agrees with $\delta([a])$ by [GS06, Remark 3.2.4] and is independent of r since different choices of r yield cohomologous cocycles. Further, δ_r induces a twisted action of Γ_K on $\overline{K}[t]/(t^n - 1)$ as follows: an element $\sigma \in \Gamma_K$ acts by the K-automorphism $\delta_{r,\sigma}$ defined by $\delta_{r,\sigma}(\beta) = \sigma(\beta)$ for all $\beta \in \overline{K}$ and $\delta_{r,\sigma}(t) = \delta_r(\sigma)t$. Consider the isomorphism of $\overline{K}$-algebras

$$\psi_r : \overline{K}[t]/(t^n - 1) \to \overline{K}[t]/(t^n - a^{-1}), \qquad t \to \zeta_n^r \alpha t. \tag{2}$$

Then the diagram

$$\begin{array}{ccc} \overline{K}[t]/(t^n - 1) & \xrightarrow{\psi_r} & \overline{K}[t]/(t^n - a^{-1}) \\ \Big\downarrow {\scriptstyle \delta_{r,\sigma}} & & \Big\downarrow {\scriptstyle \sigma} \\ \overline{K}[t]/(t^n - 1) & \xrightarrow{\psi_r} & \overline{K}[t]/(t^n - a^{-1}) \end{array}$$

commutes. Thus, the ring of invariants of the twisted Galois action corresponding to δ_r is given by $K_{a^{-1}} = K[t]/(t^n - a^{-1})$, which is the required K-form of $\mu_{n,K}$ (cf. [Ser02, Sect. III.1.3] or Example 3). On the level of group schemes, Spec $K_{a^{-1}}$ is a μ_n-torsor over K with the torsor structure given explicitly by

$$K[t]/(t^n - a^{-1}) \to K[t]/(t^n - a^{-1}) \otimes_K K[u]/(u^n - 1), \qquad t \mapsto t \otimes u.$$

□

Remark 3 Let $\zeta_n \in \overline{K}$ be a fixed n-th root of 1. For every $a \in K$, the extension $K(\zeta_n) \subseteq K(a^{\frac{1}{n}}, \zeta_n)$ is Galois, and hence uniquely determined by a, rather than by the choice of an n-th root of a. We denote by $(\star)$ the set of subfields of $\overline{K}$ of the form $K(a^{1/n}, \zeta_n)$ together with the $\mu_n(\overline{K})$-action that sends $a^{\frac{1}{n}}$ to $\zeta_n^{\frac{n}{d}} a^{\frac{1}{n}}$, where $d = [K(a^{1/n}, \zeta_n) : K(\zeta_n)]$. We observe that there is a surjective function

$$K^\times/(K^\times)^n \to (\star), \qquad [a] \mapsto K(a^{1/n}, \zeta_n). \tag{3}$$

The Galois group $\mathrm{Gal}(K(a^{\frac{1}{n}}, \zeta_n)/K(\zeta_n))$ acts on the roots of a by multiplication by d-th roots of unity where $d = [K(a^{\frac{1}{n}}, \zeta_n) : K(\zeta_n)]$. Let $a, b \in K^\times$ such that $K(a^{1/n}, \zeta_n) = K(b^{1/n}, \zeta_n)$. Let $\alpha, \beta \in \overline{K}$ such that $\alpha^n = a$ and $\beta^n = b$ and the $\mu_n(\overline{K})$ action is given by $\zeta_n \cdot \alpha = \zeta_n^{n/d}\alpha$ and $\zeta_n \cdot \beta = \zeta_n^{n/d}\beta$. Let $d = [K(\alpha, \zeta_n) : K(\zeta_n)]$. Since $d \mid n$ and $\alpha^d, \beta^d \in K(\zeta_n)$ by [Lan02, Theorem VI.6.2], $\beta, \beta^2, \dots, \beta^d$ is a basis of $K(\beta, \zeta_n)$ over $K(\zeta_n)$. Write $\alpha = \sum_{i=1}^d b_i\beta^i$ with $b_1, \dots, b_d \in K(\zeta_n)$. Since each element of $\mathrm{Gal}(K(a^{\frac{1}{n}}, \zeta_n)/K(\zeta_n))$ acts on both α and β by multiplication by the same power of $\zeta_n^{\frac{n}{d}}$, we get that $b_2 = \cdots = b_d = 0$ and $b_1 \neq 0$. Let $\gamma = b_1$, then $\gamma \in K(\zeta_n)^\times$ and $\gamma^n = a/b \in K$. We conclude that the function (3) factors through $K(\zeta_n)^\times/(K(\zeta_n)^\times)^n$ and that (3) is injective if and only if the group homomorphism

$$K^\times/(K^\times)^n \to K(\zeta_n)^\times/(K(\zeta_n)^\times)^n \tag{4}$$

induced by the inclusion $K \subseteq K(\zeta_n)$ is injective. Indeed, the injectivity of the induced map from $K(\zeta_n)^\times/(K(\zeta_n)^\times)^n$ to $(\star)$ follows from Kummer Theory (see for instance [Neu99, Corollary IV.3.6]). If $K = \mathbb{Q}$ and $n = 3$, the homomorphism (4) is injective, as $(c + d\zeta_3)^3 \in \mathbb{Q}$ if and only if $cd = 0$ or $c = d$, and in the latter case, we get $(c + c\zeta_3)^3 = (-c)^3$. If $K = \mathbb{Q}$ and $n = 4$, the homomorphism (4) is not injective, as $-4 = (1 + \zeta_4)^4$ is a 4-th power in $\mathbb{Q}(\zeta_4)$ but not in $\mathbb{Q}$. More generally, since the map (4) is the restriction map on cohomology, its kernel is $H^1(\mathrm{Gal}(K(\zeta_n)/K), \mu_n)$, by the inflation restriction sequence ([NSW08, Corollary 2.4.2]).

Remark 4 Note that $\mathrm{Spec}\, K(a^{\frac{1}{n}}, \zeta_n) \to \mathrm{Spec}\, K$ does not need to be a μ_n-torsor if $\zeta_n \notin K$, because a trivialization is given by $\overline{K} \otimes_K K(a^{\frac{1}{n}}, \zeta_n) \cong (\overline{K})^{dr}$ where $d = [K(a^{\frac{1}{n}}, \zeta_n) : K(\zeta_n)]$ and $r = [K(\zeta_n) : K]$, while the trivial torsor is given by $\overline{K}[t]/(t^n - 1) \cong (\overline{K})^n$.

Remark 5 We refer the more categorically inclined reader to [Ols16, Example 4.5.8] for a description of the category of μ_n- torsors over a scheme S in which n is invertible.

2.4 *The Stack $B(\mathbb{Z}/n\mathbb{Z})$*

2.4.1 The Constant Group Scheme $\underline{\mathbb{Z}/n\mathbb{Z}}$

Given a positive integer n, we denote by $B(\mathbb{Z}/n\mathbb{Z})$ the classifying stack of the constant group scheme $\underline{\mathbb{Z}/n\mathbb{Z}}$ defined by the finite cyclic group $\mathbb{Z}/n\mathbb{Z}$. Then $\underline{\mathbb{Z}/n\mathbb{Z}} = \mathrm{Spec}\,\mathbb{Z}^n$, where $\mathbb{Z}^n$ is the free $\mathbb{Z}$-module of rank n with a fixed basis $\{e_i : i \in \mathbb{Z}/n\mathbb{Z}\}$ and ring structure given by componentwise addition and multiplication, i.e.,

$$\sum_{i\in\mathbb{Z}/n\mathbb{Z}} a_i e_i + \sum_{i\in\mathbb{Z}/n\mathbb{Z}} b_i e_i = \sum_{i\in\mathbb{Z}/n\mathbb{Z}} (a_i + b_i)e_i, \qquad \left(\sum_{i\in\mathbb{Z}/n\mathbb{Z}} a_i e_i\right)\left(\sum_{i\in\mathbb{Z}/n\mathbb{Z}} b_i e_i\right) = \sum_{i\in\mathbb{Z}/n\mathbb{Z}} (a_i b_i)e_i,$$

for all $a_i, b_i \in \mathbb{Z}$ and $i \in \mathbb{Z}/n\mathbb{Z}$. We observe that the ring structure on $\mathbb{Z}^n \otimes_{\mathbb{Z}} \mathbb{Z}^n \cong \mathbb{Z}^{n^2}$ is given by componentwise addition and multiplication for the basis $\{e_i \otimes e_j : i, j \in \mathbb{Z}/n\mathbb{Z}\}$.

Lemma 1 *The group scheme structure on $\underline{\mathbb{Z}/n\mathbb{Z}}$ is induced by*

$$\varphi : \mathbb{Z}^n \to \mathbb{Z}^n \otimes_{\mathbb{Z}} \mathbb{Z}^n, \qquad \sum_{i \in \mathbb{Z}/n\mathbb{Z}} a_i e_i \mapsto \sum_{i,j \in \mathbb{Z}/n\mathbb{Z}} a_{i+j} e_i \otimes e_j.$$

Proof If K is a field then $\underline{\mathbb{Z}/n\mathbb{Z}}(K) \cong \mathbb{Z}/n\mathbb{Z}$. If we denote by $P_\alpha \in \underline{\mathbb{Z}/n\mathbb{Z}}(K)$ the preimage under this map of $\alpha \in \mathbb{Z}/n\mathbb{Z}$, the group scheme structure is given by

$$\underline{\mathbb{Z}/n\mathbb{Z}}(K) \times \underline{\mathbb{Z}/n\mathbb{Z}}(K) \to \underline{\mathbb{Z}/n\mathbb{Z}}(K), \qquad (P_\alpha, P_\beta) \mapsto P_{\alpha+\beta}.$$

For $i, j \in \mathbb{Z}/n\mathbb{Z}$, let $c_{i,j} = 0$ if $i = j$ and $c_{i,j} = 1$ if $i \neq j$. In the chosen notation, for $\alpha \in \mathbb{Z}/n\mathbb{Z}$ the ideal of the point P_α is $\{\sum_{i \in \mathbb{Z}/n\mathbb{Z}} a_i e_i : a_\alpha = 0\} \subseteq K^n$ and it is generated by $\sum_{i \in \mathbb{Z}/n\mathbb{Z}} c_{i,\alpha} e_i$. Fix $\alpha, \beta \in \mathbb{Z}/n\mathbb{Z}$, then the ideal of $(P_\alpha, P_\beta) \in \underline{\mathbb{Z}/n\mathbb{Z}}(K) \times \underline{\mathbb{Z}/n\mathbb{Z}}(K)$ is the ideal generated by the two elements $v_\alpha := \sum_{i,j \in \mathbb{Z}/n\mathbb{Z}} c_{i,\alpha} e_i \otimes e_j$ and $v_\beta := \sum_{i,j \in \mathbb{Z}/n\mathbb{Z}} c_{j,\beta} e_i \otimes e_j$. We observe that $\varphi\left(\sum_{i \in \mathbb{Z}/n\mathbb{Z}} c_{i,\alpha+\beta} e_i\right) = \sum_{i,j \in \mathbb{Z}/n\mathbb{Z}} c_{i+j,\alpha+\beta} e_i \otimes e_j$ belongs to the ideal generated by v_α and v_β. Indeed,

$$v_\alpha + v_\beta - v_\alpha v_\beta = \sum_{i,j \in \mathbb{Z}/n\mathbb{Z}} (c_{i,\alpha} + c_{j,\beta} - c_{i,\alpha} c_{j,\beta}) e_i \otimes e_j,$$

as $c_{i+j,\alpha+\beta} = 0$ if $i = \alpha$ and $j = \beta$ and $c_{i,\alpha} + c_{j,\beta} - c_{i,\alpha} c_{j,\beta} = 0$ if and only if $i = \alpha$ and $j = \beta$. Since the ideal generated by $\sum_{i \in \mathbb{Z}/n\mathbb{Z}} c_{i,\alpha+\beta} e_i$ is a maximal ideal of K^n, we conclude that it is equal to $\varphi^{-1}((v_\alpha, v_\beta))$. □

2.4.2 $\underline{\mathbb{Z}/n\mathbb{Z}}$-Torsors

Since $\underline{\mathbb{Z}/n\mathbb{Z}}_K = \operatorname{Spec}(K^n)$ is finite and étale over K, so is every $\underline{\mathbb{Z}/n\mathbb{Z}}$-torsor over K (see [Mil80, Proposition III.4.2]). Thus, torsors under $\underline{\mathbb{Z}/n\mathbb{Z}}$ are Galois coverings of Spec K with Galois group $\mathbb{Z}/n\mathbb{Z}$ in the sense of [Mil80, Remark I.5.4]. In the following theorem, we give some equivalent ways of describing this set.

Theorem 2 *Let K be a field of characteristic zero with a fixed algebraic closure $\overline{K}$. Then there is a bijection between the following objects:*

(1) the set of K-points of the stack $B(\mathbb{Z}/n\mathbb{Z})$;
(2) the set of classes of continuous 1-cocycles $\operatorname{Hom}_{cts}(\Gamma_K, \mathbb{Z}/n\mathbb{Z})$, where Γ_K is equipped with the profinite topology and $\mathbb{Z}/n\mathbb{Z}$ is a discrete topological space;
(3) the set of isomorphism classes of finite étale $\mathbb{Z}/n\mathbb{Z}$-algebras L/K .

Proof As in the proof of Theorem 1, since $\underline{\mathbb{Z}/n\mathbb{Z}}$ is smooth over K, the isomorphism classes of torsors under $\underline{\mathbb{Z}/n\mathbb{Z}}$ over K are classified by $H^1_{\acute{e}t}(K, \underline{\mathbb{Z}/n\mathbb{Z}}) =$

$H^1(\Gamma_K, \mathbb{Z}/n\mathbb{Z})$. Since the Γ_K-module structure on $\mathbb{Z}/n\mathbb{Z}$ is given by the trivial action of Γ_K, the 1-cocycles are just continuous group homomorphisms $\Gamma_K \to \mathbb{Z}/n\mathbb{Z}$ and the coboundaries are trivial (cf. [Ser02, Sect. I.5.1]). Thus $H^1(\Gamma_K, \mathbb{Z}/n\mathbb{Z}) = \mathrm{Hom}_{cts}(\Gamma_K, \mathbb{Z}/n\mathbb{Z})$ is in bijection with $B(\mathbb{Z}/n\mathbb{Z})(K)$. This establishes the bijection between (1) and (2).

The map between (2) and (3) is given by

$$(\varphi : \Gamma_K \to \mathbb{Z}/n\mathbb{Z}) \mapsto \prod_{r \in \operatorname{coker} \varphi} \overline{K}^{\ker \varphi}.$$

Suppose the image of φ is isomorphic to $m\mathbb{Z}/n\mathbb{Z}$ for some $m \in \mathbb{Z}$, whence its cokernel is isomorphic to $\mathbb{Z}/m\mathbb{Z}$. Then $K_\varphi := \overline{K}^{\ker \varphi}$ is Galois over K with Galois group $m\mathbb{Z}/n\mathbb{Z}$, and this action on K_φ together with the permutation action of $\mathbb{Z}/m\mathbb{Z}$ on the factors induces a $\mathbb{Z}/n\mathbb{Z}$-action on the étale algebra $L := \prod_{r \in \mathbb{Z}/m\mathbb{Z}} K_\varphi$, thus giving it the structure of an étale $\mathbb{Z}/n\mathbb{Z}$-algebra. If $\varphi \neq \varphi'$ such that $\ker \varphi = \ker \varphi'$, then $K_\varphi = K_{\varphi'}$ as subsets of $\overline{K}$, and the images $\mathrm{Im}(\varphi)$ and $\mathrm{Im}(\varphi')$ are isomorphic to $m\mathbb{Z}/n\mathbb{Z}$ for some $m \in \mathbb{Z}$. However, the induced actions of $\mathrm{Im}(\varphi)$ and $\mathrm{Im}(\varphi')$ on L are different and correspond to picking different generators of $m\mathbb{Z}/n\mathbb{Z}$. Thus, they give different maps $\mathbb{Z}/n\mathbb{Z} \hookrightarrow \mathrm{Aut}_K(L)$, establishing that the map between (2) and (3) is an injection. The surjectivity of the map follows from the definition of a $\mathbb{Z}/n\mathbb{Z}$-algebra: since $\mathbb{Z}/n\mathbb{Z}$ must act transitively on the idempotents of the étale algebra, it must be of the form $\prod_{r \in \mathbb{Z}/m\mathbb{Z}} \overline{K}^H$ for some $m|n$ and H a subgroup of Γ_K of index n/m. The Galois correspondence [Lan02, Theorem VI.1.1, Sect. VI.14] then finishes the proof. This proof is a special case of that of [Woo10, Lemma 2.6].

We now describe explicitly the map between (3) and (1). Denote by $\{e_i\}_{i \in \mathbb{Z}/n\mathbb{Z}}$ the standard basis vectors of $\overline{K}^n \cong \overline{K} \otimes_{\mathbb{Z}} \mathbb{Z}^n$. Given $\varphi \in \mathrm{Hom}_{cts}(\Gamma_K, \mathbb{Z}/n\mathbb{Z})$, let $K_\varphi = \overline{K}^{\ker \varphi}$ and write $[K_\varphi : K] = n/m$ as before. The étale algebra $\prod_{r \in \mathbb{Z}/m\mathbb{Z}} K_\varphi$ is the ring of invariant elements of $\overline{K}^n$ under the twisted Galois action

$$\Gamma_K \times \overline{K}^n \to \overline{K}^n, \qquad (\sigma, v) \mapsto \varphi(\sigma)(\sigma(v)),$$

where $\varphi(\sigma)$ acts on $\overline{K}^n$ by sending $e_i \mapsto e_{i+\varphi(\sigma)}$. This describes the étale algebra as a twist of the trivial torsor in the sense of Example 3. In order to describe the action of the group scheme $\underline{\mathbb{Z}/n\mathbb{Z}}$, let $[\cdot] : \mathbb{Z}/n\mathbb{Z} \to \mathbb{Z}/m\mathbb{Z}$ be the reduction modulo m. The elements of $\prod_{r \in \mathbb{Z}/m\mathbb{Z}} K_\varphi$ can be written uniquely as $\sum_{r \in \mathbb{Z}/m\mathbb{Z}} v_r$, where $v_r \in K_\varphi$ and $v_r = \sum_{i \in \mathbb{Z}/n\mathbb{Z}, [i]=r} a_i e_i$ as an element of $\overline{K}^n$. For every $j \in \mathbb{Z}/n\mathbb{Z}$, the shift $v_r(j) = \sum_{i \in \mathbb{Z}/n\mathbb{Z}, [i]=r} a_{i+j} e_i$ is an element of K_φ. Then $\prod_{r \in \mathbb{Z}/m\mathbb{Z}} K_\varphi$ defines a torsor under $\mathbb{Z}/n\mathbb{Z}$ with the following torsor structure:

$$\prod_{r \in \mathbb{Z}/m\mathbb{Z}} K_\varphi \to \prod_{r \in \mathbb{Z}/m\mathbb{Z}} K_\varphi \otimes K^n, \qquad \sum_{r \in \mathbb{Z}/m\mathbb{Z}} v_r \mapsto \sum_{r \in \mathbb{Z}/m\mathbb{Z}} \sum_{j \in \mathbb{Z}/n\mathbb{Z}} v_r(j) \otimes e_j.$$

□

2.5 Comparison Between Rational Points on $B\mu_n$ and $B\mathbb{Z}/n\mathbb{Z}$

Let K be a field of characteristic zero with a fixed algebraic closure $\overline{K}$. Let $\zeta_n \in \overline{K}$ be a primitive n-th root of 1.

Lemma 2 *If $\zeta_n \in K$, then $B\mu_n \cong B(\mathbb{Z}/n\mathbb{Z})$ over K.*

Proof We observe that if $\zeta_n \in K$, then there is an isomorphism of group schemes $\underline{\mathbb{Z}/n\mathbb{Z}} \to \mu_{n,K}$ given by

$$K[t]/(t^n-1) \to K^n, \qquad t^s \mapsto \sum_{i\in\mathbb{Z}/n\mathbb{Z}} \zeta_n^{si} e_i,$$

with inverse

$$K^n \to K[t]/(t^n-1), \qquad \sum_{i\in\mathbb{Z}/n\mathbb{Z}} a_i e_i \mapsto \frac{1}{n}\sum_{s=0}^{n-1}\sum_{i\in\mathbb{Z}/n\mathbb{Z}} a_i\zeta_n^{-si} t^s. \tag{5}$$

Indeed, using the notation in Sect. 2.4, the isomorphism is compatible with the group structures as

$$\frac{1}{n}\sum_{r=0}^{n-1}\sum_{i\in\mathbb{Z}/n\mathbb{Z}} a_i\zeta_n^{-ri}(u\otimes v)^r = \frac{1}{n^2}\sum_{r=0}^{n-1}\sum_{s=0}^{n-1}\sum_{i,j\in\mathbb{Z}/n\mathbb{Z}} a_{i+j}\zeta_n^{-ri-sj} u^r\otimes v^s$$

holds in $K[u]/(u^n-1)\otimes_K K[v]/(v^n-1)$ for all choices of $a_i \in K$ thanks to the fact that $\sum_{j\in\mathbb{Z}/n\mathbb{Z}} \zeta_n^{(r-s)j}$ equals n if $r = s$ and vanishes if $r \neq s$. Hence, the statement of the lemma holds. In particular, they have the same sets of K-points. □

If $\zeta_n \notin K$, then no connected μ_n-torsor over K is a $\underline{\mathbb{Z}/n\mathbb{Z}}$-torsor over K and no connected $\underline{\mathbb{Z}/n\mathbb{Z}}$-torsor over K is a μ_n-torsor over K. Indeed, connected torsors are fields, and field extensions L/K that are torsors under μ_n are not Galois, while field extensions L/K that are torsors under $\underline{\mathbb{Z}/n\mathbb{Z}}$ are always Galois extensions, as follows from the descriptions in Theorems 1 and 2.

We now spell out the difference between the classifying groups of K-forms and torsors for $\mu_{n,K}$ and $\underline{\mathbb{Z}/n\mathbb{Z}}$ over a field K of characteristic 0. By [Ser02, Sect. III.1.3], the K-forms of $\mu_{n,K}$ are classified by the cohomology group $H^1(\Gamma_K, \mathrm{Aut}(\overline{K}[t]/(t^n-1)))$, where the automorphism group $\mathrm{Aut}(\overline{K}[t]/(t^n-1))$ (in the category of $\overline{K}$-algebras) is the symmetric group S_n on the set of n elements, and the Γ_K-module structure is given by conjugation of automorphisms for the Γ_K-action on $\mu_n(\overline{K})$ induced by the base change from K to $\overline{K}$. The K-forms of $\underline{\mathbb{Z}/n\mathbb{Z}}$ are classified by the cohomology group $H^1(\Gamma_K, \mathrm{Aut}(\overline{K}^n))$, where the automorphism group $\mathrm{Aut}(\overline{K}^n)$ (in the category of $\overline{K}$-algebras) is the symmetric group S_n with Γ_K-module structure given by conjugation of automorphisms for the Γ_K-action on $\underline{\mathbb{Z}/n\mathbb{Z}}$ induced by the base change from K to $\overline{K}$. Notice that in this second case the Γ_K-module structure on S_n is trivial.

Since $\mu_{n,\overline{K}} \cong \mathbb{Z}/n\mathbb{Z}$ over $\overline{K}$, the two groups have the same set of K-forms. We identify $\mathbb{Z}/n\mathbb{Z}$ as the subgroup of S_n induced by the group operation on $\mathbb{Z}/n\mathbb{Z}$, where an element $r \in \mathbb{Z}/n\mathbb{Z}$ is identified with the permutation

$$r : \mathbb{Z}/n\mathbb{Z} \to \mathbb{Z}/n\mathbb{Z}, \quad i \mapsto r + i. \tag{6}$$

We observe that the $\mu_{n,K}$-torsors are classified by the cohomology of the subgroup $\mu_n(\overline{K}) \cong \mathbb{Z}/n\mathbb{Z} < S_n$, as the Γ_K-module structure on $\mathbb{Z}/n\mathbb{Z}$ induced by $\mathrm{Aut}(\overline{K}[t]/(t^n - 1))$ coincides with the Γ_K-module structure on $\mu_n(\overline{K})$ induced by the base change from K to $\overline{K}$. Similarly, the $\mathbb{Z}/n\mathbb{Z}$-torsors over K are classified by the cohomology of the subgroup $\mathbb{Z}/n\mathbb{Z} < S_n$ with trivial Γ_K-module structure.

Example 4 If $\zeta_n \notin K$, then some non-connected $\mu_{n,K}$-torsors are not $\mathbb{Z}/n\mathbb{Z}$-torsors and vice versa. For example, $\mu_{n,K}$ is not a $\mathbb{Z}/n\mathbb{Z}$-torsor as the polynomial $t^n - 1$ factors over K as a product of irreducible polynomials of different degrees, where at least one has degree 1 and another one is the minimal polynomial of ζ_n, which has degree > 1.

On the other hand, some non-connected $\mu_{n,K}$-torsors are $\mathbb{Z}/n\mathbb{Z}$-torsors. For example, if $K = \mathbb{Q}(\zeta_8)$, $n = 16$ and $a = -1$, the K-algebra $K_{-1} = K[t]/(t^{16} + 1)$ is both a μ_{16}-torsor and a $\mathbb{Z}/16\mathbb{Z}$-torsor. Indeed, it is clearly a μ_{16}-torsor by Theorem 1. To show that it is a $\mathbb{Z}/16\mathbb{Z}$-torsor, let $A_i = K[t]/(t^4 - \zeta_8^i)$ for $1 \leq i \leq 8$, and note that there are isomorphisms of K-algebras

$$K[t]/(t^{16} + 1) \cong A_1 \times A_3 \times A_5 \times A_7 \cong A_1^4,$$

where the first isomorphism is given by factoring the polynomial $t^{16} + 1$ over K, and the second isomorphism is given by

$$(t_1, t_3, t_5, t_7) \mapsto (t_1, t_3^3, \zeta_8 t_5, \zeta_8 t_7^3)$$

with inverse

$$(u_1, u_3, u_5, u_7) \mapsto (u_1, u_3^{11}, \zeta_8^{-1} u_5, \zeta_8^{-3} u_7^{11}).$$

The K-algebra A_1 is a $\mathbb{Z}/4\mathbb{Z}$-torsor as in Theorem 2. We show that A_1^4 is a $\mathbb{Z}/16\mathbb{Z}$-torsor. Consider the isomorphism

$$\varphi : \prod_{i=1}^{4} \overline{K}^4 \to \prod_{i=1}^{4} \overline{K}[t]/(t^4 - \zeta_8), \quad \sum_{i=1}^{4} \sum_{j \in \mathbb{Z}/4\mathbb{Z}} a_{i,j} e_{i,j} \mapsto \left(\frac{1}{4} \sum_{s=0}^{3} \sum_{j \in \mathbb{Z}/4\mathbb{Z}} a_{i,j} \zeta_4^{-js} \zeta_{32}^{-s} t^s \right)_{1 \leq i \leq 4}$$

obtained by combining the isomorphisms (5) with (2) for $r = 0$ on each of the four components. Here for each $i \in \{1, \ldots, 4\}$, $e_{i,1}, \ldots, e_{i,4}$ denotes a fixed basis of $\overline{K}^4$. Then φ induces the cocycle

$$\widetilde{\varphi} : \Gamma_K \to \mathrm{Aut}(\overline{K}^{16}) = S_{16}, \quad \sigma \mapsto \varphi^{-1} \circ \sigma \circ \varphi \circ \sigma^{-1},$$

given by

$$\widetilde{\varphi}(\sigma) : \prod_{i=1}^{4} \overline{K}^4 \to \prod_{i=1}^{4} \overline{K}^4, \quad \sum_{i=1}^{4} \sum_{j \in \mathbb{Z}/4\mathbb{Z}} a_{i,j} e_{i,j} \mapsto \sum_{i=1}^{4} \sum_{j \in \mathbb{Z}/4\mathbb{Z}} a_{i,j} e_{i,j+v_\sigma}$$

for $\sigma \in \Gamma_K$ and $v_\sigma \in \mathbb{Z}/4\mathbb{Z}$ such that $\sigma(\zeta_{32}) = \zeta_4^{v_\sigma} \zeta_{32}$. Using the exact sequence

$$0 \longrightarrow \mathbb{Z}/4\mathbb{Z} \xrightarrow{4\cdot} \mathbb{Z}/16\mathbb{Z} \longrightarrow \mathbb{Z}/4\mathbb{Z} \longrightarrow 0$$

and (6), we conclude that the cocycle $\widetilde{\varphi}$ is the continuous homomorphism

$$\Gamma_K \to \mathbb{Z}/16\mathbb{Z}, \quad \sigma \mapsto 4v_\sigma.$$

3 Heights on Classifying Stacks

As described in the previous sections, étale algebras with specified Galois groups are in bijection with points on certain stacks. Further, since Hermite's theorem [Neu99, Sect. III.2] implies that the number of such number fields with bounded discriminant is finite, the discriminant can be thought of as a height function satisfying the Northcott property on the stack BG. This idea has been fleshed out in much greater detail in recent work [ESZB23, Dar21, DY22a], building on previous work of Yasuda [Yas06] and Yasuda–Wood [WY15].

In this section we collect a number of approaches to defining heights of points on BG. In Sect. 3.1, we compute the discriminant of the étale algebras $K[t]/(t^n - a)$. In Sects. 3.2.1, 3.2.2 and 3.2.3, we describe heights on BG following the framework of [ESZB23], [Dar21] and [DY22a] respectively.

3.1 Discriminants of the Points on $B\mu_n$

For $a \in K^\times$, we compute the discriminant $\Delta(K_a/K)$ of the étale algebras $K_a = K[t]/(t^n - a)$ following [Dar21, Sect. 9].

Proposition 1 *Let K be a field such that n is invertible in K. Let $a \in K^\times$, and $K_a = K[t]/(t^n - a)$. Then there is a positive constant $C_{K,n}$ independent of a such that*

$$\Delta(K_a/K) = \prod_{v \nmid n, v \mid a} \mathfrak{p}_v^{n-d_v} \prod_{v \mid n} \mathfrak{p}_v^{e(a,v)},$$

where $\mathfrak{p}_v$ *is the prime ideal of* $\mathcal{O}_K$ *corresponding to the finite place* v, $d_v = \gcd(v(a), n)$, *and* $0 \leq e(a, v) \leq C_{K,n}$.

If $K = \mathbb{Q}$ *and* $a \in \mathbb{Z}$ *is coprime to* n, *then* $\Delta(K_a/K) = \prod_{p|a} p^{n-\gcd(v_p(a),n)} \prod_{p|n} p^{e(a,p)}$. *If* $a \in \mathbb{Z}$ *is squarefree and coprime to* n, *then* K_a *is a field and* $\Delta(K_a/K) = a^{n-1} \prod_{p|n} p^{e(a,p)}$.

Examples of complete computations of the discriminant can be found in [Was82, Proposition 2.7] for cyclotomic fields, and in [DP96, Lemma 3] for the case where n is prime and $\zeta_n \in K$.

Proof Let v be a finite place of K and K_v the completion of K at v with ring of integers $\mathcal{O}_v$ and uniformizer π. If $v \nmid n$,

$$\Delta(K_v(a^{\frac{1}{n}})/K_v) = \pi^{[K_v(a^{\frac{1}{n}}):K_v](1-\frac{d_v}{n})}\mathcal{O}_v$$

by [Dar21, Lemma 9.1.2.2], and as in [Dar21, Proposition 9.1.2.3] we get $\Delta(K_v[t]/(t^n - a)/K_v) = \pi^{n-d_v}\mathcal{O}_v$. If $v \mid n$, then $v(\Delta(K_v(a^{\frac{1}{n}})/K_v))$ is bounded in terms of $[K_v(a^{\frac{1}{n}}) : K_v] \leq n$ and v (cf. [Neu99, Theorem III.2.6]), and hence $e(a, n) = v(\Delta(K_v[t]/(t^n - a)/K_v))$ is bounded in terms of $[K_v(a^{\frac{1}{n}}) : K_v] \leq n$ and v. By [Neu99, Corollary III.2.11] we get $\Delta(K_a/K) = \prod_{v|a, v\nmid n} \mathfrak{p}_v^{n-d_v} \prod_{v|n} \mathfrak{p}_v^{e(a,v)}$ where $\mathfrak{p}_v$ is the prime ideal of $\mathcal{O}_K$ corresponding to the finite place v.

If $K = \mathbb{Q}$ and $a \in \mathbb{Z}$ is squarefree and coprime to n, then K_a is a field and $\Delta(K_a/K) = a^{n-1} \prod_{p|n} p^{e(a,p)}$. If $a \in \mathbb{Z}$ is coprime to n, then $\Delta(K_a/K) = \prod_{p|a} p^{n-\gcd(v_p(a),n)} \prod_{p|n} p^{e(a,p)}$. □

3.2 *Heights on BG*

In this subsection we describe the two approaches taken in [ESZB23, Dar21] to talk about heights on BG.

3.2.1 Heights with Respect to a Vector Bundle

If V is a projective variety over K with a very ample line bundle L, one can define the Weil height with respect to L by extending V and L to a model $\overline{V}$ over Spec $\mathcal{O}_K$, pulling L back to a point Spec $\mathcal{O}_K \to \overline{V}$ and computing its "degree". See for instance [HS00, Part B]. Such a height is geometric in nature, in that the asymptotic growth of the number of points of bounded height reflect the geometry of V. See for instance [BM90]. In [ESZB23], the authors generalize this idea to stacks by defining heights with respect to vector bundles on (proper, Artin) stacks.[1] This approach unifies counting rational points on varieties and counting number fields under the same umbrella.

[1] We do not define these technical notions in this article. The stacks we are interested in do fall into this category. For definitions, we refer the reader to [Ols16] or [Alp24].

In particular, on classifying stacks, the height in [ESZB23] specializes to some well known invariants used to count number fields, such as the discriminant.

Definition 3 Let S be an affine scheme and G a smooth, finite group scheme over S. A vector bundle on $[S/G]$ is the data of a vector bundle V on S along with a G-action that is compatible with the action of G on S. If $S = \operatorname{Spec} K$ for a field K and G is constant, then a vector bundle on BG is a finite dimensional G-representation over K. More generally, if G is a constant group scheme, and the action of G on S is trivial, then a vector bundle on $[S/G] = BG_S$ corresponds to a family of G-representations indexed by S.

Remark 6 There is a more general definition of vector bundles on stacks, which specializes to Definition 3 for quotient stacks. For the general definition, we refer the reader to [Alp24, Sect. 6.1], and for a proof that quasicoherent sheaves on the étale site of BG_S correspond to $\mathcal{O}_S$-modules with left G-action, we refer the reader to [Ols16, Example 9.1.19].

Let $\mathcal{X}$ be a nice[2] stack over the ring of integers $\mathcal{O}_K$ of a number field K, equipped with a vector bundle $\mathcal{V}$. A K-point of $\mathcal{X}$, $x : \operatorname{Spec} K \to \mathcal{X}$ doesn't necessarily extend to an $\mathcal{O}_K$ point of $\mathcal{X}$ due to failure of the valuative criterion of properness for stacks. However, it does extend to a $\mathcal{C}$-point $\bar{x} : \mathcal{C} \to \mathcal{X}$, where $\mathcal{C}$ is a stack over $\mathcal{O}_K$ that is in some sense "minimally ramified" over $\mathcal{O}_K$. This is called a tuning stack ([ESZB23, Sect. 2.1]) and is birational to $\operatorname{Spec} \mathcal{O}_K$ (in fact, $\mathcal{C} \to \operatorname{Spec} \mathcal{O}_K$ is a coarse space map, [Ols16, Definition 11.1.1]). More precisely, a tuning stack fits into a diagram of the form

$$\begin{tikzcd} \operatorname{Spec} K \arrow[r] \arrow[rr, bend left, "x"] \arrow[dr] & \mathcal{C} \arrow[r, "\bar{x}"] \arrow[d, "\pi"] & \mathcal{X} \\ & \operatorname{Spec} \mathcal{O}_K & \end{tikzcd} \tag{7}$$

where π is a birational coarse space map. The authors of [ESZB23] use this diagram to define the height of x as follows.

Definition 4 ([ESZB23, Definition 2.11]) Let $\mathcal{V}$ be a vector bundle on $\mathcal{X}$. Then the height of a point $x : \operatorname{Spec} K \to \mathcal{X}$ with respect to the vector bundle $\mathcal{V}$ is defined as

$$\operatorname{ht}_{\mathcal{V}}(x) = -\deg(\pi_* \overline{x}^* (\mathcal{V}^\vee)),$$

where deg refers to the Arakelov degree of the determinant of $\mathcal{V}$ equipped with any metrization as in [ESZB23, Sect. A3-A4]. The height is independent of the choice of the tuning stack $\mathcal{C}$.

[2] By a "nice" stack will mean a normal, proper, Artin stack with finite diagonal. The reader may just keep $\mathcal{X} = BG$ in mind for the rest of this article.

Remark 7 Definition 4 also extends to the case when K is the global function field of a smooth, proper curve C with $\operatorname{Spec} \mathcal{O}_K$ replaced by C. In that case, the degree in Definition 4 is the usual notion of the degree of a divisor on the curve C. It is only when K is a number field that one needs to account for the infinite place and therefore consider the Arakelov degree. For a quick introduction to Arakelov theory, we refer the reader to [Ara74].

The height in Definition 4 satisfies the Northcott property for appropriate choices of $\mathcal{X}$ and $\mathcal{V}$. In this article we only work with heights on BG which, as we will see in Propositions 1 and 2, agree with the discriminant at all but finitely many places, and in particular, satisfy the Northcott property.

Remark 8 Let G be a finite group. Let $\rho : G \hookrightarrow S_n$ be a degree n permutation representation of G. If $x \in BG(K)$ is a connected torsor, then since G_K is a constant group scheme, via the identification $BG(K) \cong H^1(\Gamma_K, G) \cong \operatorname{Hom}_{cts}(\Gamma_K, G)$, x corresponds to a surjection $x : \Gamma_K \twoheadrightarrow G$. When composed with $\rho : G \twoheadrightarrow S_n$, this gives a homomorphism $\rho_x : \Gamma_K \twoheadrightarrow S_n$. The composition thus defines an étale G-algebra $\tilde{L} = (\overline{K})^{\ker \rho_x}$, which in particular is a Galois extension of K. Let $H \subset S_n$ be the stabilizer of $1 \in \{1, 2, \ldots, n\}$ in $\rho(G)$. Then the fixed set $\tilde{L}^{\rho^{-1}(H)}$ is an intermediate extension L/K with Galois closure $\tilde{L}$. If the torsor $x \in BG(K)$ is not connected, then the induced homomorphism $\Gamma_K \to G$ may not be surjective. In this case, one can still generalize this construction by taking $\tilde{L} = \prod_{r \in \operatorname{coker}(x)} (\overline{K})^{\ker \rho_x}$ and L to be the corresponding product of fixed fields of 1-point stabilizer subgroups. In what follows, we take L to be this étale G-algebra.

Proposition 2 ([ESZB23, Proposition 3.1]) *Let G be a finite group, and $\underline{G}_{\mathcal{O}_K}$ the associated constant group scheme over* $\operatorname{Spec} \mathcal{O}_K$. *Let $\mathcal{V}$ be the vector bundle on $BG_{\operatorname{Spec} \mathcal{O}_K}$ corresponding to a degree n permutation representation $\rho : G \hookrightarrow S_n$ (equipped with the trivial metric described in [ESZB23, Sect. A.2]). Let $x \in BG(K)$. The composition $\Gamma_K \xrightarrow{x} G \xrightarrow{\rho} S_n$ defines a degree n étale algebra L/K as in Remark 8. Then*

$$\operatorname{ht}_{\mathcal{V}}(x) = \frac{1}{2} \log |\Delta(L/K)|,$$

where $|\Delta(L/K)|$ denotes the absolute norm of the discriminant of the extension L/K.

In [ESZB23], the authors prove Proposition 2 by using a "local decomposition" of the height, and observing that the local factors are related to the Artin conductor of a certain representation, as in [WY15, Sect. 4]. We briefly outline the way to go from the setting in [ESZB23] to [WY15].

Proof Let $x \in BG(K)$ be a G-torsor over K and $\rho : G \to S_n$ as in the statement of the proposition. One can extend ρ to a map $G \to \mathrm{GL}_n(\mathcal{O}_K)$ via permutation matrices. We will denote this by ρ as well. Let $V := \mathcal{O}_K^{\oplus n}$ equipped with the trivial metric, let $\mathcal{V} := (V, \rho)$ be the vector bundle on $BG_{\operatorname{Spec} \mathcal{O}_K}$ corresponding to ρ as in Definition 3, and let L and $\tilde{L}$ be as in Remark 8.

In the setup of (7), let $\mathcal{X} = BG_{\mathrm{Spec}\,\mathcal{O}_K}$ and take $\mathcal{C} = [\mathrm{Spec}\,\mathcal{O}_{\tilde{L}}/G]$. As a consequence, we may extend (7) to the following commutative diagram,

$$\begin{array}{ccccc} & & \mathrm{Spec}\,\mathcal{O}_{\tilde{L}} & \xrightarrow{\overline{x}_{\tilde{L}}} & \mathrm{Spec}\,\mathcal{O}_K \\ & & \big\downarrow{\scriptstyle p} & & \big\downarrow{\scriptstyle q} \\ \mathrm{Spec}\,K & \longrightarrow & \mathcal{C} & \xrightarrow{\overline{x}} & BG_{\mathrm{Spec}\,\mathcal{O}_K} \\ & & \big\downarrow{\scriptstyle \pi} & & \\ & & \mathrm{Spec}\,\mathcal{O}_K & & \end{array} \tag{8}$$

where p, q are quotient maps. In particular, q is the quotient by the trivial action of G. Note that $V = q^*\mathcal{V}$ and for every place v of K, we get a local version of this diagram.

By [ESZB23, Definition 2.21], the discussion following it and by [ESZB23, Proposition B.10], the height of x with respect to $\mathcal{V}$ satisfies

$$\mathrm{ht}_{\mathcal{V}}(x) = -\deg\left(\pi_*\overline{x}^*(\mathcal{V}^\vee)\right) = -\deg\left(\overline{x}^*\mathcal{V}^\vee\right) + \sum_{\substack{v \in M_K \\ \text{finite}}} \delta_{\mathcal{V};v}(x),$$

where the local discrepancies $\delta_{\mathcal{V};v}(x)$, which essentially measure the difference between the vector bundle on the tuning stack and its pushforward to $\mathrm{Spec}\,\mathcal{O}_K$, can be computed as

$$\delta_{\mathcal{V};v}(x) = \frac{1}{\#G} \log \left| \frac{p^*\overline{x}^*\mathcal{V}_v^\vee}{\pi_*\left(\overline{x}^*\mathcal{V}_v^\vee\right) \otimes_{\mathcal{O}_{K_v}} \mathcal{O}_{\tilde{L}_v}} \right|.$$

Here $\tilde{L}_v$ denotes the étale algebra $\prod_{w|v} \tilde{L}_w$. Further, since $V = q^*\mathcal{V}$ is the trivial vector bundle, $\deg(\overline{x}^*\mathcal{V}^\vee) = 0$. Therefore, the height decomposes into a product of local factors, and we may work locally. For notational convenience, we drop the localizing subscripts from now on: we replace K_v, L_v and $\tilde{L}_v$ by K, L and $\tilde{L}$ in what follows.

Write $L = K(\alpha)$ and let $\alpha_1 := \alpha, \alpha_2, \ldots, \alpha_n$ be the Galois conjugates of α. Note that by the way L is defined, the permutation action of G on the α_i's is determined by the embedding ρ, and vice-versa. Denote by $\sigma_i \in G$ an element that takes $\alpha_1 \mapsto \alpha_i$. If σ_i and σ_i' are two choices sending $\alpha_1 \mapsto \alpha_i$, then $\sigma_i^{-1}\sigma_i'$ stabilizes α_1 and therefore belongs to $\mathrm{Stab}_{\rho(G)}(1)$. In particular, if $\beta \in L$, then $\sigma_i(\beta) = \sigma_i'(\beta)$.

Since each σ_i fixes K,

$$V = \mathcal{O}_K^{\oplus n} \cong \oplus_{i=1}^n \sigma_i(\mathcal{O}_K)$$

as $\mathcal{O}_K$ modules. Similarly, by the commutativity of the diagram (8), we have the following isomorphisms of $\mathcal{O}_{\tilde{L}}$ modules:

$$p^*\overline{x}^*\mathcal{V} = \overline{x}_{\tilde{L}}^* V \cong V \otimes_{\mathcal{O}_K} \mathcal{O}_{\tilde{L}} \cong \mathcal{O}_{\tilde{L}}^{\oplus n} \cong \oplus_{i=1}^n \sigma_i(\mathcal{O}_{\tilde{L}}),$$

where the $\mathcal{O}_{\tilde{L}}$-module structure on $\oplus_{i=1}^n \sigma_i(\mathcal{O}_{\tilde{L}})$ is given by $x \cdot (y_1, y_2, \ldots, y_n) = (\sigma_1(x)y_1, \sigma_2(x)y_2, \ldots, \sigma_n(x)y_n)$.

Let $g \in G$. If $g(\alpha_i) = \alpha_j$ (that is, $\rho(g)(i) = j$), then $g^{-1}\sigma_j(\alpha_1) = g^{-1}(\alpha_j) = \alpha_i = \sigma_i(\alpha_1)$. Therefore, $g^{-1}\sigma_j$ and σ_i belong to the same coset of $\mathrm{Stab}_{\rho(G)}(1)$. Thus we have a G action on $p^*\overline{x}^*\mathcal{V} = \mathcal{V} \otimes \mathcal{O}_{\tilde{L}}$ given by

$$g \cdot (\sigma_1(x_1), \sigma_2(x_2), \ldots, \sigma_n(x_n)) = \left(g^{-1}\sigma_{\rho(g)(1)}(x_1), \ldots, g^{-1}\sigma_{\rho(g)(n)}(x_n)\right),$$

whose invariants are given by tuples $(\sigma_1(x_1), \sigma_2(x_2), \ldots, \sigma_n(x_n))$ such that $\sigma_i(x_i) = g^{-1}\sigma_{\rho(g)(i)}(x_i)$ for each i, that is, $x_i \in L$. In other words, the G-invariant submodule of $p^*\overline{x}^*\mathcal{V}$ is precisely the tuning module as described in the proof of [WY15, Theorem 4.8].

Now, since $p^*\overline{x}^*\mathcal{V}$ is the pullback of a vector bundle along the quotient map p, it is G-equivariant by definition. Thus, by descent theory (see for instance [Vis05, Sect. 4.4]):

$$\left(p_*p^*\overline{x}^*\mathcal{V}\right)^G \cong \overline{x}^*\mathcal{V}.$$

Taking π_* on both sides tells us that $\pi_*\overline{x}^*\mathcal{V}$ is the tuning module considered as an $\mathcal{O}_K$ module. We may now apply the proof of [WY15, Theorem 4.8] to conclude that $\delta_{\mathcal{V};v}(x) = \frac{1}{2}v(\Delta(L/K))\log(q_v)$, where q_v is the size of the residue field at v. □

Next, we consider the case of $B\mu_n$ and recall the approach of [ESZB23, Sect. 3.2] for computing heights on $B\mu_n$. While the height in the case of the constant group scheme $\mathbb{Z}/n\mathbb{Z}$ falls into the purview of the previous section, i.e., the height is exactly the discriminant, the height in the non-constant case may differ from the discriminant at finitely many places. In particular, this height is a quasi-discriminant in the sense of [Dar21]. We will explore quasi-discriminants in a later section.

Theorem 3 *Let $\chi : \mu_n \to \mathbb{G}_m$ be a primitive character. Let $\mathcal{V} = \left(\mathcal{O}_K^{\oplus n}, \oplus_{a=0}^{n-1}\chi^a\right)$ be the corresponding rank n vector bundle on $B\mu_{n,\mathcal{O}_K}$. Let $x \in B\mu_n(K)$ correspond to an étale algebra L. Assume that L/K does not contain any intermediate cyclotomic extensions, i.e., L and $K(\zeta_n)$ are linearly disjoint over K. Then, there is a constant C depending only on n and K, and $c_x \in \mathbb{R}$ satisfying $|c_x| < C$ such that*

$$\mathrm{ht}_{\mathcal{V}}(x) = c_x + \frac{1}{2}\log|\Delta(L/K)|.$$

Proof We claim that over $K(\zeta_n)$, the representation

$$\oplus_{a=0}^{n-1}\chi^a : \mu_n \to \oplus_{a=0}^{n-1}\mathbb{G}_m \hookrightarrow \mathrm{GL}_n(\mathcal{O}_K) \tag{9}$$

is isomorphic to a permutation representation of $\mathbb{Z}/n\mathbb{Z}$ acting on itself. Indeed, if χ sends $\zeta_n \mapsto \zeta_n^k$ for some positive integer k with $\gcd(k, n) = 1$, then the composition (9) sends

$$\zeta_n \mapsto (\zeta_n^0, \zeta_n^k, \zeta_n^{2k}, \ldots, \zeta_n^{(n-1)k}) \mapsto [e_1, e_k, e_{2k}, \ldots, e_{(n-1)k}],$$

where e_i is the standard i-th basis vector, where we take i modulo n. This is a permutation matrix of order n. By Lemma 2, the group schemes $\mathbb{Z}/n\mathbb{Z}$ and μ_n are isomorphic over $K(\zeta_n)$. Further, the action described above is the same as $\mathbb{Z}/n\mathbb{Z}$ acting on itself by $a \cdot b = ka + b$. In particular, over $K(\zeta_n)$ this becomes a permutation representation associated to a constant group scheme, thus falling into the setting of Proposition 2. Let y denote the base change of x to $K(\zeta_n)$ and let $L(\zeta_n)$ denote the corresponding étale algebra.

By Proposition 2,

$$\mathrm{ht}_{\mathcal{V}_{K(\zeta_n)}}(y) = \frac{1}{2}\log|\Delta(L(\zeta_n)/K(\zeta_n))|.$$

Let $\mathcal{S}$ denote the set of primes ramified in $K(\zeta_n)/K$. By [Neu99, Corollary III.2.12], it is precisely the set of primes dividing n. By the theory of local discrepancies in [ESZB23, Sect. 2] and the fact that $\mathcal{V}$ is the trivial vector bundle on Spec $\mathcal{O}_K$, one can decompose the height on $(B\mu_n)_K$ as

$$\mathrm{ht}_{\mathcal{V}}(x) = \sum_{v \in \mathcal{S}} \delta_{\mathcal{V},v}(x) + \sum_{v \notin \mathcal{S}} \delta_{\mathcal{V};v}(x).$$

Let $\mathcal{V}_n$ denote the pullback of $\mathcal{V}$ to $B\mu_{n,K(\zeta_n)}$. Then by [ESZB23, Proposition 2.24], over $K(\zeta_n)$

$$\begin{aligned}
\mathrm{ht}_{\mathcal{V}_n}(y) &= \sum_{w|v, v\in\mathcal{S}} \delta_{\mathcal{V}_n;w}(y) + [K(\zeta_n):K]\sum_{v\notin\mathcal{S}} \delta_{\mathcal{V};v}(x) \\
&= \sum_{w|v, v\in\mathcal{S}} \delta_{\mathcal{V}_n;w}(y) + [K(\zeta_n):K]\left(\mathrm{ht}_{\mathcal{V}}(x) - \sum_{v\in\mathcal{S}} \delta_{\mathcal{V},v}(x)\right).
\end{aligned}$$

Since $\mathcal{S}$ is a fixed finite set depending only on n and K, by [ESZB23, Proposition 2.25] there is a constant C depending only on n and K, such that

$$\mathrm{ht}_{\mathcal{V}}(x) + c_x = \frac{1}{[K(\zeta_n):K]}\mathrm{ht}_{\mathcal{V}_n}(y) = \frac{1}{2[K(\zeta_n):K]}\log|\Delta(L(\zeta_n)/K(\zeta_n))|,$$

and $|c_x| < C$. Further, by properties of discriminants in towers and compositums [Neu99, Sect. III.2],

$$\mathrm{Nm}_{K(\zeta_n)/K}(\Delta(L(\zeta_n)/K(\zeta_n))) = \Delta(L/K)^{[K(\zeta_n):K]},$$

where $\mathrm{Nm}_{K(\zeta_n)/K}$ denotes the norm. We see that $\mathrm{ht}_{\mathcal{V}}(x)$ is equal to $\frac{1}{2}\log|\Delta(L/K)|$ upto a constant that is bounded on $B\mu_n(K)$. □

Remark 9 In [ESZB23, Sect. 3.2], the authors describe the height on $B\mu_{n,\mathbb{Q}}$ with respect to the line bundle χ itself (as opposed to the vector bundle used in Theorem 3 above). Suppose for simplicity that χ comes from the standard representation, i.e., $\chi(\zeta_n) = \zeta_n$. By considering $B\mu_n$ as a zero-dimensional weighted projective stack,[3] the authors show that if $a \in \mathbb{Q}^*/(\mathbb{Q}^*)^n$ is an n-th power free integer, then the height of the corresponding μ_n-torsor over $\mathbb{Q}$ is $\log |a|^{1/n}$ up to a bounded function. While this is not directly relevant to this article, it gives a nice description of that height in terms of the torsor, and moreover shows ways in which one can use sums and powers of χ to construct various height functions on $B\mu_n$, i.e., number field counting invariants.

3.2.2 Darda's Height

In this subsection, we briefly describe the height defined in [Dar21]. In his work, Darda defines the notion of a quasi-toric height on a weighted projective stack. He then specializes the height to the case of $B\mu_n$, viewing it as the 0-dimensional weighted projective stack $\mathbb{P}(n)$.

As before, let K be a number field and let v denote a place of K. Let r denote the smallest prime factor of n.

Definition 5 ([Dar21, Lemma 9.1.3.1]) For every $n \in \mathbb{N}$ and every place v of K, we define a function $f_v^{\Delta,n} : K_v^\times \to \mathbb{R}_{\geq 0}$ as follows:

$$f_v^{\Delta,n}(a) = \begin{cases} |a|_v^{1/n} \left|(\Delta((K_v[t]/(t^n - a))/K_v))^{1/(n^2-n^2/r)}\right| & \text{if } v \text{ finite,} \\ |a|_v^{1/n} & \text{otherwise,} \end{cases}$$

where $|\cdot|_v = q_v^{-v(\cdot)}$ denotes the v-adic absolute value if v is finite, the real absolute value if v is real, and the square of the complex absolute value if v is complex, and $|\cdot|$ denotes the absolute ideal norm.

Lemma 3 *For $x \in B\mu_n(K)$, define the discriminant height as*

$$H^\Delta(x) := \prod_{v \in M_K} f_v^{\Delta,n}(x).$$

Then $H^\Delta(x)^{n^2-n^2/r}$ is precisely the discriminant of the torsor K_a corresponding to x.

Proof Let $a \in K^\times$ be an element such that K_a represents the class of $x \in B\mu_n(K) \cong K^\times/(K^\times)^n$. By the product formula, the multiplicativity of the ideal norm and the local decomposition of the discriminant, we have that $H^\Delta(x)$ is $|(\Delta((K[t]/(t^n - a))/K))^{1/(n^2-n^2/r)}|$. □

[3] See [Dar21, Chapter 3] for a detailed definition of weighted projective stacks.

In his PhD thesis [Dar21], Darda defined quasi-discriminant heights as functions that agree with the discriminant height at all but finitely many places.[4] He shows that any quasi-discriminant height has the Northcott property, and gives a general framework to use height zeta functions for counting points of bounded quasi-discriminant height on weighted projective stacks.

3.2.3 The Darda–Yasuda Perspective

In their work [DY22a] on a stacky version of the Batyrev–Manin conjecture, Darda and Yasuda define the height of a point on a stack $\mathcal{X}$ with respect to a "raised line bundle", which is the data of a line bundle on the coarse space of X, along with a "raising function", that captures the stackiness of $\mathcal{X}$. In this subsection, we briefly outline their definition of height on $B\mu_n$ as in [DY22a, DY23], and describe its relation to the discriminant.

Definition 6 ([DY22a, Definition 2.3]) Let $\mathcal{X}$ be a tame Deligne-Mumford stack.[5] Define the stack of 0-jets as

$$\mathcal{J}_0\mathcal{X} = \bigsqcup_{l>0} \underline{\mathrm{Hom}}_K^{rep}\left(B\mu_{lK}, \mathcal{X}\right),$$

where $\underline{\mathrm{Hom}}_K^{rep}$ denotes the stack of representable K-morphisms. We refer the reader to [Ols16, Definition 5.1.5] for a definition of representability. However for this article, one may think of this as requiring that the induced maps on automorphism groups of the points be injections (see for instance [Ces17, Lemma 3.2.2 (b)]). Henceforth, we will only keep track of the connected components of this stack, $\pi_0(\mathcal{J}_0\mathcal{X})$. Following [DY22a] we denote by $\pi_0{}^*(\mathcal{J}_0\mathcal{X})$ the set of twisted sectors, i.e., elements of $\pi_0(\mathcal{J}_0\mathcal{X})$ coming from $l > 1$.

When $\mathcal{X} = B\mu_n$, the disjoint union ranges over $l|n$, and for each such l, $\pi_0(\underline{\mathrm{Hom}}_K^{rep}(B\mu_{l,K}, B\mu_{n,K}))$ can be identified with the group of group homomorphisms $\mu_l \to \mu_n$ that are isomorphisms onto their image. This group is isomorphic to $(\mathbb{Z}/l\mathbb{Z})^\times$. Thus $\pi_0(\mathcal{J}_0 B\mu_n)$ can be identified with $\mathbb{Z}/n\mathbb{Z}$. For more on this, see [DY22a, Example 2.15].

For $B\mu_n$ over a number field K, one may think of the set of twisted sectors as keeping track of the possible (tame) inertia groups of μ_n-torsors over K. In fact, this is true more generally for BG, where G is a finite étale group scheme.

Definition 7 A raising datum, which by abuse of notation we will call a raising function, on $BG(K)$ is a collection $(c_v)_{v\in M_K}$ of continuous functions

$$c_v\colon BG(K_v) \to \mathbb{R}_{\geq 0}$$

[4] More precisely, a quasi-discriminant height coming from a degree k family is a product of local functions f_v that agree with $\left(f^{\Delta,n}\right)^k$ at all but finitely many places.

[5] See [BV24, Proposition 2.1] for the definition of a tame stack.

such that for all but finitely many $v \in M_K$, the function c_v factors through $BG(K_v) \xrightarrow{res_v} \pi_0(\mathcal{J}_0 BG) \xrightarrow{c} \mathbb{R}$, for some function c satisfying $c(BG) = 0$. Here res_v is a residue map that takes a point $x_v \in BG(K_v)$ to the induced map $B\mu_l \to BG$ as in the discussion following [DY22a, Lemma 2.16].

Each residue map essentially assigns to a local G extension its (tame) inertia group. In particular, the maps res_v are only defined outside of a finite set of places $v \in M_K$. While the definition of a raising function might appear nebulous at first glance, it performs a very natural function: it assigns local values at v to an étale algebra, depending on the tame inertia groups above the place v.

Definition 7 is related to counting functions in [Woo10] and class functions in [EV05].

Definition 8 Maintaining the notation above, the height corresponding to a raising function $c = (c_v)_{v \in M_K}$ is defined as

$$\mathrm{Ht}_c(x) = \prod_{v \in M_K} q_v^{c_v(x_v)},$$

where q_v is the size of the residue field at v if v is finite, and Euler's number e if it is infinite.

In particular, on BG for a constant group scheme G, by taking the local factor c_v to be v-adic valuation of the local discriminant at the place v, one obtains a raising function c for which $\mathrm{Ht}_c(x)$ is the discriminant of the corresponding étale algebra. See for instance [DY23, Sect. 2.5.3].

Remark 10 For a more general stack $\mathcal{X}$, the height defined in [DY22a] has a component coming from the coarse space of $\mathcal{X}$. However, in the case of BG_K, the coarse space is just a point, so we do not see this contribution.

Remark 11 One of the advantages of the Darda–Yasuda framework is that it allows for height functions that do not necessarily arise from a vector bundle. For instance, [DY22a, Example 4.11] defines a raising function on $B\mathbb{Z}/p\mathbb{Z}$ whose associated height cannot agree with the height with respect to any vector bundle in the sense of [ESZB23]. On the other hand, the Darda–Yasuda framework currently only works for tame stacks, while the Ellenberg–Satriano–Zureick-Brown framework works for wild stacks as well. For general stacks, the work in [Lan24] illustrates that there are heights studied widely that do not fit into either framework: it is shown that the Faltings height on the moduli stack of elliptic curves in characteristic three does not come from a vector bundle in the sense of [ESZB23], but this stack is not tame and so does not fit into the Darda–Yasuda framework.

4 Malle's Conjecture

In this section, we recall Malle's conjecture in Sect. 4.1 and we compare it with the recent stacky conjectures of [ESZB23] in Sect. 4.2.1 and [DY22a] in Sect. 4.2.2.

4.1 Classical Malle's Conjecture

In [Mal02, Mal04], Malle formulated a conjecture on the order of growth of the number of number fields of degree n over a given base number field K with given Galois group $G \hookrightarrow S_n$ ordered by discriminant. According to this conjecture the order of growth is expected to depend on two constants associated to the group G, denoted by $a(G)$ and $b(K, G)$ and defined below.

We view G as a subgroup of the permutation group S_n such that G acts transitively on $\{1, \ldots, n\}$. For $g \in G$, let

$$n(g) := \text{number of orbits under the action of } g \text{ on } \{1, \ldots, n\}.$$

Note that $n(g)$ depends only on the conjugacy class of g. The number $n - n(g)$ is called the *index* of g. Define

$$a(G) = (\min\{n - n(g) : g \in G, g \neq 1\})^{-1}.$$

Note that $a(G)$ depends on the embedding $G \hookrightarrow S_n$, and is equal to 1 if and only if the image of G contains a transposition.

The constant $b(K, G)$ may depend also on the base number field K. Let $\Gamma = \mathrm{Gal}(\overline{\mathbb{Q}}/\mathbb{Q})$ denote the absolute Galois group of $\mathbb{Q}$. Let $\chi : \Gamma \to \hat{\mathbb{Z}}^\times$ denote the cyclotomic character, i.e., for every $\sigma \in \Gamma$ and $m \in \mathbb{N}$, define $\chi(\sigma)$ by $\sigma \cdot \zeta_m = \zeta_m^{\chi(\sigma) \bmod m}$ for any primitive m-th root of unity ζ_m. Since for any m, $\chi(\sigma)$ is invertible modulo m, the image of χ does indeed land inside $\hat{\mathbb{Z}}^\times$. Define the action of Γ on the set $\mathrm{Cl}(G)$ of conjugacy classes of G as follows. For $\sigma \in \Gamma$ and $g \in G$, set $\sigma \cdot g := g^{\chi(\sigma)}$. This gives an action on $\mathrm{Cl}(G)$ and further, g and $g^{\chi(\sigma)}$ have the same index. The subgroup $\Gamma_K = \mathrm{Gal}(\overline{\mathbb{Q}}/K) \leq \Gamma$ thus also acts on $\mathrm{Cl}(G)$, and we can define

$$b(K, G) = \# \left(\{[g] \in \mathrm{Cl}(G) : n - n(g) = a(G)^{-1}\} / \Gamma_K \right),$$

the number of Γ_K-orbits of conjugacy classes of G with minimal index. In [Mal04, Lemma 2.2], Malle showed that if $a(G)$ is 1, then $b(K, G)$ is also 1 for all number fields K. The dependence of the constant $b(K, G)$ on the field K highlights the effect of having roots of unity in the base field K. If $\zeta_m \in K$, then for $\sigma \in \Gamma_K$, $\chi(\sigma) \equiv 1 \bmod m$. So if $m | \#G$, then Γ_K fixes more elements and thus $b(K, G)$ is smaller. For a more precise explanation, see [Mal04, Example 2.1] or Example 6 below.

Conjecture 1 (Weak Malle's conjecture [Mal02]) *Let K be a number field with algebraic closure $\overline{K}$. Let G be a subgroup of the group S_n of permutations of n elements such that G acts transitively on $\{1, \dots, n\}$. Let $M(K, G; B)$ be the number of finite field extensions L/K of degree n contained in $\overline{K}$ such that the Galois closure of L has Galois group G and the norm of the discriminant of L/K is bounded above by $B > 0$. Then for all $\varepsilon > 0$, there exist positive real constants $c_1(K, G)$ and $c_2(K, G, \varepsilon)$ such that*

$$c_1(K, G)B^{a(G)} \leq M(K, G; B) \leq c_2(K, G, \varepsilon)B^{a(G)+\varepsilon}. \tag{10}$$

A stronger version of the conjecture, stated in [Mal04], asserts the following:

Conjecture 2 (Strong Malle's conjecture [Mal04]) *Under the same assumptions,*

$$M(K, G; B) \sim c(K, G)B^{a(G)}(\log B)^{b(K,G)-1}, \tag{11}$$

for a positive constant $c(K, G)$.

To our knowledge there is no general conjecture for the value of the leading constant $c(K, G)$. A conjectural formula for $c(K, S_n)$ has been given by Bhargava in [Bha07], which is compatible with the known values for $n \leq 5$. In the cases where G is abelian and $G = \mathbb{Z}/n\mathbb{Z}$, the exact values of $c(K, G)$ are computed in [Wri89, Dar21], respectively.

Several variations of Malle's conjecture have been studied in the last few decades. For instance, one may generalize Malle's conjecture to finite extensions of $\mathbb{F}_q(t)$ in the case when q is coprime to $|G|$. In this case, the discriminant is replaced by q raised to the degree of the ramification divisor. See [EV05] for a description of this problem in terms of counting $\mathbb{F}_q$ points on Hurwitz spaces, as well as a heuristic argument for why it must hold. One can formulate Malle-type conjectures for invariants other than the discriminant, such as the conductor or any counting function ([Woo10]). The work in this area is vast and growing, but for the purpose of this article we will restrict our attention to discriminants. In the counting function $M(K, G; B)$, one may instead count isomorphism classes of extensions. This would change the leading constant of the asymptotic formula by a factor $\# \operatorname{Aut}(G)$. In [Mal04], Malle formulates the weak conjecture also for number fields that are unramified outside a finite set S. The a-invariant in this case is the same as before, and is independent of S.

Before proceeding to what is known about this conjecture, we give some examples of what it says in some cases of interest in this article.

Example 5 (*Regular representation*) If $n = \#G$ and the inclusion $G \subseteq S_n$ is induced by the action of G on itself by the group operation, then $M(K, G; x)$ is the number of degree n Galois extensions of K of Galois group G and discriminant bounded by x. In this case, $a(G) = \frac{1}{n-n/r}$ where r is the smallest prime number dividing n, i.e., the smallest order of a non-zero element of G.

Example 6 (*Cyclic groups*) If $G = \mathbb{Z}/n\mathbb{Z}$ with the inclusion in S_n given by the group operation as in (6), then Malle's conjecture concerns Galois field extensions of degree n, i.e., connected K-points of $B(\mathbb{Z}/n\mathbb{Z})$. As in Example 5, $a(G) = \frac{1}{n-n/r}$, where r is the smallest prime number dividing n. To compute $b(K, G)$, note that $\mathbb{Z}/n\mathbb{Z}$ contains $\phi(r) = r - 1$ elements of order r, all of which have the same index. If K contains all the r-th roots of unity, then all of these elements (viewed as conjugacy classes) are fixed by Γ_K and so $b(K, G) = 1$. If $K = \mathbb{Q}$, then $b(K, G) = r - 1$. A similar argument shows that in general, $b(K, G) = (r - 1)/[K(\zeta_r) : K]$.

Remark 12 Conjectures 10 and 2 concern only K-rational points on BG_K that correspond to connected torsors, where G is a constant group scheme over Spec K. One may instead set up a different counting function

$$T(K, G; B) := \#\{x \in BG(K) \mid |\Delta(x)| < B\},$$

which for constant G would count the number of isomorphism classes of étale G-algebras of bounded absolute discriminant. While the space of irreducible polynomials is Zariski dense in the space of all polynomials, it is generally not true that $M(K, G; B)$ makes up 100% of $T(K, G; B)$. This only happens in the case when G has prime exponent (see [Woo10, Sect. 2.1]).[6]

4.1.1 Known Results and Counter-Results

The strong version of Malle's conjecture is known to be true in some cases and false in some others. Some of the earliest work in this area dates back to work of Davenport–Heilbronn [DH71] for $G = S_3$, and Mäki [Mak85] for abelian G with the regular representation and $K = \mathbb{Q}$. In [Wri89], Wright proved Malle's conjecture for (Galois) abelian extensions over arbitrary number fields and for function fields with characteristic coprime to $|G|$. In recent work, Darda [Dar21] proved the following result.

Theorem 4 ([Dar21, Corollaries 9.2.5.9, 9.2.5.10]) *Let H be a quasi discriminant height coming from a degree n family, i.e., one whose local factors differ from $(H^{\Delta})^n$ (Lemma 3) at finitely many places. Let r be the smallest prime dividing n. Then, the number of K points of $B\mu_n$ with height bounded by B grows asymptotically like*

$$C_{K,H,r} B(\log B)^{r-2},$$

where $C_{K,H,r} > 0$ is a positive constant. In particular, in light of Lemma 3,

$$T(K, \mu_n; B) \sim C_{K,H^{\Delta},r} B^{1/(n-n/r)}(\log B)^{r-2}.$$

---

[6] In the terminology of [Woo10], the discriminant is a "fair" counting function only when G has prime exponent.

When the base field contains all the n-th roots of unity, μ_n and $\mathbb{Z}/n\mathbb{Z}$ are isomorphic as group schemes and the powers of B and $\log B$ in [Wri89, Dar21] agree. In light of Remark 12, note that the constants in front in general, do not.

Malle's conjecture has been proved in several non-abelian cases, including work of Bhargava [Bha05, Bha10] and Bhargava–Shankar–Wang [BSW15] for S_4 and S_5, Wang [Wan21] for certain compositums of fields, Cohen–Diaz-y-Diaz–Olivier [CDyDO02] for D_4 extensions, among many others. Since it is impossible to do justice to the vast body of literature in this area within the scope of this article, we refer the reader to [Wan21, Alb21], and [Woo10] for a comprehensive survey. The weak form of Malle's conjecture was proven by Klüners and Malle [KM04] for nilpotent groups with the regular representation; the strong upper bound was proved by Klüners [Klü22], and a recent result of Koymans and Pagano [KP23] proves the strong form of Malle's conjecture for a large class of nilpotent groups. In [Alb20], Alberts proved a general upper bound in the case of solvable groups.

Remark 13 Let K_a/K denote the étale algebra described in Sect. 2. Then $\mathrm{Gal}(K_a/K)$ is contained in some extension E of $\mathbb{Z}/n\mathbb{Z}$ by $(\mathbb{Z}/n\mathbb{Z})^\times$ and is in particular solvable. There is a finite-to-finite relation between étale algebras of the form K_a and étale algebras with Galois group contained in E. Further, there are finitely many possibilities for E. Thus one may at the very least derive weak bounds on counts of étale algebras of the form K_a from the known results about solvable groups.

Remark 14 If K is a number field, we can assume that a is an algebraic integer. By Eisenstein's criterion, if the polynomial $t^n - a$ is reducible, then a is a squareful number, that is, every prime appearing in the factorization of the ideal (a) into prime ideals has exponent at least 2. The sharp criterion for irreducibility can be found in [Lan02, Theorem VI.9.1]. If $K = \mathbb{Q}$, identifying $\mathbb{Q}^\times/(\mathbb{Q}^\times)^n$ with the set of n-th power free integers (positive if n is odd), the set of irreducible polynomials $t^n - a$ with constant term of size at most B has growth $\gg B$, while the set of reducible polynomials $t^n - a$ with constant term of size at most B grows at most as $O(B^{\frac{1}{2}})$. Hence, according to this measure, most $\mathbb{Q}$-points of $B\mu_n$ correspond to étale algebras K_a that are fields.

Of course, the elephant in the room for any discussion on Malle's conjecture is the fact that it is known to be false in some cases. In [Klü05], Klüners showed that for $G = C_3 \wr C_2 \hookrightarrow S_6$, the strong form of Malle's conjecture is incorrect. In this case, the conjecture predicts that $M(\mathbb{Q}, G; B) \asymp B^{1/2}$. However, Klüners shows that one actually obtains $B^{1/2} \log B$. In particular, only the weak form of the conjecture still holds. He also points out that if one instead counts $C_3 \wr C_2$ fields that do not contain intermediate cyclotomic extensions, one does indeed obtain the asymptotic predicted in Conjecture 2. Following this, Türkelli [Tür15] suggested a modified version of the conjecture over function fields $\mathbb{F}_q(t)$ in the case $\gcd(q, |G|) = 1$. Containing intermediate cyclotomic extensions in the number field case is analogous to having non-trivial constant subextension in the function field case. Türkelli's idea was to keep track of the constant subextensions of the extensions of function fields and modify the b-constant accordingly. See [Tür15, Sec. 2, 5]. The counterexamples to

Malle's conjecture are believed to stem from the existence of "accumulating" sets (akin to breaking thin sets in [DY22a]): these are "small" sets that contribute more than expected to the counting function.

4.1.2 Malle-Type Conjectures for Torsors

One useful perspective on counting number fields with a given Galois group (and indeed this is a perspective used in several papers mentioned above), is that this amounts to counting the number of continuous surjections $\Gamma_K \to G$. Similiarly, counting étale algebras amount to counting continuous homormorphisms. Note that if Γ_K acts on G trivially, then $H^1(\Gamma_K, G) = \mathrm{Hom}(\Gamma_K, G)$. This leads to the question of counting classes in the Galois cohomology group $H^1(\Gamma_K, G)$ for more general group actions and is related to counting points on BG for non-constant group schemes. In this context, there have been several attempts at formulating a "twisted" version of Malle's conjecture for non-constant group schemes G. Some classical approaches and results can be found in [Alb21] or [AO21], while a more stacky approach is taken in [DY23] or [DY22b].

4.2 The Stacky Batyrev–Manin–Malle Conjecture

We end our article with a discussion on Malle-type conjectures for torsors, i.e., for points on the stack BG. When G is a constant group scheme, these conjectures specialize to the classical versions of Malle's conjecture described above in Sect. 4.1. When G is non-constant, the story is more complicated: in [DY23], the authors give such a conjecture, although in this case the height of a point does not admit a straightforward description as the discriminant of an étale algebra.

In this section, we focus our attention on the constant group scheme $\underline{\mathbb{Z}/n\mathbb{Z}}$. We briefly describe work of [ESZB23, DY22a] in the context of Malle's conjecture, recalling the definitions of the invariants that they define and showing that they agree with the invariants for Malle's conjecture for $B(\mathbb{Z}/n\mathbb{Z})$. We remark that their conjectures apply to more general stacks, but that is beyond the scope of this article.

In what follows, we assume that the base field K contains a primitive n-th root of unity, i.e., that $B\mu_n$ and $B(\mathbb{Z}/n\mathbb{Z})$ are isomorphic over K. For the conjectures for general finite G see [DY23].

4.2.1 The Vector Bundle Perspective

For a "nice" stack $\mathcal{X}$, the authors in [ESZB23] define a notion of being "generically" bounded below, that helps them establish finiteness properties.

Definition 9 A function $f : \mathcal{X}(\overline{K}) \to \mathbb{R}$ is said to be generically bounded below if there is some proper substack $Z \hookrightarrow \mathcal{X}$ and a constant B such that

$$\{x \in \mathcal{X}(\overline{K}) \mid f(x) < [K(x) : K] \cdot B\} \subset Z(\overline{K}),$$

where $K(x)$ is the field of definition of x.

When $\mathcal{X} = BG_{\mathcal{O}_K}$ for a smooth finite constant group scheme G over a number field K, $\mathcal{X}$ has no proper closed substacks, and this notion reduces to a more common notion of being bounded below.

Lemma 4 *A function $f : BG(\overline{K}) \to \mathbb{R}$ is generically bounded below as in Definition 9 if and only if there is a constant B such that for all $x \in BG(\overline{K})$, $f(x) \geq [K(x) : K]B$.*

Proof Every closed substack of $BG_{\mathcal{O}_K}$ is of the form $[Z/G]$ for a closed subscheme Z of Spec $\mathcal{O}_K$ by [Alp24, Exercise 3.3.9]. If L/K is a finite field extension, an L-point on $[Z/G]$ is given by a G-torsor $Y \to \operatorname{Spec} L$ with a morphism of schemes $Y \to Z$. Since G is affine, $Y = \operatorname{Spec} A$ for an étale L-algebra A that is also an algebra over $\mathcal{O}_K/I$, where $Z = \operatorname{Spec} \mathcal{O}_K/I$. If $I \neq (0)$, then the ring homomorphism $\mathbb{Z} \to \mathcal{O}_K/I$ is not injective, but the ring homomorphism $\mathbb{Z} \to A$ is injective. Thus such a scheme Y doesn't exist, and $[Z/G](\overline{K}) = \emptyset$. □

Let K be a number field, G a constant group scheme over K, and $x \in BG(K)$. If $\mathcal{C} \to \operatorname{Spec} \mathcal{O}_K$ is a tuning stack corresponding to x, define the expected deformation dimension to be

$$\operatorname{edd}(x) := \sum_{\substack{v \in M_K, \\ \text{ramified in } \mathcal{C}}} \log q_v,$$

where q_v denotes the size of the residue field at v. This is a particular case of a more general definition, and seeks to imitate the anti-canonical height of a point. See [ESZB23, Sect. 4.1].

Given a metrized vector bundle $\mathcal{V}$ on $BG_{\mathcal{O}_K}$ and $a' \in \mathbb{R}$, define

$$D_{a'}(x) = a' \operatorname{ht}_{\mathcal{V}}(x) - \operatorname{edd}(x).$$

Definition 10 Define

$$a(BG, \mathcal{V}) := \inf\{a' \in \mathbb{R} \mid D_{a'}(x) \text{ is generically bounded below on } BG(\overline{K})\}.$$

If this infimum doesn't exist, then $a(BG, \mathcal{V}) := -\infty$.

Conjecture 3 (Weak stacky Batyrev–Manin–Malle conjecture [ESZB23, Conjecture 4.14]) *Let $\mathcal{X} = BG_{\mathcal{O}_K}$, equipped with a vector bundle $\mathcal{V}$. Consider for any open substack $\mathcal{U} \subset \mathcal{X}$, the counting function*

$$\mathcal{N}_{\mathcal{U},\mathcal{V}}(B) = \#\{x \in \mathcal{X}(K) \mid \operatorname{Ht}_{\mathcal{V}}(x) < B\}.$$

Then there is a choice of $\mathcal{U}$ such that for every $\epsilon > 0$, there is a constant c_ϵ satisfying

$$c_\epsilon^{-1} B^{a(\mathcal{X},\mathcal{V})} \ll \mathcal{N}_{\mathcal{U},\mathcal{V}}(B) \ll c_\epsilon B^{a(\mathcal{X},\mathcal{V})+\epsilon}.$$

Of course, [ESZB23, Conjecture 4.1.] is stated for more general $\mathcal{X}$, namely for proper Artin stacks. In this article, we compute the invariant $a(B\mu_n, \mathcal{V})$. For a heuristic argument for BG for more general G, we refer the reader to [ESZB23].

Proposition 3 *Let K be a number field containing the n-th roots of unity, let $\mathcal{X} = B\mu_n$ and let $\mathcal{V}$ be the vector bundle corresponding to the regular representation as in Proposition 2. Let r be the smallest prime dividing n. Then*

$$a(B\mu_n, \mathcal{V}) = \frac{2}{n - n/r}.$$

Note that the factor 2 appears here because the height in Proposition 2 is the square root of the discriminant, rather than the discriminant.

Proof Let $x \in B\mu_n(K)$ be a point corresponding to the K-algebra K_a for some non-zero element $a \in \mathcal{O}_K$. Combining Proposition 2, [ESZB23, §4.4] and Proposition 1, we get

$$\begin{aligned} D_{a'}(x) &= a' \operatorname{ht}_{\mathcal{V}}(x) - \operatorname{edd}(x) \\ &= \frac{a'}{2} \log |\Delta(K_a/K)| - \sum_{v|\Delta(K_a/K)} \log q_v \\ &= \frac{a'}{2} \left(\sum_{v|a, v\nmid n} (n - d_v) \log q_v + \sum_{v|n} e(a, v) \log q_v \right) - \sum_{v|\Delta(K_a/K)} \log q_v \\ &= \sum_{v|\Delta(K_a/K), v\nmid n} \left(\frac{a'}{2}(n - d_v) - 1 \right) \log q_v + \sum_{v|\Delta(K_a/K), v|n} (e(a, v) - 1) \log q_v. \end{aligned}$$

Note that since $0 \le e(a, v) \le C_{K,n}$ as in Proposition 1 and n is fixed, the second part of this sum is uniformly bounded by a constant for all x. We will show that if $\frac{a'}{2}(n - d_v) - 1 < 0$ for some $a' \in \mathbb{R}$, some non-zero $a \in \mathcal{O}_K$ and some $v \nmid n$ with $v \mid a$, then the first sum can be made arbitrarily negative on $B\mu_n(K)$.

Observe that if $a = b^d$ with $d \mid n$ and $b \in \mathbb{Z}$ square free, coprime to n, and such that every prime number that divides b splits completely in $\mathcal{O}_K$, then $n - d_v = n - d$ for every place of K that divides a but not n.

If $\frac{a'}{2}(n - d_v) - 1 < 0$ for some a', a, and v as above, choose d such that $d = d_v = \gcd(v(a), n)$. By the Chebotarev density theorem [Tsc26], there are infinitely many prime numbers that split completely in K. So we can produce a sequence of integers b as above where each element has more prime factors than the previous one, thus making the first part of $D_{a'}(x)$ arbitrarily negative using x coming from b^d. Therefore, the function $D_{a'}$ is not bounded from below on $B\mu_n(K)$. Thus $D_{a'}$ is generically bounded below on $B\mu_n(\overline{K})$ if and only if

$$\frac{a'}{2}(n - d_v) - 1 \geq 0$$

for all $v \mid \Delta(K_a/K)$ such that $v \nmid n$ and for all $a \in \mathcal{O}_K \setminus \{0\}$, where $d_v = \gcd(v(a), n)$. Since the infimum is the largest lower bound,

$$a(B\mu_n, \mathcal{V}) = \max_{a \in \mathcal{O}_{K,\neq 0}} \max_{v \mid \Delta(K_a/K), v \nmid n} \frac{2}{n - d_v}.$$

We observe that if $d_v \neq n$, then

$$n - d_v \geq n - n/r.$$

While if $d_v = n$, then $n - d_v = 0$, so that $v \nmid \Delta(K_a/K)$. Moreover, if $a = b^{n/r}$ with $b \in \mathbb{Z}$ square free, coprime to n, and such that every prime number that divides b splits completely in $\mathcal{O}_K$, then $n - d_v = n - n/r$ for every place of K that divides a. Thus, $a(B\mu_n, \mathcal{V}) = 2/(n - n/r)$. □

4.2.2 The Raising Function Perspective

The current state-of-the-art conjecture for the b-invariant in the stacky analogue of Malle's conjecture comes from work of Darda and Yasuda in [DY22a, DY23]. In this section, we discuss the computation of the a and b invariants from their work.

Recall from Sect. 3.2.3, that Darda and Yasuda define the height of a point on BG via raising data as in Definition 7, which captures inertia data of the corresponding étale G-algebra. The raising function $c : \pi_0(\mathcal{J}_0\mathcal{X}) \to \mathbb{R}_{\geq 0}$ plays an important role in their version of Malle's conjecture. More precisely, define,

$$a(c) := \max\{c(\mathcal{Y})^{-1} \mid \mathcal{Y} \in \pi_0{}^*(\mathcal{J}_0\mathcal{X})\}$$
$$b(c) := \#\{\mathcal{Y} \in \pi_0{}^*\mathcal{J}_0\mathcal{X} \mid c(\mathcal{Y}) = a(c)^{-1}\}.$$

In [DY22a, Proposition 9.25], the authors show that these invariants agree with the definition of the a and b invariants for more general stacks via the effective cone defined in [DY22a, Definitions 9.3, 9.7].

By [DY23, Lemma 2.5.5], if the local factors c_v are defined as the valuations of the local discriminants, then the global function c is precisely the index as in Malle's conjecture. Since in the case where $\zeta_n \in K$, the set of sectors $\pi_0(\mathcal{J}_0\mathcal{X})$ can be identified with $\mathbb{Z}/n\mathbb{Z}$, which in turn can be identified with the set of conjugacy classes in $\mathbb{Z}/n\mathbb{Z}$, $a(c)$ is precisely the minimal index $1/(n - n/r)$.

Remark 15 The proof of [DY22a, Proposition 9.24] gives an intuitive reason for why one might expect the definition of the a-invariant in [ESZB23] to be given via the condition of being generically bounded below. The latter condition imitates the condition of belonging to the effective cone as in Manin's conjecture.

Conjecture 4 ([DY22a, Conjecture 9.10], [DY23, Conjecture 2.6.5]) *Let $\mathcal{X} = BG_K$. Then there is a thin set T as in [DY22a, Sect. 9.1] and a positive constant A such that*

$$\#\{x \in BG(K) \setminus T \mid \mathrm{Ht}_c(x) < B\} \sim A \cdot B^{a(c)} \log(B)^{b(c)-1}.$$

Acknowledgements The authors would like to thank Brandon Alberts and Remy van Dobben de Bruyn for useful conversations about parts of the paper. Thanks go also to Aaron Landesman for useful comments, and to the anonymous referee for suggestions that improved the exposition and for coming up with the example that is now in Remark 2.18. The writing of this article was initiated at the Women in Numbers 6 workshop at the Banff International Research Station in Banff, Alberta, Canada. The authors are grateful for having the opportunity to participate in this workshop. SA was partially supported by the NSF award DMS-2327098. JP was partially supported by the NSF award DMS-2152182. MP was partially supported by the Dutch Research Council (NWO) grants VI.Vidi.213.019 and OCENW.XL21.XL21.011. SS was partially supported by the Dutch Research Council (NWO) grant OCENW.XL21.XL21.011. For the purpose of open access, a CC BY public copyright license is applied to any Author Accepted Manuscript version arising from this submission.

References

[Alb20] Brandon Alberts. The weak form of Malle's conjecture and solvable groups. *Res. Number Theory*, 6(1):Paper No. 10, 23, 2020.

[Alb21] Brandon Alberts. Statistics of the first Galois cohomology group: A refinement of Malle's conjecture. *Algebra and Number Theory*, 15(10):2513–2569, 2021.

[Alp24] Jarod Alper. Stacks and moduli. https://sites.math.washington.edu/~jarod/moduli-versions/moduli-5-21-24.pdf, 2024. Notes by Jarod Alper-working draft. Version May 21, 2024.

[AO21] Brandon Alberts and Evan O'Dorney. Harmonic analysis and statistics of the first Galois cohomology group. *Res. Math. Sci.*, 8(3):Paper No. 50, 16, 2021.

[Ara74] Suren Ju. Arakelov. An intersection theory for divisors on an arithmetic surface. *Izv. Akad. Nauk SSSR Ser. Mat.*, 38:1179–1192, 1974.

[Bha05] Manjul Bhargava. The density of discriminants of quartic rings and fields. *Ann. of Math. (2)*, 162(2):1031–1063, 2005.

[Bha07] Manjul Bhargava. Mass formulae for extensions of local fields, and conjectures on the density of number field discriminants. *Int. Math. Res. Not. IMRN*, (17):Art. ID rnm052, 20, 2007.

[Bha10] Manjul Bhargava. The density of discriminants of quintic rings and fields. *Ann. of Math. (2)*, 172(3):1559–1591, 2010.

[BM90] Victor V. Batyrev and Yuri I. Manin. Sur le nombre des points rationnels de hauteur borné des variétés algébriques. *Math. Ann.*, 286(1-3):27–43, 1990.

[BSW15] Manjul Bhargava, Arul Shankar, and Xiaoheng Wang. Geometry of numbers methods over global fields i: Prehomogeneous vector spaces. arXiv:1512.03035, 2015.

[BV24] Giulio Bresciani and Angelo Vistoli. An arithmetic valuative criterion for proper maps of tame algebraic stacks. *Manuscripta Math.*, 173(3-4):1061–1071, 2024.

[Ces17] Kęstutis Česnavičius. A modular description of $\mathcal{X}_0(n)$. *Algebra Number Theory*, 11(9):2001–2089, 2017.

[CDyDO02] Henri Cohen, Francisco Diaz y Diaz, and Michel Olivier. Enumerating quartic dihedral extensions of $\mathbb{Q}$. *Compositio Math.*, 133(1):65–93, 2002.

[Dar21] Ratko Darda. Rational points of bounded height on weighted projective stacks. https://theses.hal.science/tel-03682761/document, 2021.

[DH71] Harold Davenport and Hans Heilbronn. On the density of discriminants of cubic fields. II. *Proc. Roy. Soc. London Ser. A*, 322(1551):405–420, 1971.

[DP96] Mario Daberkow and Michael E. Pohst. Computations with relative extensions of number fields with an application to the construction of Hilbert class fields of prime degree. In Henri Cohen, editor, *Proceedings of the 2nd International Symposium (ANTS-II) held at the Université Bordeaux I, Talence, May 18–23, 1996*, volume 1122 of *Lecture Notes in Computer Science*, pages 68–76. Springer-Verlag, Berlin, 1996.

[DY22a] Ratko Darda and Takehiko Yasuda. The Batyrev-Manin conjecture for DM stacks. *J. Eur. Math. Soc., to appear*, https://doi.org/10.48550/arXiv.2207.03645, 2022.

[DY22b] Ratko Darda and Takehiko Yasuda. Quantitative inverse Galois problem for semi-commutative finite group schemes. *J. Théor. Nombres Bordeaux, to appear*, https://doi.org/10.48550/arXiv.2210.01495, 2022.

[DY23] Ratko Darda and Takehiko Yasuda. Torsors for finite group schemes of bounded height. *J. Lond. Math. Soc. (2)*, 108(3):1275–1331, 2023.

[ESZB23] Jordan S. Ellenberg, Matthew Satriano, and David Zureick-Brown. Heights on stacks and a generalized Batyrev-Manin-Malle conjecture. *Forum Math. Sigma*, 11:Paper No. e14, 54, 2023.

[EV05] Jordan S. Ellenberg and Akshay Venkatesh. *Counting extensions of function fields with bounded discriminant and specified Galois group*, pages 151–168. Birkhäuser Boston, Boston, MA, 2005.

[Fan01] Barbara Fantechi. Stacks for everybody. In *European Congress of Mathematics, Vol. I (Barcelona, 2000)*, volume 201 of *Progr. Math.*, pages 349–359. Birkhäuser, Basel, 2001.

[FMT89] Jens Franke, Yuri I. Manin, and Yuri Tschinkel. Rational points of bounded height on Fano varieties. *Invent. Math.*, 95(2):421–435, 1989.

[GS06] Philippe Gille and Tamás Szamuely. *Central simple algebras and Galois cohomology*, volume 101 of *Cambridge Studies in Advanced Mathematics*. Cambridge University Press, Cambridge, 2006.

[HS00] Marc Hindry and Joseph H. Silverman. *Height Functions*, pages 168–256. Springer New York, New York, NY, 2000.

[Klü05] Jürgen Klüners. A counterexample to Malle's conjecture on the asymptotics of discriminants. *C. R. Math. Acad. Sci. Paris*, 340(6):411–414, 2005.

[Klü22] Jürgen Klüners. The asymptotics of nilpotent Galois groups. *Acta Arith.*, 204(2):165–184, 2022.

[KM04] Jürgen Klüners and Gunter Malle. Counting nilpotent Galois extensions. *J. Reine Angew. Math.*, 572:1–26, 2004.

[KP23] Peter Koymans and Carlo Pagano. On Malle's conjecture for nilpotent groups. *Trans. Amer. Math. Soc. Ser. B*, 10:310–354, 2023.

[Lan02] Serge Lang. *Algebra*, volume 211 of *Graduate Texts in Mathematics*. Springer-Verlag, New York, third edition, 2002.

[Lan24] Aaron Landesman. Stacky heights on elliptic curves in characteristic 3. *Ann. Inst. Fourier (Grenoble)*, 74(5):1881–1894, 2024.

[Mak85] Sirpa Mäki. On the density of abelian number fields. *Ann. Acad. Sci. Fenn. Ser. A I Math. Dissertationes*, (54):104, 1985.

[Mal02] Gunter Malle. On the distribution of Galois groups. *J. Number Theory*, 92(2):315–329, 2002.

[Mal04] Gunter Malle. On the distribution of Galois groups. II. *Experiment. Math.*, 13(2):129–135, 2004.

[Mil80] James S. Milne. *Étale cohomology*, volume 33 of *Princeton Mathematical Series*. Princeton University Press, Princeton, N.J., 1980.

[Neu99] Jürgen Neukirch. *Algebraic number theory*, volume 322 of *Grundlehren der mathematischen Wissenschaften [Fundamental Principles of Mathematical Sciences]*. Springer-Verlag, Berlin, 1999. Translated from the 1992 German original and with a note by Norbert Schappacher, With a foreword by G. Harder.

[NSW08] Jürgen Neukirch, Alexander Schmidt, and Kay Wingberg. *Cohomology of number fields*, volume 323 of *Grundlehren der mathematischen Wissenschaften [Fundamental Principles of Mathematical Sciences]*. Springer-Verlag, Berlin, second edition, 2008.

[Ols16] Martin Olsson. *Algebraic spaces and stacks*, volume 62 of *American Mathematical Society Colloquium Publications*. American Mathematical Society, Providence, RI, 2016.

[Ser02] Jean-Pierre Serre. *Galois cohomology*. Springer Monographs in Mathematics. Springer-Verlag, Berlin, english edition, 2002. Translated from the French by Patrick Ion and revised by the author.

[Sko01] Alexei Skorobogatov. *Torsors and rational points*, volume 144 of *Cambridge Tracts in Mathematics*. Cambridge University Press, Cambridge, 2001.

[Sta23] The Stacks project authors. The stacks project. https://stacks.math.columbia.edu, 2023.

[Tsc26] Nikolai Tschebotareff. Die Bestimmung der Dichtigkeit einer Menge von Primzahlen, welche zu einer gegebenen Substitutionsklasse gehören. *Math. Ann.*, 95(1):191–228, 1926.

[Tür15] Seyfi Türkelli. Connected components of Hurwitz schemes and Malle's conjecture. *Journal of Number Theory*, 155:163–201, 2015.

[Vis05] Angelo Vistoli. Grothendieck topologies, fibered categories and descent theory. In *Fundamental algebraic geometry*, volume 123 of *Math. Surveys Monogr.*, pages 1–104. Amer. Math. Soc., Providence, RI, 2005.

[Voi04] John Voight. Introduction to algebraic stacks. https://math.dartmouth.edu/~jvoight/notes/moduli-red-harvard.pdf, 2004.

[Wan21] Jiuya Wang. Malle's conjecture for $S_n \times A$ for $n = 3, 4, 5$. *Compositio Mathematica*, 157(1):83–121, 2021.

[Was82] Lawrence C. Washington. *Introduction to cyclotomic fields*, volume 83 of *Graduate Texts in Mathematics*. Springer-Verlag, New York, 1982.

[Woo10] Melanie Matchett Wood. On the probabilities of local behaviors in abelian field extensions. *Compos. Math.*, 146(1):102–128, 2010.

[Wri89] David J. Wright. Distribution of discriminants of abelian extensions. *Proc. London Math. Soc. (3)*, 58(1):17–50, 1989.

[WY15] Melanie Machett Wood and Takehiko Yasuda. Mass formulas for local Galois representations and quotient singularities. I: a comparison of counting functions. *Int. Math. Res. Not. IMRN*, (23):12590–12619, 2015.

[Yas06] Takehiko Yasuda. Motivic integration over Deligne–Mumford stacks. *Advances in Mathematics*, 207(2):707–761, 2006.

On a Slice of the Cubic 2-Adic Mandelbrot Set

Jacqueline Anderson, Emerald Stacy, and Bella Tobin

Abstract Consider the one-parameter family of cubic polynomials defined by $f_t(z) = -\frac{3}{2}t(-2z^3 + 3z^2) + 1, t \in \mathbb{C}_2$. This family corresponds to a slice of the parameter space of cubic polynomials in $\mathbb{C}_2[z]$. We investigate which parameters in this family belong to the cubic 2-adic Mandelbrot set, a p-adic analog of the classical Mandelbrot set. When $t = 1$, $f_t(z)$ is post-critically finite with a strictly preperiodic critical orbit. We establish that this is a non-isolated boundary point on the cubic 2-adic Mandelbrot set and show asymptotic self-similarity of the Mandelbrot set near this point. Subsequently, we investigate the Julia set for a polynomial on the boundary and demonstrate a similarity between the Mandelbrot set at this point and the Julia set, similar to what is seen in the classical complex case.

1 Introduction

The *Mandelbrot* set, perhaps the most famous set in complex dynamics, is a subset of the complex plane determined by the dynamical properties of quadratic polynomials of the form $f_c(z) = z^2 + c$. The Mandelbrot set may be defined in terms of the forward orbit of the critical point 0 under iteration of f_c. The (forward) orbit of a point z under iteration of a map f is denoted as follows:

$$\mathcal{O}_f(z) = \{f^n(z) : n \in \mathbb{Z}_{\geq 0}\},$$

J. Anderson
Mathematics Department, Bridgewater State University, Bridgewater, MA, USA
e-mail: jacqueline.anderson@bridgew.edu

E. Stacy
Department of Mathematics and Computer Science, Washington College, Chestertown, MD, USA
e-mail: estacy2@washcoll.edu

B. Tobin (✉)
Department of Mathematics, Agnes Scott College, Decatur, GA, USA
e-mail: btobin@agnesscott.edu

S. Akhtari et al. (eds.), *Research Directions in Number Theory*, Association for Women in Mathematics Series 39, https://doi.org/10.1007/978-3-032-11183-8_2

where $f^n = f \circ f \circ \cdots \circ f$ denotes the n-fold composition of f with itself. The Mandelbrot set is then defined as

$$\mathcal{M} = \{c \in \mathbb{C} : \mathcal{O}_{f_c}(0) \text{ is bounded for } f_c(z) = z^2 + c\}.$$

A *Misiurewicz point* is a point in the Mandelbrot set for which the orbit of 0 is strictly preperiodic. These points all appear on the boundary of the Mandelbrot set, and in fact are dense in the boundary of $\mathcal{M}$. In 1989, Tan Lei proved that the Mandelbrot set zoomed in on a Misiurewicz point c and the corresponding Julia set for f_c zoomed in on c at the same scale were asymptotically similar about c [L90]. Moreover, both are asymptotically self-similar as well. The aim of this paper is to explore the extent to which similar properties exist in the 2-adic setting for cubic polynomials.

The Mandelbrot set is of great interest in complex dynamics and has been an inspiration for analogues in the p-adic setting. One might notice that looking at the Mandelbrot set in the p-adic setting yields less interesting results than in the complex setting. In particular, the Mandelbrot set for the family of maps $z^2 + c$ is the unit disk in $\mathbb{C}_p$. A natural extension of the Mandelbrot set is to look at polynomials of higher degrees.

We say a map f is *post-critically bounded*, or PCB, if all of its critical points have bounded forward orbits under iteration of f. Similarly, we say f is *post-critically finite*, or PCF, if all of its critical points have finite forward orbits.

Let $\mathcal{P}_{d,p}$ denote the $(d-1)$-dimensional parameter space of degree d polynomials f defined over $\mathbb{C}_p$, normalized so that f is monic and $f(0) = 0$. (Note that every polynomial can be conjugated so that it is in this normal form, but that representation is not unique in that there exist different polynomials in $\mathcal{P}_{d,p}$ that are conjugates of each other.) We can define the degree d p-adic Mandelbrot set, $\mathcal{M}_{d,p}$, as the set of parameters in $\mathcal{P}_{d,p}$ that yield a PCB polynomial. When $p \geq d$, this generalized p-adic Mandelbrot set is similar to what we see in the quadratic case: a polynomial in $\mathcal{P}_{d,p}$ is PCB if and only if all of its coefficients belong to the unit disk in $\mathbb{C}_p$.

When $p < d$, however, the p-adic Mandelbrot set $\mathcal{M}_{d,p}$ has an interesting structure more akin to that of the complex Mandelbrot set; in particular, this set has boundary points. In [A10] the first author found a point on the boundary of $\mathcal{M}_{3,2}$. She completed this by investigating the one-parameter family of cubic polynomials of the form $g_t(z) = z^3 - \frac{3}{2}tz^2$ for $t \in \mathbb{C}_2$. Note that g_t has two critical points, 0 and t, and that 0 is fixed. When $t = 1$ it is easy to see that both critical points have a finite forward orbit. In [A10] it was shown that $g_1(z) = z^3 - \frac{3}{2}z^2$ is a boundary point on the cubic 2-adic Mandelbrot set.

In this paper we explore a different slice of the cubic 2-adic Mandelbrot set, the one-parameter family defined as follows:

$$f_t(z) = -\tfrac{3}{2}t(-2z^3 + 3z^2) + 1.$$

The polynomials in this family have critical points 0 and 1, and $f_t(0) = 1$, so there is just one critical orbit to keep track of. Note that we are using a different normal form than the one used in the definition of $\mathcal{P}_{d,p}$, as these polynomials are not monic and zero is not fixed. Rather, in this family we choose to normalize so the critical points are 0 and 1, following the normal form for bicritical polynomials as used in [AMT20, T19]. Using this family, we can find another point on the boundary of $\mathcal{M}_{3,2}$.

Main Theorem *The point associated to* $f_1(z) = -\frac{3}{2}(-2z^3 + 3z^2) + 1$ *is on the boundary of the cubic 2-adic Mandelbrot set.*

In the complex quadratic setting, Tan Lei showed that there are similarities between the Mandelbrot set near a Misiurewicz point and the Julia set for the corresponding quadratic polynomial. Inspired by this result, we explore the Julia set of $f_1(z)$ in Sect. 4 to see if there is any such relationship in the p-adic setting for the boundary point given in the Main Theorem.

1.1 Outline

We begin by providing background and necessary definitions in Sect. 2. In Sect. 3 we prove the Main Theorem by identifying one sequence of disks approaching $t = 1$ in which the parameters belonging to those disks are not in the Mandelbrot set, and other sequences of disks approaching $t = 1$ whose contents are in the Mandelbrot set. Then, in Sect. 4 we discuss the 2-adic Julia set for $f_1(t) = -\frac{3}{2}(-2z^3 + 3z^2) + 1$, demonstrating that there appears to be similarities between the Mandelbrot set and this Julia set. Those similarities can be seen when one compares the results in Theorems 1 and 3 on the Mandelbrot set to those in Propositions 1 and 2 on the Julia set for f_1.

2 Preliminaries

2.1 Basic Definitions in Arithmetic Dynamics

As we are interested in the behavior of points under iteration of rational maps, we begin this section by defining terms for the possible behavior. Let K be a field and $f(z) \in K(z)$. We say a point α is *periodic* for f if there exists some positive integer n such that $f^n(\alpha) = \alpha$. Note that if $n = 1$ then α is a *fixed point* for f. If n is the smallest positive integer such that $f^n(\alpha) = \alpha$ then α is periodic of *exact period n*. If there exist distinct positive integers n, m such that $f^n(\alpha) = f^m(\alpha)$, we say α is a *preperiodic point* for f. If α is preperiodic but not periodic, then we say it is *strictly preperiodic*. If α is not periodic or preperiodic, so it has an infinite forward orbit under f, we say α is a *wandering point*.

Periodic points can have an effect on the dynamical behavior of nearby points.

Definition 1 Let α be a point of exact period n, so $f^n(\alpha) = \alpha$ and $f^m(\alpha) \neq \alpha$ for any $0 < m < n$. Then the *multiplier* of α for f is

$$\lambda_f(\alpha) := (f^n)'(\alpha).$$

The multiplier is invariant under iteration, so all points in the orbit of a periodic point will have the same multiplier.

We say that α is repelling for f if $|\lambda_f(\alpha)| > 1$, attracting if $|\lambda_f(\alpha)| < 1$, and neutral if $|\lambda_f(\alpha)| = 1$. One can think of the multiplier of α as an indicator of the behavior of nearby points under iteration of f. For example, if α is a repelling periodic point of exact period n, then there exists a neighborhood of α in which the points move away from α under iteration of f^n, and if α is attracting, then there exists a neighborhood of α in which the points move closer to α with each iteration of f^n.

We say that $f(z) \in K(z)$ is *conjugate* to a rational map g if there exists $\varphi(z) \in \mathrm{PGL}_2(\bar{K})$ such that

$$g = f^\varphi := \varphi^{-1} \circ f \circ \varphi.$$

Conjugation commutes with iteration so if α is a periodic point for f then $\varphi^{-1}(\alpha)$ is periodic point (of the same period) for f^φ. Similarly, critical points for f correspond to critical points for f^φ. In particular, if f is a PCF (or PCB) polynomial than f^φ is PCF (resp. PCB) for any $\varphi \in \mathrm{PGL}_2(\bar{K})$. In this way conjugation respects many dynamical properties and allows us to investigate the dynamics of a map by choosing a convenient representative from its conjugacy class. For example, every quadratic polynomial is conjugate to exactly one polynomial of the form $z^2 + c$, which is why polynomials of that form are used in the definition of the classical Mandelbrot set.

2.2 Julia Sets

For a rational map $f \in \mathbb{C}(z)$, the Julia set of f has many equivalent definitions. Among others, the *Julia set* of f is characterized as

- The closure of the repelling periodic points.
- The smallest closed set, $\mathcal{J}$, in $\mathbb{C}$ containing at least 3 points that is completely invariant under f. That is, if $f(\alpha) \in \mathcal{J}$ and $f^{-1}(\alpha) \in \mathcal{J}$ for any $\alpha \in \mathcal{J}$.

For a polynomial f, the *filled Julia set* is the set of points with bounded forward orbits, and the Julia set is the boundary of the filled Julia set. Thus, a point is in the Julia set if its forward orbit is bounded but there are arbitrarily close points whose orbits tend to infinity. The parameters c that belong to the complex (degree 2) Mandelbrot set are also precisely those for which the Julia set of $f_c(z) = z^2 + c$ is connected.

2.3 Notation

Throughout the paper we will write $|\cdot|$ to denote the normalized 2-adic absolute value and $\overline{D}(a, \delta)$ to denote the closed disk in $\mathbb{C}_2$ centered at a of radius δ. We recall that $z \in \overline{D}(a, 2^{-r})$ if and only if $z \equiv a \mod 2^r$.

3 A Point on the Cubic 2-Adic Mandelbrot Set

In this section we analyze the parameter space associated to the one-parameter family of cubic polynomials of the form $f_t(z) = -\frac{3}{2}t(-2z^3 + 3z^2) + 1$ in $\mathbb{C}_2[z]$. Recall that in this family, the critical points are 0 and 1, and $f_t(0) = 1$, so there is only one critical orbit to consider. We are concerned with which parameters t belong to the Mandelbrot set for this family; in other words, for which t values is the critical orbit bounded? We first note that if $|t| > 1$, the critical orbit is always unbounded, as $|f_t(1)| = |1 - \frac{3}{2}t| = 2|t| > 2$, and all points of absolute value greater than 2 have unbounded orbits. If $|t| \leq \frac{1}{2}$, on the other hand, the coefficients of f_t are all inside the 2-adic unit disk, and so all points in the unit disk stay bounded under iteration of f_t. So, all t with $|t| \leq \frac{1}{2}$ belong to the Mandelbrot set for this family, while all t with $|t| > 1$ do not belong to the Mandelbrot set. We examine the case where $|t| = 1$.

The proof of the main theorem lies in the statements of Theorems 1 and 2 below. Theorem 1 establishes that there are values of t arbitrarily close to $t = 1$ such that the map $f_t(z)$ is post-critically bounded, while Theorem 2 establishes the reverse: that there are values of t arbitrarily close to $t = 1$ for which the map $f_t(z)$ is post-critically unbounded. In Fig. 1 below, we show the structure of the slice of the 2-adic, degree 3 Mandelbrot set for this family. Each node corresponds to a disk of t values centered at the given value with radius 2^{-r}, where r is indicated by the power of 2 in the modulus in the right column. The nodes are ovals if the critical orbit of the corresponding polynomials is known to be unbounded, and squares if they are known to be bounded.

We begin with a lemma in which we build on the fact that $-\frac{1}{2}$ is a repelling fixed point for $f_1(z)$, which implies that points sufficiently close to $-\frac{1}{2}$ move away from $-\frac{1}{2}$ by a fixed amount with each iteration of f_1. For parameters t sufficiently close to 1, we are able to show that the distance between certain points and $-\frac{1}{2}$ increases by a factor of 4 upon each iteration of f_t.

Lemma 1 *Suppose that* $t \in \overline{D}\left(1 + 2^{2n-1}, 2^{-2n}\right)$ *for some* $n \geq 1$ *and assume* $|z + \frac{1}{2}| = 2^k$*, where* $-2n < k < 1$*. Then* $|f_t(z) + \frac{1}{2}| = 2^{k+2}$*.*

Proof First, note that $f_t(z) + \frac{1}{2} = (z + \frac{1}{2})(3tz^2 - 6tz + 3t) - \frac{3}{2}(t - 1)$. This gives us that

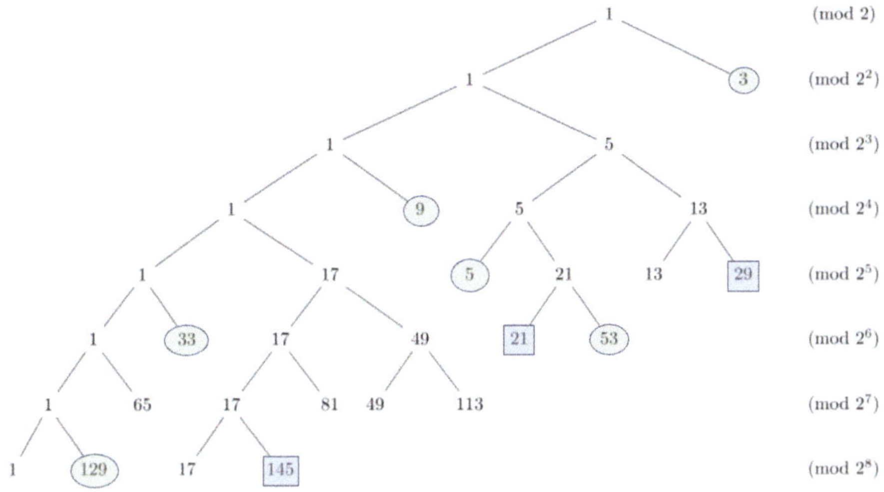

Fig. 1 The structure of the slice of the 2-adic, degree 3 Mandelbrot set given by f_t

$$\begin{aligned}\left|f_t(z)+\tfrac{1}{2}\right| &\le \max\{|z+\tfrac{1}{2}|\cdot|3t|\cdot|z^2-2z+1|, |\tfrac{3}{2}|\cdot|t-1|\}\\ &= \max\{2^k|z-1|^2, 2\cdot 2^{-(2n-1)}\},\end{aligned}$$

with equality if the values for which we are taking the maximum of are distinct. Since $|z+\frac{1}{2}| = 2^k < 2$, it must be the case that $|z| = 2$ and therefore $|z-1| = 2$, so we have that

$$\max\{2^k|z-1|^2, 2\cdot 2^{-(2n-1)}\} = \max\{2^{k+2}, 2^{-2n+2}\} = 2^{k+2} > 2^{-2n+2}.$$

Thus, by the strong triangle inequality, we have that $\left|f_t(z)+\frac{1}{2}\right| = 2^{k+2}$, as desired. □

In Theorem 1 we establish a repeating pattern for 2-adic disks arbitrarily close to $t = 1$ yielding post-critically unbounded polynomials. This pattern is highlighted in Fig. 2 with the highlighted disks representing disks of t-values for which f_t is post-critically unbounded.

Theorem 1 *If* $t \in \overline{D}(1+2^{2n-1}, 2^{-2n})$ *for* $n \ge 1$ *then* $f_t(z) = -\frac{3}{2}t(-2z^3+3z^2)+1$ *is post-critically unbounded.*

Proof Since $f_t(1)+\frac{1}{2} = \frac{3}{2}(1-t)$ then for $t \in \overline{D}(1+2^{2n-1}, 2^{-2n})$ we have that

$$|f_t(1)+\tfrac{1}{2}| = 2|1-t| = 2^{1-(2n-1)} = 2^{-2n+2}.$$

Then, repeatedly applying Lemma 1 shows the following (for $m \le n+1$):

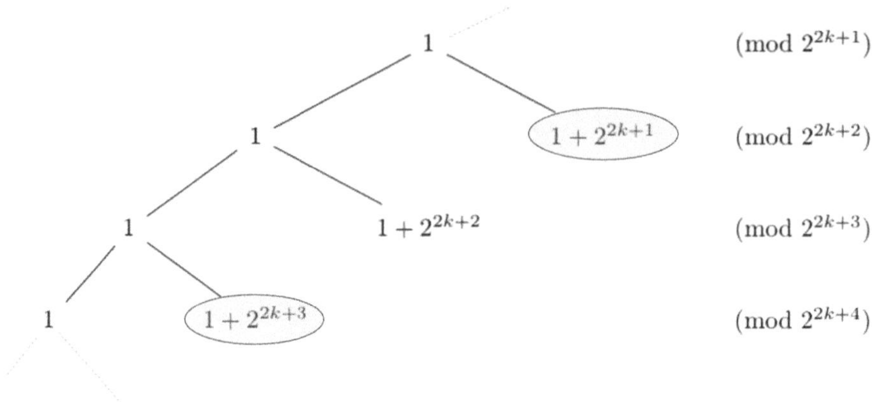

Fig. 2 Repeating pattern of unbounded disks in a slice of the 2−adic degree 3 Mandelbrot set

$$\left|f_t^m(1) + \tfrac{1}{2}\right| = 2^{-2n+2m}.$$

In particular, $|f_t^n(1) + \frac{1}{2}| = 1$, and $|f_t^{n+1}(1) + \frac{1}{2}| = 4$. This implies that $|f^{n+1}(1)| = 4 > 2$. Any point of absolute value greater than 2 will have an orbit that escapes to infinity, and so the orbit of the critical point 1 is unbounded. □

The previous theorem gives a sequence of disks getting arbitrarily close to $t = 1$ whose elements all correspond to polynomials with unbounded critical orbits. The next theorem shows that there is a different sequence of disks, also getting arbitrarily close to $t = 1$, whose elements all correspond to PCB polynomials. This establishes $t = 1$ as a boundary point (and not an isolated point) in the Mandelbrot set for this family.

Theorem 2 *There exists a sequence* $\{t_n\}_{n=3}^{\infty}$ *such that* $\lim_{n\to\infty} t_n = 1$ *and* f_{t_n} *is PCF with a periodic critical orbit of length n. Moreover, for* $t \in \overline{D}(t_n, 2^{8n/3+2})$ *the polynomial* $f_t(z)$ *is PCB.*

Proof Consider the polynomial in s defined by $g_n(s) = f_{s+1}^n(0)$. The roots of g_n are parameters s for which 0 is periodic of period n under iteration of f_{s+1}. We'll show that the Newton polygon for g_n has vertices at $(0, -1)$ and $(1, 3 - 2n)$, which implies that there exists a root s_n of g_n of absolute value 2^{4-2n}. Setting $t_n = s_n + 1$, we have a sequence of t values approaching 1 for which the critical orbit is periodic, and therefore bounded.

We will prove this by induction. First, take $n = 3$. In this case, we can explicitly compute $g_3(s)$:

$$g_3(s) = -\tfrac{1}{2} - \tfrac{93}{8}s - \tfrac{243}{8}s^2 - \tfrac{243}{8}s^3 - \tfrac{81}{8}s^4.$$

The leftmost segment of the Newton polygon for this polynomial has slope -2, as desired. This implies that g_3 has a root s_3 of absolute value $\frac{1}{4}$. Take $t_3 = 1 + s_3$.

For the inductive step, suppose that $g_n(s)$ is of the following form:

$$g_n(s) = -\tfrac{1}{2} + \sum_{i=1}^{3n-5} a_i s^i,$$

where $|a_1| = 2^{2n-3}$ and $|a_i| \leq 2^{1-(4-2n)i}$ for $i > 1$. This is what is necessary to produce the desired Newton polygon properties. We must show that $g_{n+1}(s)$ is of the form

$$g_{n+1}(s) = -\tfrac{1}{2} + \sum_{i=1}^{3n-2} b_i s^i,$$

where $|b_1| = 2^{2n-1}$ and $|b_i| \leq 2^{1-(2-2n)i}$ for $i > 1$.

We compute $g_{n+1}(s)$ from g_n:

$$\begin{aligned} g_{n+1}(s) &= -\tfrac{3}{2}(s+1)\left(-\tfrac{1}{2} + \sum_{i=1}^{3n-5} a_i s^i\right)^2 \left(3 - 2\left(-\tfrac{1}{2} + \sum_{i=1}^{3n-5} a_i s^i\right)\right) + 1 \\ &= -\tfrac{3}{2}(s+1)\left(-\tfrac{1}{2} + \sum_{i=1}^{3n-5} a_i s^i\right)^2 \left(4 - \sum_{i=1}^{3n-5} 2a_i s^i\right) + 1 \\ &= (-3s-3)\left(-\tfrac{1}{2} + \sum_{i=1}^{3n-5} a_i s^i\right)^2 \left(2 - \sum_{i=1}^{3n-5} a_i s^i\right) + 1. \end{aligned}$$

By expanding the above, we see that the constant term is $-\frac{1}{2}$, and the degree 1 term is

$$b_1 = \tfrac{27}{4}a_1 - \tfrac{3}{2},$$

which has the desired absolute value.

It remains to argue that for $2 \leq m \leq 3n-2$, the coefficient b_m has absolute value less than or equal to $2^{1-(2-2n)m}$. Carefully considering the expansion of the expression for $g_{n+1}(s)$ above, we see that the terms of degree m have coefficients of the form $ca_ia_ja_k$, where $i+j+k=m$ or $i+j+k=m-1$ and $|c| \leq 4$. This implies

$$|b_m| \leq 4 \cdot 2^{1-(4-2n)i} \cdot 2^{1-(4-2n)j} \cdot 2^{1-(4-2n)k} \leq 2^{5-(4-2n)m} \leq 2^{1-(2-2n)m},$$

as desired.

Now that we have established the existence of a sequence $\{t_n\}_{n=3}^{\infty}$ with $\lim_{n\to\infty} t_n = 1$ such that f_{t_n} has a periodic critical orbit of length n, we will show that values of t in disks centered at t_n will correspond to PCB polynomials. Specifically, by tracking the precision of a small disk around t_n as we iterate, we see that if $|t - t_n| \leq 2^{-(\frac{8}{3}n+2)}$, then the critical orbit for f_t will be bounded.

Write $t = t_n + 2^m v$, where $m \geq \frac{8}{3}n + 2$ and $|v| \leq 1$. Let $\{0, 1, c_2, c_3, \ldots, c_{n-1}\}$ denote the periodic critical orbit for f_{t_n}. Letting k be the integer nearest to $\frac{2}{3}n$, we will show that f_t^n will map a disk centered at 0 of radius 2^{-k} into itself, and thus the critical orbit is bounded. If $z \in \bar{D}(0, 2^{-k})$, then $z = 2^k u$ for some u with $|u| \leq 1$. Then,

$$f_t(z) = 1 - \tfrac{3}{2}(t_n + 2^m v)(z^2)(3 - 2z) = 1 - \tfrac{3}{2}(t_n + 2^m v)(2^{2k}u^2)(3 - 2^{k+1}u)$$

$$\equiv 1 \pmod{2^{2k-1}}.$$

So $f_t(\bar{D}(0, 2^{-k})) \subset \bar{D}(1, 2^{-(2k-1)})$.

Looking at the next iterate, we see that $f_t(\bar{D}(1, 2^{-(2k-1)})) \subset \bar{D}(c_2, 2^{-(4k-1)})$ by taking $z = 1 + 2^{2k-1}u$ for some u such that $|u| \leq 1$ and computing $f_t(z)$:

$$\begin{aligned} f_t(1 + 2^{2k-1}u) &= 1 - \tfrac{3}{2}(t_n + 2^m v)(1 + 2^{2k}u + 2^{4k-2}u^2)(1 - 2^{2k}u) \\ &= 1 - \tfrac{3}{2}(t_n + 2^m v)(1 - 2^{4k}u^2 + 2^{4k-2}u^2 - 2^{6k-2}u^3) \\ &\equiv 1 - \tfrac{3}{2}t_n \pmod{2^{4k-1}}. \end{aligned}$$

Since $c_2 = f_t(1) = 1 - \frac{3}{2}t_n$, we have that $f_t(1 + 2^{2k-1}u) \equiv c_2 \pmod{2^{4k-1}}$.

Continuing to track iterates from here, the radius of the disk will expand by a factor of 4 each time. We can see this by showing that if $|z - z_n| \leq 2^{-r}$, $|z| \leq 2$ and $r \leq m - 1$ then $|f_t(z) - f_{t_n}(z_n)| \leq 2^{-(r-2)}$. Write $z = z_n + 2^r u$ for some u such that $|u| \leq 1$. Then,

$$f_t(z_n + 2^r u) = 1 - \tfrac{3}{2}(t_n + 2^m v)(z_n^2 + 2^{r+1}z_n u + 2^{2r}u^2)(3 - 2z_n - 2^{r+1}u)$$

$$\equiv 1 - \tfrac{3}{2}t_n(z_n^2)(3 - 2z_n) \equiv f_{t_n}(z_n) \pmod{2^{r-2}}.$$

Applying this $n - 2$ times to $D(c_2, 2^{-(4k-1)})$ gives the following:

$$f_t^n(\bar{D}(0, 2^{-k})) \subset \bar{D}(0, 2^{-(4k-1-2(n-2))}).$$

Note that $4k - 2n + 3 > k$, because this is equivalent to $k > \frac{2}{3}n - 1$, and k was chosen to be the nearest integer to $\frac{2}{3}n$. So, $f_t^n(\bar{D}(0, 2^{-k})) \subset \bar{D}(0, 2^{-k})$, and the proof is complete. □

In addition to the sequence of disks belonging to the Mandelbrot set described in the previous theorem, in Theorem 3 we identify two sequences of disks approaching $t = 1$ in which all $\mathbb{Q}_2$ points belong to the Mandelbrot set.

Theorem 3 *For all integers $n \geq 3$, define the following two sequences of disks:*

$$D_{a,n} = \overline{D}\left(1 + 5 \cdot 2^{2n}, 2^{-(2n+3)}\right)$$

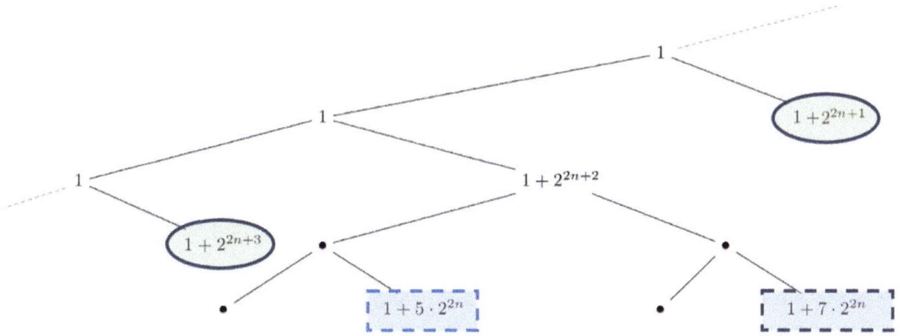

Fig. 3 The self-similar pattern of disks in the slice of the 2−adic cubic Mandelbrot set given by f_t

$$D_{b,n} = \overline{D}\left(1 + 7 \cdot 2^{2n}, 2^{-(2n+3)}\right).$$

Then any $t \in \mathbb{Q}_2$ belonging to one of these disks belongs to the Mandelbrot set.

Figure 3 shows the self-similar pattern of disks in the slice of the cubic Mandelbrot set described in Theorems 1 and 3. The shaded ovals represent disks of t-values for which f_t is post-critically unbounded, as in Theorem 1. The rectangles represent disks whose $\mathbb{Q}_2$ points described in Theorem 3 correspond to t-values for which f_t is post-critically bounded, and thus are in the cubic 2-adic Mandelbrot set.

Remark 1 In Theorem 3 above, we must restrict to $t \in \mathbb{Q}_2$ to guarantee that the critical orbit for f_t is bounded. This is because for these t values, the critical orbit eventually ends up near a repelling 2-cycle. Since that cycle is repelling, there are arbitrarily close points whose orbits escape to infinity, but it can be shown that all $\mathbb{Q}_2$ points near that repelling 2-cycle have bounded orbits.

To prove Theorem 3, we will need a few lemmas. The first establishes that for t values sufficiently close to 1, a $\mathbb{Q}_2$ point whose orbit eventually enters one of two particular disks will have a bounded orbit under iteration of f_t.

Lemma 2 *Let $t \in \mathbb{Q}_2, t \equiv 1 \pmod{32}$. Then if $z \in \mathbb{Q}_2$ and either $z \in \overline{D}\left(\frac{19}{2}, 2^{-4}\right)$ or $z \in \overline{D}\left(\frac{27}{2}, 2^{-4}\right)$, $\mathcal{O}_{f_t}(z)$ is bounded.*

Proof The proof is a straightforward computation of iteration of disks. First we show that a point in $\overline{D}\left(\frac{27}{2}, 2^{-4}\right)$ maps into $\overline{D}\left(\frac{19}{2}, 2^{-4}\right)$ after three iterations. Let $z = \frac{27}{2} + 16k_0$, where $k_0 \in \mathbb{Z}_2$. Then applying f_t to z, we get

$$f_t\left(\frac{27}{2} + 16k_0\right) = 1 + 6561t + 24300k_0t + 29952k_0^2t + 12288k_0^3t \in \overline{D}\left(2, 2^{-2}\right).$$

Next, we apply f_t to a point of the form $2 + 4k_1$, $k_1 \in \mathbb{Z}_2$:

$$f_t\left(2+4k_1\right)=1+6t+72k_1t+216k_1^2t+192k_1^3t\in\overline{D}\left(3,2^{-2}\right).$$

Finally, we apply f_t to a point of the form $3+4k_2$, $k_2\in\mathbb{Z}_2$:

$$f_t\left(3+4k_2\right)=1+\frac{81}{2}t+216k_2t+360k_2^2t+192k_2^3t\equiv\frac{19}{2}+8k_2(k_2+1)\pmod{16}.$$

Note that since $k_2\in\mathbb{Z}_2$, we know $k_2(k_2+1)\equiv 0\pmod 2$, and so this point lies in the disk $\overline{D}\left(\frac{19}{2},2^{-4}\right)$.

Next, we show that points in $\overline{D}\left(\frac{19}{2},2^{-4}\right)$ map back into the disk $\overline{D}\left(3,2^{-2}\right)$, and so $\mathbb{Q}_2$ points in these disks have bounded orbits under iteration of f_t.

Let $z=\frac{19}{2}+16k_3$, $k_3\in\mathbb{Z}_2$. Then we have

$$f_t\left(z\right)=1+2166t+11628k_3t+20736k_3^2t+12288k_3^3t\in\overline{D}\left(3,2^{-2}\right),$$

as desired. □

In the next lemma, we show that if t and t_0 are sufficiently close together, then f_t and f_{t_0} will map disks of a certain type to the same image.

Lemma 3 *Let $\alpha\in\mathbb{C}_2$ such that $\left|\alpha-\frac{1}{2}\right|\leq 1$ and let $r\leq 1$. Suppose that f_{t_0} maps $\overline{D}\left(\alpha,r\right)$ to $\overline{D}\left(\beta,4r\right)$, where $\beta=f_{t_0}(\alpha)$. Then if $|t-t_0|\leq r$, f_t will also map $\overline{D}\left(\alpha,r\right)$ to $\overline{D}\left(\beta,4r\right)$.*

Proof Write $t=t_0+\epsilon$, where $|\epsilon|\leq r$. Let $z=\alpha+\delta$, where $|\delta|\leq r$. Then we have

$$f_t(z)=-\tfrac{3}{2}(t_0+\epsilon)(-2z^3+3z^2)+1=f_{t_0}(z)-\tfrac{3}{2}\epsilon(-2z^3+3z^2).$$

This will lie in $\overline{D}\left(\beta,4r\right)$ if $\left|\frac{3}{2}\epsilon(-2z^3+3z^2)\right|\leq 4r$. Since $z\equiv\frac{1}{2}\pmod 1$, we know that $|-2z^3+3z^2|\leq 2$, and thus

$$\left|\tfrac{3}{2}\epsilon(-2z^3+3z^2)\right|\leq 2\cdot|\epsilon|\cdot 2\leq 4r,$$

as desired. □

We are now ready to prove Theorem 3.

Proof of Theorem 3 We will show that if $t\in D_{a,n}\cap\mathbb{Q}_2$ or if $t\in D_{b,n}\cap\mathbb{Q}_2$, then $f_t^n(1)$ will lie in either $\overline{D}\left(\frac{19}{2},2^{-4}\right)$ or $\overline{D}\left(\frac{27}{2},2^{-4}\right)$. Then, Lemma 2 will imply that the critical orbit is bounded.

Let $n=3$. Then the first disk in each sequence is

$$D_{a,3}=\overline{D}\left(321,2^{-9}\right),$$

$$D_{b,3}=\overline{D}\left(449,2^{-9}\right).$$

First consider $t \in D_{a,3} \cap \mathbb{Q}_2$. In this case, we can show by direct computation that $f_t^3(1) \in \overline{D}\left(\frac{19}{2}, 2^{-4}\right)$:

$$f_t(1) = 1 - \tfrac{3}{2}t \equiv -\tfrac{961}{2} \pmod{2^8}$$

$$\implies f_t^2(1) \equiv f_t\left(-\tfrac{961}{2}\right) \equiv \tfrac{111}{2} \pmod{2^6}$$

$$\implies f_t^3(1) \equiv f_t\left(\tfrac{111}{2}\right) \equiv \tfrac{19}{2} \pmod{2^4}.$$

Similarly, we can show that if $t \in D_{b,3} \cap \mathbb{Q}_2$, then $f_t^3(1) \in \overline{D}\left(t\frac{27}{2}, 2^{-4}\right)$.

$$f_t(1) = 1 - \tfrac{3}{2}t \equiv -\tfrac{1345}{2} \pmod{2^8}$$

$$\implies f_t^2(1) \equiv f_t\left(-\tfrac{1345}{2}\right) \equiv \tfrac{79}{2} \pmod{2^6}$$

$$\implies f_t^3(1) \equiv f_t\left(\tfrac{79}{2}\right) \equiv \tfrac{27}{2} \pmod{2^4}.$$

Next, let $n > 3$. We will show that if $t \in D_{a,n}$ and $t_0 \in D_{b,n-1}$, then $f_t^3(1) \in \overline{D}\left(f_{t_0}^2(1), 2^{-(2n-2)}\right)$. Similarly, if $t \in D_{b,n}$ and $t_0 \in D_{a,n-1}$, then $f_t^3(1) \in \overline{D}\left(f_{t_0}^2(1), 2^{-(2n-2)}\right)$. From there, using Lemma 3 we can argue that t is in the Mandelbrot set if t_0 is in the Mandelbrot set. The theorem will then follow by induction.

First, we track the first three iterates of 1 under f_t if $t \in D_{a,n}$. Let $t \equiv 1 + 5 \cdot 2^{2n} \pmod{2^{2n+3}}$. Then,

$$f_t(1) = 1 - \tfrac{3}{2}t \equiv -\tfrac{1}{2} + 2^{2n-1} \pmod{2^{2n+2}}$$

$$\implies f_t^2(1) \equiv f_t\left(-\tfrac{1}{2} + 2^{2n-1}\right) \equiv -\tfrac{1}{2} + 7 \cdot 2^{2n-3} \pmod{2^{2n}}$$

$$\implies f_t^3(1) \equiv f_t\left(-\tfrac{1}{2} + 7 \cdot 2^{2n-3}\right) \equiv -\tfrac{1}{2} + 5 \cdot 2^{2n-5} \pmod{2^{2n-2}}.$$

We next do the same for $t \in D_{b,n}$, so $t \equiv 1 + 7 \cdot 2^{2n} \pmod{2^{2n+3}}$:

$$f_t(1) = 1 - \tfrac{3}{2}t \equiv -\tfrac{1}{2} + 3 \cdot 2^{2n-1} \pmod{2^{2n+2}}$$

$$\implies f_t^2(1) \equiv -\tfrac{1}{2} + 5 \cdot 2^{2n-3} \pmod{2^{2n}}$$

$$\implies f_t^3(1) \equiv -\tfrac{1}{2} + 7 \cdot 2^{2n-5} \pmod{2^{2n-2}}.$$

Note by the above that if $t \in D_{a,n}$ and $t_0 \in D_{b,n-1}$, then $f_t^3(1) \equiv f_{t_0}^2(1) \pmod{2^{2n-2}}$. Note also that $|t - t_0| = 2^{-(2n-2)}$. Thus, we can repeatedly apply Lemma 3 to conclude that for $3 \le m \le n$, $|f_t^m(1) - f_{t_0}^{m-1}(1)| \le 2^{-(2n-2m+4)}$. In particular, if $f_{t_0}^{n-1}(1)$ lies in either $\overline{D}\left(\frac{19}{2}, 2^{-4}\right)$ or $\overline{D}\left(\frac{27}{2}, 2^{-4}\right)$, then $f_t^n(1)$ will as well. Thus, by induction, the critical orbit is bounded.

The same argument works if $t \in D_{b,n}$ and $t_0 \in D_{a,n-1}$ by noting that $f_t^3(1) \equiv f_{t_0}^2(1) \pmod{2^{2n-2}}$. □

Remark 2 As Theorems 1 and 3 are given in terms of closed 2-adic disks, it is natural to consider an interpretation of these results in Berkovich space. For the interested reader, please see [BR10, B19, T15] for background on Berkovich space. Our results show that the type I point at $t = 1$ is on the boundary of the slice of the degree 3 2-adic Mandelbrot set. Theorem 1 shows that every type II point centered at $1 + 2^{2n-1}$ with radius 2^{-2n} for $n \geq 1$ is outside of the Mandelbrot set, along with all points below these type II points. Theorem 3 shows that for $n \geq 3$ every $\mathbb{Q}_2$ point below a type II point centered at $1 + 5 \cdot 2^{2n}$ or at $1 + 7 \cdot 2^{2n}$ with radius $2^{-(2n+3)}$ is in the Mandelbrot set.

4 The Julia Set for f_1

In this section, we explore the filled Julia set for the following map in our family:

$$f_1(z) = -\tfrac{3}{2}(-2z^3 + 3z^2) + 1 = 3z^3 - \tfrac{9}{2}z^2 + 1.$$

In particular, we look in a neighborhood of the repelling fixed point $z = -\frac{1}{2}$, to examine the structural similarities with the Mandelbrot set near this parameter.

First we note that if $|z| > 2$, then $|f_1(z)| = |z|^3$, and so $|f_1^n(z)| = |z|^{3^n}$ and the orbit of z tends to infinity. Thus, the filled Julia set for f_1 is contained in the closed disk centered at 0 of radius 2. Moreover, the filled Julia set therefore consists of all points z for which $\mathcal{O}_{f_1}(z)$ is contained inside the disk $\overline{D}(0, 2)$. This bound for the radius of the filled Julia set is sharp, since $z = -\frac{1}{2}$ is a repelling fixed point, and therefore in the Julia set. Note that both critical points 0 and 1, along with $\frac{3}{2}$, are in the backward orbit of $-\frac{1}{2}$ and are thus in the Julia set as well. (We have $f_1(0) = 1$, $f_1\left(\frac{3}{2}\right) = 1$,and $f_1(1) = -\frac{1}{2}$.)

Similarly, if $1 < |z| < 2$ or if $2^{-1/2} < |z| < 1$, the orbit of z will be unbounded: if $1 < |z| < 2$, then $|f_1(z)| = 2|z|^2 > 2$, and if $2^{-1/2} < |z| < 1$, then $|f_1(z)| = 2|z|^2$, and $|f_1^2(z)| = 2(2|z|^2)^2 = 8|z|^4 > 2$. Thus the filled Julia set consists of points of absolute value 2, points of absolute value 1, and points of absolute value less than or equal to $2^{-1/2}$.

Since $z = -\frac{1}{2}$ is a fixed point with multiplier $|\lambda| = |f_1'\left(-\frac{1}{2}\right)| = |\frac{27}{4}| = 4$, we know it is repelling and therefore in the Julia set. This implies that there are points arbitrarily close to $-\frac{1}{2}$ with bounded orbits, and there are also points arbitrarily close to $-\frac{1}{2}$ with unbounded orbits. We explore a tree of points centered at $-\frac{1}{2}$ to determine if there are any patterns in how these points with bounded and unbounded orbits are distributed as we zoom in on our repelling fixed point.

Propositions 1 and 2 show that we have a tree pattern as shown in Fig. 4 with the pattern shown in Fig. 5 repeating every two levels as we zoom in on $-\frac{1}{2}$. In Figs. 4 and 5, a disk is labeled with a rectangle if all $\mathbb{Q}_2$ points in that disk belong to the

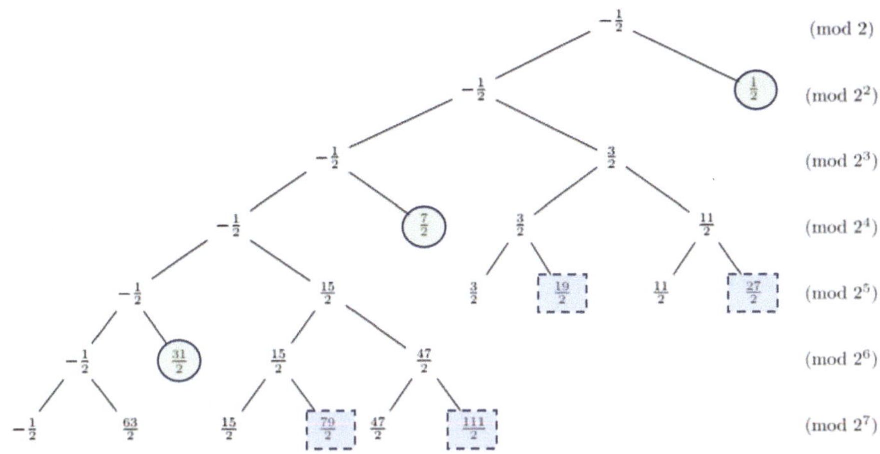

Fig. 4 Disks of points with bounded (blue) and unbounded (green) orbits for f_1 in a neighborhood of the repelling fixed point $z = -\frac{1}{2}$

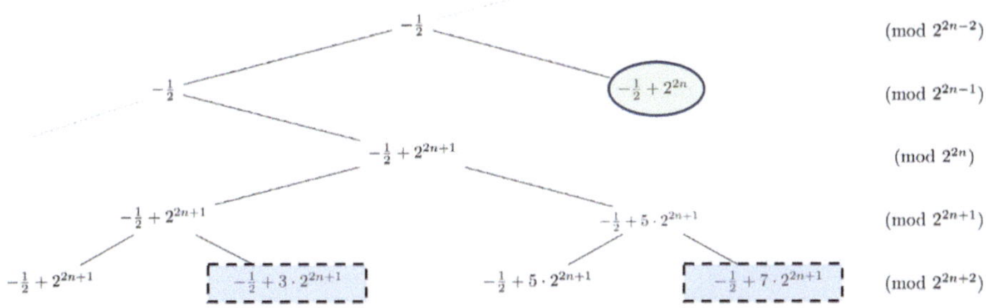

Fig. 5 The self-similar pattern of disks in the Julia set for f_1 centered at $z = -\frac{1}{2}$

filled Julia set, and with a circle if no points in that disk belong to the filled Julia set, i.e., if all points have unbounded orbits.

The following proposition gives a sequence of disks decreasing in radius and approaching $-\frac{1}{2}$ consisting entirely of points with unbounded orbits. It establishes a pattern of disks outside the filled Julia set for f_1 similar to the pattern of disks outside the Mandelbrot set in the parameter space that we observed in Theorem 1.

Proposition 1 *All points contained in the disk* $D_n = \overline{D}\left(-\frac{1}{2} + 2^{2n}, 2^{-(2n+1)}\right)$ *for* $n \in \mathbb{Z}, n \geq 0$ *have unbounded orbits under iteration of* f_1.

Proof First note that if $z \in D_0$, then $z = \frac{1}{2} + 2k$ for some k such that $|k| \leq 1$. Then, $|f_1(z)| > 2$ and thus the orbit of z is unbounded:

$$\left|f_1\left(\tfrac{1}{2} + 2k\right)\right| = \left|3\left(\tfrac{1}{2} + 2k\right)^3 - \tfrac{9}{2}\left(\tfrac{1}{2} + 2k\right)^2 + 1\right| = \left|\tfrac{1}{4} - \tfrac{9}{2}k + 24k^3\right| = 4.$$

We complete the proof by noting that for $n > 1$, $f_1(D_n) \subseteq D_{n-1}$. If $z \in D_n$ then $z = -\frac{1}{2} + 2^{2n} + 2^{2n+1}k$ for some k such that $|k| \leq 1$. Then, computing $f_1(z)$, we get the following:

$$f_1\left(-\tfrac{1}{2} + 2^{2n}(1+2k)\right) = 3\left(-\tfrac{1}{2} + 2^{2n}(1+2k)\right)^3 - \tfrac{9}{2}\left(-\tfrac{1}{2} + 2^{2n}(1+2k)\right)^2 + 1$$

$$= -\tfrac{1}{2} + 27 \cdot 2^{2n-2}(1+2k) - 9 \cdot 2^{4n}(1+2k)^2 + 3 \cdot 2^{6n}(1+2k)^3 \equiv -\tfrac{1}{2} + 2^{2n-2} \pmod{2^{2n-1}},$$

and so it is contained in D_{n-1}.

Since each D_n maps into D_{n-1}, after n iterations, every point in D_n maps to a point of absolute value 4 and thus has unbounded orbit. □

In the next proposition, we give two sequences of disks decreasing in radius and approaching $-\frac{1}{2}$ in which all of the $\mathbb{Q}_2$ points have bounded orbits. This proposition gives a pattern of points with bounded orbits that is exactly analogous to the pattern we saw in the Mandelbrot set in Theorem 3. One can see the similarities between the Julia set for f_1 zoomed in to a particular level on $z = -\frac{1}{2}$ and the slice of the Mandelbrot set for f_t zoomed in to the same level on $t = 1$ by comparing Figs. 2 and 5. This similarity between the Mandelbrot set and the corresponding Julia set near a boundary point of the Mandelbrot set is analogous to what one sees for Misiurewicz points in the complex Mandelbrot set.

Proposition 2 *Let $D_{a,n}$ and $D_{b,n}$ be sequences of disks defined as follows for $n \in \mathbb{Z}$, $n \geq 0$:*

$$D_{a,n} = \overline{D}\left(-\tfrac{1}{2} + 5 \cdot 2^{2n+1}, 2^{-(2n+4)}\right),$$

$$D_{b,n} = \overline{D}\left(-\tfrac{1}{2} + 7 \cdot 2^{2n+1}, 2^{-(2n+4)}\right).$$

Then, all of the $\mathbb{Q}_2$ points in these disks have bounded orbits under iteration of f_1.

Proof First consider $n = 0$. Then, $D_{a,0} = \overline{D}\left(\frac{19}{2}, 2^{-4}\right)$ and $D_{b,0} = \overline{D}\left(\frac{27}{2}, 2^{-4}\right)$. We will show that the $\mathbb{Q}_2$ points in $D_{b,0}$ map to $\mathbb{Q}_2$ points in $D_{a,0}$ after two iterations of f_1, and that the $\mathbb{Q}_2$ points in $D_{a,0}$ map into $D_{a,0}$ after two iterations of f_1. Thus, all of these points have bounded orbit.

By explicitly computing the iteration of disks under f_1 below, we will see the following:

$$\overline{D}\left(\tfrac{27}{2}, 2^{-4}\right) \to \overline{D}\left(2, 2^{-2}\right) \to \overline{D}\left(3, 2^{-2}\right) \leftarrow \overline{D}\left(\tfrac{19}{2}, 2^{-4}\right).$$

Moreover, we will show that the $\mathbb{Q}_2$ points in $\overline{D}\left(3, 2^{-2}\right)$ map into $\overline{D}\left(\frac{19}{2}, 2^{-4}\right)$. So, these $\mathbb{Q}_2$ points simply move back and forth between these two disks every time we apply f_1, and thus their orbits stay bounded.

Applying f_1 to a point of the form $\frac{19}{2} + 16k$, $|k| \leq 1$, we get the following:

$$f_1\left(\tfrac{19}{2} + 16k\right) = 2167 + 11628k + 20736k^2 + 12288k^3 \equiv 3 \pmod{4}.$$

Doing the same with $z = \frac{27}{2} + 16k$,

$$f_1\left(\tfrac{27}{2} + 16k\right) = 6562 + 24300k + 29952k^2 + 12288k^3 \equiv 2 \pmod 4.$$

Next, with $z = 2 + 4k$,

$$f_1(2 + 4k) = 7 + 72k + 216k^2 + 192k^3 \equiv 3 \pmod 4.$$

Finally, applying f_1 to a point of the form $3 + 4k$, $k \in \mathbb{Q}_2$, we get the following:

$$f_1(3 + 4k) = \tfrac{83}{2} + 216k + 360k^2 + 192k^3 \equiv \tfrac{19}{2} + 8k + 8k^2 \pmod{16}.$$

Note that for $k \in \mathbb{Q}_2$, we have $8k + 8k^2 = 8k(k+1) \equiv 0 \pmod{16}$, and so $f_1(3 + 4k)$ lies in the disk $\overline{D}\left(\frac{19}{2}, 2^{-4}\right)$.

For $n \geq 1$, we will show that if $z \in D_{a,n}$, then $f_1(z) \in D_{b,n-1}$, and if $z \in D_{b,n}$, then $f_1(z) \in D_{a,n-1}$. This will complete the proof that all $\mathbb{Q}_2$ points in these disks have bounded orbits. First, suppose $z \in D_{a,n}$. Then we can write $z = -\frac{1}{2} + 2^{2n+1}(5 + 8k)$, where $|k| \leq 1$. Applying f_1 to z, we obtain

$$\begin{aligned} f_1\left(-\tfrac{1}{2} + 2^{2n+1}(5+8k)\right) &= -\tfrac{1}{2} + 27 \cdot 2^{2n-1}(5+8k) - 9 \cdot 2^{4n+2}(5+8k)^2 + 3 \cdot 2^{6n+3}(5+8k)^3 \\ &\equiv -\tfrac{1}{2} + 7 \cdot 2^{2n-1} \pmod{2^{2n+2}} \in D_{b,n-1}. \end{aligned}$$

Next, consider $z \in D_{b,n}$. Then we can write $z = -\frac{1}{2} + 2^{2n+1}(7 + 8k)$, where $|k| \leq 1$. Applying f_1 to z, we get the following:

$$\begin{aligned} f_1\left(-\tfrac{1}{2} + 2^{2n+1}(7+8k)\right) &= -\tfrac{1}{2} + 27 \cdot 2^{2n-1}(7+8k) - 9 \cdot 2^{4n+2}(7+8k)^2 + 3 \cdot 2^{6n+3}(7+8k)^3 \\ &\equiv -\tfrac{1}{2} + 5 \cdot 2^{2n-1} \pmod{2^{2n+2}} \in D_{a,n-1}. \end{aligned}$$

Thus, for all $n \geq 0$, all $\mathbb{Q}_2$ points in $D_{a,n}$ or $D_{b,n}$ eventually map to $\mathbb{Q}_2$ points in $\bar{D}\left(\frac{19}{2}, 2^{-4}\right)$, and therefore have bounded orbits. □

In the proposition above, the $\mathbb{Q}_2$ points in the disks $\overline{D}(3, 2^{-2})$ and $\overline{D}\left(\frac{19}{2}, 2^{-4}\right)$ map to each other under f_1. This is because there is a (repelling) 2-cycle in these disks, namely the roots of $6z^2 - 3z - 1$. The roots of this polynomial, call them α_1 and α_2, map to each other under f_1. A Newton polygon analysis shows that one of these points, say α_1, has absolute value 1 and the other has absolute value 2. Using Hensel's Lemma, we can see that $\alpha_1 \in \mathbb{Z}_2$ with $\alpha_1 \equiv 7 \pmod{16}$. So, the multiplier for this cycle is

$$|\lambda| = |f_1'(\alpha_1) f_1'(\alpha_2)| = |\alpha_1||\alpha_1 - 1||\alpha_2||\alpha_2 - 1| = 1 \cdot \tfrac{1}{2} \cdot 2 \cdot 2 = 2.$$

Since this cycle is repelling, there will be $\mathbb{C}_2$ points that are not in $\mathbb{Q}_2$ in these disks whose orbits will escape to infinity, while every $\mathbb{Q}_2$ point in these disks has a bounded orbit. This is precisely the same as what we observed in Theorem 3.

Acknowledgements This work was initiated during the Women in Numbers 6 workshop at Banff International Research Station (BIRS) in 2023. The authors thank BIRS and the WIN6 organizers for providing accommodations that allowed us to focus on this work and the opportunity to collaborate on this project.

References

[A13] Jacqueline Anderson. Bounds on the radius of the p-adic Mandelbrot set. *Acta Arith.*, 158(3):253–269, 2013.

[A10] Jacqueline Anderson. *p-adic Analogues of the Mandelbrot Set.* PhD thesis, Brown University, 2013.

[AMT20] Jacqueline Anderson, Michelle Manes, and Bella Tobin. Cubic post-critically finite polynomials defined over $\mathbb{Q}$. *Open Book Series*, 4(1):23–38, 2020.

[BR10] Matthew Baker and Robert S Rumely. *Potential theory and dynamics on the Berkovich projective line*. Number 159. American Mathematical Soc., 2010.

[B19] Robert L Benedetto. *Dynamics in one non-archimedean variable*, volume 198. American Mathematical Soc., 2019.

[L90] Tan Lei. Similarity between the mandelbrot set and julia sets. *Communications in Mathematical Physics*, 134(3):587–617, 1990.

[S07] Joseph H. Silverman. *The arithmetic of dynamical systems*, volume 241 of *Graduate Texts in Mathematics*. Springer, New York, 2007.

[T15] Michael Temkin. Introduction to berkovich analytic spaces. *Berkovich spaces and applications*, pages 3–66, 2015.

[T19] Bella Tobin. *Belyi Maps and Bicritical Polynomials*. PhD thesis, University of Hawai'i at Mānoa, 2019.

Towards a Classification of p^2-Discriminant Ideal Twins Over Number Fields

Alyson Deines, Asimina S. Hamakiotes, Andreea Iorga, Changningphaabi Namoijam, Manami Roy, and Lori D. Watson

Abstract Isogenous elliptic curves have the same conductor but not necessarily the same minimal discriminant ideal. In this article, we explicitly classify all p^2-isogenous elliptic curves defined over a number field with the same minimal discriminant ideal for odd prime p where $X_0(p^2)$ has genus 0, i.e., $p = 3$ or 5. As a consequence, we give a list of all p^2-isogenous discriminant (ideal) twins over $\mathbb{Q}$ for such p.

2010 Mathematics Subject Classification Primary 11G05 · 14K02 · 14H10 · 14H52

A. Deines
Center for Communications Research, San Diego, CA, USA

A. S. Hamakiotes
Department of Mathematics, Fordham University, 414 E 52nd Street, New York, NY 10022, USA
e-mail: ahamakiotes@fordham.edu

A. Iorga
Department of Mathematics, Cornell University, 212 Garden Ave, Ithaca, NY 14853, USA
e-mail: ai324@cornell.edu

C. Namoijam
Department of Mathematics, Fairfield University, 1073 North Benson Road, Fairfield, CT 06824, USA
e-mail: cnamoijam@fairfield.edu

M. Roy (✉)
Department of Mathematics, Lafayette College, Easton, PA, USA
e-mail: royma@lafayette.edu

L. D. Watson
Department of Mathematics, Trinity College, Hartford, CT, USA
e-mail: lori.watson@trincoll.edu

S. Akhtari et al. (eds.), *Research Directions in Number Theory*, Association for Women in Mathematics Series 39, https://doi.org/10.1007/978-3-032-11183-8_3

1 Introduction

Isogenous elliptic curves defined over a number field have the same conductor; however, their minimal discriminant ideals might be different. We are interested in understanding when two (cyclic) isogenous elliptic curves also have the same minimal discriminant ideal, i.e., when they are discriminant ideal twins. This work continues the classification of discriminant (ideal) twins begun by the first author in [D18] and continued in [BBDHR24]. Discriminant ideal twins serve as an obstruction to determining the optimal quotient of the modular parameterization of an elliptic curve by a modular or Shimura curve using the algorithm given by Ribet and Takahashi in [RT97]. In [D18], the first author determines all such obstructions for elliptic curves E defined over $\mathbb{Q}$ with at least one prime of multiplicative reduction, providing a list of all j-invariants corresponding to semistable isogenous[1] discriminant twins. This result is extended to (non-semistable) p-isogenous elliptic curves defined over arbitrary number fields for $p \in \{3, 5, 7, 13\}$ by Barrios, Brucal-Hallara, Deines, Harris, and Roy in [BBDHR24]. With a goal of completing the classification for other isogeny degrees we further extend this work by determining p^2-isogenous discriminant ideal twins over arbitrary number fields in the case where p is an odd prime and $X_0(p^2)$ has genus 0, i.e., $p^2 \in \{9, 25\}$.

If E_1 and E_2 are (cyclic) isogenous discriminant ideal twins, then we can create infinitely many more by twisting at primes ℓ coprime to 2 and the conductor of the curves. Note that twisting preserves j-invariants and so, even though twisting by such ℓ creates infinitely many pairs of isogenous elliptic curves with the same discriminant ideals, it does not create new twist families.

It is easy to check the Cremona database (via SageMath [D20] or LMFDB [LMFDB21]) and find all twist families of isogenous discriminant ideal twins over $\mathbb{Q}$. For each prime isogeny degree, all the j-invariant pairs found in the Cremona database are given in [BBDHR24, Table 5.1]. For each p^2-isogeny degree, all the j-invariant pairs found in the Cremona database are given in Table 1.

Table 1 j-invariant pairs for p^2-isogenous elliptic curves

Deg	j_1	j_2
4	–1/15	56667352321/15
	35937/17	82483294977/17
	287496	287496
9	–50357871050752/19	32768/19
	4096000/37	727057727488000/37
25	–52893159101157376/11	–4096/11
	190705121216/71	39225406342464307 81376/71

[1] Throughout, we will restrict to cases of cyclic isogenies.

The next question to ask is, are there more j-invariant pairs of p or p^2-isogenous discriminant ideal twins? The Cremona tables in LMFDB contain curves with conductor up to 299,996,953 and are complete for curves with conductor up to 500,000. With more data, would we find more discriminant ideal twins? For isogeny degrees $p = 2, 3, 5, 7, 13$ over $\mathbb{Q}$, the answer is no. This was first examined by the first author [D18] for elliptic curves with at least one prime with multiplicative reduction, and the more general prime isogeny case for elliptic curves defined over number fields was completed by Barrios, Brucal-Hallara, Deines, Harris, and Roy [BBDHR24]. Similarly, the first author [D18] also examined the cases $p^2 = 4, 9, 25$ for curves with at least one prime of multiplicative reduction.

In this paper, we continue this exploration and finish certain cases over $\mathbb{Q}$ for curves without multiplicative reduction; we also work with curves with both multiplicative and non-multiplicative reduction over number fields, in general.

As in previous work, all isogenies are cyclic. Specifically, we examine when isogenous discriminant ideal twins defined over a number field K occur for p^2-isogeny degrees coming from genus 0 modular curves for odd p, i.e., for isogeny degrees $n = 9, 25$. In [B23], Barrios gives a parameterization of p^2-isogenous elliptic curves with the property that their j-invariants are not both 0 and 1728. In particular, for $p = 3, 5$, if E_1 and E_3 are two p^2-isogenous elliptic curves then there exist $t \in K, d \in K^{\times}/K^{\times 2}$ such that $E_i \cong \mathcal{C}_{p^2,i}(t, d)$ (see Table 3 for precise definition of $\mathcal{C}_{p^2,i}(t, d)$). Using this notation, we state our main results.

Theorem 1 *Let E_1 and E_3 be two 9-isogenous elliptic curves over a number field K such that their j-invariants are not equal. Suppose further that $E_i \cong \mathcal{C}_{9,i}(t, d)$, where $t \in \mathcal{O}_K$ and $d \in \mathcal{O}_K$ given in Table 3. Then E_1 and E_3 are discriminant ideal twins if and only if for each prime $\mathfrak{p}$ of $\mathcal{O}_K$,*

$$\nu_{\mathfrak{p}}(t-3) = 3k_{\mathfrak{p}}, \quad \text{for } 0 \leq k_{\mathfrak{p}} \leq \nu_{\mathfrak{p}}(3). \tag{1}$$

Moreover, the two curves are discriminant twins if and only if t satisfies (1) *and* $(t-3)^8 \in \mathcal{O}_K^{12}$.

Theorem 2 *Let E_1 and E_3 be two 25-isogenous elliptic curves over a number field K such that their j-invariants are not equal. Suppose further that $E_i \cong \mathcal{C}_{25,i}(t, d)$, where $t \in \mathcal{O}_K$ and $d \in \mathcal{O}_K$ given in Table 3. Then E_1 and E_3 are discriminant ideal twins if and only if for each prime $\mathfrak{p}$ of $\mathcal{O}_K$,*

$$\nu_{\mathfrak{p}}(t-1) = k_{\mathfrak{p}}, \quad \text{for } 0 \leq k_{\mathfrak{p}} \leq \nu_{\mathfrak{p}}(5). \tag{2}$$

In fact, the two curves are discriminant twins if t satisfies (2).

We prove Theorem 1 (see Theorems 12 and 13) in Sect. 4 and Theorem 2 (see Theorems 14 and 15) in Sect. 5. The proofs of these theorems rely on the fact that we can take $t \in \mathcal{O}_K$, not just in K, which is proved in Lemma 9 of Sect. 3. The rest of this paper is organized as follows. In Sect. 2, we recall preliminary definitions and results about elliptic curves. Section 3 discusses the cases where the curves have the same j-invariant. We treat the case when the two curves have j-invariants both equal to 0 or 1728 separately. As a consequence of our main theorems, in Sect. 6 it is confirmed that, over $\mathbb{Q}$, the Cremona database contains all the j-invariant pairs associated to p^2-isogenies, for odd p. Examples and code are available on github [DHINRW24].

2 Preliminaries

In this section we will recall some basic definitions and facts about elliptic curves; see Silverman [S09] and [BBDHR24, Sect. 2] for details. To start, let K be a number field or a local field of characteristic 0. An elliptic curve E defined over K is given by a Weierstrass model

$$E : y^2 + a_1xy + a_3y = x^3 + a_2x^2 + a_4x + a_6, \tag{3}$$

with $a_i \in K$. The b_i and c_i-invariants are further defined as

$$\begin{aligned} &b_2 = a_1^2 + 4a_2, && b_4 = 2a_4 + a_1a_3, && b_6 = a_3^2 + 4a_6,\\ &c_4 = b_2^2 - 24b_4, && c_6 = -b_2^3 + 36b_2b_4 - 216b_6,\\ &\Delta = (c_4^3 - c_6^2)/1728, && j = c_4^3/\Delta. \end{aligned}$$

We say E is given by an *integral Weierstrass model* if each a_i is in the ring of integers of K for $i \in \{1, 2, 3, 4, 6\}$.

An elliptic curve E' is K-isomorphic to E if there is an *admissible change of variables* $\tau : E \to E'$ defined by $(x, y) \mapsto (u^2x + r, u^3y + su^2x + w)$, where $u, r, s, w \in K$, with $u \neq 0$. We write $\tau = [u, r, s, w]$. Consequently,

$$j' = j, \qquad \Delta' = u^{-12}\Delta, \qquad c_4' = u^{-4}c_4, \qquad c_6' = u^{-6}c_6.$$

If any of $u, r, s, w \notin K$, then E and E' are isomorphic over an algebraic closure of K. When this occurs, i.e., when τ is defined over an extension of K, we say that E' is a *twist* of E.

In the case that K is a number field with ring of integers $\mathcal{O}_K$, let $\mathfrak{p}$ be a prime ideal of K and $\nu_\mathfrak{p}$ be the normalized valuation of the completion $K_\mathfrak{p}$ of K at $\mathfrak{p}$. Further, let $R_\mathfrak{p}$ be the ring of integers of $K_\mathfrak{p}$. By choosing an element of $\mathfrak{p} \setminus \mathfrak{p}^2$, we can find an element $\pi \in \mathcal{O}_K$ that has $\nu_\mathfrak{p}(\pi) = 1$ and π is thus a uniformizer in $R_\mathfrak{p}$.

Consider E defined over K with Weierstrass equation as in (3). We view E as being defined over $K_\mathfrak{p}$ via the inclusion $\iota : K \hookrightarrow K_\mathfrak{p}$. Via this inclusion we often conflate E being defined over K with E being defined over $K_\mathfrak{p}$. If E is not already $\mathfrak{p}$-integral, then there is at least one coefficient a_i with negative valuation. The transformation $\tau = [u^{-1}, 0, 0, 0]$ on E gives a $K_\mathfrak{p}$-isomorphic elliptic curve whose Weierstrass coefficients are u^ia_i. Thus, choosing u^{-1} to be an appropriate power of π we can find a model for E such that all coefficients u^ia_i are in $R_\mathfrak{p}$, i.e., we can always find a

$\mathfrak{p}$-integral Weierstrass model. As we chose π to be an element of $\mathcal{O}_K$, the model τE is defined over K in addition to $K_\mathfrak{p}$.

A $\mathfrak{p}$-integral Weierstrass model as constructed above will, by construction, have an integral discriminant $\nu_\mathfrak{p}(\Delta) \geq 0$. As $\nu_\mathfrak{p}$ is discrete, there will be a $\mathfrak{p}$-integral model such that $\nu_\mathfrak{p}(\Delta) \geq 0$ is minimal. This model will not be unique, but the minimal valuation of the discriminant will be.

Definition 3 A Weierstrass model for E defined over $K_\mathfrak{p}$ is called a $\mathfrak{p}$-*minimal (Weierstrass) model* if $\nu_\mathfrak{p}(\Delta)$ is minimized subject to the constraint that the model is a $\mathfrak{p}$-integral Weierstrass model. Any Δ such that $\nu_\mathfrak{p}(\Delta)$ is minimal is called a $\mathfrak{p}$-*minimal discriminant* of E, and we call $\nu_\mathfrak{p}(\Delta)$ the *valuation of the minimal discriminant* of E at $\mathfrak{p}$.

Remark 1 While we have been working prime by prime, we note that we can always find a $\mathfrak{p}$-integral model that is globally integral. By [C10, Proposition 4.7.8], we can always construct $\beta \in \mathcal{O}_K$ that has prescribed (non-negative) valuations at a finite set of prime ideals and non-negative valuation at all other primes. Using this element, we can transform E to an integral model τE that is integral and $\mathfrak{p}$-minimal at the prescribed set of primes. The caveat is that the discriminant of τE is not necessarily minimal at primes not in the prescribed set.

When working with elliptic curves over local fields, we will also consider the c_i-invariants along with the discriminant. The following definition will be used frequently in the later sections.

Definition 4 For $\mathfrak{p}$ a prime of K, we define the signature of E defined over K with respect to $\mathfrak{p}$ to be

$$\mathrm{sig}_\mathfrak{p}(E) = \left(\nu_\mathfrak{p}(c_4), \nu_\mathfrak{p}(c_6), \nu_\mathfrak{p}(\Delta)\right).$$

Next, we look at the global properties of elliptic curves related to the valuation of the minimal discriminant of E at $\mathfrak{p}$. Using the embedding ι, for each prime $\mathfrak{p}$ of K, we can view an elliptic curve E defined over K as an elliptic curve defined over $K_\mathfrak{p}$. Then, together with [BBDHR24, Lemma 2.6] and the above discussion, we have the following global definition.

Definition 5 Let E be an elliptic curve defined over K. For each prime $\mathfrak{p}$ of $\mathcal{O}_K$ we can choose a $\mathfrak{p}$-minimal model given by an integral Weierstrass model that is K-isomorphic to E. This $\mathfrak{p}$-minimal model $E_\mathfrak{p}$ has discriminant $\Delta_\mathfrak{p}$, which is defined to be the $\mathfrak{p}$-minimal discriminant of E. The *minimal discriminant ideal* of E defined over K, denoted by $\mathfrak{D}^{\min}$, is the (integral) ideal of K given by

$$\mathfrak{D}^{\min} = \prod_{\mathfrak{p}} \mathfrak{p}^{\nu_\mathfrak{p}(\Delta_\mathfrak{p})}.$$

Over a number field, it is not always possible to find a single Weierstrass equation that is simultaneously minimal for every prime $\mathfrak{p}$, as illustrated in [BBDHR24, Example 2.8]. When it is possible, we have a global minimal model as follows.

Definition 6 A *global minimal model* for E defined over K is an integral Weierstrass equation

$$y^2 + a_1xy + a_3y = x^3 + a_2x^2 + a_4x + a_6$$

such that the discriminant Δ of the equation satisfies $\mathfrak{D}^{\min} = (\Delta)$.

The conductor of an elliptic curve is closely related to its discriminant, and measures, in some sense, the arithmetic complexity of the curve. The conductor describes the possible bad reduction types of an elliptic curve. We will skip the definition here and refer to Silverman [S09, Sect. VIII.11]. Next, we define the object of study, discriminant (ideal) twins.

Definition 7 Let K be a number field, and let E and E' be elliptic curves defined over K that are not K-isomorphic. We say that E and E' are *discriminant ideal twins* if they have the same minimal discriminant ideal and the same conductor. If, additionally, for each prime $\mathfrak{p}$ there exist $\mathfrak{p}$-minimal models for E and E' defined over $\mathcal{O}_K$ such that $\Delta_\mathfrak{p} = \Delta'_\mathfrak{p}$, then we say E and E' are *discriminant twins.*

Now that we have defined discriminant (ideal) twins, we can investigate the discriminant ideals of n-isogenous elliptic curves over number fields. Two elliptic curves E_1 and E_2 defined over the same field K are *isogenous* if there exists a non-constant morphism $\varphi\colon E_1 \to E_2$ with coefficients in K mapping the neutral element of E_1 to the neutral element of E_2. An equivalent definition states that an isogeny is a morphism $\varphi\colon E_1 \to E_2$ that is surjective and has finite kernel, $\ker\varphi$. If $\ker\varphi \cong \mathbb{Z}/n\mathbb{Z}$ for a natural number n, we say E_1 and E_2 are n-isogenous. Furthermore, the isogeny φ is defined over K when $\ker\varphi$ is $\mathrm{Gal}(\overline{K}/K)$-invariant. For an n-isogeny $\varphi : E_1 \to E_2$ defined over K, the K-isomorphism class of $(E, \ker\varphi)$ is a non-cuspidal K-rational point on the classical modular curve $X_0(n)$. Note that the non-cuspidal K-rational points of $X_0(n)$ parameterize isomorphism classes of pairs (E, C) where E is an elliptic curve defined over K and C is a cyclic subgroup of E of order n such that C is $\mathrm{Gal}(\overline{K}/K)$-invariant.

In this article we focus on some n-isogenies when $X_0(n)$ has genus 0. In particular we focus on the case when $n = p^2$, for an odd prime p. If $E_1 \to E_3$ is a p^2-isogeny, then there is an elliptic curve E_2 such that $E_i \to E_{i+1}$ is a p-isogeny for $i = 1, 2$. Here, E_1, E_2 and E_3 are non-isomorphic curves. Theorem 8 gives a parameterization of such curves E_i. Before introducing Theorem 8, let us define some polynomials T_i, S_i in Table 2 that will be used frequently in this article.

Table 2 Polynomials in t that appear in models for $\mathcal{C}_{p^2,i}(t,d)$

p^2	Notation	Polynomial
9	T_1	$t-3$
	T_2	$t+6$
	T_3	t^2+3t+9
	T_4	t^3-24
	T_5	$t^3+234t^2+756t+2160$
	T_6	t^6-36t^3+216
	T_7	$t^6-504t^5-16632t^4-123012t^3-517104t^2-1143072t-1475496$
	T_8	t
	T_9	$t^2-6t+36$
	T_{10}	$t^2-6t-18$
	T_{11}	$t^4+6t^3+54t^2-108t+324$
25	S_1	$t^4+t^3+6t^2+6t+11$
	S_2	t^2+4
	S_3	$t-1$
	S_4	$t^{10}+240t^9+2170t^8+8880t^7+34835t^6+83748t^5+206210t^4+313380t^3+503545t^2+424740t+375376$
	S_5	$t^4+6t^3+21t^2+36t+61$
	S_6	$t^{10}-510t^9-13580t^8-36870t^7-190915t^6-393252t^5-1068040t^4-1508370t^3-2581955t^2-2087010t-1885124$
	S_7	$t^{10}+10t^8+35t^6-12t^5+50t^4-60t^3+25t^2-60t+16$
	S_8	t^4+3t^2+1
	S_9	$t^{10}+10t^8+35t^6-18t^5+50t^4-90t^3+25t^2-90t+76$
	S_{10}	t^2+3t+1
	S_{11}	$t^4-4t^3+11t^2-14t+31$
	S_{12}	$t^4+t^3+11t^2-4t+16$
	S_{13}	t^2-2t-4
	S_{14}	$t^4-4t^3+21t^2-34t+41$

Theorem 8 (Barrios [B23, Theorem 1]) *Let p be an odd prime such that $X_0(p^2)$ has genus 0 and let K be a number field or a local field of characteristic 0. Let E_1 and E_3 be elliptic curves defined over K such that the j-invariants of E_1 and E_3 are not both identically 0 or 1728. Suppose further that E_1 and E_3 are p^2-isogenous elliptic curves over K so that there is an elliptic curve E_2 that is p-isogenous to both E_1 and E_3. Then there are $t \in K$ and $d \in \mathcal{O}_K$ such that E_i are K-isomorphic to $\mathcal{C}_{p^2,i}(t,d)$ given in Table 3 for $i \in \{1,2,3\}$, respectively.*

Table 3 The elliptic curve $\mathcal{C}_{p^2,i}(t,d) : y^2 = x^3 + d^2 A_{p^2,i}(t)x + d^3 B_{p^2,i}(t)$

p^2	i	$A_{p^2,i}(t)$	$B_{p^2,i}(t)$
9	1	$-3 \cdot T_2 \cdot T_5$	$-2 \cdot T_7$
	2	$-3^5 \cdot T_2 \cdot T_8 \cdot T_9$	$-2 \cdot 3^6 \cdot T_{10} \cdot T_{11}$
	3	$-3^9 \cdot T_4 \cdot T_8$	$-2 \cdot 3^{12} \cdot T_6$
25	1	$-3^3 \cdot S_2 \cdot S_4$	$-2 \cdot 3^3 \cdot S_2^2 \cdot S_5 \cdot S_6$
	2	$-3^3 \cdot 5^4 \cdot S_2 \cdot S_{10} \cdot S_{11} \cdot S_{12}$	$-2 \cdot 3^3 \cdot 5^6 \cdot S_2^2 \cdot S_5 \cdot S_8 \cdot S_{13} \cdot S_{14}$
	3	$-3^3 \cdot 5^8 \cdot S_2 \cdot S_7$	$-2 \cdot 3^3 \cdot 5^{12} \cdot S_2^2 \cdot S_8 \cdot S_9$

Remark 2 We note that Theorem 8 differs slightly from [B23, Theorem 1] in that the latter is stated more generally for any field K of characteristic relatively prime to $6p^2$, and $d \in K^\times/(K^\times)^2$. In this article, we will only consider number fields or local fields of characteristic 0. In this setting, if $d \in K^\times/(K^\times)^2$ and $d \notin \mathcal{O}_K$, then we may take a different representative $d' \in \mathcal{O}_K$ such that $\mathcal{C}_{p,i}(t,d)$ is K-isomorphic to $\mathcal{C}_{p,i}(t,d')$. The curves in [B23, Theorem 1] $\mathcal{C}_{n,i}(t,d)$ are given in the form $y^2 = x^3 + d^2 A_{n,i}(t)x + d^3 B_{n,i}(t)$. Let $\{\mathfrak{p}_j\}$ be the finite set of primes such that $\nu_{\mathfrak{p}_j}(d) = -e_j$ where $e_j > 0$. By [C10, Proposition 4.7.8], we can take $\beta \in \mathcal{O}_K$ such that $\nu_{\mathfrak{p}_j}(\beta) = e_j$. The isomorphism $\tau = [1/\beta, 0, 0, 0]$ yields $\tau\mathcal{C}_{n,i}(t,d) : y^2 = x^3 + d^2\beta^4 A_{n,i}(t)x + d^3\beta^6 B_{n,i}(t)$. Thus setting $d' = d\beta^2$ gives $d' \in \mathcal{O}_K$ and $d' = d$ in $K^\times/(K^\times)^2$.

Next, Table 4 gives the j-invariants, c_i-invariants, and discriminants of $\mathcal{C}_{p^2,i}(t,d)$ for $i \in \{1, 2, 3\}$, which will be helpful in the work that follows. As d is a twisting parameter, we see that $j_{p,i}(t) = j_{p,i}(t,d)$ does not depend on d.

Table 4 The quantities $d^{-2}c_{4,i}(t,d)$, $d^{-3}c_{6,i}(t,d)$, $d^{-6}\Delta_i(t,d)$, and $j_i(t)$

p^2	i	$d^{-2}c_{4,i}(t,d)$	$d^{-3}c_{6,i}(t,d)$	$d^{-6}\Delta_i(t,d)$	$j_i(t)$
9	1	$2^4 \cdot 3^2 \cdot T_2 \cdot T_5$	$2^6 \cdot 3^3 \cdot T_7$	$2^{12} \cdot 3^6 \cdot T_1^9 \cdot T_3$	$T_2^3 \cdot T_5^3 \cdot T_1^{-9} \cdot T_3^{-1}$
	2	$2^4 \cdot 3^6 \cdot T_2 \cdot T_8 \cdot T_9$	$2^6 \cdot 3^9 \cdot T_{10} \cdot T_{11}$	$2^{12} \cdot 3^{18} \cdot T_1^3 \cdot T_3^3$	$T_2^3 \cdot T_8^3 \cdot T_9^3 \cdot T_1^{-3} \cdot T_3^{-3}$
	3	$2^4 \cdot 3^{10} \cdot T_4 \cdot T_8$	$2^6 \cdot 3^{15} \cdot T_6$	$2^{12} \cdot 3^{30} \cdot T_1 \cdot T_3$	$T_4^3 \cdot T_8^3 \cdot T_1^{-1} \cdot T_3^{-1}$
25	1	$2^4 \cdot 3^4 \cdot S_2 \cdot S_4$	$2^6 \cdot 3^6 \cdot S_2^2 \cdot S_5 \cdot S_6$	$2^{12} \cdot 3^{12} \cdot S_1 \cdot S_2^3 \cdot S_3^{25}$	$S_4^3 \cdot S_1^{-1} \cdot S_3^{-25}$
	2	$2^4 \cdot 3^4 \cdot 5^4 \cdot S_2 \cdot S_{10} \cdot S_{11} \cdot S_{12}$	$2^6 \cdot 3^6 \cdot 5^6 \cdot S_2^2 \cdot S_5 \cdot S_8 \cdot S_{13} \cdot S_{14}$	$2^{12} \cdot 3^{12} \cdot 5^{12} \cdot S_1^5 \cdot S_2^3 \cdot S_3^5$	$S_{10}^3 \cdot S_{11}^3 \cdot S_{12}^3 \cdot S_1^{-5} \cdot S_3^{-5}$
	3	$2^4 \cdot 3^4 \cdot 5^8 \cdot S_2 \cdot S_7$	$2^6 \cdot 3^6 \cdot 5^{12} \cdot S_2^2 \cdot S_8 \cdot S_9$	$2^{12} \cdot 3^{12} \cdot 5^{24} \cdot S_1 \cdot S_2^3 \cdot S_3$	$S_7^3 \cdot S_1^{-1} \cdot S_3^{-1}$

3 Results for $n = p^2$ for Odd Primes p

In this section, we prove a few results for p^2-isogenous curves where $p \in \{3, 5\}$ that are used to establish our main results. Let K be a number field. Suppose that E_1 and E_3 are p^2-isogenous elliptic curves over K so that there is an elliptic curve E_2 that is p-isogenous to both E_1 and E_3. Note that the curves E_1, E_2 and E_3 are non K-isomorphic curves. The following result ensures that one can assume $t \in \mathcal{O}_K$ when using parametrized families in Theorem 8 for discriminant ideal twins.

Lemma 9 *Let p be an odd prime such that $X_0(p^2)$ has genus 0. Suppose that E_1 and E_3 are p^2-isogenous discriminant ideal twins over K. Then there exists $t \in \mathcal{O}_K$ and $d \in \mathcal{O}_K$ such that $E_i \cong \mathcal{C}_{p^2,i}(t, d)$.*

Proof Since E_1 and E_3 are p^2-isogenous, it follows from Theorem 8 that there exist elements $t \in K$ and $d \in \mathcal{O}_K$ such that $E_i \cong \mathcal{C}_{p^2,i}(t, d)$. It remains to show that $t \in \mathcal{O}_K$ when E_1 and E_3 are discriminant ideal twins.

Towards a contradiction, assume that there exists a prime $\mathfrak{p}$ of $\mathcal{O}_K$ such that $\nu_{\mathfrak{p}}(t) = -k < 0$, for some $k \in \mathbb{Z}$. Given such t, for a monic polynomial $f(t) = t^n + a_{n-1}t^{n-1} + \cdots + a_0 \in \mathbb{Z}[t]$, then $\nu_{\mathfrak{p}}(f(t)) = \min(t^n, a_{n-1}t^{n-1}, \ldots, a_0) = -kn = -k \cdot \deg(f)$. Looking at the j-invariant of E_1 when $p = 3$ in Table 4, we obtain

$$\begin{aligned}\nu_{\mathfrak{p}}(j(E_1)) &= \nu_{\mathfrak{p}}(T_2^3 \cdot T_5^3 \cdot T_1^{-9} \cdot T_3^{-1}) \\ &= 3\nu_{\mathfrak{p}}(T_2) + 3\nu_{\mathfrak{p}}(T_5) - 9\nu_{\mathfrak{p}}(T_1) - \nu_{\mathfrak{p}}(T_3) \\ &= -3k - 9k + 9k + 2k = -k,\end{aligned}$$

where the third equality follows as $T_i \in \mathbb{Z}[t]$ are monic. By similar arguments, using Table 4, for both $p = 3$ and $p = 5$ we obtain

$$\begin{aligned}\nu_{\mathfrak{p}}(j(E_1)) &= \nu_{\mathfrak{p}}(t) = -k, \\ \nu_{\mathfrak{p}}(j(E_2)) &= p\nu_{\mathfrak{p}}(t) = -pk, \\ \nu_{\mathfrak{p}}(j(E_3)) &= p^2\nu_{\mathfrak{p}}(t) = -p^2k.\end{aligned}$$

It follows that the three curves must have either multiplicative reduction or potentially multiplicative reduction at $\mathfrak{p}$. Let $u_i \in K^\times$ be such that the curve $\tau_i E_i$ is minimal at $\mathfrak{p}$ for $\tau_i = [u_i, 0, 0, 0]$. Let $\Delta(E_{i,\mathfrak{p}})$ be the minimal discriminant of E_i at $\mathfrak{p}$.

If the curves have multiplicative reduction, then from [DD15, Table 1] we know that

$$\nu_{\mathfrak{p}}(\Delta(E_{2,\mathfrak{p}})) = p\nu_{\mathfrak{p}}(\Delta(E_{1,\mathfrak{p}})) \text{ as } \nu_{\mathfrak{p}}(j(E_2)) = p\nu_{\mathfrak{p}}(j(E_1))$$

and

$$\nu_{\mathfrak{p}}(\Delta(E_{3,\mathfrak{p}})) = p\nu_{\mathfrak{p}}(\Delta(E_{2,\mathfrak{p}})) \text{ as } \nu_{\mathfrak{p}}(j(E_3)) = p\nu_{\mathfrak{p}}(j(E_2)).$$

If $\nu_{\mathfrak{p}}(\Delta(E_{1,\mathfrak{p}})) \neq 0$, then E_1 and E_3 cannot be discriminant ideal twins. However, if $\nu_{\mathfrak{p}}(\Delta(E_{1,\mathfrak{p}})) = 0$, then we could potentially have discriminant twins, so let us consider $\nu_{\mathfrak{p}}(\Delta(E_{1,\mathfrak{p}}))$. Let $\nu_{\mathfrak{p}}(2) = f$, $\nu_{\mathfrak{p}}(3) = e$, $\nu_{\mathfrak{p}}(u_i) = \alpha_i$, and $\nu_{\mathfrak{p}}(d) = \delta \geq 0$. Then, using Table 4, we get the following:

$n = p^2$	$\nu_{\mathfrak{p}}(c_{4,1,\min})$	$\nu_{\mathfrak{p}}(\Delta(E_{1,\mathfrak{p}}))$
9	$4\alpha_1 + 4f + 2e + 2\delta - 4k$	$12\alpha_1 + 12f + 6e + 6\delta - 11k$
25	$4f + 4e + 4\alpha_1 + 2\delta - 12k$	$12f + 12e + 12\alpha_1 + 6\delta - 35k$

As the curve is integral at $\mathfrak{p}$, we must have $\nu_{\mathfrak{p}}(c_{4,1,\min}) \geq 0$. Consequently,

$$\nu_{\mathfrak{p}}(\Delta(E_{i,\mathfrak{p}})) = 3\nu_{\mathfrak{p}}(c_{4,1,\min}) + k \geq k > 0,$$

so $\nu_{\mathfrak{p}}(\Delta(E_{1,\mathfrak{p}}))$ cannot be 0. Therefore E_1 and E_3 cannot be discriminant ideal twins in the multiplicative reduction case. This is a contradiction to our assumption and hence $t \in \mathcal{O}_K$ in this case.

Next, if the curves have potentially multiplicative reduction, then from [DD15, Table 1] we know that

$$\nu_{\mathfrak{p}}(\Delta(E_{2,\mathfrak{p}})) = \nu_{\mathfrak{p}}(\Delta(E_{1,\mathfrak{p}})) + (p-1)k \text{ as } \nu_{\mathfrak{p}}(j(E_2)) = p\nu_{\mathfrak{p}}(j(E_1))$$

and

$$\nu_{\mathfrak{p}}(\Delta(E_{3,\mathfrak{p}})) = \nu_{\mathfrak{p}}(\Delta(E_{2,\mathfrak{p}})) + p(p-1)k \text{ as } \nu_{\mathfrak{p}}(j(E_3)) = p\nu_{\mathfrak{p}}(j(E_2)).$$

Therefore,

$$\nu_{\mathfrak{p}}(\Delta(E_{3,\mathfrak{p}})) = \nu_{\mathfrak{p}}(\Delta(E_{1,\mathfrak{p}})) + (p^2-1)k \neq \nu_{\mathfrak{p}}(\Delta(E_{1,\mathfrak{p}})),$$

contradicting the assumption that E_1 and E_3 are discriminant ideal twins. It thus follows that $t \in \mathcal{O}_K$. □

3.1 Equal j-Invariants

In this section we categorize all p^2-isogenous discriminant (ideal) twins for $p = 3, 5$ where the two curves share the same j-invariant. Note that the parameterized families of elliptic curves $\mathcal{C}_{p,i}(t, d)$ fail to capture the case when two p^2-isogenous elliptic curves share the same j-invariant 0 or 1728. Thus, as with [BBDHR24], we break our theorems into two categories $j = 0, 1728$ and otherwise. Recall from [BBDHR24, Lemma 3.1] that isogenous curves with the same j-invariant have complex multiplication (CM) and will be isomorphic over any field containing their CM endomorphism ring. Thus they can only possibly be discriminant (ideal) twins over number fields that do not contain their CM endomorphism ring.

Remark 3 When the discriminant of a genus one curve is 0, the curve is singular. If t_0 is such that $\mathcal{C}_{n,i}(t_0, d)$ has discriminant 0, then we call t_0 a singular value.

Theorem 10 *Let $p \in \{3, 5\}$. There are no non-isomorphic p^2-isogenous curves that both have j-invariant* 0 *or* 1728.

Proof Following [BBDHR24, Lemma 3.2], we compute valuations of the modular polynomials $\Phi_{p^2}(j, y)$ for $j = 0, 1728$ and $p = 3, 5$ at $y - j$. Only $\Phi_{25}(1728, y)$ has any appropriate factors, $\nu_{y-1728}(\Phi_{25}(1728, y)) = 2$; for code see [DHINRW24]. Thus there are no non-isomorphic 9-isogenous curves that both have j-invariant 0 or 1728 and there are no non-isomorphic 25-isogenous curves that both have j-invariant 0. This leaves us to check the 25-isogeny case when both j-invariants are 1728.

Every cyclic 25-isogeny $E_1 \to E_3$ factors as two 5-isogenies: $E_1 \to E_2 \to E_3$. If all three curves have j-invariant 1728, then by [BBDHR24, Proposition 3.4], all three curves were actually isomorphic. To find the possible j-invariants for E_2 we use the 5-isogeny parameterization found in [B23] to set $j(\mathcal{C}_{5,1}(t, d)) = 1728$. We then solve for t and determine the possible j-invariants for $\mathcal{C}_{5,2}(t, d)$. The polynomial factors as $j(\mathcal{C}_{5,1}(t, d)) - 1728 = (t^2 + 22t + 125)(t^2 - 500t - 15625)^2/t^5$ with the roots of the factor $t^2 + 22t + 125$ corresponding to singular values. Thus we look at the roots of $t^2 - 500t - 15625$, which are $t_0 = 250 \pm 125\sqrt{5}$. These give j-invariants 1728 and $\alpha, \bar{\alpha} = 22015749613248 \pm 9845745509376\sqrt{5}$ for $\mathcal{C}_{5,1}(t_0, 1)$ and $\mathcal{C}_{5,2}(t_0, 1)$ respectively. Next, we check whether the curve $\mathcal{C}_{5,2}(t_0, 1)$ is 5-isogenous to another elliptic curve with j-invariant 1728 not isomorphic to $\mathcal{C}_{5,2}(t_0, 1)$. To do so, we reuse the trick from the beginning of this proof and factor the modular polynomial $\Phi_5(\alpha, y)$ to find $\nu_{y-1728}(\Phi_5(\alpha, y)) = 1$. As any curve with j-invariant α only has one 5-isogeny to a curve with j-invariant 1728 over $\overline{K}$, there are no (non-isomorphic) 25-isogenous curves that both have j-invariant 1728. □

Theorem 11 *Let $p = 3, 5$. If E_1, E_3 are p^2-isogenous discriminant ideal twins defined over a number field K with $j = j(E_1) = j(E_3)$, but $j \neq 0$ or* 1728, *then j is as follows. If $p = 3$, then t_0 is a root of $t^2 - 6t - 18$ and $j(E_1) = j(E_3) = 14776832t_0 + 32440512$ while $j(E_2) = 1728$ with CM endomorphism ring in $\mathbb{Q}(i)$. If $p = 5$, then t_0 is a root of $t^2 - 2t - 4$ and $j(E_1) = j(E_2) = 9845745509376t_0 + 12170004103872$ and again $j(E_2) = 1728$ with CM endomorphism ring in $\mathbb{Q}(i)$.*

Proof The proof is exactly analogous to the p-isogenous case for $p = 2, 3, 5, 7, 13$ in [BBDHR24]. For $p^2 = 9, 25$, factor the parameterized j-invariants $j_1(t) - j_3(t)$. For each factor $f(t)$, we let $\mathbb{Q}(t_0) \cong \mathbb{Q}[t]/(f(t))$ be the associated number field. If t_0 is not a singular value, then compute $E_1(t_0, 1)$, $E_3(t_0, 1)$ and check if they are isomorphic over $\mathbb{Q}(t_0)$. If they are not isomorphic, then check if they are discriminant ideal twins and compute their CM field.

Following this algorithm in [DHINRW24], we find that for $p^2 = 9$ the factors $t^2 + 3t + 9$ and $t - 3$ give singular values. The factor $t^2 - 6t - 18$ gives non-isomorphic curves over $\mathbb{Q}(t_0)$ that are discriminant twins. The rest of the factors give isomorphic curves over $\mathbb{Q}(t_0)$. Similarly, for $p^2 = 25$ we find that the factors $t^4 + t^3 + 6t^2 + 6t + 11$, $t^2 + 4$ and $t - 1$ give singular values. The factor $t^2 - 2t - 4$ gives non-isomorphic curves that are discriminant twins. As in the $p^2 = 9$ case, the rest of the factors give curves isomorphic over $\mathbb{Q}(t_0)$. Note that unlike the $p = 2, 3, 5, 7, 13$ cases in [BBDHR24], there are no non-isomorphic, non-discriminant ideal twin cases. □

4 Proof for $n = 9$

In this section, we parameterize all 9-isogenous discriminant ideal twins over number fields that do not share the same j-invariant. As we classified all 9-isogenous discriminant ideal twins over number fields that do share the same j-invariant, this completes the classification. Our main tool is the 9-isogeny elliptic curve parameterization due to Barrios [B23]. Consider the polynomials T_m for $1 \leq m \leq 11$ in Table 2. Then for $i = 1, 3$ we can parameterize the two 9-isogenous curves as in Table 3. Moreover, the invariants j_i, Δ_i, $c_{4,i}$, and $c_{6,i}$ are as given in Table 4.

Theorem 12 *Let E_1 and E_3 be two 9-isogenous elliptic curves over a number field K with the property that their j-invariants are not equal. If E_1 and E_3 are discriminant ideal twins, then there exists $t \in \mathcal{O}_K$ and $d \in \mathcal{O}_K$ such that $E_i \cong C_{9,i}(t, d)$, and for each prime $\mathfrak{p}$ of $\mathcal{O}_K$,*

$$\nu_{\mathfrak{p}}(t - 3) = 3k_{\mathfrak{p}}, \qquad 0 \leq k_{\mathfrak{p}} \leq \nu_{\mathfrak{p}}(3). \tag{4}$$

Moreover, if E_1 and E_3 are discriminant twins, then t satisfies (4) *and $(t - 3)^8 \in \mathcal{O}_K^{12}$.*

Proof By Lemma 9, there exist $t \in \mathcal{O}_K$ and $d \in \mathcal{O}_K$ such that $E_i \cong C_{p^2,i}(t, d)$, and thus we can say E_i is given by a model

$$y^2 = x^3 + d^2 A_{9,i}(t)x + d^3 B_{9,i}(t),$$

with $A_{9,i}(t), B_{9,i}(t) \in \mathcal{O}_K$ as defined in Table 3. Let $\mathfrak{D}_i^{\min}$ denote the minimal discriminant ideal of E_i; then $\mathfrak{D}_1^{\min} = \mathfrak{D}_3^{\min}$ since E_1 and E_3 are discriminant ideal twins. Note that the models currently given for E_1 and E_3 are not necessarily minimal at a given prime $\mathfrak{p}$ and their discriminants Δ_1 and Δ_3 are not necessarily minimal at $\mathfrak{p}$ either. As we are interested in the minimal discriminant ideal, we proceed prime by prime to construct $\mathcal{O}_K$-integral, $\mathfrak{p}$-minimal models of E_1 and E_3 to access and compare $\nu_{\mathfrak{p}}(\mathfrak{D}_1^{\min})$ and $\nu_{\mathfrak{p}}(\mathfrak{D}_3^{\min})$.

Let $\mathfrak{p}$ be a prime such that $\nu_{\mathfrak{p}}(3) \geq 1$. Then the signature of E_3 is of the form $(\geq 10\nu_{\mathfrak{p}}(3), \geq 15\nu_{\mathfrak{p}}(3), \geq 30\nu_{\mathfrak{p}}(3))$. In this case, by [P93, Tableau III], we know that the curve is not minimal at $\mathfrak{p}$. We can therefore make the change of variable $\tau_3 = [3^2, 0, 0, 0]$ to reduce this curve. From now on, replace E_3 by $\tau_3 E_3$.

For any prime $\mathfrak{p}$, we take $E_{i,\mathfrak{p}}$ to be a local minimal model of E_i at $\mathfrak{p}$. Suppose further that $E_{i,\mathfrak{p}}$ is $\mathcal{O}_K$-integral. Consequently, $\nu_{\mathfrak{p}}(\Delta(E_{1,\mathfrak{p}})) = \nu_{\mathfrak{p}}(\Delta(E_{3,\mathfrak{p}}))$, where $\Delta(E_{i,\mathfrak{p}})$ is the discriminant of the local minimal model $E_{i,\mathfrak{p}}$. In particular, there exists $\mu_{\mathfrak{p}} \in K$ such that $\nu_{\mathfrak{p}}(\mu_{\mathfrak{p}}) = 0$ and $\Delta(E_{1,\mathfrak{p}}) = \mu_{\mathfrak{p}}\Delta(E_{3,\mathfrak{p}})$. By [S09, Proposition VII.1.3], there exists $u_{i,\mathfrak{p}} \in \mathcal{O}_K$ such that $u_{i,\mathfrak{p}}^{-12}\Delta_i = \Delta(E_{i,\mathfrak{p}})$. Consequently, $u_{1,\mathfrak{p}}^{-12}\Delta_1 = \mu_{\mathfrak{p}} u_{3,\mathfrak{p}}^{-12}\Delta_3$. Then

$$(t-3)^8 = \frac{\Delta_1}{\Delta_3} = \mu_{\mathfrak{p}} \left(\frac{u_{1,\mathfrak{p}}}{u_{3,\mathfrak{p}}}\right)^{12}. \tag{5}$$

Note that if E_1 and E_3 are discriminant twins, then $\mu_{\mathfrak{p}} = 1$.

Let $\nu_{\mathfrak{p}}(T_1) = \nu_{\mathfrak{p}}(t-3) = s_{\mathfrak{p}} \geq 0$ and $\nu_{\mathfrak{p}}(3) = \ell \geq 0$. Suppose first that E_1 and E_3 are discriminant ideal twins. Applying $\nu_{\mathfrak{p}}(\cdot)$ to both sides of (5), we have

$$8s_{\mathfrak{p}} = 12\nu_{\mathfrak{p}}(u_{1,\mathfrak{p}}/u_{3,\mathfrak{p}}).$$

Note that since the valuation takes integral values, $s_{\mathfrak{p}} = 3k_{\mathfrak{p}}$ for some $k_{\mathfrak{p}} \geq 0$. We want to show that $k_{\mathfrak{p}} \leq \ell$. In order to do this we investigate the possible values of $\nu_{\mathfrak{p}}(u_{1,\mathfrak{p}}/u_{3,\mathfrak{p}})$.

We start by showing that we can locally take the twisting parameter d up to squares. For a prime $\mathfrak{p}$ of $\mathcal{O}_K$, let $\delta = \nu_{\mathfrak{p}}(d)$, and let $\pi \in \mathfrak{p} \setminus \mathfrak{p}^2$ be a uniformizer for $R_{\mathfrak{p}}$. We may write $d = d_{\mathfrak{p}}\pi^{2i_{\mathfrak{p}}}$, where $\delta_{\mathfrak{p}} = \nu_{\mathfrak{p}}(d_{\mathfrak{p}}) \in \{0, 1\}$. Then, using the change of variables $[\pi^{i_{\mathfrak{p}}}, 0, 0, 0]$, we obtain the $\mathfrak{p}$-integral model

$$E_i' : y^2 = x^3 + d_{\mathfrak{p}}^2 A_{9,i}(t)x + d_{\mathfrak{p}}^3 B_{9,i}(t).$$

Note that if the signature of E_i is $(\nu_{\mathfrak{p}}(c_{4,i}), \nu_{\mathfrak{p}}(c_{6,i}), \nu_{\mathfrak{p}}(\Delta_i))$, then the signature of E_i' is $(\nu_{\mathfrak{p}}(c_{4,i}) - 4i_{\mathfrak{p}}, \nu_{\mathfrak{p}}(c_{6,i}) - 6i_{\mathfrak{p}}, \nu_{\mathfrak{p}}(\Delta_i) - 12i_{\mathfrak{p}})$. As we are concerned only with the signature of a minimal model, we therefore consider the signature of E_i' rather than E_i. From hereon, we replace E_i with E_i'.

First, we assume that $\ell = 0$. We will show that in this case $k_{\mathfrak{p}} = 0$. Towards the contradiction, let us assume that $k_{\mathfrak{p}} \neq 0$. This forces $\nu_{\mathfrak{p}}(T_2) = \nu_{\mathfrak{p}}(T_3) = \nu_{\mathfrak{p}}(T_5) = \nu_{\mathfrak{p}}(T_7) = 0$. Then the signatures of the two curves are

$$\begin{aligned}
&\nu_{\mathfrak{p}}(c_{4,1}) = 4\nu_{\mathfrak{p}}(2) + 2\delta_{\mathfrak{p}},\ \nu_{\mathfrak{p}}(c_{6,1}) = 6\nu_{\mathfrak{p}}(2) + 3\delta_{\mathfrak{p}},\ \nu_{\mathfrak{p}}(\Delta_1) = 12\nu_{\mathfrak{p}}(2) + 6\delta_{\mathfrak{p}} + 9s_{\mathfrak{p}},\\
&\nu_{\mathfrak{p}}(c_{4,3}) = 4\nu_{\mathfrak{p}}(2) + 2\delta_{\mathfrak{p}},\ \nu_{\mathfrak{p}}(c_{6,3}) = 6\nu_{\mathfrak{p}}(2) + 3\delta_{\mathfrak{p}},\ \nu_{\mathfrak{p}}(\Delta_3) = 12\nu_{\mathfrak{p}}(2) + 6\delta_{\mathfrak{p}} + s_{\mathfrak{p}}.
\end{aligned}$$

If $\nu_{\mathfrak{p}}(2) = 0$, then the two curves are already minimal at $\mathfrak{p}$. If $\nu_{\mathfrak{p}}(2) \neq 0$, we can use [P93, Tableau V] to see that E_1 and E_3 are minimal at $\mathfrak{p}$. In both cases, we observe that the two curves cannot be discriminant ideal twins (as $s_{\mathfrak{p}} = 3k_{\mathfrak{p}} \neq 0$). This is a contradiction, hence $k_{\mathfrak{p}} = 0$.

Next, we consider the case when $\ell \geq 1$. We want to show that $k_{\mathfrak{p}} \leq \ell$, i.e., $s_{\mathfrak{p}} \leq 3\ell$. Towards a contradiction, assume that $3\ell < s_{\mathfrak{p}}$. In this case, using Table 6, we observe that the signatures of the two curves are

$$\nu_{\mathfrak{p}}(c_{4,1}) = 12\ell + 2\delta_{\mathfrak{p}},\ \nu_{\mathfrak{p}}(c_{6,1}) = 18\ell + 3\delta_{\mathfrak{p}},\ \nu_{\mathfrak{p}}(\Delta_1) = 9\ell + 6\delta_{\mathfrak{p}} + 9s_{\mathfrak{p}},$$
$$\nu_{\mathfrak{p}}(c_{4,3}) = 4\ell + 2\delta_{\mathfrak{p}},\ \ \nu_{\mathfrak{p}}(c_{6,3}) = 6\ell + 3\delta_{\mathfrak{p}},\ \ \nu_{\mathfrak{p}}(\Delta_3) = 9\ell + 6\delta_{\mathfrak{p}} + s_{\mathfrak{p}}.$$

If we reduce the curve E_3 by $\tau_3 = [\pi^i, 0, 0, 0]$ (for some integer $i \geq 0$) and the curve E_1 by $\tau_1 = [\pi^{i+2\ell}, 0, 0, 0]$, then the new reduced curves will have the following signatures:

$$\nu_{\mathfrak{p}}(c_{4,1}) = 4\ell + 2\delta_{\mathfrak{p}} - 4i,\ \nu_{\mathfrak{p}}(c_{6,1}) = 6\ell + 3\delta_{\mathfrak{p}} - 6i,\ \nu_{\mathfrak{p}}(\Delta_1) = 9\ell + 3k_{\mathfrak{p}} + 6\delta_{\mathfrak{p}} - 12i + 24(k_{\mathfrak{p}} - \ell),$$
$$\nu_{\mathfrak{p}}(c_{4,3}) = 4\ell + 2\delta_{\mathfrak{p}} - 4i,\ \nu_{\mathfrak{p}}(c_{6,3}) = 6\ell + 3\delta_{\mathfrak{p}} - 6i,\ \nu_{\mathfrak{p}}(\Delta_3) = 9\ell + 3k_{\mathfrak{p}} + 6\delta_{\mathfrak{p}} - 12i.$$

Observe that $\tau_1 E_1$ is an integral curve if and only if $\tau_3 E_3$ is an integral curve. Moreover, if we consider the value i for which one of these curves is minimal, then using [P93, Tableau III], we observe that the other curve must also be minimal. It follows that these reductions give us minimal models $E_{1,\mathfrak{p}}$ and $E_{3,\mathfrak{p}}$. By noting that $\nu_{\mathfrak{p}}(\Delta(E_{1,\mathfrak{p}})) = 24(k_{\mathfrak{p}} - \ell) + \nu_{\mathfrak{p}}(\Delta(E_{3,\mathfrak{p}})) > \nu_{\mathfrak{p}}(\Delta(E_{3,\mathfrak{p}}))$ (since $3k_{\mathfrak{p}} = s_{\mathfrak{p}} > 3\ell$), we conclude that in this case the two curves cannot be discriminant ideal twins, reaching a contradiction.

Hence we conclude that, for all primes $\mathfrak{p}$ of $\mathcal{O}_K$, t satisfies

$$\nu_{\mathfrak{p}}(t-3) = 3k_{\mathfrak{p}}, \quad \text{for } 0 \leq k_{\mathfrak{p}} \leq \nu_{\mathfrak{p}}(3).$$

Lastly, suppose that E_1 and E_2 are discriminant twins. Then for each prime $\mathfrak{p}$ of $\mathcal{O}_K$, we can take $\mu_{\mathfrak{p}} = 1$ in (5). Consequently, we have that for $\mathfrak{p}$ with $\nu_{\mathfrak{p}}(3) \neq 0$, we can write

$$(t-3) = \left(u_{\mathfrak{p}}\pi\right)^3, \tag{6}$$

where $u_{\mathfrak{p}}$ is a unit in $\mathcal{O}_{K_{\mathfrak{p}}}$ and π is a uniformizer. For prime $\mathfrak{p}$ with $\nu_{\mathfrak{p}}(3) = 0$, it follows from (5) that there is a unit $u_{\mathfrak{p}}$ in $\mathcal{O}_{K_{\mathfrak{p}}}$ such that

$$(t-3)^8 = u_{\mathfrak{p}}^{12}. \tag{7}$$

Now consider the inclusion $K \hookrightarrow K_{\mathfrak{p}}$. Then $(t-3)^8 \in \mathcal{O}_{K_{\mathfrak{p}}}^{12}$ for each prime $\mathfrak{p}$ of K. We conclude by the Grunwald-Wang Theorem [S50, Corollary 2] that $(t-3)^8 \in \mathcal{O}_K^{12}$. □

Theorem 13 *Let E_1 and E_3 be two 9-isogenous elliptic curves over a number field K such that their j-invariants are not the same. If there exists $t \in \mathcal{O}_K$ and $d \in \mathcal{O}_K$ such that E_i is K-isomorphic to the elliptic curve $C_{9,i}(t,d)$, and for each prime $\mathfrak{p}$ of $\mathcal{O}_K$,*

$$\nu_{\mathfrak{p}}(t-3) = 3k_{\mathfrak{p}}, \qquad 0 \leq k_{\mathfrak{p}} \leq \nu_{\mathfrak{p}}(3), \tag{8}$$

then E_1 and E_3 are discriminant ideal twins. Moreover, if t satisfies (8) *and* $(t-3)^8 \in \mathcal{O}_K^{12}$, *then E_1 and E_3 are discriminant twins.*

Proof Let E_1 and E_3 be two 9-isogenous elliptic curves over a number field K. Suppose that there exist $t \in \mathcal{O}_K$ and and $d \in \mathcal{O}_K$ such that $E_i \cong C_{9,i}(t,d)$ with t

satisfying

$$\nu_{\mathfrak{p}}(t-3) = 3k_{\mathfrak{p}} = s_{\mathfrak{p}},$$

for some integer $0 \le k_{\mathfrak{p}} \le \nu_{\mathfrak{p}}(3)$, for all primes $\mathfrak{p}$ of $\mathcal{O}_K$. Using the definitions of the invariants of the curve E_3 and [P93, Tableau III], we note that we can reduce E_3 at primes $\mathfrak{p}$ above 3. Thus, replace E_3 by $\tau_3 E_3$ where $\tau_3 = [3^2, 0, 0, 0]$. Let $E_{i,\mathfrak{p}}$ is a minimal model of E_i at $\mathfrak{p}$.

Let $\mathfrak{p}$ be a prime of $\mathcal{O}_K$. Let $\delta = \nu_{\mathfrak{p}}(d) \ge 0$. Note that we do not assume that $\delta \in \{0, 1\}$. If $\nu_{\mathfrak{p}}(T_1) = \nu_{\mathfrak{p}}(t-3) > 0$, then by (8) we have $\nu_{\mathfrak{p}}(3) > 0$. In order to prove our statement, we consider all the cases given in Table 6. This table computes the valuations of the polynomials T_m given in Table 2, deriving the signatures and the valuation of the j-invariants of the two curves. Note that the j-invariants satisfy $\nu_{\mathfrak{p}}(j_1) \ge 0$ and $\nu_{\mathfrak{p}}(j_3) \ge 0$ for all the cases in Table 6 except for the case when $k_{\mathfrak{p}} = \ell/2$. Moreover, in all these cases, $\nu_{\mathfrak{p}}(\Delta_1) = \nu_{\mathfrak{p}}(\Delta_3) + 24k_{\mathfrak{p}}$, i.e., $\nu_{\mathfrak{p}}(\Delta_1) \equiv \nu_{\mathfrak{p}}(\Delta_3) \mod 12$. Then by [BBDHR24, Lemma 4.4], we conclude that the $\nu_{\mathfrak{p}}(\Delta(E_{1,\mathfrak{p}})) = \nu_{\mathfrak{p}}(\Delta(E_{3,\mathfrak{p}}))$.

Next, we assume $k_{\mathfrak{p}} = \ell/2$. Let $\delta = \nu_{\mathfrak{p}}(d)$. Then, from Table 6,

$$\begin{aligned} &\nu_{\mathfrak{p}}(c_{4,1}) = 8\ell + 2\delta,\ \nu_{\mathfrak{p}}(c_{6,1}) = 12\ell + 3\delta,\ \nu_{\mathfrak{p}}(\Delta_1) = 9\ell + 6\delta + \delta_3 + 9s_{\mathfrak{p}},\\ &\nu_{\mathfrak{p}}(c_{4,3}) = 4\ell + 2\delta,\ \nu_{\mathfrak{p}}(c_{6,3}) = 6\ell + 3\delta,\quad \nu_{\mathfrak{p}}(\Delta_3) = 9\ell + 6\delta + \delta_3 + s_{\mathfrak{p}}.\end{aligned}$$

Here, δ_3 is a non-negative integer. If we reduce the curve E_3 by $\tau_3 = [\mathfrak{p}^i, 0, 0, 0]$ and the curve E_1 by $\tau_1 = [\mathfrak{p}^{i+2k_{\mathfrak{p}}}, 0, 0, 0]$ (for some integer $i \ge 0$), then the new reduced curves will have equal signatures, given as follows:

$$(4\ell + 2\delta - 4i,\quad 6\ell + 3\delta - 6i,\quad 9\ell + 6\delta + \delta_3 + s_{\mathfrak{p}} - 12i).$$

Consider the value of i for which $\tau_3 E_3$ is minimal. Then $\tau_1 E_1$ is also minimal (since they have the same signature). We conclude that $\nu_{\mathfrak{p}}(\Delta(E_{1,\mathfrak{p}})) = \nu_{\mathfrak{p}}(\Delta(E_{3,\mathfrak{p}}))$ in this case.

Lastly, we assume that $\nu_{\mathfrak{p}}(T_1) = \nu_{\mathfrak{p}}(t-3) = 0$. We have the following cases:

(a) Assume $\nu_{\mathfrak{p}}(3) \ne 0$. In this case, the curves E_1 and E_3 have the same signature, so they reduce in the same way. It follows that $\nu_{\mathfrak{p}}(\Delta(E_{1,\mathfrak{p}})) = \nu_{\mathfrak{p}}(\Delta(E_{3,\mathfrak{p}}))$.

(b) Assume $\nu_{\mathfrak{p}}(3) = 0$ and $\nu_{\mathfrak{p}}(T_3) \ne 0$. Again, in this case, the curves E_1 and E_3 have the same signature, so they reduce in the same way. It follows that $\nu_{\mathfrak{p}}(\Delta(E_{1,\mathfrak{p}})) = \nu_{\mathfrak{p}}(\Delta(E_{3,\mathfrak{p}}))$.

(c) Assume $\nu_{\mathfrak{p}}(3) = 0$ and $\nu_{\mathfrak{p}}(T_3) = 0$. If $\nu_{\mathfrak{p}}(2) = 0$, then E_1 and E_3 are already minimal at $\mathfrak{p}$. If $\nu_{\mathfrak{p}}(2) \ne 0$, then using [P93, Tableau V], we can see that E_1 and E_3 are minimal at $\mathfrak{p}$. Thus, $\nu_{\mathfrak{p}}(\Delta(E_{1,\mathfrak{p}})) = \nu_{\mathfrak{p}}(\Delta(E_{3,\mathfrak{p}}))$.

We conclude that E_1 and E_3 are discriminant ideal twins when t satisfies (8).

It remains to show that if t satisfies (8) and $(t-3)^8 \in \mathcal{O}_K^{12}$, then E_1 and E_3 are discriminant twins. To this end, for a prime $\mathfrak{p}$ of K, let $E_{i,\mathfrak{p}}$ be a $\mathfrak{p}$-minimal model of E_i. By [S94, Proposition VII.1.3] there exists $u_{i,\mathfrak{p}} \in \mathcal{O}_K$ such that $u_{i,\mathfrak{p}}^{-12}\Delta_i = \Delta(E_{i,\mathfrak{p}})$.

Moreover, since E_1 and E_3 are discriminant ideal twins (as they satisfy (8)), it is the case that there is a $\mu_\mathfrak{p} \in K$ with $\nu_\mathfrak{p}(\mu_\mathfrak{p}) = 0$ such that $\Delta(E_{1,\mathfrak{p}}) = \mu_\mathfrak{p}\Delta(E_{3,\mathfrak{p}})$. It follows that

$$(t-3)^8 = \frac{\Delta_1}{\Delta_3} = \mu_\mathfrak{p}\left(\frac{u_{1,\mathfrak{p}}}{u_{3,\mathfrak{p}}}\right)^{12} \quad\Longrightarrow\quad \mu_\mathfrak{p} = (t-3)^8\left(\frac{u_{3,\mathfrak{p}}}{u_{1,\mathfrak{p}}}\right)^{12}.$$

By our assumption, we get $\mu_\mathfrak{p} \in \mathcal{O}_K^{12}$. In particular, there is a $\kappa_\mathfrak{p} \in \mathcal{O}_K$ such that $\kappa_\mathfrak{p}^{12} = \mu_\mathfrak{p}$ and $\nu_\mathfrak{p}(\kappa_\mathfrak{p}) = 0$. Now let $E'_{1,\mathfrak{p}}$ be the elliptic curve obtained from $E_{1,\mathfrak{p}}$ via the isomorphism $[\kappa_\mathfrak{p}, 0, 0, 0]$. Then $E'_{1,\mathfrak{p}}$ is an $\mathfrak{p}$-minimal model with $\Delta(E'_{1,\mathfrak{p}}) = \kappa_\mathfrak{p}^{-12}\Delta(E_{1,\mathfrak{p}})$. Hence $\Delta(E_{1,\mathfrak{p}}) = \mu_\mathfrak{p}\Delta(E'_{1,\mathfrak{p}})$, which, in turn, yields that $\Delta(E'_{1,\mathfrak{p}}) = \Delta(E_{3,\mathfrak{p}})$. This shows that for each $\mathfrak{p}$, there are $\mathfrak{p}$-minimal models of E_1 and E_3 having equal discriminants. Therefore E_1 and E_3 are discriminant twins. □

5 Proof for $n = 25$

In this section, we parameterize all 25-isogenous discriminant ideal twins over number fields that do not share the same j-invariant. Again, as we have already classified all 25-isogenous discriminant ideal twins over number fields that do share the same j-invariant, this completes the classification. Our main tool is the 25-isogeny elliptic curve parameterization due to Barrios [B23]. Consider the polynomials S_m for $1 \leq m \leq 14$ in Table 2. Then for $i = 1, 3$ we can parameterize the two 25-isogenous curves as in Table 3. Moreover, we can define the invariants j_i, Δ_i, $c_{4,i}$, and $c_{6,i}$ as in Table 4.

Theorem 14 *Let E_1 and E_3 be two 25-isogenous elliptic curves over K with the property that their j-invariants are not the same. If E_1 and E_3 are discriminant ideal twins then there exists $t \in \mathcal{O}_K$ and $d \in \mathcal{O}_K$ such that $E_i \cong C_{25,i}(t, d)$ and for each prime $\mathfrak{p}$ of K, t satifies*

$$\nu_\mathfrak{p}(t-1) = k_\mathfrak{p}, \qquad 0 \leq k_\mathfrak{p} \leq \nu_\mathfrak{p}(5). \tag{9}$$

Proof By Lemma 9, there exist $t \in \mathcal{O}_K$ and $d \in \mathcal{O}_K$ such that $E_i \cong C_{p^2,i}(t, d)$, and thus E_i is given by an $\mathcal{O}_K$-integral model. Now let $\mathfrak{D}_i^{\min}$ denote the minimal discriminant ideal of E_i. Then $\mathfrak{D}_1^{\min} = \mathfrak{D}_3^{\min}$ since E_1 and E_3 are discriminant ideal twins.

Note that the curve E_3 is not minimal at the primes $\mathfrak{p}$ above 5, so we can make the change of variables $\tau_3 = [5^2, 0, 0, 0]$ to reduce this curve. From hereon, replace E_3 by $\tau_3 E_3$.

For any prime $\mathfrak{p}$, let $E_{i,\mathfrak{p}}$ denote a local minimal model of E_i at $\mathfrak{p}$. Suppose further that $E_{i,\mathfrak{p}}$ is $\mathcal{O}_K$-integral. Consequently, $\nu_\mathfrak{p}(\Delta(E_{1,\mathfrak{p}})) = \nu_\mathfrak{p}(\Delta(E_{3,\mathfrak{p}}))$. In particular, there exists $\mu_\mathfrak{p} \in K$ such that $\nu_\mathfrak{p}(\mu_\mathfrak{p}) = 0$ and $\Delta(E_{1,\mathfrak{p}}) = \mu_\mathfrak{p}\Delta(E_{3,\mathfrak{p}})$. By [S94,

Proposition VII.1.3], there exists $u_{i,\mathfrak{p}} \in \mathcal{O}_K$ such that $u_{i,\mathfrak{p}}^{-12}\Delta_i = \Delta(E_{i,\mathfrak{p}})$. Consequently, $u_{1,\mathfrak{p}}^{-12}\Delta_1 = \mu_{\mathfrak{p}} u_{2,\mathfrak{p}}^{-12}\Delta_3$. Then

$$(t-1)^{24} = \frac{\Delta_1}{\Delta_3} = \mu_{\mathfrak{p}} \left(\frac{u_{1,\mathfrak{p}}}{u_{3,\mathfrak{p}}}\right)^{12}. \tag{10}$$

Note that if E_1 and E_3 are discriminant twins, then $\mu_{\mathfrak{p}} = 1$.

Let $\nu_{\mathfrak{p}}(S_3) = \nu_{\mathfrak{p}}(t-1) = k_{\mathfrak{p}} \geq 0$ and $\nu_{\mathfrak{p}}(5) \geq 0$. Suppose first that E_1 and E_3 are discriminant ideal twins. Applying $\nu_{\mathfrak{p}}(\cdot)$ to both sides of (10) implies that

$$2k_p = 2\nu_{\mathfrak{p}}(t-1) = \nu_{\mathfrak{p}}(u_{1,\mathfrak{p}}/u_{3,\mathfrak{p}}).$$

Now, we will show that $\nu_{\mathfrak{p}}(t-1) \leq \nu_{\mathfrak{p}}(5)$. For this we will investigate possible values of $\nu_{\mathfrak{p}}(u_{1,\mathfrak{p}}/u_{3,\mathfrak{p}})$ over all primes $\mathfrak{p}$ of $\mathcal{O}_K$.

As in the proof of Theorem 12, we may write $d = d_{\mathfrak{p}}\pi^{2i_{\mathfrak{p}}}$ for an uniformizer $\pi \in \mathfrak{p} \setminus \mathfrak{p}^2$ for $R_{\mathfrak{p}}$ and $\delta_{\mathfrak{p}} = \nu_{\mathfrak{p}}(d_{\mathfrak{p}}) \in \{0,1\}$. Just as in the proof of Theorem 12, from hereon we may replace E_i by the isomorphic $\mathfrak{p}$-integral model E_i' using the change of variables $[\pi^{i_{\mathfrak{p}}}, 0, 0, 0]$. From hereon, we can thus replace $\delta = \nu_{\mathfrak{p}}(d)$ by $\delta_{\mathfrak{p}} \in \{0,1\}$.

If $\nu_{\mathfrak{p}}(\Delta_3) = 0$, then $\nu_{\mathfrak{p}}(\Delta_1) = 0$. So in this case $\nu_{\mathfrak{p}}(u_{1,\mathfrak{p}}/u_{3,\mathfrak{p}}) = 0$ and $k_{\mathfrak{p}} = 0$. We can therefore assume that $\nu_{\mathfrak{p}}(\Delta_3) > 0$.

We have two cases, depending on whether or not $\nu_{\mathfrak{p}}(5) = 0$. Firstly, assume that $\nu_{\mathfrak{p}}(5) = 0$. We want to show that $k_{\mathfrak{p}} = 0$. Towards the contradiction, let us assume $k_{\mathfrak{p}} \neq 0$. Then $\nu_{\mathfrak{p}}(S_1) = \nu_{\mathfrak{p}}(S_2) = \nu_{\mathfrak{p}}(S_4) = \nu_{\mathfrak{p}}(S_5) = \nu_{\mathfrak{p}}(S_6) = \nu_{\mathfrak{p}}(S_7) = \nu_{\mathfrak{p}}(S_8) = \nu_{\mathfrak{p}}(S_9) = 0$. Then the signatures of the two curves are

$$\mathrm{sig}_{\mathfrak{p}}(E_1) = (4\nu_{\mathfrak{p}}(2) + 4\nu_{\mathfrak{p}}(3) + 2\delta_{\mathfrak{p}}, 6\nu_{\mathfrak{p}}(2) + 6\nu_{\mathfrak{p}}(3) + 3\delta_{\mathfrak{p}}, 12\nu_{\mathfrak{p}}(2) + 12\nu_{\mathfrak{p}}(3) + 6\delta_{\mathfrak{p}} + 25k_{\mathfrak{p}}),$$
$$\mathrm{sig}_{\mathfrak{p}}(E_3) = (4\nu_{\mathfrak{p}}(2) + 4\nu_{\mathfrak{p}}(3) + 2\delta_{\mathfrak{p}}, 6\nu_{\mathfrak{p}}(2) + 6\nu_{\mathfrak{p}}(3) + 3\delta_{\mathfrak{p}}, 12\nu_{\mathfrak{p}}(2) + 12\nu_{\mathfrak{p}}(3) + 6\delta_{\mathfrak{p}} + k_{\mathfrak{p}}).$$

If $\nu_{\mathfrak{p}}(2) = \nu_{\mathfrak{p}}(3) = 0$, then $\nu_{\mathfrak{p}}(c_{4,1}) = \nu_{\mathfrak{p}}(c_{4,3}) = 2\delta_p$, so E_1 and E_3 are minimal at primes $\mathfrak{p}$ with $\nu_{\mathfrak{p}}(2) = \nu_{\mathfrak{p}}(3) = \nu_{\mathfrak{p}}(5) = 0$. If $\nu_{\mathfrak{p}}(2) > 0$, then using [P93, Tableau V], we observe that the models of E_1 an E_3 are minimal at $\mathfrak{p}$ in this case as well. If $\nu_{\mathfrak{p}}(3) > 0$, then $\nu_{\mathfrak{p}}(\Delta_1) = \nu_{\mathfrak{p}}(\Delta_3) + 24k_{\mathfrak{p}}$, $\nu_{\mathfrak{p}}(c_{4,1}) = \nu_{\mathfrak{p}}(c_{4,3})$, and $\nu_{\mathfrak{p}}(c_{6,1}) = \nu_{\mathfrak{p}}(c_{6,3})$. Note that if we reduce the curve E_3 by $\tau_3 = [\mathfrak{p}^i, 0, 0, 0]$ and E_1 by $\tau_1 = [\mathfrak{p}^{i+2k_{\mathfrak{p}}}, 0, 0, 0]$ (for some integer $i \geq 0$), we obtain two new curves with the property that $\tau_1 E_1$ is an integral model if and only if $\tau_3 E_3$ is an integral model. Choose i for which $\tau_3 E_3$ is a minimal model. By [P93, Tableau III], we observe that $\tau_1 E_1$ is also a minimal model. In all these cases, we observe that the minimal models satisfy the fact that $\nu_{\mathfrak{p}}(\Delta(E_{1,\mathfrak{p}})) = \nu_{\mathfrak{p}}(\Delta(E_{3,\mathfrak{p}})) + 24k_{\mathfrak{p}}$, so the two curves cannot be discriminant ideal twins (since $k_p \neq 0$), which is a contradiction.

Now, assume that $\ell = \nu_{\mathfrak{p}}(5) > 0$. In this case, $k_{\mathfrak{p}} \geq 0$. We want to show that $k_{\mathfrak{p}} \leq \ell$. Towards the contradiction, let us assume that $\ell < k_{\mathfrak{p}}$. Then by Table 7,

$$\nu_{\mathfrak{p}}(c_{4,1}) = 10\ell + 2\delta_{\mathfrak{p}},\ \nu_{\mathfrak{p}}(c_{6,1}) = 15\ell + 3\delta_{\mathfrak{p}},\ \nu_{\mathfrak{p}}(\Delta_1) = 5\ell + 6\delta_{\mathfrak{p}} + 25k_{\mathfrak{p}},$$
$$\nu_{\mathfrak{p}}(c_{4,3}) = 2\ell + 2\delta_{\mathfrak{p}},\quad \nu_{\mathfrak{p}}(c_{6,3}) = 3\ell + 3\delta_{\mathfrak{p}},\quad \nu_{\mathfrak{p}}(\Delta_3) = 5\ell + 6\delta_{\mathfrak{p}} + k_{\mathfrak{p}}.$$

Note that, in order to obtain minimal models, E_3 can be reduced by $\tau_3 = [\mathfrak{p}^i, 0, 0, 0]$ if and only if E_1 can be reduced by $\tau_1 = [\mathfrak{p}^{i+2\ell}, 0, 0, 0]$ for some $i \geq 0$. In this case the minimal discriminants of E_1 and E_3 are not equal, since $k_\mathfrak{p} \neq 0$. This is a contradiction. Hence we obtain that t satisfies

$$\nu_\mathfrak{p}(t-1) = k_\mathfrak{p}, \quad \text{for } 0 \leq k_\mathfrak{p} \leq \nu_\mathfrak{p}(5).$$

□

Theorem 15 *Let E_1 and E_3 be two 25-isogenous elliptic curves over K such that their j-invariants are not the same. If there exists $t \in \mathcal{O}_K$ and $d \in \mathcal{O}_K$ such that E_i is K-isomorphic to the elliptic curve $C_{25,i}(t,d)$, and for each prime $\mathfrak{p}$ of $\mathcal{O}_K$,*

$$\nu_\mathfrak{p}(t-1) = k_\mathfrak{p}, \qquad 0 \leq k_\mathfrak{p} \leq \nu_\mathfrak{p}(5), \tag{11}$$

then E_1 and E_3 are discriminant ideal twins. Moreover, E_1 and E_3 are discriminant twins.

Proof Let E_1 and E_3 be two 25-isogenous elliptic curves over a number field K. Assume moreover that there exist $t \in \mathcal{O}_K$ and $d \in \mathcal{O}_K$ such that $E_i \cong C_{25,i}(t,d)$ with t satisfying the equation

$$\nu_\mathfrak{p}(t-1) = k_\mathfrak{p},$$

for some integer $0 \leq k_\mathfrak{p} \leq \nu_\mathfrak{p}(5)$. The curve E_3 is not minimal at the primes dividing 5, so we can replace E_3 by $\tau_3 E_3$, where $\tau_3 = [5^2, 0, 0, 0]$. Let $E_{i,\mathfrak{p}}$ be a local minimal model of E_i at $\mathfrak{p}$.

Let $\mathfrak{p}$ be a prime of K. If $k_\mathfrak{p} = \nu_\mathfrak{p}(S_3) > 0$, then $\nu_\mathfrak{p}(5) > 0$. In order to prove the statement, we consider the cases given in Table 7. This table computes the signatures of the curves E_1 and E_3, using the valuations of the polynomials S_j in Table 2. By looking at the signatures in each case of Table 7, we observe that in order to get minimal models $E_{1,\mathfrak{p}}$ and $E_{3,\mathfrak{p}}$, the curve E_3 can be reduced by $\tau_3 = [\mathfrak{p}^i, 0, 0, 0]$ if and only if the curve E_1 can be reduced by $\tau_1 = [\mathfrak{p}^{i+2k_\mathfrak{p}}, 0, 0, 0]$. Then, using the fact that $\nu_\mathfrak{p}(\Delta_1) = 24k_\mathfrak{p} + \nu_\mathfrak{p}(\Delta_3)$, we conclude that the $\nu_\mathfrak{p}(\Delta(E_{1,\mathfrak{p}})) = \nu_\mathfrak{p}(\Delta(E_{3,\mathfrak{p}}))$.

If $\nu_\mathfrak{p}(S_3) = k_\mathfrak{p} = 0$, then we have the following cases:

(a) Assume $\nu_\mathfrak{p}(S_3) = 0$ and $\nu_\mathfrak{p}(S_2) > 0$. If $\nu_\mathfrak{p}(2) = \nu_\mathfrak{p}(3) = \nu_\mathfrak{p}(5) = 0$, then the two curves are already reduced at $\mathfrak{p}$, so $\nu_\mathfrak{p}(\Delta(E_{1,\mathfrak{p}})) = \nu_\mathfrak{p}(\Delta(E_{3,\mathfrak{p}}))$. It remains to consider the other cases.

   (i) Assume $\nu_\mathfrak{p}(2) > 0$ or $\nu_\mathfrak{p}(3) > 0$. In this case, $\nu_\mathfrak{p}(c_{4,1}) = \nu_\mathfrak{p}(v_{4,3})$, $\nu_\mathfrak{p}(c_{6,1}) = \nu_\mathfrak{p}(c_{6,3})$ and $\nu_\mathfrak{p}(\Delta_1) = \nu_\mathfrak{p}(\Delta_3)$, so the two curves reduce in the same way. Thus, $\nu_\mathfrak{p}(\Delta(E_{1,\mathfrak{p}})) = \nu_\mathfrak{p}(\Delta(E_{3,\mathfrak{p}}))$.

   (ii) Assume $\nu_\mathfrak{p}(5) > 0$. Then $\nu_\mathfrak{p}(\Delta_1) = \nu_\mathfrak{p}(\Delta_3) = 3\nu_\mathfrak{p}(S_2)$, $\nu_\mathfrak{p}(c_{4,1}) = \nu_\mathfrak{p}(c_{4,3}) = \nu_\mathfrak{p}(S_2)$, and $\nu_\mathfrak{p}(c_{6,1}) \geq 2\nu_\mathfrak{p}(S_2)$, $\nu_\mathfrak{p}(c_{6,3}) \geq 2\nu_\mathfrak{p}(S_2)$, so the two curves reduce in the same way. Thus, $\nu_\mathfrak{p}(\Delta(E_{1,\mathfrak{p}})) = \nu_\mathfrak{p}(\Delta(E_{3,\mathfrak{p}}))$.

(b) Assume $\nu_\mathfrak{p}(S_3) = \nu_\mathfrak{p}(S_2) = 0$ and $\nu_\mathfrak{p}(S_1) > 0$. If $\nu_\mathfrak{p}(5) = 0$, then the two curves are minimal at $\mathfrak{p}$. On the other hand, if $\nu_\mathfrak{p}(5) > 0$, then $\nu_\mathfrak{p}(S_3) > 0$, reaching a contradiction. Thus, $\nu_\mathfrak{p}(\Delta(E_{1,\mathfrak{p}})) = \nu_\mathfrak{p}(\Delta(E_{3,\mathfrak{p}}))$.

(c) Assume $\nu_\mathfrak{p}(S_1) = \nu_\mathfrak{p}(S_2) = \nu_\mathfrak{p}(S_3) = 0$, but $\nu_\mathfrak{p}(2) > 0$ or $\nu_\mathfrak{p}(3) > 0$. In this case, the two curves are minimal at $\mathfrak{p}$ and $\nu_\mathfrak{p}(\Delta(E_{1,\mathfrak{p}})) = \nu_\mathfrak{p}(\Delta(E_{3,\mathfrak{p}}))$.

Hence, we obtained that E_1 and E_3 are discriminant ideal twins when (11) is satisfied.

It remains to show that E_1 and E_3 are discriminant twins. To this end, for a prime $\mathfrak{p}$ of K, let $E_{i,\mathfrak{p}}$ be a $\mathfrak{p}$-minimal model of E_i. By [S94, Proposition VII.1.3], there exists $u_{i,\mathfrak{p}} \in \mathcal{O}_K$ such that $u_{i,\mathfrak{p}}^{-12}\Delta_i = \Delta(E_{i,\mathfrak{p}})$. Moreover, since E_1 and E_3 are discriminant ideal twins it is the case that there is a $\mu_\mathfrak{p} \in K$ such that $\nu_\mathfrak{p}(\mu_\mathfrak{p}) = 0$ and $\Delta(E_{1,\mathfrak{p}}) = \mu_\mathfrak{p}\Delta(E_{3,\mathfrak{p}})$. It follows that

$$(t-1)^{24} = \frac{\Delta_1}{\Delta_3} = \mu_\mathfrak{p}\left(\frac{u_{1,\mathfrak{p}}}{u_{3,\mathfrak{p}}}\right)^{12} \quad \Longrightarrow \quad \mu_\mathfrak{p} = (t-1)^{24}\left(\frac{u_{3,\mathfrak{p}}}{u_{1,\mathfrak{p}}}\right)^{12}.$$

Note that $\mu_\mathfrak{p} \in \mathcal{O}_K^{12}$ since $(t-1)^{24} \in \mathcal{O}_K^{12}$. In particular, there is a $\kappa_\mathfrak{p} \in \mathcal{O}_K$ such that $\kappa_\mathfrak{p}^{12} = \mu_\mathfrak{p}$ and $\nu_\mathfrak{p}(\kappa_\mathfrak{p}) = 0$. Now let $E'_{1,\mathfrak{p}}$ be the elliptic curve obtained from $E_{1,\mathfrak{p}}$ via the isomorphism $[\kappa_\mathfrak{p}, 0, 0, 0]$. Then $E'_{1,\mathfrak{p}}$ is a $\mathfrak{p}$-minimal model with $\Delta(E'_{1,\mathfrak{p}}) = \kappa_\mathfrak{p}^{-12}\Delta(E_{1,\mathfrak{p}})$. Hence $\Delta(E_{1,\mathfrak{p}}) = \mu_\mathfrak{p}\Delta(E'_{1,\mathfrak{p}})$, which, in turn, yields that $\Delta(E'_{1,\mathfrak{p}}) = \Delta(E_{3,\mathfrak{p}})$. This shows that for each $\mathfrak{p}$, there are $\mathfrak{p}$-minimal models of E_1 and E_3 having equal discriminants. Therefore E_1 and E_3 are discriminant twins. □

6 Explicit Results over $\mathbb{Q}$

Proposition 16 *Up to twists, there are finitely many p^2-isogenous discriminant ideal twins over $\mathbb{Q}$, for odd p for which $X_0(p^2)$ has genus 0. These elliptic curves, E_1 and E_3, are given in Table 5 by their LMFDB label. The table also lists their j-invariant and minimal discriminant.*

Proof Suppose E_1 and E_3 are two p^2-isogenous elliptic curves that are discriminant ideal twins over $\mathbb{Q}$, for $p = 3, 5$. By Theorem 10, we can assume that the two curves have j-invariants not both identically 0 or 1728. By Lemma 9, we know that there exists $t \in \mathbb{Z}$ and $d \in \mathbb{Z}$ such that $E_i \cong C_{p^2,i}(t, d)$.

Table 5 p^2-isogenous discriminant ideal twins over $\mathbb{Q}$

p^2	E_1	E_3	$j(E_1)$	$j(E_3)$	$\Delta_{E_1}^{\min}$	$\Delta_{E_3}^{\min}$
9	37.b1	37.b2	$\frac{727057727488000}{37}$	$\frac{4096000}{37}$	37	37
	171.b1	171.b3	$\frac{-50357871050752}{19}$	$\frac{32768}{19}$	−13851	−13851
25	11.a1	11.a3	$\frac{-52893159101157376}{11}$	$\frac{-4096}{11}$	−11	−11
	18176.e1	18176.e3	$\frac{3922540634246430781376}{71}$	$\frac{190705121216}{71}$	36352	36352

If $p = 3$, then Theorem 1 says that $\nu_q(t-3) = 3k_q$, for $0 \le k_q \le \nu_q(3)$, for all rational primes q. This means that $t-3$ is either a unit in $\mathbb{Q}$ or a unit multiplied by 3^3. Since the only units in $\mathbb{Q}$ are ± 1, it follows that $t-3 = \pm 1$ or $t-3 = \pm 3^3$, which implies that $t = 2, 4, -24, 30$. Note that $j(\mathcal{C}_{9,1}(\pm 1+3, d)) = j(\mathcal{C}_{9,3}(\pm 27+3, d))$, so $\mathcal{C}_{9,3}(\pm 27+3, d)$ is a twist of $\mathcal{C}_{9,1}(\pm 1+3, d)$. Therefore, we only have to consider the pairs for which $t = \pm 1 + 3$. As d is a twisting parameter, for $t = 4$, we obtain the curves 37.b1 and 37.b3; for $t = 2$, we obtain the curves 171.b1 and 171.b3.

If $p = 5$, then Theorem 2 says that $\nu_q(t-1) = k_q$, for $0 \le k_q \le \nu_q(5)$, for all rational primes q. This means that $t-3$ is either a unit in $\mathbb{Q}$ or a unit multiplied by 5. Since the only units in $\mathbb{Q}$ are ± 1, it follows that $t-1 = \pm 1$ or $t-1 = \pm 5$, which implies that $t = 0, 2, -4, 6$. Note that $j(\mathcal{C}_{25,1}(\pm 1+1, d)) = j(\mathcal{C}_{25,3}(\pm 5+1, d))$, so $\mathcal{C}_{25,3}(\pm 5+1, d)$ is a twist of $\mathcal{C}_{25,1}(\pm 1+1, d)$. Therefore, we only have to consider the pairs for which $t = \pm 1 + 1$. As d is a twisting parameter, for $t = 0$, we obtain the curves 11.a1 and 11.a3; for $t = 2$, we obtain the curves 18176.e1 and 18176.e3. □

Acknowledgements The authors thank the Banff International Research Station (BIRS) and Women in Numbers 6 (23w5175) for the opportunity to initiate this collaboration. The authors would like to thank the referee for valuable suggestions and comments, and for identifying an error in an earlier version of the paper. CN was partially supported by a Colby College Research Grant. The AMS Simons Travel Grant program partially supported MR during this work.

7 Appendix: Tables of Valuations

Let E_1 and E_3 be two p^2-isogenous elliptic curves, for $p = 3, 5$. It follows from the work of Barrios [B23] that there exist elements $t, d \in K$ such that $E_i \cong \mathcal{C}_{p^2,i}(t, d)$, where the curves $\mathcal{C}_{p^2,i}(t, d)$ are defined in Table 3. For each $p = 3, 5$, let $\mathfrak{p}$ be a prime of K. Let $\delta = \nu_{\mathfrak{p}}(d)$. Moreover, let

$$\ell = \begin{cases} \nu_{\mathfrak{p}}(3) & \text{if } p = 3 \\ \nu_{\mathfrak{p}}(5) & \text{if } p = 5 \end{cases}$$

and

$$\begin{aligned} s &= \nu_{\mathfrak{p}}(t-3) & \text{if } p = 3 \\ k &= \nu_{\mathfrak{p}}(t-1) & \text{if } p = 5 \end{aligned}$$

Note that $k = k_{\mathfrak{p}}$ in Theorems 14 and 15 and $s = s_{\mathfrak{p}}$ in Theorems 12 and 13. Assume that $k > 0$, $s > 0$, and $\ell > 0$ and define the following tables. Note that in these tables, the values δ_i (for $1 \le i \le 8$) are integers with the property that $\delta_i \ge 0$, for all i.

Table 6 Cases for $n = 9$

Poly.	$s<\ell/3$	$s=\ell/3$	$\ell/3<s<\ell/2$	$s=\ell/2$	$\ell/2<s<\ell$
T_1	s	s	s	s	s
T_2	s	s	s	s	s
T_3	$2s$	$2s$	$2s$	$2s$	$2s$
T_4	$3s$	$3s+\delta_4$	ℓ	ℓ	ℓ
T_5	$3s$	$3s$	$3s$	$3s$	$3s$
T_6	$6s$	$6s$	$6s$	$3\ell+\delta_6$	3ℓ
T_7	$6s$	$6s$	$6s$	$6s$	$6s$
T_8	s	s	s	s	s
$c_{4,1}$	$2\ell+4s+2\delta$	$10s+2\delta$	$2\ell+4s+2\delta$	$8s+2\delta$	$2\ell+4s+2\delta$
$c_{6,1}$	$3\ell+6s+3\delta$	$15s+3\delta$	$3\ell+6s+3\delta$	$12s+3\delta$	$3\ell+6s+3\delta$
Δ_1	$6\ell+11s+6\delta$	$29s+6\delta$	$6\ell+11s+6\delta$	$23s+6\delta$	$6\ell+11s+6\delta$
j_1	s	s	s	s	s
$c_{4,3}$	$2\ell+4s+2\delta$	$10s+\delta_4+2\delta$	$3\ell+s+2\delta$	$7s+2\delta$	$3\ell+s+2\delta$
$c_{6,3}$	$3\ell+6s+3\delta$	$15s+3\delta$	$3\ell+6s+3\delta$	$12s+\delta_6+3\delta$	$6\ell+3\delta$
Δ_3	$6\ell+3s+6\delta$	$21s+6\delta$	$6\ell+3s+6\delta$	$15s+6\delta$	$6\ell+3s+6\delta$
j_3	$9s$	$9s+3\delta_4$	3ℓ	$6s$	3ℓ
Poly.	$\ell=s$	$\ell<s<3\ell/2$	$s=3\ell/2$	$3\ell/2<s<2\ell$	$s=2\ell$
T_1	s	s	s	s	s
T_2	s	s	s	s	$2\ell+\delta_2$
T_3	$2s$	$2s$	$3\ell+\delta_3$	3ℓ	3ℓ
T_4	ℓ	ℓ	ℓ	ℓ	ℓ
T_5	$3s$	$3s$	$3s$	$3s$	$3s$
T_6	3ℓ	3ℓ	3ℓ	3ℓ	3ℓ
T_7	$6s$	$6s$	$6s$	$6s$	$6s$
T_8	$\ell+\delta_8$	ℓ	ℓ	ℓ	ℓ
$c_{4,1}$	$6s+2\delta$	$2\ell+4k+2\delta$	$8\ell+2\delta$	$2\ell+4s+2\delta$	$10\ell+\delta_2+2\delta$
$c_{6,1}$	$9s+3\delta$	$3\ell+6s+3\delta$	$12\ell+3\delta$	$3\ell+6s+3\delta$	$15\ell+3\delta$
Δ_1	$17s+6\delta$	$6\ell+11s+6\delta$	$9\ell+9s+\delta_3+6\delta$	$9\ell+9s+6\delta$	$27\ell+6\delta$
j_1	s	s	$s-\delta_3$	$3s-3\ell$	$3\ell+3\delta_2$
$c_{4,3}$	$4s+\delta_8+2\delta$	$4\ell+2\delta$	$4\ell+2\delta$	$4\ell+2\delta$	$4\ell+2\delta$
$c_{6,3}$	$6s+3\delta$	$6\ell+3\delta$	$6\ell+3\delta$	$6\ell+3\delta$	$6\ell+3\delta$
Δ_3	$9s+6\delta$	$6\ell+3s+6\delta$	$9\ell+s+\delta_3+6\delta$	$9\ell+s+6\delta$	$11\ell+6\delta$
j_3	$3s+3\delta_8$	$6\ell-3s$	$s-\delta_3$	$3\ell-s$	ℓ
Poly.	$2\ell<s<5\ell/2$	$s=5\ell/2$	$5\ell/2<s<8\ell/3$	$s=8\ell/3$	$8\ell/3<s$
T_1	s	s	s	s	s
T_2	2ℓ	2ℓ	2ℓ	2ℓ	2ℓ
T_3	3ℓ	3ℓ	3ℓ	3ℓ	3ℓ

(continued)

Table 6 (continued)

Poly.	$s < \ell/3$	$s = \ell/3$	$\ell/3 < s < \ell/2$	$s = \ell/2$	$\ell/2 < s < \ell$
T_4	ℓ	ℓ	ℓ	ℓ	ℓ
T_5	$3s$	$3s$	$3s$	$8\ell + \delta_5$	8ℓ
T_6	3ℓ	3ℓ	3ℓ	3ℓ	3ℓ
T_7	$6s$	$15\ell + \delta_7$	15ℓ	15ℓ	15ℓ
T_8	ℓ	ℓ	ℓ	ℓ	ℓ
$c_{4,1}$	$4\ell + 3s + 2\delta$	$4\ell + 3s + 2\delta$	$4\ell + 3s + 2\delta$	$12\ell + \delta_5 + 2\delta$	$12\ell + 2\delta$
$c_{6,1}$	$3\ell + 6s + 3\delta$	$18\ell + \delta_7 + 3\delta$	$18\ell + 3\delta$	$18\ell + 3\delta$	$18\ell + 3\delta$
Δ_1	$9\ell + 9s + 6\delta$	$9\ell + 9s + 6\delta$	$9\ell + 9s + 6\delta$	$9\ell + 9s + 6\delta$	$9\ell + 9s + 6\delta$
j_1	3ℓ	3ℓ	3ℓ	$3\ell + 3\delta_5$	$27\ell - 9s$
$c_{4,3}$	$4\ell + 2\delta$	$4\ell + 2\delta$	$4\ell + 2\delta$	$4\ell + 2\delta$	$4\ell + 2\delta$
$c_{6,3}$	$6\ell + 3\delta$	$6\ell + 3\delta$	$6\ell + 3\delta$	$6\ell + 3\delta$	$6\ell + 3\delta$
Δ_3	$9\ell + s + 6\delta$	$9\ell + s + 6\delta$	$9\ell + s + 6\delta$	$9\ell + s + 6\delta$	$9\ell + s + 6\delta$
j_3	$3\ell - s$	$3\ell - s$	$3\ell - s$	$3\ell - s$	$3\ell - s$

Table 7 Cases for $n = 25$

Poly.	$\ell < k$	$\ell = k$	$k < \ell < 2k$	$\ell = 2k$	$2k < \ell < 10k$	$\ell = 10k$	$10k < \ell$
S_1	2ℓ	2ℓ	2ℓ	$2\ell + \delta_1$	$4k$	$4k$	$4k$
S_2	ℓ	$\ell + \delta_2$	k	k	k	k	k
S_3	k	k	k	k	k	k	k
S_4	9ℓ	9ℓ	$10k$	$10k$	$10k$	$10k$	$10k$
S_5	3ℓ	$3\ell + \delta_5$	$2k + \ell$	$4k + \delta_5$	$4k$	$4k$	$4k$
S_6	10ℓ	$10\ell + \delta_6$	$10k$	$10k$	$10k$	$10k$	$10k$
S_7	ℓ	ℓ	ℓ	ℓ	ℓ	$\ell + \delta_7$	$10k$
S_8	ℓ	ℓ	ℓ	$\ell + \delta_8$	$2k$	$2k$	$2k$
S_9	0	0	0	0	0	0	0
$c_{4,1}$	$10\ell + 2\delta$	$10k + \delta_2 + 2\delta$	$11k + 2\delta$	$11k + 2\delta$	$11k + 2\delta$	$11k + 2\delta$	$11k + 2\delta$
$c_{6,1}$	$15\ell + 3\delta$	$15k + 2\delta_2 + \delta_5 + \delta_6 + 3\delta$	$14k + \ell + 3\delta$	$16k + \delta_5 + 3\delta$	$16k + 3\delta$	$16k + 3\delta$	$16k + 3\delta$
Δ_1	$5\ell + 25k + 6\delta$	$30k + 3\delta_2 + 6\delta$	$2\ell + 28k + 6\delta$	$32k + \delta_1 + 6\delta$	$32k + 6\delta$	$32k + 6\delta$	$32k + 6\delta$
j_1	$25\ell - 25k$	0	$5k - 2\ell$	$k - \delta_1$	k	k	k
$c_{4,3}$	$2\ell + 2\delta$	$2k + \delta_2 + 2\delta$	$k + \ell + 2\delta$	$3k + 2\delta$	$k + \ell + 2\delta$	$11k + \delta_7 + 2\delta$	$11k + 2\delta$
$c_{6,3}$	$3\ell + 3\delta$	$3k + 2\delta_2 + 3\delta$	$2k + \ell + 3\delta$	$4k + \delta_8 + 3\delta$	$4k + 3\delta$	$4k + 3\delta$	$4k + 3\delta$
Δ_3	$5\ell + k + 6\delta$	$6k + 3\delta_2 + 6\delta$	$2\ell + 4k + 6\delta$	$8k + \delta_1 + 6\delta$	$8k + 6\delta$	$8k + 6\delta$	$8k + 6\delta$
j_3	$\ell - k$	0	$\ell - k$	$k - \delta_1$	$3\ell - 5k$	$25k + 3\delta_7$	$25k$

References

[B23] Alexander J. Barrios, *Explicit classification of isogeny graphs of rational elliptic curves*, Int. J. Number Theory **19** (2023), no. 4, 913–936. MR4555392.

[BBDHR24] Alexander J. Barrios, Maila Brucal-Hallare, Alyson Deines, Piper Harris, and Manami Roy, *Prime isogenous discriminant ideal twins*, https://arxiv.org/abs/2402.19183v1, 2024.

[C10] Henri Cohen, *A course in computational algebraic number theoyr*, Graduate Texts in Mathematics, vol. 138, Springer Berlin, Heidelberg, 2010.

[D18] Alyson Deines, *Discriminant twins*, Women in Numbers Europe II, Assoc. Women Math. Ser., vol. 11, Springer, Cham, 2018, pp. 83–106. MR3882707.

[DHINRW24] Alyson Deines, Asimina S. Hamakiotes, Andreea Iorga, Changningphaabi Namoijam, Manami Roy, and Lori D. Watson, *Code for prime squared isogenous discriminant ideal twins*, https://github.com/adeines/DiscriminantTwins/tree/master/WIN6, 2024.

[D20] The Sage Developers, *Sagemath, the Sage Mathematics Software System (version 9.1)*, 2020, http://www.sagemath.org.

[DD15] Tim Dokchitser and Vladimir Dokchitser, *Local invariants of isogenous elliptic curves*, Trans. Amer. Math. Soc. **367** (2015), no. 6, 4339–4358. MR3324930.

[LMFDB21] The LMFDB Collaboration, *The L-functions and modular forms database*, http://www.lmfdb.org, 2021, [Online; accessed 18 October 2021].

[P93] Ioannis Papadopoulos, *Sur la classification de Néron des courbes elliptiques en caractéristique résiduelle* 2 *et* 3, J. Number Theory **44** (1993), no. 2, 119–152. MR1225948.

[RT97] Kenneth A. Ribet and S. Takahashi, *Parametrizations of elliptic curves by Shimura curves and by classical modular curves*, Proceedings of the National Academy of Sciences of the United States of America **94 21** (1997), 11110–4.

[S94] Joseph H. Silverman, *Advanced topics in the arithmetic of elliptic curves*, Graduate Texts in Mathematics, vol. 151, Springer-Verlag, New York, 1994. MR1312368.

[S09] Joseph H. Silverman, *The arithmetic of elliptic curves*, second ed., Graduate Texts in Mathematics, vol. 106, Springer, Dordrecht, 2009. MR2514094.

[S50] Shianghaw Wang, *On Grunwald's theorem*, Ann. of Math. (2) **51** (1950), 471–484. MR33801.

Generalized Eckardt Points on Del Pezzo Surfaces of Degree 1

Julie Desjardins, Yu Fu, Kelly Isham, and Rosa Winter

Abstract We study intersections of exceptional curves on del Pezzo surfaces of degree 1, motivated by questions in arithmetic geometry. Outside characteristics 2 and 3, at most 10 exceptional curves can intersect in a point. We classify the different ways in which 10 exceptional curves can intersect, construct a new family of surfaces with 10 exceptional curves intersecting in a point, and discuss strategies for finding more such examples.

1 Introduction

Del Pezzo surfaces are smooth projective algebraic surfaces with ample anticanonical divisor, and can be classified by their *degree*, which is defined as the self-intersection of the canonical divisor; an integer between 1 and 9. Both the geometric as well as the arithmetic complexity of del Pezzo surfaces goes up as the degree goes down (see also the *Motivation from the rational points perspective* below). Over an algebraically closed field, a del Pezzo surface contains a fixed number of exceptional curves (which we often call *lines*), depending on the degree; the lower the degree, the more lines on the surface.

Del Pezzo surfaces of degree 1 contain 240 lines over an algebraically closed field, and at most 10 lines intersect in one point outside characteristic 2 and 3 (at

J. Desjardins
University of Toronto at Mississauga, Mississauga, ON, Canada
e-mail: julie.desjardins@utoronto.ca

Y. Fu
Caltech Pasadena, Pasadena, CA, USA
e-mail: Yufu@caltech.edu

K. Isham
Colgate University, Hamilton, NY, USA
e-mail: kisham@colgate.edu

R. Winter (✉)
Albert-Ludwigs-Universität Freiburg, Freiburg im Breisgau, Germany
e-mail: rosa.winter@math.uni-freiburg.de

S. Akhtari et al. (eds.), *Research Directions in Number Theory*, Association for Women in Mathematics Series 39, https://doi.org/10.1007/978-3-032-11183-8_4

most 16 and 12 in those characteristics, respectively) [vLW23]. Besides this, little is known about the configurations of these lines. This is in sharp contrast with del Pezzo surfaces of higher degree. For example, del Pezzo surfaces of degree 3 (the smooth cubic surfaces in $\mathbb{P}^3$) contain 27 lines, and any point on the surface is contained in at most 3 lines; such a point is called an *Eckardt point*. There are at most 45 Eckardt points on a cubic surface in characteristic 2, and at most 18 otherwise. A del Pezzo surface of degree 2 contains 56 lines, a point on it is contained in at most 4 lines (a *generalized Eckardt point*), and there are at most 126 generalized Eckardt points on such a surface.

Following the terminology established for del Pezzo surfaces of degrees 2 and 3, we call a point on a del Pezzo surface of degree 1 over a field with characteristic unequal to 2 and 3, that is contained in the intersection of 10 lines, a *generalized Eckardt point*. For characteristics 2 and 3 we define this to be the intersection point of 16 and 12 lines, respectively. The knowledge we have about del Pezzo surfaces of degree at least 2 prompts the following questions.

Question 1 *How many generalized Eckardt points can a del Pezzo surface of degree 1 contain?*

Question 2 *How likely is it for a del Pezzo surface of degree 1 to contain a generalized Eckardt point?*

It is worth mentioning that there are only a handful of examples in the literature of (families of) del Pezzo surfaces of degree 1 with a generalized Eckardt point. Moreover, all these examples show only 1 such point on the surface [SvL14, vLW23].

In this paper we start an investigation toward answering Questions 1 and 2. Our main contribution is the following result (c.f. Theorem 16).

Theorem 3 *Outside characteristic 2, there are 9 possible configurations for 10 exceptional curves on a del Pezzo surface of degree 1 to intersect in one point, and they are listed in Figs. 2 and 3. In characteristic 0, at most 7 of the configurations (1, 2, 3, 5, 6, 8 in Fig. 2 and the configuration in Fig. 3) can correspond to a generalized Eckardt point on a del Pezzo surface of degree 1.*

Theorem 16 is a first step in answering Questions 1 and 2, as it tells us which configurations of intersecting lines to count. As mentioned, only two examples of (families of) del Pezzo surfaces of degree 1 with a generalized Eckardt points can be found in the current literature. In [SvL14, Example 4.1], the authors construct a family of del Pezzo surfaces of degree 1 that all contain 10 exceptional curves that intersect in one point, whose intersection graph is isomorphic to number 5 in Fig. 2. In [vLW23, Example 5.2] the authors construct a del Pezzo surface of degree 1 over a field of characteristic unequal to 2 with 10 exceptional curves that intersect in one point, whose intersection graph is isomorphic to the one in Fig. 3. We extend this to a whole family of such examples in Theorem 29.

The strategy used in order to obtain the examples of Theorem 29 is explained in detail with Algorithm 26. This strategy combined with combinatorial computations proves that there are no del Pezzo surface of degree 1 with a generalized Eckardt point corresponding to the configurations in Fig. 2 over the finite field $\mathbb{F}_p$ with $p \in \{1, \ldots, 17, 23\}$ (Theorems 34 and 35) and allows us to find an example of a del Pezzo surface of degree 1 with a generalized Eckardt point corresponding to configuration 8 over $\mathbb{F}_{19}$ (Theorem 36). We are quite confident that our strategy will lead to more results of the kind we are presenting in Sects. 5 and 6.

Motivation from the Rational Points Perspective: One motivation to study Questions 1 and 2 comes from the arithmetic of del Pezzo surfaces. Over algebraically closed fields, these surfaces are *rational* (i.e., birationally equivalent to the projective plane). Over non-algebraically closed fields, this is not true in general, and the set $X(k)$ of rational points on a del Pezzo surface X over such a field k is far from fully understood. It is generally believed that del Pezzo surfaces that contain a rational point have "many" rational points, and that these rational points are "well distributed". There are several results that support this belief. Del Pezzo surfaces of degree at least 3 with a rational point are *unirational* (i.e., dominated by the projective plane) [Seg43, Seg51, Man74, Kol02, Pie12], for del Pezzo surfaces of degree 2 the same holds if they contain a rational point outside a closed subset [STVA14], and for del Pezzo surfaces of degree 1 unirationality is known when the surface contains a conic bundle [KM17]. If the field k is infinite, then unirationality over k implies that the set $X(k)$ is *Zariski dense* (i.e., not contained in a closed subset). There are several partial results on Zariski density of $X(k)$ for a del Pezzo surface of degree 1 [Ula08, Ula07, UT10, Jab12, SvL14, Des18, DW22]. Another measure of "many" and "well distributed" rational points for a variety X over a number field k is *weak approximation* (i.e., the set $X(k)$ of rational points is dense in the product of local points $\prod_{v\in\Omega_k} X(k_v)$, where Ω_k is the set of places of k). Weak approximation is satisfied for all del Pezzo surfaces of degree at least 5 with a rational point, and shown not to hold in general for del Pezzo surfaces of lower degree. However, it is shown that del Pezzo surfaces of degree 3 and 4 with a rational point satisfy *weak weak approximation* (i.e., $X(k)$ is dense in $\prod_{v\in\Omega_k\setminus S} X(k_v)$ for a finite set S) [SS91, SD01], and the same holds for del Pezzo surfaces of degree 2 with a general enough rational point [DSW24] and del Pezzo surfaces of degree 1 with a conic bundle under an extra condition [DS23].

The configuration of lines often come up when proving these arithmetic results. For example, the results in [STVA14, DSW24] require the existence of a rational point that is not contained in the intersection of 4 lines on a del Pezzo surface of degree 2. For del Pezzo surfaces of degree 1, the result in [DS23] requires that no 4 of the exceptional curves over the algebraic closure intersect in one point, the result in [SvL14] requires the existence of a point not contained in 6 or more lines, and in [DW22] one needs a point which is not torsion on its fiber of a corresponding elliptic surface, which implies that it cannot be contained in 9 or more lines, as was shown in [DW].

Outline of the Paper

In Sect. 2 we give the necessary background. In Sect. 3 we determine the different configurations in which 10 exceptional curves on a del Pezzo surface of degree 1 can pairwise intersect (Theorem 16 and Remark 17). Section 4 contains a discussion of how to determine which of these configurations can correspond to 10 exceptional curves that all intersect in *one point*, i.e., which configurations can correspond to generalized Eckardt points (Algorithm 26). In Sect. 5 we construct a family of surfaces that all contain a generalized Eckardt point (Theorem 29), and we study the same questions over finite fields in Sect. 6 (Theorems 34, 35, and 36). We end with a short discussion on open questions in Sect. 7.

Computations were done in `magma` [BCP97]. The code can be found at [DFIW].

2 Background

In this section, we give the necessary background for the rest of this paper, namely the geometric properties of del Pezzo surfaces of degree 1, and the weighted graph on exceptional classes. We show how this graph and its automorphism group come into play when studying intersecting exceptional curves.

2.1 Del Pezzo Surfaces of Degree One and Exceptional Curves

Definition 4 Let $1 \leq n \leq 8$ be an integer and let $P_1, \ldots, P_n$ be points in $\mathbb{P}^2$. We say that $P_1, \ldots, P_n$ are in *general position* if there is no line containing three of the points, no conic containing six of the points, and no cubic containing eight of the points with a singularity at one of them.

A del Pezzo surface over a field k is a smooth, projective, geometrically integral surface with ample anticanonical divisor. The degree of a del Pezzo surface is the self-intersection number of the canonical divisor, which is an integer between 1 and 9. Moreover, a del Pezzo surface over an algebraically closed field is either isomorphic to $\mathbb{P}^1 \times \mathbb{P}^1$ when $d = 8$ or to $\mathbb{P}^2$ blown up at $9 - d$ points in general position otherwise [Man74, Theorem 24.4]. Let X be a del Pezzo surface of degree 1 over an algebraically closed field k, then X is isomorphic to the blow up of $\mathbb{P}^2$ in eight points $P_1, \ldots, P_8$ in general position. Let $\text{Pic}(X)$ be the Picard group of X. Let K_X be a fixed divisor in the canonical class of X.

Definition 5 Define an *exceptional curve* on X to be an irreducible projective curve $C \subset X$ such that

$$C \cdot C = -1, C \cdot K_X = -1.$$

The corresponding class $[C] \in \text{Pic}(X)$ is called an *exceptional class* with

$$[C] \cdot [C] = -1, [C] \cdot [K_X] = -1.$$

For $i \in \{1, \ldots, 8\}$, let $[E_i]$ be the exceptional class of the exceptional curve above P_i, and $[L]$ be the class of the pullback of a line $\ell \subset \mathbb{P}^2$ not passing through any of the P_i. Once we realize $\mathrm{Pic}(X)$ as a lattice isomorphic to $\mathbb{Z}^9$, we can write down a basis as $\{[L], [E_1], \ldots, [E_8]\}$ [Man74, Proposition 20.9.1]. Moreover, for $i, j \in \{1, \ldots, 8\}$, $i \neq j$, we have

$$[E_i] \cdot [E_i] = -1, [E_i] \cdot [E_j] = 0,\ [L] \cdot [E_i] = 0, [L] \cdot [L] = 1.$$

Remark 6 We often use the word *lines* in order to speak about exceptional curves. This comes from the fact that for del Pezzo surfaces of degree $d \geq 3$, the exceptional curves are lines on the image of the surface under the anticanonical embedding in $\mathbb{P}^d$.

Manin proved that every exceptional class in $\mathrm{Pic}(X)$ contains exactly one exceptional curve on X and gave a geometrical description of the exceptional classes.

Theorem 7 ([Man74, Theorem 26.2]) *The exceptional curves (therefore exceptional classes) on the del Pezzo surface X which is obtained by blowing up 8 points $P_1, \ldots, P_8$ in general position are*

- *the exceptional curves E_i above the points P_i, for $i \in \{1, \ldots, 8\}$,*

and the strict transforms of the following curves in $\mathbb{P}^2$;

- *the lines $\ell_{i,j}$ passing through the points P_i and P_j for $i \neq j$;*
- *the conics passing through any five of the points;*
- *the cubics $c_{i,j}$ not passing through P_i, passing once through points P_k with $k \neq i, j$, passing twice through P_j;*
- *the quartics $q_{i,j,k}$ passing through the eight points with a double point in P_i, P_j and P_k for i, j, k distinct;*
- *the quintics passing through the eight points with double points at six of them;*
- *the sextics passing through the eight points with double points at seven of them, and a triple point at one of them.*

One can write the exceptional classes in the form of linear combinations of the basis

$$a[L] - \sum_{i=1}^{8} b_i[E_i]$$

where a is the degree of the corresponding curve in $\mathbb{P}^2$ and b_i is the multiplicity of P_i. In this way, we obtain a vector $(a, b_1, \ldots, b_8)$ in $\mathbb{Z}^9$. Let A be the set of 240 vectors obtained in this way, and let C be the set of all exceptional classes in $\mathrm{Pic}(X)$. There is a one-to-one correspondence between A and C induced by the basis $L, E_1, \ldots, E_8$ [vLW23, Lemma 2.4, Remark 2.5]. Therefore we are able to study the intersection of exceptional curves (and classes) by carrying out computations in some vector spaces, either by `magma` or by hand.

2.2 The Weighted Graph on Exceptional Classes

Definition 8 Define a *graph* to be a pair (V, D), where V is a set of elements called vertices, and D a subset of the power set of V such that every element in D has cardinality 2; elements in D are called edges, and the size of the graph is the cardinality of V. A graph (V, D) is complete if for every two distinct vertices $v_1, v_2 \in V$, the pair $\{v_1, v_2\}$ is in D.

Define a *weighted graph* to be a graph (V, D) with a map $\psi : D \longrightarrow A$, where A is any set, whose elements we call weights; for any element d in D we call $\psi(d)$ its weight. If (V, D) is a weighted graph with weight function ψ, then we define a weighted subgraph of (V, D) to be a graph $\left(V', D'\right)$ with map ψ', where V' is a subset of V, while D' is a subset of the intersection of D with the power set of V', and ψ' is the restriction of ψ to D'.

An *isomorphism* between two weighted graphs (V, D) and $\left(V', D'\right)$ with weight functions $\psi : D \longrightarrow A$ and $\psi' : D' \longrightarrow A'$, respectively, consists of a bijection f between the sets V and V' and a bijection g between the sets A and A', such that for any two vertices $v_1, v_2 \in V$, we have $\{v_1, v_2\} \in D$ with weight w if and only if $\{f(v_1), f(v_2)\} \in D'$ with weight $g(w)$. We call the map f an *automorphism* of (V, D) if $(V, D) = \left(V', D'\right)$, $\psi = \psi'$, and g is the identity on A.

Definition 9 A *clique* of a weighted graph is a complete weighted subgraph. A *maximal clique* is a maximal complete weighted subgraph with respect to the number of vertices.

Let X be a del Pezzo surface of degree 1 with canonical divisor K_X, and let $K_X^{\perp}$ be the orthogonal complement of K_X in $\text{Pic}(X)$. Let $\langle\cdot,\cdot\rangle$ be the *negative* of the intersection pairing on $\text{Pic}(X)$ and note that $\langle\cdot,\cdot\rangle$ induces an inner product on $\mathbb{R} \otimes_{\mathbb{Z}} \text{Pic}(X)$ which realizes $K_X^{\perp}$ as Euclidean space. Vectors in $K_X^{\perp}$ with length $\sqrt{2}$, i.e., classes in $K_X^{\perp}$ with self-intersection -2, form a root system of type E_8 [Man74, Theorem 23.9]:

$$E = \{D \in \text{Pic}(X) \mid \langle D, D\rangle = 2;\, D \cdot K_X = 0\}.$$

There is a bijection between the set C of 240 exceptional classes in $\text{Pic}(X)$ and E:

$$F : C \longrightarrow E, c \longmapsto c + K_X.$$

For $c_1, c_2 \in C$ we have $\langle c_1 + K_X, c_2 + K_X\rangle = 1 - c_1 \cdot c_2$. As a consequence of this bijection, the group of permutations of C that preserve the intersection pairing is isomorphic to the Weyl group W_8, which is the group of permutations of E_8 generated by the reflections in the hyperplanes orthogonal to the roots [Man74, Theorem 23.9].

Denote by Γ the complete weighted graph whose vertex set is the set of simple roots in E, and whose weight function is induced by the inner product on the Euclidean space. Denote by G the complete weighted graph whose vertex set is C, and whose weight function is the intersection pairing in $\text{Pic}(X)$. Then there is an isomorphism f between G and Γ, which sends an edge $d = \{c_1, c_2\}$ in G with weight ω to the edge $\tilde{d} = \{c_1 + K_X, c_2 + K_X\}$ of weight $1 - \omega$ in Γ.

Note that, since the action of W_8 on the set C preserves the intersection pairing, the group W_8 is also the automorphism group of G (and of Γ).

We aim to study the intersection of exceptional curves in X. If two such curves intersect, then the corresponding vertices in G are connected by edges of positive weights. The possible positive weights that occur in G are 1, 2, and 3, so the corresponding weights that occur in Γ are 0, -1, and -2. Therefore, the size of maximal cliques in Γ with edges of these weights gives an upper bound for the maximal possible number of lines on X that go through one point. Note that this upper bound does not need to be sharp.

Definition 10 We call a set of exceptional curves *concurrent* if there is a point on X contained in all of them.

As mentioned in the introduction, the maximal number of concurrent exceptional curves on a del Pezzo surface of degree 3 is 3, and for a del Pezzo surface of degree 2, the number of exceptional curves that are concurrent in a point is at most 4. In both these cases, this number is the same as the maximal size of complete subgraphs of the intersection graph on the exceptional curves. But for del Pezzo surface of degree 1, things are more complicated. The maximal size of cliques in the graph G is 16, hence this is also the maximal number of exceptional curves that are concurrent in a point on X [vLW23, Proposition 2.8]. However, this is only sharp in characteristic 2; for char $k \neq 2$, the number of exceptional curves that are concurrent in a point is not given by the maximal size of cliques in G; it is at most 10 outside characteristics 2 and 3, and at most 12 in characteristic 3 [vLW23, Theorems 1.1, 1.2]. Both of these upper bounds are shown to be sharp except possibly in characteristic 5.

To answer Questions 1 and 2, we study maximal cliques of size ten in Γ, especially those with weights $\{-1, 0\}$ and $\{-2, 0\}$, as we will explain now.

Definition 11 By $\Gamma_{\{-2,0\}}$ and $\Gamma_{\{-1,0\}}$ we denote the subgraph of Γ whose vertices are all the vertices of Γ, and whose edges are only those in Γ of weights $\{-2, 0\}$ and $\{-1, 0\}$ respectively.

Lemma 12 ([vLW23, Lemma 2.7])

(i) Let e be an exceptional class in $\text{Pic}(X)$. *Then there is exactly one exceptional class f with $e \cdot f = 3$, there are 56 exceptional classes f with $e \cdot f = 0$, there are 126 exceptional classes f with $e \cdot f = 1$, and 56 exceptional classes f with $e \cdot f = 2$.*

(ii) *For* e_1, e_2 *two exceptional classes in* $\mathrm{Pic}(X)$ *with* $e_1 \cdot e_2 = 3$*, and* f *a third exceptional class, we have* $e_1 \cdot f + e_2 \cdot f = 2$.

The bi-anticanonical divisor $-2K_X$ induces a morphism

$$\varphi : X \to \mathbb{P}^3 \tag{1}$$

such that X can be realized as a double cover of a cone Q in $\mathbb{P}^3$, ramified over a sextic curve [CO99, Proposition 2.3]. Moreover, by [CO99, Proposition 2.6], φ maps an exceptional curve on X bijectively to a smooth conic which is the intersection of Q with a hyperplane in $\mathbb{P}^3$ not containing the vertex of Q. Conversely, for a hyperplane section H of Q not containing the vertex of Q and tri-tangent to the smooth sextic curve, the pullback $\varphi^* H = e_1 + e_2$ under φ is the sum of two exceptional curves intersecting with multiplicity 3. We call e_1 the *partner* of e_2. Thus every point on H has two preimages, except for the points with preimages in $e_1 \cap e_2$.

Remark 13 We conclude from the above that if a set of exceptional curves concurrent in P contains an exceptional curve and its partner, then P is contained in the ramification locus of φ. Conversely, if a set of exceptional curves is concurrent at a point P that lies on the ramification locus, then the set of partners of the exceptional curves are also concurrent at P. Moreover, by Lemma 12, a set of exceptional curves that is concurrent at one point on the ramification curve forms a clique in G with only edges of weights 1 and 3, and a set of exceptional curves concurrent at one point outside the ramification curve forms a clique in G with only edges of weights 1 and 2.

Cliques in G with only edges of weights 1 and 3 or 1 and 2 correspond to cliques in $\Gamma_{\{-2,0\}}$ and in $\Gamma_{\{-1,0\}}$, respectively. This motivates us to study those two graphs, especially the maximal cliques in them.

3 Configurations of Ten Concurrent Lines

A naive approach to give an upper bound for the answer to Question 1 is to count all cliques of size 10 in the graph G on the 240 exceptional curves on a del Pezzo surface of degree 1. Although this does give an upper bound for the number of generalized Eckardt points on a del Pezzo surface of degree 1, it is very large (only counting those cliques with edges of weights 1 and 2 we obtain already 107,700,096 cliques of size 10, which is the sum of the orbits in Table 1), and far from sharp. For example, we know that some cliques in G never correspond to exceptional curves that are concurrent in a point [vLW23, Propositions 3.6, 4.6]. A better first step in answering Questions 1 and 2 is therefore to determine which cliques to count, that is, we want to determine in which way 10 concurrent lines on a del Pezzo surface of degree 1 can be concurrent. From Lemma 12 we know that exceptional curves can intersect each other with multiplicities 1, 2, or 3. Moreover, from Remark 13 we know how to distinguish between lines that are concurrent in a point on the ramification curve of the map (1), and lines that are concurrent in a point outside the ramification curve. In Theorem 16 we list all cliques of size 10 that possibly correspond to concurrent exceptional curves. Moreover, we show that cliques of the same isomorphism type form one orbit under the action of the Weyl group on the set of these cliques.

Let G be the weighted graph on the 240 exceptional curves of a del Pezzo surface of degree 1, as in the previous section. Recall that the automorphism group of G is the Weyl group W_8.

Definition 14 We say that a clique K in G can be *realized as a generalized Eckardt point*, or simply *realized*, if there exists a del Pezzo surface of degree 1 with an isomorphism between the intersection graph on its 240 exceptional curves and G, such that K corresponds to a set of exceptional curves concurrent in a generalized Eckardt point under this isomorphism.

Note that W_8 acts on the set of cliques of the same isomorphism type in G. The following lemma tells us that we only have to check one representative per orbit under this action in order to know if all cliques in the orbit can be realized.

Lemma 15 *For two cliques in G that are conjugate under the action of the Weyl group, one can be realized as a generalized Eckardt point if and only if the other one can as well.*

Proof Let $K_1 = \{c_1, \ldots, c_n\}$ be a clique that cannot be realized as a generalized Eckardt point, and let $K_2 = \{w(c_1), \ldots, w(c_n)\}$ be another clique in G, which is the image of K_1 under the element $w \in W_8$. Assume by contradiction that K_2 can be realized as a generalized Eckardt point. Then there is a del Pezzo surface X of degree 1, with an isomorphism ψ between G and the graph on the 240 exceptional curves on X, such that $\psi(K_2)$ corresponds to a set of exceptional curves that are concurrent on X. But then the map $\psi \circ w$ gives an isomorphism between G and the graph of 240 exceptional curves on X, such that the image $\psi(w(K_1)) = \psi(K_2)$ of K_1 under this isomorphism corresponds to a set of exceptional curves that are concurrent on X. This contradicts the fact that K_1 is not realizable as a generalized Eckardt point. We conclude that K_1 is realizable if and only if K_2 is. □

Recall that we distinguish between generalized Eckardt points on a del Pezzo surface of degree 1 that are on the ramification curve of the map (1) and those that are outside the ramification curve, by only considering cliques in G with edges of weights 1 and 3, or of weights 1 and 2, respectively.

Theorem 16 *(i) There are 8 orbits of cliques of size 10 in G under the action of the Weyl group with only edges of weights 1 and 2, and their isomorphism types are listed in Fig. 2. In characteristic 0, at most 6 of these (1, 2, 3, 5, 6, 8) correspond to cliques in G that can be realized as generalized Eckardt points.*

Fig. 1 Maximal clique T of size twelve in G

(ii) *Outside characteristic 2, there is only one orbit of cliques of size 10 in G under the action of the Weyl group with only edges of weights 1 and 3 that can be realized as a generalized Eckardt point. Cliques in this orbit are of the form depicted in Fig. 3: a set of vertices* $\{c_1, d_1, \ldots, c_5, d_5\}$, *where* $c_i \cdot c_j = 1$ *for* $i \neq j$, *and* $c_i \cdot d_i = 3$ *for all* i.

Proof (i). Recall that cliques in G with only edges of weights 1 and 2 correspond to cliques in $\Gamma_{\{-1,0\}}$. The table in [WvL21, Appendix A] gives the complete list of orbits of the maximal cliques in $\Gamma_{\{-1,0\}}$, and we see that there are 6 orbits of maximal cliques of size 10. Translating this to cliques in G, a direct computation by hand shows that these are orbits $1-3$, $5-7$ in Fig. 2. We are left with determining the orbits of the non-maximal cliques of size 10 in G with only edges of weights 1 and 2. By [vLW23, Proposition 19], there are no maximal cliques in $\Gamma_{\{-1,0\}}$ of size 11, and there is one orbit of maximal cliques of size 12 under the action of W_8, which is the maximal size of cliques in $\Gamma_{\{-1,0\}}$. When translating this to the graph G, a maximal clique of size 12 in this orbit has the form where edges drawn correspond to edges of weight 2 and every two vertices not connected by a drawn edge are connected by an edge of weight 1. Non-maximal cliques of size 10 in G are contained in a clique of size 12. After removing two vertices from T in Fig. 1, we get a clique of size 10 either isomorphic to clique 4 or clique 8 in Fig. 2. We are left to show that all cliques of type 4 in G form one orbit under the action of W_8, and the same for type 8.

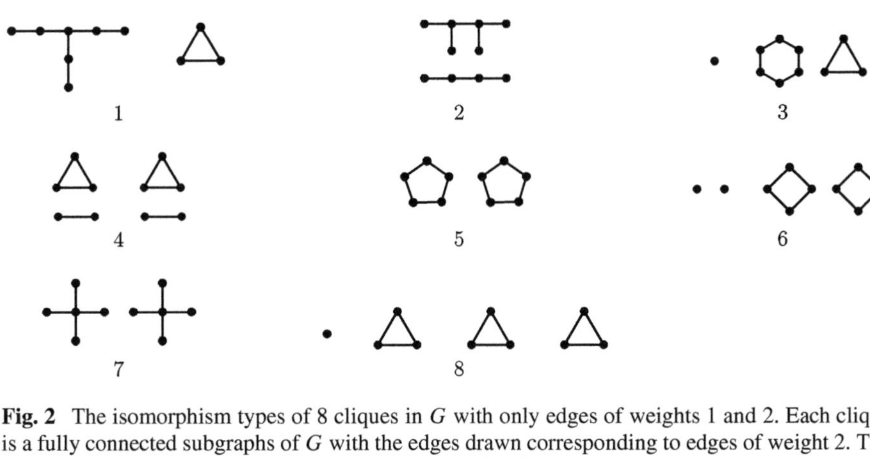

Fig. 2 The isomorphism types of 8 cliques in G with only edges of weights 1 and 2. Each clique is a fully connected subgraphs of G with the edges drawn corresponding to edges of weight 2. The edges not drawn correspond to edges of weight 1 in the clique

Fig. 3 The isomorphism type of a clique of size 10 in G with only edges of weights 1 and 3. The clique is a fully connected subgraph of G with the edges drawn corresponding to edges of weight 3. The edges not drawn correspond to edges of weight 1 in the clique

We obtain a clique of type 4 by removing two vertices from T that are connected by an edge of weight 1. Let $H = \mathrm{Stab}_{W_8}(T)$ be the stabilizer of T inside W_8. By [WvL21, Proposition 20 (ii)] (after translating from the graph $\Gamma_{\{-1,0\}}$ to G), the group H acts transitively on the set

$$S_4 = \{(e_1, e_2, e_3, e_4) \in T^4 \mid e_i \cdot e_j = 1 \text{ for all } i \neq j\}.$$

Since we have a surjective map $S_4 \longrightarrow S_2$, with

$$S_2 = \{(e_1, e_2) \in T^2 \mid e_1 \cdot e_2 = 1\},$$

we conclude that H also acts transitively on S_2. But every clique of type 4 in T is the complement of an element in S_2, so H acts transitively on the set of cliques of type 4 in T as well. Since W_8 acts transitively on the set of cliques of size 12 of isomorphism type T in G, we conclude that W_8 acts transitively on the set of cliques of type 4 in G.

We obtain a clique of type 8 by removing two vertices e_1, e_2 from T that are connected by an edge of weight 2. By [WvL21, Corollary 15], the stabilizer H of T in W_8 acts transitively on the clique T. Let e_3 be the vertex in T with $e_1 \cdot e_3 = e_2 \cdot e_3 = 2$. Then for every other pair $\{c_1, c_2\}$ of vertices in T intersecting with multiplicity 2, the element of H that sends e_3 to the unique vertex c_3 with $c_1 \cdot c_3 = c_2 \cdot c_3 = 2$ sends the pair $\{e_1, e_2\}$ to $\{c_1, c_2\}$. Therefore, H acts transitively on the set

$$S = \{\{e_1, e_2\} \mid e_1, e_2 \in T,\ e_1 \cdot e_2 = 2\}.$$

Since the cliques of type 8 in T are complements of elements in S, we conclude that H acts transitively of the cliques of type 8 in T, and, as before, we conclude that W_8 acts transitively on the set of cliques of type 8 in G.

Now that we have proved that there are 8 orbits of cliques of size 10 in G with only edges of weights 1 and 2, we are left to show that in characteristic 0, at most 6 of them can be realized as a generalized Eckardt point. In [vLW23, Proposition 4.6] it is shown that a clique of type 4 cannot be realized as a generalized Eckardt point outside characteristic 3, and the same is shown for a subset of a clique of type 7 in [DW, Proposition 5.4] in characteristic 0. From Lemma 15 we conclude that no cliques of type 4 or 7 can be realized as a generalized Eckardt point outside characteristic 3 and in characteristic 0, respectively. This concludes part (i).

(ii) Outside characteristic 2, the maximal size of a clique in G with only edges of weights 1 and 3 corresponding to concurrent exceptional curves is 10 by [vLW23, Theorem 1.1]. Moreover, from Remark 13 it follows that such a clique should be of the form as claimed, since, for every element in such a clique, its partner intersects the same points on the ramification curve of φ. There is only one orbit of such cliques, which follows from [WvL21, Proposition 16]. □

Remark 17 In characteristic 2, the maximal number of concurrent exceptional curves in a point on the ramification curve of the map (1) is 16, and a clique in G corresponding to such curves is of the form $K = \{c_1, d_1, \ldots, c_8, d_8\}$, where $c_i \cdot c_j = 1$ for $i \neq j$, and $c_i \cdot d_i = 3$ for all i. Moreover, there are no maximal cliques in G with only edges of weights 1 and 3 of size smaller than 16, so any clique corresponding to 10 exceptional curves that are concurrent on the ramification curve is a subgraph of K. We see that there are 4 different isomorphism types, depending on how many pairs c_i, d_i we remove from K: a clique of size 10 is left with 2, 3, 4, or 5 such pairs. With `magma`, we compute the number of orbits of the set of all cliques of size 10 in G with only edges of weights 1 and 3 under the action of W_8. Since every such clique contains two vertices connected by an edge of weight 1, and W_8 acts transitively on the set of pairs of exceptional curves intersecting with multiplicity 1 [WvL21, Proposition 8], we can reduce computations and still find representatives for each orbit, by fixing two exceptional curves e_1, e_2 with $e_1 \cdot e_2 = 1$ and computing all cliques of size 10 with edges of weights 1 and 3 containing e_1 and e_2. Interestingly enough, in this case there are multiple orbits with the same isomorphism type: those cliques of size 10 with 3 pairs c_i, d_i as well as those with 4 such pairs split into two orbits. The other isomorphism types form a full orbit each. This gives a total of 6 orbits of cliques of size 10 with only edges of weights 1 and 3 in G. Note that this is also true outside characteristic 2, but in that case we already know that only one of these orbits contains cliques that can be realized as a generalized Eckardt point (the cliques consisting of 5 pairs, as described in Theorem 16 (ii)).

Corollary 18 *If 10 lines on a del Pezzo surface of degree 1 over a field of characteristic 0 are concurrent in a point outside the ramification curve of the map* (1), *their intersection graph equals one of 1, 2, 3, 5, 6, 8 in Fig. 2.*

To our knowledge there is no example in the literature of 10 concurrent exceptional curves on a del Pezzo surface of degree 1 whose intersection graph is isomorphic to one of cliques 1, 2, 3, 6, or 8 in Fig. 2. In Sect. 4 we discuss strategies for finding such examples, or for proving that more cliques are not realizable.

We finish this section with a brief discussion on the relation between Question 1 and the maximal number of generalized Eckardt points on del Pezzo surfaces of degree 2, a connection that was suggested to us by Bianca Viray. Over an algebraically closed field, every del Pezzo surface of degree 1 is the blow-up of a (non-unique) del Pezzo surface of degree 2, so another step in determining an upper bound for the number of generalized Eckardt points on a del Pezzo surface of degree 1 is to use what we know about del Pezzo surfaces of degree 2. A first question to answer is, which of the cliques in Fig. 2 correspond to exceptional curves on a del Pezzo surface of degree 1 that, after blowing down an exceptional curve, could form a generalized Eckardt point on a del Pezzo surface of degree 2? In the following proposition we give a partial answer to this question.

Proposition 19 *Let $e_1, \ldots, e_{10}$ be 10 exceptional curves on a del Pezzo surface X of degree 1. Assume there is another exceptional curve c on X such that, after blowing down c, there is a subset of 4 exceptional curves among $e_1, \ldots, e_{10}$ whose images on the resulting del Pezzo surface Y of degree 2 are concurrent in a point. Then the intersection graph of $e_1, \ldots, e_{10}$ is not isomorphic to clique 4, 5, or 8 in Fig. 2.*

Proof Assume that the set $\{e_1, \ldots, e_{10}\}$ contains a subset of 4 curves $e_{i_1}, e_{i_2}, e_{i_3}, e_{i_4}$ that correspond to 4 concurrent exceptional curves on Y after blowing down c. Since the exceptional curve on Y are exactly the images of the 56 exceptional curves on X that are disjoint from c, it follows that $e_{i_1}, e_{i_2}, e_{i_3}, e_{i_4}$ are all disjoint from c. Choose a basis $[L], [E_1], \ldots, [E_8]$ for the exceptional classes in Pic X as in Sect. 2. We can write every exceptional curve as a vector $(a, b_1, \ldots, b_8)$ with respect to this basis as denoted in the discussion below Theorem 7. Using this notation, observe that the set S of 10 exceptional curves

$$\{(1, 0, 0, 0, 0, 0, 0, 1, 1), (1, 0, 0, 1, 1, 0, 0, 0, 0), (1, 1, 1, 0, 0, 0, 0, 0, 0), (1, 0, 0, 0, 0, 1, 1, 0, 0),$$

$$(4, 2, 1, 2, 1, 2, 1, 1, 1), (4, 1, 2, 2, 1, 1, 1, 1, 2), (4, 1, 1, 2, 1, 1, 2, 2, 1), (4, 1, 2, 1, 2, 1, 2, 1, 1),$$

$$(4, 2, 1, 1, 2, 1, 1, 2, 1), (4, 2, 1, 1, 1, 1, 2, 1, 2)\}$$

has intersection graph isomorphic to graph 4 in Fig. 2. It is a quick check with `magma` that no exceptional curve on X is disjoint to any 4 curves in S simultaneously, so no 4 curves in S will correspond to concurrent exceptional curves after blowing down c. Since every clique in G with isomorphism type that of clique 4 in Fig. 2 is conjugate under the action of W_8 to the clique S by Theorem 16, and the action of W_8 respects the intersection multiplicities, we conclude that the intersection graph of $\{e_1, \ldots, e_{10}\}$ is not isomorphic to graph 4 in Fig. 2. Cliques 5 and 8 are excluded analogously. □

Remark 20 From Proposition 19 it follows that concurrent exceptional curves on a del Pezzo surface of degree 1 whose intersection graph looks like graphs 4, 5, or 8 in Fig. 2 do not come from generalized Eckardt points on a del Pezzo surface of degree 2 (by blowing up a point). Therefore, we cannot deduce an upper bound for the number of generalized Eckardt points realized by such cliques by using upper bounds for generalized Eckardt points on del Pezzo surfaces of degree 2. From Theorem 16 we know that clique 4 in Fig. 2 can never be realized. We will show that clique 8 can be realized over $\mathbb{F}_{19}$ (Theorem 36). We also know that clique 5 can be realized, as mentioned in the introduction. It could be the case that, outside characteristic 3, a generalized Eckardt point on a del Pezzo surface of degree 1 outside the ramification curve can never come from a generalized Eckardt point on a del Pezzo surface of degree 2; for example, if it turns out that no clique except cliques 5 and 8 in Fig. 2 can be realized as a generalized Eckardt point.

4 Strategies—Generalized Eckardt Points Outside the Ramification Curve

From Theorem 16 it follows that, outside characteristics 2 and 3 (where a generalized Eckardt point corresponds to 16 and 12 concurrent exceptional curves, respectively), we have a generalized Eckardt point outside the ramification curve of (1) if and only if it is possible to realize one of the cliques 1–8 in Fig. 2, where we can exclude clique 4, and clique 7 in characteristic 0. However, apart from clique 5, it is unclear if such realizations are possible. In this section, we provide strategies to produce examples of 10 concurrent lines corresponding to a given clique, or to show that no examples can exist.

Definition 21 By a *representative* of a clique in the graph G we mean an element from its orbit under the action of the Weyl group W_8.

For each clique in Fig. 2, we can use a representative of the corresponding orbit, compute the size of the stabilizer of that clique in the group W_8 with `magma`, and divide the order of the group W_8 (which is 696,729,600) by this number to obtain the size of its orbit under the action of W_8. The results are in the columns "Stabilizer Size" and "Orbit Size" of Table 1. The sizes of these orbits are very large, and obtaining a full list of all representatives would take a very long time with `magma`. In order to still obtain a list with many different representatives, we compute for each clique a "sub-orbit" as follows. Since the maximal size of cliques in G with only edges of weight 2 is 3 [WvL21, Lemma 7], every clique in Fig. 2 contains two vertices with an edge of weight 1. We fix two such vertices and compute with `magma` a list of only those cliques in each orbit that contain these two vertices. The size of these sub-orbits is found in Table 1, and the database containing all elements in all sub-orbits is available at [DFIW].

Since there are many representatives in each sub-orbit to choose from, a first step is determining which representatives will be best to work with. As we will discuss, there is not one best option, and we will need to consider trade-offs between them.

Fix a field k, and suppose we have points $P_1, P_2, \ldots, P_8$ in $\mathbb{P}^2_k$ in general position. For a del Pezzo surface X of degree 1 obtained by blowing up these points, we denote representatives of cliques in the intersection graph G on exceptional curves on X by sets of curves in $\mathbb{P}^2$, using the correspondence in Theorem 7. We will use the following notation to simplify our discussion, letting i, j, k be distinct indices in $\{1, 2, \ldots, 8\}$.

Table 1 The stabilizer size, orbit size, and sub-orbit size for each clique in Fig. 2

Clique	Stabilizer size	Orbit size	Sub-orbit size
1	18	38,707,200	92,160
2	32	21,772,800	51,840
3	36	19,353,600	46,080
4	72	9,676,800	23,680
5	100	6,967,296	16,128
6	128	5,443,200	13,320
7	192	3,628,800	8,880
8	324	2,150,400	5,120

- Let $\ell_{i,j}$ denote the line through points P_i and P_j.
- Let $b_{i,j,k}$ denote the conic through points P_m for $m \in \{1, \dots, 8\} \setminus \{i, j, k\}$.
- Let $c_{i,j}$ denote the cubic through points P_m for $m \in \{1, 2, \dots, 8\} \setminus \{i\}$ that is singular at point P_j.
- Let $q_{i,j,k}$ denote the quartic through points $P_1, P_2, \dots, P_8$ that is singular at points P_i, P_j, and P_k.
- Let $k_{i,j}$ denote the quintic through points $P_1, P_2, \dots, P_8$ that is singular at all P_m so that $m \in \{1, 2, \dots, 8\} \setminus \{i, j\}$.
- Let s_i denote the sextic through points $P_1, P_2, \dots, P_8$ that passes through P_i three times and every other point twice.

4.1 Clique Representatives

First, it is always possible to pick a representative for cliques 1–8 containing four lines.

Proposition 22 *For each clique 1–8, there exists a representative containing the lines $\ell_{1,2}$, $\ell_{3,4}$, $\ell_{5,6}$, and $\ell_{7,8}$.*

Proof We find such a representative for each clique in the list of sub-orbits produced with `magma`, available at [DFIW]. □

Example 23 Clique 1 has several representatives containing these four lines. We give two examples of such representatives below.

Representative 1 has the above four lines together with the following exceptional curves

- the cubic $c_{2,1}$,
- the quartics $q_{2,3,5}$, $q_{2,4,7}$, $q_{2,6,8}$, $q_{3,6,7}$, $q_{4,5,8}$.

Representative 2 has the above four lines together with the following exceptional curves

- the quartics $q_{1,3,5}$, $q_{1,4,7}$, $q_{1,6,8}$, $q_{3,6,7}$, $q_{4,5,8}$,
- the sextic s_2.

In this case, up to projective equivalence, we can fix four points in general position, say $P_1 = (0:1:1)$, $P_3 = (1:0:1)$, $P_5 = (1:1:1)$, $P = (0:0:1)$, where P is the point we wish to be concurrent to all 10 exceptional curves. Observe that this determines the following lines

$$\ell_{1,2} : x = 0,$$
$$\ell_{3,4} : y = 0,$$
$$\ell_{5,6} : x = y.$$

Since $\ell_{7,8}$ contains P and is distinct from the other three lines, then the line determined by $\ell_{7,8}$ must be of the form $x = gy$ for some $g \in k \setminus \{0\}$. Thus we can fix the following nine points for **point set-up A**

$$\begin{aligned} P_1 &= (0:1:1), & P_5 &= (1:1:1),\\ P_2 &= (0:1:a), & P_6 &= (1:1:c),\\ P_3 &= (1:0:1), & P_7 &= (d:1:e),\\ P_4 &= (1:0:b), & P_8 &= (d:1:f),\\ P &= (0:0:1), \end{aligned}$$

where $a, b, c, d, e, f \in k$. We construct a del Pezzo surface of degree 1 by choosing $P_1, P_2, \ldots, P_8$ in general position, and then blowing up the points.

An example of a del Pezzo surface of degree 1 with a generalized Eckardt point can be constructed by finding eight points in general position that pass through a given set of 10 concurrent exceptional curves and then blowing up these points. In order to determine whether such an example exists, we must determine whether it is possible to find eight points in general position of the above form so that these points determine 10 exceptional curves that pass through P. Since these eight points lie on lines $\ell_{1,2}$, $\ell_{3,4}$, $\ell_{5,6}$, and $\ell_{7,8}$ by construction, which are concurrent in P, we only need to check 6 more exceptional curves.

Notice that in this example there are several high-degree curves that must pass through the same point. As we will see later in this section, it may be computationally infeasible to compute such examples.

As an alternative, we search for representatives with fewer lines, but with more curves of low degree.

Proposition 24 *For each clique 1–8, there exists a representative containing the lines $\ell_{1,2}$ and $\ell_{3,4}$ and no other $\ell_{i,j}$.*
Proof Every representative in the list of sub-orbits contains lines $\ell_{1,2}$ and lines $\ell_{3,4}$ by construction. By going through the list of sub-orbits we find a representative without any other line for each clique. □
Example 25 Consider clique 1. We have the following examples of representatives with two lines.

Representative 1 is described by the two lines above together with

- the conics $b_{1,4,7}, b_{2,3,7}, b_{2,4,8}$,
- the cubic $c_{5,8}, c_{6,7}$,
- the quartics $q_{2,4,5}, q_{2,6,7}$,
- the sextic s_3.

Representative 2 is described by the two lines above together with

- the conics $b_{1,3,4}, b_{2,4,8}$,
- the cubics $c_{2,8}, c_{3,4}, c_{5,6}, c_{6,7}, c_{7,5}$,
- the sextic s_2.

As in the previous case, we can fix four points $P_1 = (0:1:1)$, $P_3 = (1:0:1)$, $P_5 = (1:1:1)$, and $P = (0:0:1)$ in general position up to projective equivalence. We determine the lines

$$\ell_{1,2} : x = 0,$$
$$\ell_{3,4} : y = 0.$$

Since we know that P_5, P_6, P_7, and P_8 do not lie on lines $\ell_{1,2}$ and $\ell_{3,4}$, we know that their first coordinate cannot be equal to 0. Thus we can scale it to be 1. We then construct the points for **point set-up B**

$$\begin{aligned} P_1 &= (0:1:1), & P_5 &= (1:1:1),\\ P_2 &= (0:1:a), & P_6 &= (1:c:d),\\ P_3 &= (1:0:1), & P_7 &= (1:e:f),\\ P_4 &= (1:0:b), & P_8 &= (1:g:h),\\ P &= (0:0:1), \end{aligned}$$

where $a, b, c, d, e, f, g, h \in k$ and P is the point of intersection of lines $\ell_{1,2}$ and $\ell_{3,4}$. As before, we can find examples of generalized Eckardt points by constructing eight points in general position so that their blow-up produces a del Pezzo surface of degree 1 with 10 concurrent lines. Since we have already determined that lines $\ell_{1,2}$ and $\ell_{3,4}$ pass through P, we need to produce eight more exceptional curves that pass through P.

4.2 Strategy for Computing Examples

We now discuss a strategy for determining whether it is possible to have 10 concurrent lines on a del Pezzo surface of degree 1 in a fixed configuration.

The first step is to determine an equation that must be satisfied for each exceptional curve in terms of the variables $a, b, \ldots, h$ used in the point set-up A or B. We illustrate the process using conics and cubics. The higher degree curves are similar.

A general equation of a conic in $\mathbb{P}^2$ is given by $t_1x^2 + t_2xy + t_3xz + t_4y^2 + t_5yz + t_6z^2$. Observe that a conic $b_{i,j,k}$ is determined by five points of $P_1, P_2, \ldots, P_8$. As we want this conic to pass through P, we plug the points in $\{P_1, \ldots, P_8, P\} \setminus \{P_i, P_j, P_k\}$ into the general equation for the conic and obtain a system of six linear homogeneous equations in six unknowns. There is at least one i with $t_i \neq 0$; by fixing such an i, we can rewrite the system of equations to obtain a single equation of the form $Ft_i = 0$ where $F \in k(a, b, c, d, e, f, g, h)$. Since $t_i \neq 0$, then $F = 0$.

Observe that F is a rational function, so our steps may be invalid if the denominator is 0. If the goal is to find an example of a generalized Eckardt point, we can find solutions to the numerator of F equals 0 first, then check that the solutions we found make the denominator non-zero. Thus in order to find examples, we can ignore the denominator at first, then perform a simple check at the end. In order to show that no examples of a generalized Eckardt point realized by a given clique exist, the denominator must be considered.

The process for cubics is similar. A general equation of a cubic in $\mathbb{P}^2$ is given by $t_1x^3 + t_2x^2y + t_3x^2z + t_4xy^2 + t_5xz^2 + t_6xyz + t_7y^3 + t_8y^2z + t_9yz^2 + t_{10}z^3$. The cubic $c_{i,j}$ passes through $\{P_1, P_2, \ldots, P_8\} \setminus \{P_i\}$ and passes through P_j twice. Since we want the cubic to pass through P, we first get eight equations from forcing each of the eight points to lie on the cubic. Since the cubic is singular at P_j, we construct two more equations using two of the three partial derivatives such that the remaining coordinate of the point is non-zero. The third partial derivative will be equal to 0 if we force the other two partials to be 0. This gives us 10 linear homogeneous equations into 10 unknowns. We then solve as we did in the conic case.

We can construct a single rational homogeneous equation for each exceptional curve in a similar way. Once we have an equation for each curve, we need to understand whether the variety formed by the numerators of these rational functions contains any k-points. The rational points of the corresponding variety are possible examples of 10 concurrent lines provided that no denominator is equal to 0. However, the last step is to ensure that the 8 points determined by the variety are in general position. If they are, we have produced an example of a del Pezzo surface of degree 1 with a generalized Eckardt point.

Algorithm 26

(1) Input point set-up A or B and a representative of a clique.
(2) For each exceptional curve in the clique, construct a general equation for such a curve along with a square system of homogeneous equations that determine the coefficients of this equation.

(3) For each exceptional curve in the clique, what remains is a single rational function equal to 0. Construct a variety over k from the numerators of these equations.
(4) Compute k-points on this variety and then find which rational points produce non-zero denominators for each rational function from Step 3.
(5) For each remaining rational point from Step 4, check that the corresponding points $P_1, \ldots, P_8$ are in general position.
(6) If an example of eight points in general position exists, output the corresponding del Pezzo surface of degree 1.

Example 27 Using point set-up A, consider the conic $b_{2,4,8}$. By forcing points P_1, P_3, P_5, P_6, P_7, and P to lie on this conic, we arrive at the equations

$$\begin{aligned} t_4 + t_5 + t_6 &= 0 \\ t_1 + t_3 + t_6 &= 0 \\ t_1 + t_2 + t_3 + t_4 + t_5 + t_6 &= 0 \\ t_1 + t_2 + ct_3 + t_4 + ct_5 + c^2 t_6 &= 0 \\ d^2 t_1 + dt_2 + det_3 + t_4 + et_5 + e^2 t_6 &= 0 \\ t_6 &= 0. \end{aligned}$$

Substituting $t_6 = 0$, $t_4 = -t_5$, $t_1 = -t_3$, we arrive at the simpler system

$$\begin{aligned} t_2 &= 0 \\ (c-1)t_3 + (c-1)t_5 &= 0 \\ (de - d^2)t_3 + (e-1)t_5 &= 0. \end{aligned}$$

We solve $t_5 = -t_3$ and arrive at the single equation

$$(de - d^2 - e + 1)t_3 = 0.$$

Since $t_3 \neq 0$, in order for all six points to lie on this conic, we need $de - d^2 - e + 1 = 0$. Notice that in order to solve for t_5 in the last step, we divided by $c - 1$. This solution could be incorrect if $c = 1$, but in this case $c = 1$ is impossible since the points $P_1, \ldots, P_8$ should be in general position.

Since this equation is linear in e, one simplification we could make it to solve is

$$e = \frac{d^2 - 1}{d - 1} = d + 1.$$

This is valid as long as $d \neq 1$.

There are two main points to make about Algorithm 26. First, since we are ignoring denominators, showing that the variety has no k-points does not automatically mean that there cannot be an example of the given clique. Instead, more work must be done to show what happens if any of the denominators from Step 3 equal 0.

The next main point is that even in magma, computing the variety formed from Algorithm 26 is very challenging. In fact, we have not been able to successfully compute the variety for any of the cliques 1–8. Thus in order to make progress toward this question, we need to make further simplifications.

Example 27 gives one possible simplification. Whenever an equation is linear in a variable, we could solve for that variable and make a substitution in the other exceptional curve equations. This reduces the dimension of the affine space over which the variety is defined at the cost of making the equations more complicated.

Another simplification is that we can fix more of the points in the beginning set-up. However, there is no guarantee that examples will exist in this special case, and we will not be able to prove rigorously that examples do not exist by using this method.

A third simplification is to factor the polynomials that define the equations and use a single factor for each polynomial when defining the variety. This works if the polynomials factor, such as the polynomial in Example 27 which factors as

$$de - d^2 - e + 1 = (d - 1)(e - d - 1)$$

though many of the polynomials defined will be irreducible. Further, this splits the computation into several pieces, one for each combination of factors from each polynomial.

Example 28 To illustrate how difficult computing these varieties can be, we write the equation that must be satisfied in order for points from point set B to lie on the cubic $c_{7,5}$, omitting the variable t_i which we know must be non-zero. We have

$$\begin{aligned}
0 =(&- abc^3gh + abc^3g + abc^3h - abc^3 + abc^2dgh - abc^2dg - abc^2dh + abc^2d \\
&+ abc^2gh - abc^2g - abc^2h + abc^2 + abcdg^3 - abcdg^2h - abcdg^2 + abcdg \\
&+ abcdh - abcd - abcg^3 + abcg^2h + abcg^2 - abcgh - abdg^3 + abdg^2h \\
&+ abdg^2 - abdgh + abg^3 - abg^2h - abg^2 + abgh + ac^3gh - ac^3g - ac^3h^2 \\
&+ ac^3h - ac^2dgh + ac^2dg + ac^2dh^2 - ac^2dh - ac^2gh + ac^2g + ac^2h^2 \\
&- ac^2h - acdg^3 + acdg^2h + acdg^2 - acdg - acdh^2 + acdh + acg^3 - acg^2h \\
&- acg^2 + acgh + ad^2g^3 - ad^2g^2h - ad^2g^2 + ad^2gh - adg^3 + adg^2h + adg^2 \\
&- adgh + bc^2dgh - bc^2dg - bc^2dh + bc^2d - bc^2gh + bc^2g + bc^2h - bc^2 \\
&- bcd^2gh + bcd^2g + bcd^2h - bcd^2 - bcdg^2h + bcdg^2 + bcdgh^2 - bcdg \\
&- bcdh + bcd + bcg^2h - bcg^2 - bcgh^2 + bcgh + bdg^2h - bdg^2 - bdgh^2 + bdgh \\
&- bg^2h + bg^2 + bgh^2 - bgh - c^2dgh + c^2dg + c^2dh^2 - c^2dh + c^2gh - c^2g
\end{aligned}$$

$$-c^2h^2 + c^2h + cd^2gh - cd^2g - cd^2h^2 + cd^2h + cdg^2h - cdg^2 - cdgh^2 + cdg$$
$$+cdh^2 - cdh - cg^2h + cg^2 + cgh^2 - cgh - d^2g^2h + d^2g^2 + d^2gh^2 - d^2gh$$
$$+dg^2h - dg^2 - dgh^2 + dgh)/(bcd - bc - bd + b - cd + c + d^2 - d).$$

The numerator of this rational function is irreducible.

5 Surfaces with Generalized Eckardt Points on the Ramification Curve

As mentioned before, Example 5.2 in [vLW23] contains a del Pezzo surface of degree 1 over any field of characteristic unequal to $2, 3, 5, 7, 11, 13, 17, 19$ with 10 concurrent exceptional curves on the ramification curve. In this section we extend this example to a larger family of surfaces with a generalized Eckardt point. For simplicity we work over a field of characteristic 0.

Let k be any field of characteristic 0. For $a, b, c, d, e, f \in k$, define the following eight points in $\mathbb{P}_k^2$ as in point set-up A.

$$P_1 = (0:1:1), \qquad P_5 = (1:1:1),$$
$$P_2 = (0:1:a), \qquad P_6 = (1:1:c),$$
$$P_3 = (1:0:1), \qquad P_7 = (d:1:e),$$
$$P_4 = (1:0:b), \qquad P_8 = (d:1:f).$$

It is easy to compute with `magma` the determinants of the matrices in [vLW23, Lemma 3.4] that determine whether the points are in general position. This gives a set S of polynomials in $k[a, b, c, d, e, f]$, and if none of these polynomials vanish the points $P_1, \ldots P_8$ are in general position. In that case, the blow-up of $\mathbb{P}_k^2$ in these points is a del Pezzo surface of degree 1, which we denote by X. We define the following four lines in $\mathbb{P}^2$.

The line ℓ_1 through P_1 and P_2, which is given by $x = 0$,
the line ℓ_2 through P_3 and P_4, which is given by $y = 0$,
the line ℓ_3 through P_5 and P_6, which is given by $x = y$,
the line ℓ_4 through P_7 and P_8, which is given by $x = dy$.

Note that $\ell_1, \ldots, \ell_4$ all contain the point $P = (0:0:1)$. Let $c_{7,8}$ be the unique cubic through $P_1, P_3, \ldots, P_6, P_8$ that is singular in P_8, and $c_{8,7}$ be the unique cubic through $P_1, \ldots, P_7$ that is singular in P_7. Define $e_1, \ldots, e_4$ to be the strict transforms on X of $\ell_1, \ldots, \ell_4$, and e_5, e_5' the strict transforms on X of $c_{7,8}$ and $c_{8,7}$, respectively. These are exceptional curves on X. Now assume that both $c_{7,8}$ and $c_{8,7}$ contain P. Then the six exceptional curves $e_1, \ldots, e_5, e_5'$ are concurrent in a point $Q \in X$ that blows down to $P \in \mathbb{P}^2$. Let ϕ be the morphism

$$\phi : X \to \mathbb{P}^3$$

given by the linear system $|-2K_S|$ as in (1). By Theorem 7, we can write

$$e_5 = 3L - \sum_{i=1}^{6} E_i - 2E_8$$

and

$$e_5' = 3L - \sum_{i=1}^{6} E_i - 2E_7,$$

where E_i are exceptional curves on X above the point P_i for $1 \le i \le 8$. It is easy to see that $e_5 \cdot e_5' = 3$, therefore Q is contained in the ramification curve of ϕ and for $1 \le i \le 4$, the partners of e_i are concurrent in Q by Remark 13, hence there are 10 exceptional curves on X concurrent in Q.

We now describe a family of del Pezzo surfaces of degree 1 in terms of the variables a, b, c, d, e, f for which these curves are concurrent. The assumption that $c_{7,8}$ contains P gives the polynomial equation $F_1 \cdot s = 0$, where s is a product of elements in S, and

$$\begin{aligned} F_1(a,b,c,d,e,f) = a(&bd^2f - 2bdf + bf - bd^3 + bd^2 + bd - b - cf + c - df^2 + 2f^2 + d^2f - 2f \\ &- d^2 + d) + (f-d)(bcd^2 - 2bdf + bf + bd - b - cdf - cf + cd + c + 2f^2 - 2f). \end{aligned}$$

We disregard the factor s, since we are interested in configurations of $P_1, \ldots, P_8$ in general position such that $c_{7,8}$ contains P. Write

$$F_1(a,b,c,d,e,f) = a \cdot p(b,c,d,e,f) + q(b,c,d,e,f).$$

We claim that, if $P_1, \ldots, P_8$ are in general position and $F_1(a,b,c,d,e,f) = 0$, then we have $p(b,c,d,e,f) \neq 0$ and we can write

$$a = -q(b,c,d,e,f)/p(b,c,d,e,f).$$

To prove this, assume that we have $p(b,c,d,e,f) = 0$. Since we assume $F_1(a,b,c,d,e,f) = 0$, this implies $q(b,c,d,e,f) = 0$. But then we have

$$0 = (f-d)p(b,c,d,e,f) + q(b,c,d,e,f) = d(d-f)(c+d-f-1)(bd-f+1),$$

and the latter four factors are all contained in S, as they correspond to configurations of $P_1, \ldots, P_8$ where P_1, P_2, P_7, P_8 are collinear, P_3, P_5, P_8 are collinear, P_3, P_6, P_8

are collinear, and P_1, P_4, P_8 are collinear, respectively. Therefore, if $P_1, \ldots, P_8$ are in general position we have $p(b, c, d, e, f) \neq 0$.

Writing $a = -q(b, c, d, e, f)/p(b, c, d, e, f)$, we compute with magma the polynomial equation that corresponds to $c_{8,7}$ containing P, and we find that it factors as $F_2 \cdot t = 0$, where t is a product of elements in S, and

$$\begin{aligned}F_2 = {}& b^2cd^4 - 2b^2cd^3 + b^2cd^2 + 2b^2d^3ef - 5b^2d^2ef + 4b^2def - b^2ef - 2b^2d^4e + 3b^2d^3e + b^2d^2e \\ & - 3b^2de + b^2e - 2b^2d^4f + 3b^2d^3f + b^2d^2f - 3b^2df + b^2f + 2b^2d^5 - 3b^2d^4 + 2b^2d - b^2 \\ & + bc^2d^3 - bc^2d^2 + bcd^2ef - 3bcdef + 2bcef - 2bcd^3e + 2bcd^2e + 2bcde - 2bce - 2bcd^3f \\ & + 2bcd^2f + 2bcdf - 2bcf + bcd^4 - 2bcd^2 - bcd + 2bc - 2bd^2e^2f + 4bde^2 - 2be^2f + 2bd^3e^2 \\ & - 2bd^2e^2 - 2bde^2 + 2be^2 - 2bd^2ef^2 + 4bdef^2 - 2bef^2 + 4bd^3ef - bd^2ef - 7bdef + 4bef \\ & - 2bd^4e - bd^3e + 4bd^2e + bde - 2be + 2bd^3f^2 - 2bd^2f^2 - 2bdf^2 + 2bf^2 - 2bd^4f - bd^3f \\ & + 4bd^2f + bdf - 2bf + 3bd^4 - 3bd^3 - bd^2 + bd - c^2def - c^2ef + c^2de + c^2e + c^2df + c^2f \\ & - c^2d - c^2 + 2ce^2f - 2ce^2 + 2cef^2 + cd^2ef - cdef - 4cef - cd^2e + cde + 2ce - 2cf^2 - cd^2f \\ & + cdf + 2cf + cd^2 - cd + 2de^2f^2 - 4e^2f^2 - 2d^2e^2f + 4e^2f + 2d^2e^2 - 2de^2 - 2d^2ef^2 \\ & + 4ef^2 + 2d^3ef + 2d^2ef - 2def - 4ef - 2d^3e + 2de + 2d^2f^2 - 2df^2 - 2d^3f + 2df + 2d^3 - 2d^2.\end{aligned}$$

Let $\mathbb{A}^5$ be affine space with coordinate ring $k[b, c, d, e, f]$. Points in $\mathbb{A}^5$ correspond to configurations of eight points $P_1, \ldots, P_8$ in $\mathbb{P}^2$. From what we have shown, it follows that all points in the algebraic set $V_1 = Z(F_2)$ defined by $F_2 = 0$ that are not contained in the algebraic set $V_2 = \cup_{s \in S} Z(s)$ correspond to a del Pezzo surface of degree 1 with a generalized Eckardt point on the ramification curve of the map (1). We know that there are points in the set $V_1 \setminus V_2$, as [vLW23, Example 5.2] gives an example: the point $Q = \left(-1, \frac{5}{4}, -1, \frac{1}{2}, -\frac{1}{2}\right)$. We can extend this point to an infinite family as follows.

Theorem 29 *The intersection of V_1 with the hyperplanes $b = -1$, $c = \frac{5}{4}$, and $d = -1$ gives an elliptic curve of rank 1 in $\mathbb{A}^5$, that contains infinitely many points corresponding to a del Pezzo surface of degree 1 with a generalized Eckardt point.*

Proof With magma we check that E is an elliptic curve of rank 1. Since it contains the point Q, which does not lie in V_2, we know that E is not contained in V_2 (in fact, a quick point search with magma gives that all points on E up to height 100 (which are 6) are not in V_2). Since E is irreducible, we conclude that it can only intersect V_2 in finitely many points, hence there are infinitely many rational points on E corresponding to a del Pezzo surface of degree 1 with a generalized Eckardt point. □

6 Finite Fields

In this section we discuss the analogous problem of finding generalized Eckardt points over finite fields. When the characteristic of k is 2 or 3, it is possible to have more than 10 concurrent lines.

Theorem 30 ([vLW23, Theorem 1.1]) *Let $P \in X(k)$ be a point on the ramification curve of φ in (1). The number of exceptional curves that go through P is at most ten if char $k \neq 2$, and at most sixteen if char $k = 2$.*

Theorem 31 ([vLW23, Theorem 1.2]) *Let $P \in X(k)$ be a point outside of the ramification curve of φ in (1). The number of exceptional curves that go through P is at most ten if char $k \neq 3$, and at most twelve if char $k = 3$.*

Further, van Luijk and the fourth author produced examples of del Pezzo surfaces of degree 1 to show that the upper bounds in these theorems are sharp except possibly in char $k = 5$ outside the ramification curve. This leaves the following open question.

Question 32 *Does there exist a del Pezzo surface of degree 1 in characteristic 5 with 10 concurrent lines outside the ramification curve?*

In Theorem 16, we showed that there are exactly 8 possible configurations of 10 concurrent lines on a del Pezzo surface of degree 1 outside the ramification curve.

Question 33 *Which of the eight cliques can be realized over a finite field?*

Answering this question will provide a full classification of all possible generalized Eckardt points over finite fields outside characteristics 2 and 3 (for the latter two characteristics it is known that there is one possible configuration for a generalized Eckardt point, and that these are realizable—see [vLW23]; generalized Eckardt points on the ramification curve outside characteristic 2 are described in Theorem 16 (ii)).

6.1 Results for $\mathbb{F}_p$, *p* Prime

Let p be a prime and consider the finite field $\mathbb{F}_p$. In order to search for examples of generalized Eckardt points, we first need to understand for which characteristics del Pezzo surfaces of degree 1 over $\mathbb{F}_p$ can exist.

Theorem 34 *For $p < 17$, there does not exist a set of eight points in general position in $\mathbb{P}^2_{\mathbb{F}_p}$.*

Proof We prove this theorem using `magma`. Up to projective equivalence, we can fix four points in general position. We then run Algorithm 26 to compute all possible collections of four additional points, and then determines whether these eight points are in general position. We find that for $p < 17$, there are no such eight points. □

For $p \geq 17$, it makes sense to consider Question 33. We use point set-up A to construct the points

$$\begin{aligned} P_1 &= (0:1:1), & P_5 &= (1:1:1),\\ P_2 &= (0:1:a), & P_6 &= (1:1:c),\\ P_3 &= (1:0:1), & P_7 &= (d:1:e),\\ P_4 &= (1:0:b), & P_8 &= (d:1:f),\\ P &= (0:0:1) \end{aligned}$$

as in Sect. 4 and then use a brute force algorithm that (1) determines whether $P_1, P_2, \ldots, P_8$ are in general position and then (2) determines whether 10 lines pass through P.

Theorem 35 *There are no examples of a generalized Eckardt point coming from cliques 1–8 over $\mathbb{F}_p$ for primes $p \in \{17, 23\}$.*

Interestingly, we do find examples of a generalized Eckardt point coming from clique 8 in $\mathbb{F}_{19}$.

Theorem 36 *Clique 8 can be realized over $\mathbb{F}_{19}$.*

Proof We use `magma` to run the brute force computation described above. We find several examples of 8 points in general position over $\mathbb{F}_{19}$ that pass through 10 concurrent exceptional curves. □

Example 37 Our code fixes one representative of clique 8, with the exceptional curves

- the lines $\ell_{12}, \ell_{34}, \ell_{56}, \ell_{78}$,
- the quartics $q_{135}, q_{238}, q_{246}, q_{147}, q_{257}, q_{168}$.

We find the the following points $P_1, \ldots, P_8$ are in general position and determine the 10 exceptional curves above. All 10 exceptional curves pass through P.

$$\begin{aligned} P_1 &= (0:1:1), & P_5 &= (1:1:1),\\ P_2 &= (0:1:2), & P_6 &= (1:1:16),\\ P_3 &= (1:0:1), & P_7 &= (7:1:18),\\ P_4 &= (1:0:4), & P_8 &= (7:1:16),\\ P &= (0:0:1). \end{aligned}$$

Remark 38 As shown in Table 1, there are many representatives of clique 8 that we can choose. Thus many more examples of realizations of clique 8 over $\mathbb{F}_{19}$ are likely to exist by changing the representative.

It seems difficult to push this further using a brute force method. The calculation took 1 h for $p = 17$, 2 h for $p = 19$, and overnight for $p = 23$.

6.2 General $\mathbb{F}_q$

Over $\mathbb{F}_p$, we showed that there are no points in general position for $p < 17$ and we only found realizations of one clique in $\mathbb{F}_{19}$, but no realizations of any cliques when $p \in \{17, 23\}$. However, there are examples in the literature of generalized Eckardt points both on and outside the ramification curve in every characteristic (except characteristic 5 outside the ramification curve). Techniques to find such examples are similar to the technique described in the previous section; one must first choose a field extension of $\mathbb{F}_p$ by adding the roots of an irreducible polynomial. This is done in [vLW23, Sect. 5]; for example, Example 5.1 shows 16 concurrent exceptional curves over a del Pezzo surface of degree 1 over $\mathbb{F}_{32}$, and Example 5.3 shows 10 concurrent exceptional curves on a del Pezzo surface of degree 1 on the ramification curve in characteristics 3, 5, 7, 11, 13, 17, and 19. One can see from this example that for $p = 5, 7, 11, 13, 17, 19$, examples already exist over a quadratic extension of $\mathbb{F}_p$, while in characteristic 3 it was needed to go to a cubic extension. Similarly, Example 5.4 exhibits 12 concurrent exceptional curves in a del Pezzo surface of degree 1 outside the ramification curve over $\mathbb{F}_{27}$.

7 Further Questions

In this paper, we make progress toward answering Questions 1 and 2. We conclude with some questions that can help us make further partial progress.

It is known that cliques 4 and 7 cannot be realized outside characteristic 3 and in characteristic 0, respectively. Clique 5 can be realized at least outside characteristic 5, and the clique in Fig. 3 can be realized in all characteristics; Theorem 29 gives a family of del Pezzo surfaces of degree 1 with such a generalized Eckardt point. However, it remains open if the remaining cliques can be realized over any field, with the exception of clique 8 over $\mathbb{F}_{19}$ (Theorem 36).

Question 39 *Is it possible to find a generalized Eckardt point that realizes clique 1, 2, 3, 6, or 8 over any field that is not* $\mathbb{F}_{19}$*?*

While this paper focused on 10 concurrent lines over any field, 10 concurrent lines do not produce a generalized Eckardt point in characteristics 2 or 3. In these characteristics it is known that there is only one configuration of maximal cliques of exceptional curves, which can be realized as a generalized Eckardt point. It might therefore be easier to answer Question 1 for these cases.

Question 40 *How many generalized Eckardt points can a del Pezzo surface of degree 1 over a field of characteristic 2 or 3 contain?*

Finally, since all known examples give a single generalized Eckardt point, a more attainable question leading to Question 1 would be the following.

Question 41 *Does there exist a del Pezzo of degree 1 that has more than 1 generalized Eckardt point?*

Acknowledgements This project originated during the workshop "Women in Numbers 6", held in March 2023 at BIRS in Banff, Canada. We are grateful for the organizers and the organization of BIRS. We thank Bianca Viray for useful discussions. We are grateful to the anonymous referee for useful comments that improved the quality of the paper. Julie Desjardins was partially supported by an NSERC Discovery grant. Rosa Winter was supported by UKRI fellowship MR/T041609/2.

References

[BCP97] W. Bosma, J. Cannon, and C. Playoust, *The Magma algebra system. I. The user language*, J. Symbolic Comput. **24** (1997), no. 3–4, 235–265, Computational algebra and number theory (London, 1993). MRMR1484478.

[CO99] P. Cragnolini and P. A. Oliverio, *Lines on del Pezzo surfaces with $K_S^2 = 1$ in characteristic $\neq 2$*, Comm. Algebra **27** (1999), no. 3, 1197–1206. MR1669140.

[Des18] J. Desjardins, *On the variation of the root number of the fibers of families of elliptic curves*, J. London Math. Soc. **00** (2018), no. 2, 1–37.

[DFIW] J. Desjardins, Y. Fu, K. Isham, and R. Winter, Magma code and output, available at https://github.com/rosawinter/Magma-Gen_Eckardt_Points_dP1.

[DS23] J. Demeio and S. Streeter, *Weak approximation for del Pezzo surfaces of low degree*, Int. Math. Res. Not. IMRN (2023), no. 13, 11549–11576. MR4609792.

[DSW24] J. Demeio, S. Streeter, and R. Winter, *Weak weak approximation and the hilbert property for degree-two del Pezzo surfaces*, Proc. London Math. Soc. **128** (2024), no. 5, e12601.

[DW] J. Desjardins and R. Winter, *Torsion points and concurrent exceptional curves on del Pezzo surfaces of degree one*, Preprint, arXiv:2210.11659.

[DW22] J. Desjardins and R. Winter, *Density of rational points on a family of del Pezzo surfaces of degree one*, Adv. Math. **405** (2022), 108489, With an appendix by Jean-Louis Colliot-Thélène (in French).

[Jab12] E. Jabara, *Rational points on some elliptic surfaces*, Acta Arith. **153** (2012), no. 1, 93–108.

[KM17] J. Kollár and M. Mella, *Quadratic families of elliptic curves and unirationality of degree 1 conic bundles*, Amer. J. Math. **139** (2017), no. 4, 915–936.

[Kol02] J. Kollár, *Unirationality of cubic hypersurfaces*, J. Inst. Math. Jussieu **1** (2002), no. 3, 467–476.

[Man74] Yu. I. Manin, *Cubic forms: algebra, geometry, arithmetic*, North-Holland Publishing Co., Amsterdam-London; American Elsevier Publishing Co., New York, 1974, Translated from the Russian by M. Hazewinkel, North-Holland Mathematical Library.

[Pie12] M. Pieropan, *On the unirationality of del Pezzo surfaces over an arbitrary field*, 2012, Algant master's thesis at Concordia University and Université Paris-Sud 11, Available at https://spectrum.library.concordia.ca/974864/6/Pieropan_MA_F2012.pdf.

[SD01] P. Swinnerton-Dyer, *Weak approximation and R-equivalence on cubic surfaces*, Rational points on algebraic varieties, Progr. Math., vol. 199, Birkhäuser, Basel, 2001, pp. 357–404. MR1875181.

[Seg43] B. Segre, *A note on arithmetical properties of cubic surfaces*, J. London Math. Soc. **18** (1943), no. 1, 24–31.

[Seg51] B. Segre, *On the rational solutions of homogeneous cubic equations in four variables*, Math. Notae **11** (1951), 1–68.

[SS91] P. Salberger and A. N. Skorobogatov, *Weak approximation for surfaces defined by two quadratic forms*, Duke Math. J. **63** (1991), no. 2, 517–536. MR1115119.

[STVA14] C. Salgado, D. Testa, and A. Várilly-Alvarado, *On the unirationality of del Pezzo surfaces of degree two*, J. London Math. Soc. **90** (2014), 121–139.

[SvL14] C. Salgado and R. van Luijk, *Density of rational points on del Pezzo surfaces of degree one*, Adv. Math. **261** (2014), 154–199.

[Ula07] M. Ulas, *Rational points on certain elliptic surfaces*, Acta Arith. **129** (2007), no. 2, 167–185.

[Ula08] M. Ulas, *Rational points on certain del Pezzo surfaces of degree one*, Glasg. Math. J. **50** (2008), no. 3, 557–564.

[UT10] M. Ulas and A. Togbé, *On the Diophantine equation* $z^2 = f(x)^2 \pm f(y)^2$, Publ. Math. Debrecen **76** (2010), no. 1–2, 183–201.

[vLW23] R. van Luijk and R. Winter, *Concurrent lines on del Pezzo surfaces of degree one*, Math. Comp. **92** (2023), 451–481.

[WvL21] R. Winter and R. van Luijk, *The action of the Weyl group on the* E_8 *root system*, Graphs Combin. **37** (2021), no. 6, 1965–2064.

Period-Index in Top Cohomology Over Semiglobal Fields

Sarah Dijols, Raman Parimala, Sujatha Ramdorai, and Charlotte Ure

Abstract We prove a common slot lemma for symbols in top cohomology classes over semiglobal fields. Furthermore, we prove that period and index agree for general top cohomology classes over such fields. We discuss applications to quadratic forms and related open problems.

2020 Mathematics Subject Classification Primary 12G05 · Secondary 11E04 · 11E81 · 13F25 · 14F22 · 14H25

1 Introduction

Let F be a field and fix a separable closure $\overline{F}$ of F. The Galois group $\mathrm{Gal}\left(\overline{F}/F\right)$ will be denoted by G_F. Let μ_m be the group of m-th roots of unity in $\overline{F}$, considered as a module over G_F, and denote by $\mu_m^{\otimes i}$ its i-fold tensor product. The Galois cohomology groups $H^i\left(G_F, \mu_m^{\otimes i}\right)$ will be denoted $H^i\left(F, \mu_m^{\otimes i}\right)$ throughout. For any field extension E of F, the profinite group G_E is a subgroup of G_F and so there are natural restriction maps

$$\mathrm{res} : H^i\left(F, \mu_m^{\otimes i}\right) \to H^i\left(E, \mu_m^{\otimes i}\right) \tag{1.1}$$

S. Dijols
University of British Columbia, Vancouver, BC, Canada
e-mail: sarah.dijols@math.ubc.ca

R. Parimala
Emory University Atlanta, Atlanta, GA, USA
e-mail: parimala.raman@emory.edu

S. Ramdorai
University of British Columbia, Vancouver, BC, Canada
e-mail: sujatha@math.ubc.ca

C. Ure (✉)
Illinois State University, Normal, IL, USA
e-mail: cure@ilstu.edu

S. Akhtari et al. (eds.), *Research Directions in Number Theory*, Association for Women in Mathematics Series 39, https://doi.org/10.1007/978-3-032-11183-8_5

that take a cohomology class ξ to its restriction ξ_E. If $\xi_E = 0$, we call E a splitting field of ξ.

We call an element $\xi \in H^i\left(F, \mu_m^{\otimes i}\right)$ a *symbol* if it is in the image of the i-fold cup product map

$$H^1(F, \mu_m) \times \cdots \times H^1(F, \mu_m) \to H^i\left(F, \mu_m^{\otimes i}\right).$$

If $\mu_m \subset F$, then $\mu_m^{\otimes i} \cong \mu_m$ as G_F-modules, so that there is a notion of symbols in $H^i(F, \mu_m)$. In the case that $i = m = 2$, a symbol in $H^2\left(F, \mu_2^{\otimes 2}\right)$ determines a quaternion algebra over F and vice versa. Tate's common slot lemma states that if F is a number field, then any finite set of symbols in $H^2\left(F, \mu_m^{\otimes 2}\right)$ can be split by a single extension F of degree dividing m [Tat76, Theorem 4.4 and the remarks following the theorem]. In particular, any finite set of quaternion algebras over a number field may be split by a single quadratic extension of F.

A semiglobal field F is a function field in one variable over a complete discretely valued field K. We prove an analogue of Tate's common slot lemma in the setting of semiglobal fields in Theorem 4.3.

Understanding the splitting fields of general cohomology classes, as in Tate's common slot lemma, has been an interesting problem in the study of Galois cohomology. This motivates the notion of the index of $\xi \in H^i(F, \mu_m)$ as the greatest common divisor of the degrees of field extensions E of F so that ξ splits over E. For an element ξ in $H^i(F, \mu_m)$, we can also define the period of ξ to be its order, which is denoted by $\operatorname{per}(\xi)$. It is known that the period divides the index and that they have the same prime factors ([Pie82], Proposition 14.4(b)(ii)). The period-index question asks for the smallest number r so that

$$\operatorname{ind}(\xi) \mid \operatorname{per}(\xi)^r$$

for all $\xi \in H^i(F, \mu_m)$. There has been much progress toward answering this problem over various fields. In the case that F is a local or global field, period and index agree in the Brauer group ($i = 2$) by the Albert–Brauer–Hasse–Noether Theorem [BNH32]. Recent important results have been obtained in the case of degree 3 cohomology, where [PS98] showed that period and index coincide in $H^3(F, \mu_m)$ if F is the function field of a p-adic curve or the function field of a surface over a finite field. Furthermore, this coincidence has been settled for function fields of curves over imaginary number fields in [Sur20]. In [HHK23], the authors obtain bounds on the index in terms of the period for F a semiglobal field.

Throughout this paper, F is a semiglobal field, i.e., a function field in one variable over a complete discretely valued field K with residue field k. Let m be a number coprime to the characteristic of k, and n be the largest number such that $H^n(k, \mu_m)$ does not vanish. We investigate the cohomology groups $H^{n+2}(F, \mu_m)$. Since $H^i(F, \mu_m)$ vanishes for any $i > n+2$, we call $H^{n+2}(F, \mu_m)$ the *top cohomology*. For classes $\xi \in H^{n+2}(F, \mu_m)$, Gosavi proved in [Gos22, Corollary 3.3] that $\operatorname{ind}(\xi) \mid \operatorname{per}(\xi)^2$ for m an odd prime and $\operatorname{ind}(\xi) \mid \operatorname{per}(\xi)^3$ for $m = 2$. In Theorem 4.8, we prove that period and index agree in top cohomology for semiglobal fields for any m coprime to the characteristic of k. In fact, for a class $\xi \in H^{n+2}(F, \mu_m)$, we construct a field extension L of F of degree dividing m that is a splitting field of ξ.

Let K be a complete discretely valued field with residue field k of characteristic coprime to m and suppose that k has cohomological dimension n. This paper is organized as follows. In Sect. 2, we give necessary background on cohomological dimension, the period-index problem, semiglobal fields, and the local-global principle. In Sect. 3, we explore the top cohomology groups $H^{n+1}(K, \mu_m)$ of the complete discretely valued field K. In particular, we prove Lemma 3.1, which provides a decomposition of elements in $H^i(K, \mu_m)$ that is crucial to our proofs. Using this decomposition, we show that period and index agree in the top cohomology $H^{n+1}(K, \mu_m)$ groups in Lemma 3.4. In Sect. 4, we state and prove our two main theorems. First, we prove an analogue of Tate's common slot lemma over semiglobal fields in Theorem 4.3. We show that for any finite set $\mathcal{S}$ of symbols in $H^{n+2}\left(F, \mu_m^{\otimes n+2}\right)$, there is a field extension L of F of degree dividing m so that all $s \in \mathcal{S}$ are split over L. The proof relies on the local-global principle discussed in Theorem 2.11. More precisely, we show that for any discrete valuation ω on L with restriction ν to F, the restriction s_{F_ν} splits over s_{L_ω} for all $s \in \mathcal{S}$. Finally, we show in Theorem 4.8 that period and index agree in top cohomology of $H^{n+2}(F, \mu_m)$. We conclude the paper by discussing some open questions on top cohomology and connections of period-index questions with the u-invariant of quadratic forms in Sect. 5.

2 Preliminaries

2.1 Cohomological Dimension

For a profinite group G and a G-module A, we denote by $H^i(G, A)$ the i-th profinite cohomology group. If G_F is the absolute Galois group of a field F, we also call A an F-module and write $H^i(F, A) = H^i(G_F, A)$ (see for instance [Ser73, Chap. 1 Sect. 2.2 and Chap. 2]). The cohomological dimension is an important invariant associated with a field (for more details, we refer the reader to [Ser73, Chap. 1 Sect. 3 and Chap. 2 Sect. 4]).

Definition 2.1 For a prime ℓ, the *ℓ-cohomological dimension of* F, $\mathrm{cd}_\ell(F)$, is the smallest integer R for which the ℓ-primary component of $H^{i+R}(F, A)$ is zero for any integer $i \geq 1$ and for every torsion F-module A. If no such R exists, we say that $\mathrm{cd}_\ell(F) = \infty$. Analogously, the *cohomological dimension* $\mathrm{cd}(F)$ *of* F is the smallest integer R so that $H^{i+R}(F, A) = 0$ for any $i \geq 1$ and for any torsion F-module A. If no such R exists, we say that $\mathrm{cd}(F) = \infty$.

In this paper, we are only interested in the groups $H^i(F, \mu_m)$ for $m \in \mathbb{Z}$, $m \geq 2$, $(m, \mathrm{char}(F)) = 1$. We remark that if the characteristic of F is different from ℓ, for ℓ

prime, then by [Ser73, Proposition 4, Chap. 2, Sect. 2.3] the following are equivalent for any $n \geq 1$:

(a) $\mathrm{cd}_\ell(F) \leq n$;
(b) For any finite separable extension L of F of degree prime to ℓ, $H^{n+1}(L, \mu_\ell) = 0$.

An algebraically closed field has cohomological dimension 0. Finite fields and the Laurent series field $k((t))$ with k algebraically closed of characteristic 0 have cohomological dimension 1. More generally, the cohomological dimension is invariant under finite field extensions (see Lemma 2.2 below) and increases as the transcendental degree increases (see Lemma 2.3 below).

Lemma 2.2 ([Ser73], Proposition 10, Chap. 2, Sect. 4.1) *Let L be an algebraic extension of the field F and ℓ a prime number. Assume $[L : F] < \infty$ and $\mathrm{cd}_\ell(F) < \infty$, then $\mathrm{cd}_\ell(L) = \mathrm{cd}_\ell(F)$. In particular, if $[L : F] < \infty$ and $\mathrm{cd}(F) < \infty$, then $\mathrm{cd}(L) = \mathrm{cd}(F)$.*

Lemma 2.3 ([Ser73], Proposition 11, Chap. 2, Sect. 4.2) *Let L be an extension of K, then*

$$\mathrm{cd}_\ell(L) \leq \mathrm{cd}_\ell(K) + \mathrm{trdeg}_K(L)$$

for any prime ℓ. There is equality if L is finitely generated over K, $\mathrm{cd}_\ell(K) < \infty$, and ℓ is coprime to the characteristic of K.

Let K be a complete discretely valued field with residue field k and suppose that $\mathrm{char}(k) \neq \ell$. For $n \geq 1$, we consider the following exact sequence in Galois cohomology

$$0 \to H^{n+1}(k, \mu_\ell) \to H^{n+1}(K, \mu_\ell) \to H^n(k, \mathbb{Z}/\ell\mathbb{Z}) \to 0. \tag{2.1}$$

This sequence is obtained combining the surjective homomorphism between $H^n(k, \mu_\ell)$ and $H^n(k, \mathbb{Z}/\ell\mathbb{Z})$ given in [Ser97, Sect. 4.3] and residue maps (see [GS17, Sect. 6.8]). Using this sequence, we deduce the following relationship between the cohomological dimension of K and k, respectively.

Lemma 2.4 *Let K be a complete discretely valued field with residue field k and suppose that* $\mathrm{char}(k) \neq \ell$. *If $\mathrm{cd}_\ell(k) \leq n$, then $\mathrm{cd}_\ell(K) \leq n + 1$.*

Proof Let L be a finite separable extension of K of degree coprime to ℓ with residue field l, containing k. As $\mathrm{cd}(k) \leq n$ and l is a finite extension of k of degree coprime to ℓ, we deduce from Lemma 2.2 that $\mathrm{cd}_\ell(l) = \mathrm{cd}_\ell(k) \leq n$. In particular, we see that $H^{n+1}(l, \mathbb{Z}/\ell\mathbb{Z}) = 0$. The sequence in (2.1) implies that $H^{n+2}(L, \mu_\ell) = 0$. As this holds for any finite extension L of K, we conclude that $\mathrm{cd}(K) \leq n + 1$. □

Lemma 2.5 *Let K be a complete discretely valued field with residue field k and suppose that* $\mathrm{char}(k) \neq \ell$. *If* $\mathrm{cd}_\ell(k) = n$, *then* $\mathrm{cd}_\ell(K) = n + 1$.

Proof By Lemma 2.4, it remains to show that $\mathrm{cd}_\ell(K) \geq n + 1$, or equivalently, that there is some finite field extension L of K of degree coprime to ℓ so that $H^{n+1}(L, \mu_\ell) \neq 0$.

By assumption $\mathrm{cd}_\ell(k) = n$ and so there exists a finite separable field extension l of k of degree coprime to ℓ so that $H^n(l, \mu_\ell) \neq 0$. Let ρ be a primitive ℓ-th root of unity in the algebraic closure of l. We remark that $l[\rho]$ is a separable extension of l of degree coprime to ℓ. Replacing l by $l[\rho]$, we may assume without loss of generality, that ρ is contained in l. Therefore

$$H^n(l, \mathbb{Z}/\ell\mathbb{Z}) = H^n(l, \mu_\ell) \neq 0. \tag{2.2}$$

By the primitive element theorem, there is some $\overline{\alpha} \in l$ so that $l = k(\overline{\alpha})$. Denote by $\overline{f}(x) \in k[x]$ the minimal polynomial of $\overline{\alpha}$ over k. Let $f(x) \in K[x]$ be a monic lift of $\overline{f}(x)$ and consider $L = K(\alpha) = K[x]/(f(x))$. Then L/K is a separable extension of degree coprime to ℓ that lifts l/k. The short exact sequence in (2.1) for L and l together with Eq. (2.2) implies that $H^{n+1}(L, \mu_\ell)$ does not vanish. □

Let $F = K(C)$ be the function field of an algebraic curve over a complete discretely valued field K with residue field k. Then F is a semiglobal field in the sense discussed below in Sect. 2.3. If ℓ is a prime number which is coprime to the characteristic of the residue field k, then it is also coprime to the characteristic of K and F. In particular, we may apply Lemmas 2.3 and 2.5 to F and K in this setting to deduce that if $\mathrm{cd}_\ell(k) = n$, then $\mathrm{cd}_\ell(K) = n + 1$ and $\mathrm{cd}_\ell(F) = n + 2$. Furthermore, if $\mathrm{cd}(k) = n$, then $\mathrm{cd}_\ell(K) \leq n + 1$ and $\mathrm{cd}_\ell(F) \leq n + 2$ for any prime ℓ coprime to the residue characteristic.

For a field F with $\mathrm{cd}(F) = n$ and A a torsion G-module, the *top cohomology with coefficients in* A is the cohomology group $H^n(F, A)$. Throughout this paper, A will often be the Galois module arising from roots of unity.

2.2 Period-Index Problem and Voevodsky's Theorem

Let F be a field and let m be any number coprime to the characteristic of F. Fix a separable closure $\overline{F}$ of F and denote by G_F its absolute Galois group. Consider the Kummer exact sequence of G_F-modules

$$1 \longrightarrow \mu_m \longrightarrow \overline{F}^\times \xrightarrow{a \mapsto a^m} \overline{F}^\times \longrightarrow 1\,.$$

The induced long exact sequence of cohomology in conjunction with Hilbert 90 gives an isomorphism

$$H^1(F,\mu_m) \xrightarrow{\sim} F^\times/(F^\times)^m. \tag{2.3}$$

There is an i-fold cup-product from the product of i copies of $H^1(F,\mu_m)$ to $H^i(F,\mu_m)$ for any $i > 1$

$$\cup : H^1(F,\mu_m) \times \cdots \times H^1(F,\mu_m) \longrightarrow H^i(F,\mu_m). \tag{2.4}$$

For elements $a_1,\ldots,a_i \in F^\times/(F^\times)^m$, we denote by $(a_1,\ldots,a_i) \in H^i(F,\mu_m)$ the image under the composition of the isomorphism in (2.3) and the cup product in (2.4). Such an element in $H^i(F,\mu_m)$ is called a *symbol*. Abusing notation, we will consider $a_1,\ldots,a_i \in F^\times$ throughout. The cup-product map in (2.4) is alternating and we have the following relations for any $a, b \in F^\times$ and $n \in \mathbb{Z}$

$$\begin{aligned} -(a,b) &= (b,a) \\ n(a,b) &= (a,b^n) = (a^n,b) \\ (a,a) &= (-1,a) \end{aligned} \tag{2.5}$$

(see, e.g., [GS17, Sects. 3.4 and 4.6]).

Let $K_i^M(F)$ denote the i-th Milnor K-group of F (see H. Bass's Algebraic K-theory). Let ℓ be a prime not equal to char(F). Suppose that F contains the ℓ-th roots of unity μ_ℓ. By a celebrated result of Voevodsky [Voe03], the norm-residue map determines an isomorphism between Milnor's K-groups and cohomology

$$K_i^M(F) \otimes \mathbb{Z}/\ell\mathbb{Z} \xrightarrow{\sim} H^i(F,\mu_\ell).$$

In particular, any element ξ in $H^i(F,\mu_\ell)$ may be written as a sum of symbols

$$\xi = \sum_{j=1}^{r} \left(a_{j,1},\ldots,a_{j,i}\right),$$

with $a_{j,1},\ldots,a_{j,i} \in F^\times$ for $1 \le j \le r$. Related to this presentation, the *symbol length* of the i-th cohomology group $H^i(F,\mu_\ell)$, denoted $\mathrm{symb}_i^\ell(F)$, is defined as the minimal number r such that any element $\xi \in H^i(F,\mu_\ell)$ is expressible as a sum of at most r symbols. Investigating the symbol lengths $\mathrm{symb}_i^\ell(F)$ for fields F is of interest. For more background and open problems related to the symbol lengths of fields, we refer the interested reader to [ABGV11]. We note that by definition, the symbol length $\mathrm{symb}_1^\ell(F) = 1$ for any field F. For $n = 2$, the second cohomology group $H^2(F,\mu_\ell)$ is isomorphic to the ℓ-torsion of the Brauer group $\mathrm{Br}(F)$. For a p-adic local field F, the Brauer group $\mathrm{Br}(F) \cong \mathbb{Q}/\mathbb{Z}$ [Ser97, Chap. XII] and if $\mu_\ell \subset F$, then $\mathrm{symb}_2^\ell(F) = 1$. Using this together with the Albert–Brauer–Hasse–Noether Theorem [BNH32], we deduce that $\mathrm{symb}_2^\ell(F) = 1$ for any totally imaginary global field F. In either of the two previous cases $\mathrm{cd}(F) = 2$ and so in summary

$$\mathrm{symb}_n^{\ell}(F) = \begin{cases} 1 & n = 1, 2 \\ 0 & n > 2 \end{cases}$$

for a local or a totally imaginary global field F containing μ_ℓ. Now, let F be the function field of a p-adic curve, and let ℓ be a prime which is coprime to p, then the symbol length $\mathrm{symb}_2^{\ell}(F)$ is 2 [Sal97, Sal98] and $\mathrm{symb}_3^{\ell}(F) = 1$ [PS10].

Remark 2.6 We recall that a cohomology class $\xi \in H^i(F, A)$ for a field F and a Galois module A has splitting field L if the restriction ξ_L as defined in (1.1) is trivial. For example, any symbol (a, b) in $H^2(F, \mu_m)$ is split by the degree m extension $L = F(\sqrt[m]{a})$ since

$$0 = m\left(\sqrt[m]{a}, b\right)_L \overset{(2.5)}{=} \left(\left(\sqrt[m]{a}\right)^m, b\right)_L = (a, b)_L,$$

where the first equality holds as $H^n(L, \mu_m)$ is m-torsion. More generally, a symbol $(a_1, \dots, a_i) \in H^i(F, \mu_m)$ splits over the degree m extension $F(\sqrt[m]{a_j})$ for any $1 \leq j \leq i$.

The index of a cohomology class gives a measure for splitting.

Definition 2.7 Let F be a field and $\xi \in H^i(F, A)$. The *index of* ξ is

$$\mathrm{ind}(\xi) = \gcd\{[L : F] : \xi_L \text{ splits}\}.$$

The period-index question discussed in the introduction explores the relationship between symbol lengths and the degrees of splitting field extensions for symbols. In Sect. 5.2 we will explain the connection between symbol lengths and the u-invariant of a field.

2.3 Semiglobal Fields and a Local-Global Principle

One of the main tools we build on in this paper is patching in semiglobal fields. We follow [HHK14] in the following definitions.

Definition 2.8 (*semiglobal fields*) Let K be a complete discretely valued field. A *semiglobal field* over K is a one-variable function field F over K.

Example 2.9 Examples of semiglobal fields are $F = \mathbb{Q}_p(x)$, $F = k((t))(x)$ for any field k, and any finite extension of these.

Let F be a semiglobal field, that is F is the function field of a smooth projective curve C over K. Let $\mathcal{O}_K$ be the corresponding valuation ring, $\pi \in \mathcal{O}_K$ a (fixed) uniformizing parameter and k the residue field. A *regular proper model of* C is a proper, flat, projective $\mathcal{O}_K$-scheme $\mathfrak{X}$ with function field F that is regular as a scheme. We denote its reduced closed fiber by $\mathfrak{X}_k$. By [Lip78, Main Theorem] such a regular proper model exists and by [Lip75, page 193], we may assume that the reduced closed fiber $\mathfrak{X}_k$ is a union of regular curves with normal crossings. We call such a model a *normal crossings model* of F.

$$\begin{array}{ccccc} C & \longrightarrow & \mathfrak{X} & \longleftarrow & \mathfrak{X}_k \\ \downarrow & & \downarrow & & \downarrow \\ \operatorname{Spec}(K) & \longrightarrow & \operatorname{Spec}(\mathcal{O}_K) & \longleftarrow & \operatorname{Spec}(k) \end{array}$$

For each prime divisor D, we can look at the discretely valued ring $\mathcal{O}_{D,\mathfrak{X}}$ and get a corresponding discrete valuation v_D on F. Such valuations are known as divisorial.

Definition 2.10 The *divisorial set of places of* F *corresponding to the model* $\mathfrak{X}$ is the set

$$\Omega_{F,\mathfrak{X}} = \{v_D \mid D \text{ prime divisor on } \mathfrak{X}\}.$$

We call v_D with D as above a divisorial valuation on F. The set of all divisorial valuations v_D, on all regular models $\mathfrak{X}$ of F, will be denoted by Ω_F.

A new approach to local-global principles for homogeneous varieties over function fields of curves defined over complete discretely valued fields was introduced by Harbater, Hartmann, and Krashen in [HHK14] via patching. Further, Parimala and Suresh, building on the results of Hartmann, Harbater, and Krashen, proved the following Theorem that will fundamentally be used later in our paper:

Theorem 2.11 (Theorem 4.3, [HKP21]) *Let F be a semiglobal field, ℓ be a prime number, Ω_F the set of divisorial places of F, and let r be any integer. Suppose that char$(k) \neq \ell$, and let $i \geq 2$. Then*

$$Ш^i_{\Omega_F} := \operatorname{Ker}\left(H^i\left(F, \mu_\ell^{\otimes r}\right) \to \prod_{\omega \in \Omega_F} H^i\left(F_\omega, \mu_\ell^{\otimes r}\right)\right) = 0.$$

3 Complete Discretely Valued Field of Cohomological Dimension $n + 1$

Let K be a complete discretely valued field, $\mathcal{O}_K$ its ring of integers, π its local uniformizer, and k its residue field. Throughout this section, we suppose that ℓ is a prime which is coprime to the residue characteristic and $\operatorname{cd}_\ell(k) = n$. By Lemma 2.5, we see that $\operatorname{cd}_\ell(K) = n + 1$. As a result, the expression *top cohomology* of K will refer to $H^{n+1}(K, \mu_\ell)$. The aim of this section is to prove that period and index agree in top cohomology in this setting.

We first assume that $\mu_\ell \subset k$, which implies that $\mu_\ell \subset K$ by Hensel's lemma, using that ℓ is coprime to the characteristic of k. We recall that there is an exact sequence on Galois cohomology

$$0 \longrightarrow H^{n+1}_{et}(\mathcal{O}_K, \mu_\ell) \xrightarrow{\ r\ } H^{n+1}(K, \mu_\ell) \xrightarrow{\ \delta\ } H^n(k, \mu_\ell), \qquad (3.1)$$

where r is induced by restriction and δ is a residue map (see, e.g., [GS17, Chap. 6.8]). An element ξ in $H^{n+1}(K, \mu_\ell)$ is called *unramified* if it lies in the kernel of δ, or equivalently in the image of r. Otherwise, we say that ξ is *ramified.*

We show that any cohomology class splits into a *sum of ramified and unramified parts*. We recall from Sect. 2.2 that there is a cup-product map from $\otimes^m H^1(K, \mu_\ell)$ to $H^m(K, \mu_\ell)$ and that $H^1(K, \mu_\ell) \cong K^\times/(K^\times)^\ell$.

Lemma 3.1 *Let K be a discrete valued field, $\mathcal{O}_K$ its ring of integers, and k its residue field. Suppose that $\mu_\ell \subset k$ and denote by π the local uniformizer in $\mathcal{O}_K$. Any element $\xi \in H^m(K, \mu_\ell)$ decomposes as*

$$\xi = \xi_1 + (\xi_2, \pi)$$

with $\xi_1 \in H^m(K, \mu_\ell)$ and $\xi_2 \in H^{m-1}(K, \mu_\ell)$ unramified for any $m > 0$. In the above equation (ξ_2, π) denotes the cup product of ξ_2 with the symbol $(\pi) \in H^1(K, \mu_\ell)$.

Proof Let $\xi \in H^m(K, \mu_\ell)$ and suppose first that ξ is a symbol, that is $\xi = (a_1, \ldots, a_m)$ with $a_i \in K^\times$. Write $a_i = u_i \pi^{j_i}$ with u_i a unit in $\mathcal{O}_K$ and $0 \leq j_i \leq \ell - 1$. Using the relations in (2.5), we see that

$$\begin{aligned}
\xi &= (a_1, \ldots, a_m) \\
&= \left(u_1 \pi^{j_1}, \ldots, u_m \pi^{j_m}\right) \\
&= \left(u_1, u_2 \pi^{j_2}, \ldots, u_m \pi^{j_m}\right) + \left(\pi^{j_1}, u_2 \pi^{j_2}, \ldots, u_m \pi^{j_m}\right) \\
&= \left(u_1, u_2, u_3 \pi^{j_3}, \ldots, u_m \pi^{j_m}\right) + \left(u_1, \pi^{j_2}, u_3 \pi^{j_3}, \ldots, u_m \pi^{j_m}\right) \\
&\quad + \left(\pi^{j_1}, u_2, u_3 \pi^{j_3}, \ldots, u_m \pi^{j_m}\right) + \left(\pi^{j_1}, \pi^{j_2}, u_3 \pi^{j_3}, \ldots, u_m \pi^{j_m}\right) \\
&= \left(u_1, u_2, u_3 \pi^{j_3}, \ldots, u_m^{j_m}\right) + \left(\frac{u_1^{j_2}}{u_2^{j_1}}, \pi, u_3 \pi^{j_3}, \ldots, u_m \pi^{j_m}\right) \\
&\quad + \left((-1)^{j_1 + j_2}, \pi, u_3 \pi^{j_3}, \ldots, u_m \pi^{j_m}\right) \\
&\ \ \vdots \\
&= (u_1, u_2, \ldots, u_m) + \sum_t \left(v_1^t, v_2^t, \ldots, v_{m-1}^t, \pi\right)
\end{aligned}$$

for some units v_s^t in $\mathcal{O}_K$. Since $\mu_\ell \subset k$, any cohomology class $\xi \in H^{n+1}(K, \mu_\ell)$ may be decomposed as a sum of symbols $\xi = \sum_i (a_{i,1}, \ldots, a_{i,m})$ by Voevodsky's Theorem (see Sect. 2.2). We decompose each summand of ξ as above to write

$$\xi = \sum_i (a_{i,1}, \ldots, a_{i,m}) = \sum_i (u_{i,1}, \ldots, u_{i,m}) + \sum_{i,t} (v_{i,1}^t, \ldots, v_{i,m-1}^t, \pi) = \xi_1 + (\xi_2, \pi),$$

which finishes the proof. □

Using the description from the previous Lemma, we are now able to explicitly determine a splitting field of $\xi \in H^{n+1}(K, \mu_\ell)$. We recall from Remark 2.6, that a symbol $(a_1, \ldots, a_n) \in H^n(K, \mu_\ell)$ may be split by the extension $L = K(\sqrt[\ell]{a_i})$ for any $1 \leq i \leq n$.

Lemma 3.2 *Let K be a complete discretely valued field, π its local uniformizer, and k its residue field. Suppose that $\mathrm{cd}_\ell(k) = n$. Assume additionally that $\mu_\ell \subset k$. Then any element $\xi \in H^{n+1}(K, \mu_\ell)$ is split by the degree ℓ field extension $K(\sqrt[\ell]{\pi})$ of K. In particular, it has index bounded above by ℓ.*

Proof We note that as $\mathrm{cd}_\ell(k) = n$, by Lemma 2.5, we have $\mathrm{cd}_\ell(K) = n + 1$. Let $\xi \in H^{n+1}(K, \mu_\ell)$. By Lemma 3.1, ξ can be written as $\xi = \xi_1 + (\xi_2, \pi)$ for some unramified $\xi_1 \in H_{et}^{n+1}(\mathcal{O}_K, \mu_\ell)$ and $\xi_2 \in H_{et}^n(\mathcal{O}_K, \mu_\ell)$. We note that if ξ is unramified, then $\xi = \xi_1$. Since $\mathcal{O}_K$ is Henselian, we remark that $H_{et}^m(\mathcal{O}_K, \mu_\ell) \cong H^m(k, \mu_\ell)$ for any m. Together with the fact that $\mathrm{cd}_\ell(k) = n$, we deduce that $\xi_1 = 0$. In particular, if ξ is unramified, then $\xi = \xi_1 = 0$ is already split over K so that $\mathrm{ind}(\xi) = 0 \leq \ell$. In general, we write any $\xi = (\xi_2, \pi)$ with ξ_2 as before. This cohomology class is split by the degree ℓ field extension $K(\sqrt[\ell]{\pi})$ of K so that $\mathrm{ind}(\xi)$ is again bounded above by ℓ. □

The following lemma extends Lemma 3.2 to the case $\mu_\ell \not\subset k$.

Lemma 3.3 *Let K be a complete discretely valued field, π its local uniformizer, and k its residue field. Suppose that $\mathrm{cd}_\ell(k) = n$ and assume that the characteristic of k is coprime to ℓ, but possibly $\mu_\ell \not\subset k$. Then any element $\xi \in H^{n+1}(K, \mu_\ell)$ is split by the degree ℓ field extension $K(\sqrt[\ell]{\pi})$ of K. In particular, it has index bounded above by ℓ.*

Proof We note that as $\mathrm{cd}_\ell(k) = n$, by Lemma 2.5, we have $\mathrm{cd}_\ell(K) = n + 1$. Since K does not contain μ_ℓ, let us first consider the degree $\ell - 1$ extension $E = K(\rho)$ where ρ is a generator of μ_ℓ. We remark that the characteristic of k is coprime to ℓ, so that the cyclotomic extension E over K is unramified and thus π is a local uniformizer for E as well [Cas86, Eq. (5.5) in the proof of Lemma 5.4]. By Lemma 3.2, after restricting to E, ξ is split by the extension $E(\sqrt[\ell]{\pi})$. Now consider the following diagram of field extensions

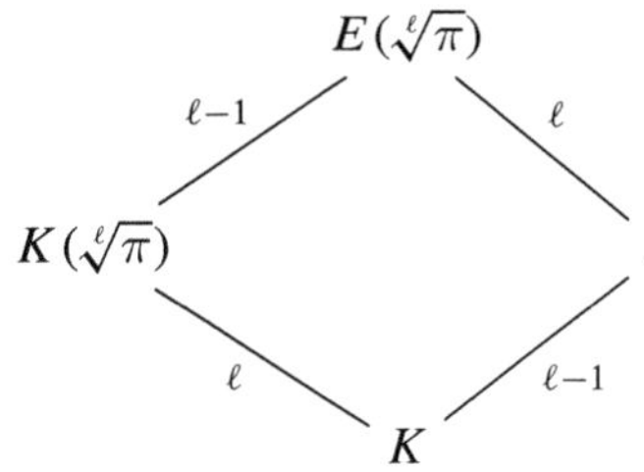

The composition

$$H^{n+1}\left(K(\sqrt[\ell]{\pi}), \mu_\ell\right) \xrightarrow{\text{res}} H^{n+1}\left(E(\sqrt[\ell]{\pi}), \mu_\ell\right) \xrightarrow{\text{cor}} H^{n+1}\left(K(\sqrt[\ell]{\pi}), \mu_\ell\right)$$

is multiplication by $[E(\sqrt[\ell]{\pi}) : K(\sqrt[\ell]{\pi})] = \ell - 1$ [NSW08, Corollary 1.5.7]. Thus

$$(\ell-1)\xi_{K(\sqrt[\ell]{\pi})} = \text{cor} \circ \text{res}\left(\xi_{K(\sqrt[\ell]{\pi})}\right) = \text{cor}(0) = 0.$$

As ℓ and $\ell - 1$ are coprime, we deduce that $\xi_{K(\sqrt[\ell]{\pi})} = 0$ as well and so $K(\sqrt[\ell]{\pi})$ is a splitting field for ξ. □

Finally, using induction on the number of prime factors, we prove the general period-index result for complete discretely valued fields.

Lemma 3.4 *Let K be a complete discretely valued field and k its residue field. Suppose that the characteristic of k is coprime to an integer $m > 1$. Suppose that $\operatorname{cd}(k) = n$. Then any element $\xi \in H^{n+1}(K, \mu_m)$ has index bounded above by m.*

Since m is coprime to the residue characteristic and $\operatorname{cd}_\ell(k) \leq \operatorname{cd}(k) = n$ for any prime divisor ℓ of m, by Lemmas 2.2 and 2.5, we remark that $\operatorname{cd}_\ell(K) \leq n+1$ and so $H^r(K, \mu_m) = 0$ for any $r > n+1$. This motivates the term *top cohomology* for $H^{n+1}(K, \mu_m)$ in this setting.

Proof We use induction on the number of prime factors of m. For $m = \ell$ prime, the result follows from Lemma 3.3. If $m = s\ell$ for some prime ℓ, consider the short exact sequence of K-modules

$$0 \longrightarrow \mu_\ell \longrightarrow \mu_m \xrightarrow{\ell} \mu_s \longrightarrow 0\,. \tag{3.2}$$

For any $\xi \in H^{n+1}(K, \mu_m)$, let $\tilde{\xi}$ be its image under the map

$$H^{n+1}(K, \mu_m) \longrightarrow H^{n+1}(K, \mu_s)$$

arising from the long exact sequence in cohomology. By the inductive hypothesis, there is an extension L_1 of K of degree at most s such that L_1 splits $\tilde{\xi}$. By exactness of the long exact sequence of cohomology induced by (3.2), the restriction ξ_{L_1} of ξ to L_1 lifts to an element $\eta \in H^{n+1}(L_1, \mu_\ell)$.

$$\begin{array}{ccccc} H^{n+1}(L_1,\mu_\ell) & \longrightarrow & H^{n+1}(L_1,\mu_m) & \longrightarrow & H^{n+1}(L_1,\mu_s) \\ \uparrow \text{res} & & \uparrow \text{res} & & \uparrow \text{res} \\ H^{n+1}(K,\mu_\ell) & \longrightarrow & H^{n+1}(K,\mu_m) & \longrightarrow & H^{n+1}(K,\mu_s) \end{array}$$

By Lemma 3.3, there is a field extension L of L_1 of degree at most ℓ so that L splits η. As a result, $\xi_L = \eta_L = 0$ and so L is a splitting field of ξ of degree

$$[L : K] = [L : L_1][L_1 : K] \leq \ell s = m.$$

The statement follows. □

4 The Main Theorems

In this section, we prove our two main theorems. First, we prove an analogue of Tate's common slot lemma over semiglobal fields. When considering coefficients in μ_ℓ for ℓ prime, we assume that $\mathrm{cd}_\ell(k) = n$. For non-prime m, there is no notion of m-cohomological dimension and therefore we assume instead that $\mathrm{cd}(k) = n$.

Let F be a semiglobal field with residue field k. We show in Theorem 4.3 that any finite set $\mathcal{S}$ of r symbols in $H^{n+2}\left(F, \mu_m^{\otimes n+2}\right)$ for m coprime to the characteristic of k is split by a common extension L of F of degree dividing m. We construct this extension explicitly in the proof of Theorem 4.3. First, we consider the case that $m = \ell$ is prime and we recall from Theorem 2.11 that

$$Ш^{n+2}_{\Omega_L} = \mathrm{Ker}\left(H^{n+2}\left(L, \mu_\ell^{\otimes r}\right) \to \prod_{\omega \in \Omega_L} H^{n+2}\left(L_\omega, \mu_\ell^{\otimes r}\right)\right) = 0$$

for any integer r. By the above, it is sufficient to show that L splits any symbol $s \in \mathcal{S}$ locally. More precisely, let $\omega \in \Omega_L$ be a discrete valuation and denote its restriction to F by ν. We show that the restriction s_{F_ν} splits over L_ω. Finally, we extend the result to the case that m is any number coprime to the characteristic of k.

If $\mu_m \subset k$, then Theorem 4.3 implies that period and index agree in the top cohomology for semiglobal fields (see Proposition 4.5 for the precise statement). In the second part of this section, we remove the assumption that $\mu_m \subset k$ while retaining the equality for period and index in top cohomology. More precisely, in Theorem 4.8 we show that, for a semiglobal field F, any class ξ in $H^{n+2}(F, \mu_m)$ has index bounded above by m.

Recall the following setting for semiglobal fields. Let K be a complete discretely valued field and denote by $\mathcal{O}_K$ its ring of valuations. Denote by C a smooth, projective curve over K, and $F = K(C)$ is the function field of C. In particular, F is a semiglobal field. Let k be the residue field of K. We discussed models and divisorial places in Sect. 2.3, but let us recall:

Definition 4.1 (*Model*) A *regular model* of the curve C is a flat, projective, regular, integral $\mathcal{O}_K$-scheme $\mathfrak{X}$ with $\mathfrak{X}_K = \mathfrak{X} \times_{\mathcal{O}_K} K = C$. We denote the reduced special fiber by $\mathfrak{X}_k$.

$$\begin{array}{ccccc} C & \longrightarrow & \mathfrak{X} & \longleftarrow & \mathfrak{X}_k \\ \downarrow & & \downarrow & & \downarrow \\ \mathrm{Spec}(K) & \longrightarrow & \mathrm{Spec}(\mathcal{O}_K) & \longleftarrow & \mathrm{Spec}(k) \end{array}$$

In the course of the proof, we will choose an appropriate model $\mathfrak{X}$ so that the given cohomology class can only ramify along a divisor with normal crossings. Recall the notation Ω_F from Sect. 2.3 for the set of divisorial places on F, which is the set of valuations on F corresponding to codimension one points on all (regular, projective) models of C. To split the cohomology class, it will suffice to find an extension of F, which splits the cohomology class locally everywhere by Theorem 2.11.

We begin with the following. Let A be a complete regular local two-dimensional ring with field of fractions E and residue field k. Let (π, δ) be the maximal ideal of A. For $m \geq 2$ and $1 \leq i \leq m$, we can write $a_i \in E^\times$ as

$$a_i = u_i \pi^{r_i} \delta^{s_i} \theta_i^\ell$$

for some units $u_i \in A$, integers $r_i, s_i \in \mathbb{Z}$ with $0 \leq r_i, s_i \leq \ell - 1$, and $\theta_i \in E$.

Lemma 4.2 *Let A be a complete regular local two-dimensional ring with field of fractions E and residue field k. Let (π, δ) be the maximal ideal of A, and take $a_i \in E^\times, 1 \leq i \leq m$. Then*

$$(a_1, \ldots, a_m) = \xi_1 + (\xi_2, \pi) + (\xi_3, \delta) + (\xi_4, \pi, \delta),$$

where $\xi_1 \in H^n_{et}(A, \mu_\ell^{\otimes m})$, $\xi_2, \xi_3 \in H^{n-1}_{et}(A, \mu_\ell^{\otimes m-1})$, $\xi_4 \in H^{n-2}_{et}(A, \mu_\ell^{\otimes m-2})$ are unramified.

Proof The proof follows along the same lines as the proof of Lemma 3.1. We remark that the assumption that $\mu_\ell \subset k$ in the proof of Lemma 3.1 is only required to prove the statement for a general cohomology class, which has to be written as a sum of symbols first. □

We now prove an analogue of Tate's common slot lemma for number fields over semiglobal fields. We remark that the roots of unity μ_m may not be contained in k for this result.

Theorem 4.3 *Let F be a semiglobal field over a complete discretely valued field K, with residue field k. Let ℓ be a prime not equal to char(k). Assume that* $\mathrm{cd}_\ell(k) = n$. *Let*

$$\mathcal{S} = \left\{ \left(a_{i,1}, a_{i,2}, \dots, a_{i,n+2}\right) : 1 \le i \le r \right\}$$

be a set of r symbols in $H^{n+2}\left(F, \mu_\ell^{\otimes n+2}\right)$*. Then there is a field extension L of F of degree* ℓ *which splits all symbols in* $\mathcal{S}$*.*

Proof We denote by $s_i = \left(a_{i,1}, \dots, a_{i,n+2}\right)$ for $1 \le i \le r$ the symbols in $\mathcal{S}$. By Theorem 2.11, it suffices to prove the existence of a field extension L of F satisfying $[L : F] \le \ell$ with the property that the restriction $(s_i)_{L_\omega}$ of s_i to L_ω is trivial for any divisorial place $\omega \in \Omega_L$. We will show the stronger statement that this is true for any discrete valuation ω of L, not just any divisorial valuation.

We now construct the desired splitting field L. Let $\mathfrak{X}$ be any model of F so that the function field of $\mathfrak{X}$ is F. We recall that $\mathfrak{X}_k$ denotes the reduced special fiber of $\mathfrak{X}$. Abusing notation, we denote a representative of $a_{i,j}$ in $F^\times$ by $a_{i,j}$ as well. As an element in the function field of $\mathfrak{X}$, $a_{i,j}$ is a rational function on $\mathfrak{X}$. Let $D_{i,j} = \operatorname{div}(a_{i,j})$ be the corresponding divisor on $\mathfrak{X}$. After blowing up, we may assume that there are regular curves C_i so that

$$\mathfrak{X}_k \cup \left(\bigcup_{i,j} \operatorname{supp}\left(\operatorname{div}(a_{i,j})\right) \right) \subseteq \bigcup_i C_i$$

and the C_i have normal crossings. We denote $\mathcal{C} = \{C_i\}$. We call a point P on $\mathfrak{X}$ a *nodal point of* $\mathcal{C}$ if it lies on both C_i and C_j for some $i \ne j$. Choose a finite set of closed points $\mathcal{P} = \{P_j\}$ that includes all nodal points of $\mathcal{C}$ and at least one point on the components of $\mathfrak{X}_k$. We note that even though the C_i may not be on the reduced special fiber $\mathfrak{X}_k$, all closed points P_j in $\mathcal{P}$ lie on $\mathfrak{X}_k$.

Since $\mathfrak{X}$ is projective, there is an affine open subscheme U of $\mathfrak{X}$ that contains the finite set of closed points $\mathcal{P} = \{P_j\}$. We denote the affine coordinate ring of U by A. Since $\mathfrak{X}$ is regular, A is regular as well. Let R be the semilocalization of A at $\mathcal{P}$. Because A is regular, we deduce that R is a semilocal regular domain and hence a unique factorization domain. Therefore each C_i defines a height one prime ideal in R, whose generator we denote by π_i. Finally, let $g = \prod_i \pi_i$ and set $L = F(\sqrt[\ell]{g})$. We aim to show that L is a splitting field for s_i for all $1 \le i \le r$. Any discrete valuation ω on L restricts to a discrete valuation ν on F. Let $\mathcal{O}_\nu$ denote the valuation ring of F at ν, F_ν the completion of F at ν and L_ω the completion of L at ω. We claim that L_ω is a splitting field for the restriction $(s_i)_{F_\nu}$ of s_i to F_ν for any $1 \le i \le r$. Throughout the remainder of the proof, we fix $1 \le i \le r$ and we denote $s = s_i$. We remark that by properness of the model $\mathfrak{X}$, any discrete valuation ν is centered on a closed point or on a point of codimension one. For any such P, we can define a regular local ring $\mathcal{O}_{\mathfrak{X},P}$. Let $m_{\mathfrak{X},P}$ be the maximal ideal of $\mathcal{O}_{\mathfrak{X},P}$, and denote by $\widehat{\mathcal{O}_{\mathfrak{X},P}}$ the completion of $\mathcal{O}_{\mathfrak{X},P}$ with respect to this maximal ideal.

- **Case 1: Suppose that ν is centered on a codimension one point P on $\mathfrak{X}$.** There are two possibilities for P, either $P \notin \mathcal{C}$ or $P \in \mathcal{C}$.

– **Case 1a: Suppose that $P \neq C_i$ for all i.** By definition of the C_i's and our hypothesis, a_{ij} is a unit in $\mathcal{O}_{\mathfrak{X},P}$ for all i and j. Thus s is unramified at the completion $\widehat{\mathcal{O}_{\mathfrak{X},P}}$ and so

$$s_{F_\nu} \in H^{n+2}_{\mathrm{nr}}\left(F_\nu, \mu_\ell^{\otimes n+2}\right) \cong H^{n+2}_{et}\left(\widehat{\mathcal{O}_{\mathfrak{X},P}}, \mu_\ell^{\otimes n+2}\right) \cong H^{n+2}\left(k_\nu, \mu_\ell^{\otimes n+2}\right), \tag{4.1}$$

where the last isomorphism holds since $\widehat{\mathcal{O}_{\mathfrak{X},P}}$ is Henselian. We remark that, since P is a codimension one point, the residue field k_ν of F_ν has transcendence degree 1 over k, and hence has the property that $\mathrm{cd}_\ell(k_\nu) = n+1$. Using this and (4.1), we have that $s_{F_\nu} = 0$ and therefore its restriction s_{L_ω} also vanishes.

– **Case 1b: Suppose that $P = C_i$ for some i.** Since g is a parameter at C_i, $F_\nu(\sqrt[\ell]{g}) = L_\omega$, by Lemma 3.3, s_{F_ν} are split by the extension L_ω.

- **Case 2: Suppose that ν is centered on a codimension two point P on $\mathfrak{X}$.** Then $\mathcal{O}_{\mathfrak{X},P} \subset \mathcal{O}_\nu$ as a local homomorphism, which is an inclusion inducing an inclusion of the completions $\widehat{\mathcal{O}_{\mathfrak{X},P}} \subset \widehat{\mathcal{O}_\nu}$. As a result, the field of fractions of $\widehat{\mathcal{O}_{\mathfrak{X},P}}$ is contained in the field F_ν, which is a complete discretely valued field.
We distinguish three cases (2a) P is a nodal point of C_i and C_j for some $i \neq j$, (2b) P is a point on exactly one of the C_i, and (2c) P does not lie on any C_i. Denote by $k(P)$ the residue field of both $\mathcal{O}_{\mathfrak{X},P}$ and its completion.

  – **Case 2a: Suppose that P is a nodal point on $\bigcup_i C_i$.** Since $\mathfrak{X}$ has normal crossings, a nodal point will be at the intersection of two distinct curves C_i and C_j. Without loss of generality, assume that P is the intersection point of C_1 and C_2. We remark that the maximal ideal of the local ring $\mathcal{O}_{\mathfrak{X},P}$ is generated by π_1 and π_2 and $\widehat{\mathcal{O}_{\mathfrak{X},P}}$ is a complete local ring of dimension 2. We also note that the residue field $k(P)$ is a finite field extension of k, thus $\mathrm{cd}_\ell(k(P)) = \mathrm{cd}_\ell(k) = n$. By Lemma 4.2, we can decompose s_{F_ν} as

$$s_{F_\nu} = \xi_1 + (\xi_2, \pi_1) + (\xi_3, \pi_2) + (\xi_4, \pi_1, \pi_2),$$

for unramified cohomology classes $\xi_1 \in H^{n+2}_{et}\left(\widehat{\mathcal{O}_{\mathfrak{X},P}}, \mu_\ell^{\otimes n+2}\right)$, $\xi_2, \xi_3 \in H^{n+1}_{et}\left(\widehat{\mathcal{O}_{\mathfrak{X},P}}, \mu_\ell^{\otimes n+1}\right)$, and $\xi_4 \in H^{n}_{et}\left(\widehat{\mathcal{O}_{\mathfrak{X},P}}, \mu_\ell^{\otimes n}\right)$. In conclusion, ξ_1, ξ_2, ξ_3 are unramified at $\mathcal{O}_{\mathfrak{X},P}$ and vanish over its completion as

$$H^{n+i}_{et}\left(\widehat{\mathcal{O}_{\mathfrak{X},P}}, \mu_\ell^{\otimes n+i}\right) \cong H^{n+i}\left(k(P), \mu_\ell^{\otimes n+i}\right)$$

for $i = 1, 2$ and $\mathrm{cd}_\ell\, k(P) = n$. We conclude that $s_{F_\nu} = (\xi_4, \pi_1, \pi_2)$. In $\mathcal{O}_{\mathfrak{X},P}$, we have that $g = u\pi_1\pi_2$ and so

$$\begin{aligned} s_{F_\nu} &= (\xi_4, \pi_1, \pi_2) \\ &= (\xi_4, -\pi_1, \pi_1\pi_2) \end{aligned}$$

$$\begin{aligned} &= \left(\xi_4, -\pi_1, u^{-1}g\right) \\ &= (\xi_4, u, -\pi_1) + (\xi_4, -\pi_1, g) \\ &= (\xi_4, -\pi_1, g)\,, \end{aligned}$$

where the last equality holds since (ξ_4, u) is unramified in

$$H^{n+1}\left(\widehat{\mathcal{O}_{\mathfrak{X},P}}, \mu_\ell^{\otimes n+1}\right) \cong H^{n+1}\left(k(P), \mu_\ell^{\otimes n+1}\right) = 0,$$

since $k(P)$ has cohomological dimension n. Using Remark 2.6, we see that $L_\omega = F_\nu\left(\sqrt[\ell]{g}\right)$ splits s_{F_ν}.

- **Case 2b: Suppose that** P **is a point on exactly one of the** C_i. Without loss of generality, we may assume that P lies on C_1. By Lemma 4.2, the class s_{F_ν} becomes
$$s_{F_\nu} = \xi_1 + (\xi_2, \pi_1)$$
for some $\xi_1 \in H^{n+2}_{et}\left(\widehat{\mathcal{O}_{\mathfrak{X},P}}, \mu_\ell^{\otimes n+2}\right)$ and $\xi_2 \in H^{n+1}_{et}\left(\widehat{\mathcal{O}_{\mathfrak{X},P}}, \mu_\ell^{\otimes n+1}\right)$. As in Case 2a, s is split by the degree ℓ extension given by attaching an ℓ-th root of $g = u\pi_1$, where π_1 is the local uniformizer at C_1 for ν.
- **Case 2c: Suppose that** P **does not lie on** C_i **for any** i. Then a_{ij} is a unit in $\mathcal{O}_{\mathfrak{X},P}$ and therefore s_{F_ν} is an unramified cohomology class in
$$H^{n+2}_{et}\left(\widehat{\mathcal{O}_{\mathfrak{X},P}}, \mu_\ell^{\otimes n+2}\right) \cong H^{n+2}\left(k(P), \mu_\ell^{\otimes n+2}\right).$$
The residue field $k(P)$ is a finite extension of k, implying that $\mathrm{cd}_\ell(k(P)) = \mathrm{cd}_\ell(k) = n$. Hence, the class $s_{\widehat{\mathcal{O}_{\mathfrak{X},P}}}$ vanishes, and so s_{F_ν} does as well. □

Corollary 4.4 *Let F be a semiglobal field over a complete discretely valued field K, with residue field k. Let $m \geq 2$ be coprime to char(k). Assume that* $\mathrm{cd}(k) = n$. *Suppose that $\mu_m \subset k$. Let*

$$\mathcal{S} = \left\{\left(a_{i,1}, a_{i,2}, \ldots, a_{i,n+2}\right) : 1 \leq i \leq r\right\}$$

be a set of r symbols in $H^{n+2}\left(F, \mu_m^{\otimes n+2}\right)$. Then there is a field extension L of F of degree ℓ which splits all symbols in $\mathcal{S}$.

Proof We note that the assumption $\mathrm{cd}(k) = n$ implies that $\mathrm{cd}_\ell(k) \leq \mathrm{cd}(k) = n$ for any prime ℓ dividing m. The proof follows from Theorem 4.3 and arguments analogous to the proof of Lemma 3.4. □

If k contains μ_m, the previous theorem implies that period and index agree in top cohomology. We later remove this assumption.

Proposition 4.5 *Let F be the function field of the smooth projective curve C over the complete discretely valued field K, with residue field k. Assume additionally that*

$\mu_m \subset k$, m is coprime to the characteristic of k, and $\mathrm{cd}(k) = n$. Then any element ξ in $H^{n+2}(F, \mu_m)$ has index bounded above by m. In fact, there is a field extension L of F of degree m so that $\xi_L = 0$.

Proof Suppose $m = \ell$ is a prime number coprime to the characteristic of k and let $\xi \in H^{n+2}(F, \mu_\ell)$. By Voevodsky's result (see Sect. 2.2), we may write

$$\xi = \sum_{i=1}^{r} \left(a_{i,1}, \ldots, a_{i,n+1}\right)$$

for some $a_{i,j} \in F^\times$. We denote

$$\mathcal{S} = \left\{\left(a_{i,1}, a_{i,2}, \ldots, a_{i,n+2}\right) : 1 \leq i \leq r\right\}.$$

By Theorem 4.3 there is some field extension L of F of degree ℓ which splits all symbols in $\mathcal{S}$. As a result, the sum ξ of the symbols in $\mathcal{S}$ also splits over L. The general case m coprime to the characteristic of k with $\mu_m \subset k$ follows arguments by induction analogous to the proof of Lemma 3.4. □

From now on, we no longer require that $\mu_m \subset k$. We first assume that $m = \ell$ is a prime number. Consider the field extension $F[\rho]$ obtained by attaching a primitive ℓ-th root of unity ρ. We denote by G the Galois group of $F[\rho]$ over F. To descend Theorem 4.7 to this setting, we first need to construct an appropriate model $\mathfrak{X}$ for a fixed cohomology class $\xi \in H^{n+2}\left(F, \mu_\ell\right)$. Consider the restriction $\xi_{F[\rho]} \in H^{n+2}\left(F[\rho], \mu_\ell\right)$. Then $\xi_{F[\rho]}$ may be written as the sum of symbols

$$\xi_{F[\rho]} = \sum_{i=1}^{r} \left(a_{i,1}, \ldots, a_{i,n+2}\right)$$

for some $a_{i,j} \in F[\rho]$.

Lemma 4.6 *With the notation as above, there is a regular proper model $\mathfrak{X}$ over $\mathcal{O}_K$ of F such that the following conditions are satisfied:*

(a) *The base change $\mathfrak{X}[\rho]$ of $\mathfrak{X}$ to $\mathcal{O}_{K[\rho]}$ is a regular proper model of $F[\rho]$;*
(b) *There are distinct irreducible curves $C_1, \ldots, C_r$ on $\mathfrak{X}[\rho]$ so that*

$$(\mathfrak{X}[\rho])_{k[\rho]} \cup \left(\bigcup_{i,j} \mathrm{supp}\left(\mathrm{div}(a_{i,j})\right)\right) \subseteq \bigcup_{\substack{1 \leq i \leq r \\ \sigma \in G}} C_i^\sigma, \tag{4.2}$$

where C_i^σ denotes the Galois conjugate of C_i;
(c) *For the projection $\pi : \mathfrak{X}[\rho] \to \mathfrak{X}$, we have that $\pi(C_i) = D_i$ for $1 \leq i \leq r$ are distinct irreducible curves on $\mathfrak{X}$; and*
(d) *$\sum_{i,\sigma} C_i^\sigma$ on $\mathfrak{X}[\rho]$ and $\sum_i D_i$ on $\mathfrak{X}$ are divisors with normal crossings.*

Proof Let $\mathfrak{X}$ be a regular proper model of F over $\mathcal{O}_K$ as discussed in Sect. 2.3 and denote by $\mathfrak{X}[\rho]$ its base change to $\mathcal{O}_{K[\rho]}$. We remark that the projection $\pi : \mathfrak{X}$
$[\rho] \to \mathfrak{X}$ is a finite étale morphism. We may choose distinct irreducible curves $D_1, \ldots, D_r$ on $\mathfrak{X}$ so that if $C_1, \ldots, C_r$ are distinct irreducible curves on $\mathfrak{X}[\rho]$ with $\pi(C_i) = D_i$, then (b) is satisfied.

We remark that the model $\mathfrak{X}$ does not necessarily satisfy (d). Let $b : \tilde{\mathfrak{X}} \to \mathfrak{X}$ be a blow up such that $b^* \left(\sum_{i=1}^r D_i\right)$ is a normal crossing divisor on $\tilde{\mathfrak{X}}$. We denote by $\tilde{\mathfrak{X}}[\rho]$ the base change, by $\tilde{\pi} : \tilde{\mathfrak{X}}[\rho] \to \tilde{\mathfrak{X}}$ the corresponding étale projection, and by $b[\rho] : \tilde{\mathfrak{X}}[\rho] \to \mathfrak{X}[\rho]$ the base change of b. Then the following diagram commutes

$$\begin{array}{ccc} \tilde{\mathfrak{X}}[\rho] & \xrightarrow{b[\rho]} & \mathfrak{X}[\rho] \\ {\scriptstyle\tilde{\pi}}\downarrow & & \downarrow{\scriptstyle\pi} \\ \tilde{\mathfrak{X}} & \xrightarrow{b} & \mathfrak{X} \end{array}$$

and therefore

$$\tilde{\pi}^* b^* \left(\sum_{i=1}^r D_i \right) = (b[\rho])^* \, \pi^* \left(\sum_{i=1}^r D_i \right) = (b[\rho])^* \left(\sum_{\substack{1 \le i \le r \\ \sigma \in G}} C_i^\sigma \right).$$

Since $b^* \left(\sum_{i=1}^r D_i\right)$ is a normal crossing divisor on $\tilde{\mathfrak{X}}$ and $\tilde{\pi} : \tilde{\mathfrak{X}}[\rho] \to \tilde{\mathfrak{X}}$ is finite étale, the pullback $(b[\rho])^* \left(\sum_{i,\sigma} C_i^\sigma\right)$ is a divisor with normal crossings on $\tilde{\mathfrak{X}}[\rho]$. This shows that the model $\tilde{\mathfrak{X}}$ satisfies the conditions (a)-(d) in the statement. □

We are now ready to split the cohomology class $\xi \in H^{n+2}(F, \mu_\ell)$ using the model constructed in the previous lemma.

Theorem 4.7 *Let F be a semiglobal field over a complete discrete-valued field K with residue field k. Let ℓ be a prime number coprime to the characteristic of the residue field k, and suppose that* $\mathrm{cd}_\ell(k) = n$. *Any element ξ in $H^{n+2}(F, \mu_\ell)$ is split in a field extension L of F of degree $\le \ell$ so that* $\mathrm{ind}(\xi)$ *divides ℓ.*

Proof In view of Proposition 4.5, we may assume that F does not contain a primitive ℓ-th root of unity. Let ρ be a primitive ℓ-th root of unity in $\bar{F}$ and remark that $[F[\rho] : F] = \ell - 1$. We write $\xi_{F[\rho]} = \sum_{1 \le i \le s} \left(a_{i,1}, \ldots, a_{i,n+2}\right)$, with $a_{i,j} \in F[\rho]$. Let $\mathfrak{X}$ be a regular proper model chosen as in the previous Lemma, i.e., satisfying conditions (a)-(d). We denote $\mathcal{C} = \{C_1, \ldots, C_r\}$ and $\mathcal{C}^\sigma = \left\{C_i^\sigma : 1 \le i \le r, \sigma \in G\right\}$, where G is the Galois group of $F[\rho]$ over F.

As in the proof of Proposition 4.5, we take $g \in F[\rho]$ so that

$$\mathrm{div}_{\mathfrak{X}[\rho]}(g) = \sum_{i,\sigma} C_i^\sigma + E, \tag{4.3}$$

where E is a divisor on $\mathfrak{X}[\rho]$ whose support does not contain any C_i^σ and no component of supp(E) passes through a nodal point of $\mathcal{C}^\sigma$. Let $N : F[\rho] \to F$ denote the norm map and we recall that $\pi : \mathfrak{X}[\rho] \to \mathfrak{X}$ denotes the projection. Let us also denote $k(C)$ (resp. $k(D)$) the residue field at the codimension one point C on the 2-dimensional scheme $\mathfrak{X}[\rho]$ (resp. $\mathfrak{X}$). Then

$$\begin{aligned} \operatorname{div}_{\mathfrak{X}[\rho]}(N(g)) &= \pi^*\pi_*\left(\operatorname{div}_{\mathfrak{X}[\rho]}(g)\right) \\ &\overset{(4.3)}{=} \pi^*\pi_*\left(\sum_{i,\sigma} C_i^\sigma + E\right) \\ &= \pi^*\left(\sum_i f_i g_i D_i\right) + \pi^*\pi_* E \\ &= (\ell-1)\sum_{i,\sigma} C_i^\sigma + \pi^*\pi_* E, \end{aligned} \tag{4.4}$$

where f_i is the degree of the residue field extension $[k(C_i) : k(D_i)]$ and g_i is the number of conjugates C_i^σ of C_i. For the last equality, we remark that

$$f_i g_i = [F[\rho] : F] = (\ell - 1).$$

We now prove that no component of $\pi^*\pi_* E$ passes through a nodal point of $\mathcal{C}^\sigma$. Each component of $\pi^*\pi_* E$ is of the form E_r^τ, with E_r a component of supp(E) and $\tau \in G$. Suppose that E_r^τ passes through a nodal point $C_i^{\sigma_1} \cap C_j^{\sigma_2}$ of $\mathcal{C}^\sigma$. Then E_r passes through $C_i^{\sigma_1\tau} \cap C_j^{\sigma_2\tau}$ which is a nodal point of $\mathcal{C}^\sigma$ contradicting that no component of E passes through a nodal point of $\mathcal{C}^\sigma$. □

As in the proof of Theorem 4.3, let $\mathcal{P}$ be a set of closed points on $\mathfrak{X}[\rho]$ that includes all nodal points of $\mathcal{C}$ and at least one point on each component. We denote the set $\mathcal{P}^\sigma = \{P^\sigma : P \in \mathcal{P}, \sigma \in G\}$. Let U be an open subscheme of $\mathfrak{X}[\rho]$ that contains $\mathcal{P}^\sigma$ and denote its affine coordinate ring by A. In the semilocalization R of A that corresponds to $U - \mathcal{P}^\sigma$ each C_i^σ defines a height one prime ideal whose generator we denote by π_i^σ. Without loss of generality, we may assume that $\sigma(\pi_i) = \pi_i^\sigma$. We denote $L = F\left(\sqrt[\ell]{N(g)}\right)$ and we claim that $L[\rho]$ splits $\xi_{F[\rho]}$. Let ω be a discrete valuation of $L[\rho]$ and denote its restriction to $F[\rho]$ by ν. If ν is centered on a codimension one point P on $\mathfrak{X}[\rho]$ so that $P \notin \mathcal{C}$, then the argument in the proof of Theorem 4.3 Case 1a applies to show that $\xi_{F[\rho]_\nu}$ splits over $\xi_{L[\rho]_\omega}$. If P is centered on C_i^σ for some i and some σ, then we remark that $N(g) = u\left(\pi_i^\sigma\right)^{f_i}$ for some unit $u \in \mathcal{O}_{\mathfrak{X}[\rho], C_i^\sigma}$. Since f_i is coprime to ℓ, the ℓ-torsion symbol (ξ_2, π_i) for $\xi_2 \in H^{n+1}\left(\widehat{\mathcal{O}_{\mathfrak{X}, C_i^\sigma}}, \mu_\ell\right)$ is split if and only if the symbol $\left(\xi_2, (\pi_i^\sigma)^{f_i}\right) = \left(\xi_2, \pi_i^\sigma\right)^{f_i}$ is split. Thus the argument in the proof of Theorem 4.3 Case 1b implies that $\xi_{F[\rho]_\nu}$ splits over $L[\rho]_\omega$. A similar reasoning shows the local splitting if ν is centered on a codimension-two point on $\mathfrak{X}[\rho]$. We conclude that $L[\rho]$ splits $\xi_{F[\rho}$. It remains to show that ξ is already split over L.

We consider the following commutative diagram

$$\begin{array}{ccc} H^{n+2}(L, \mu_\ell) & \xrightarrow{\text{res}} & H^{n+2}(L[\rho], \mu_\ell) \\ \uparrow{\scriptstyle\text{res}} & & \uparrow{\scriptstyle\text{res}} \\ H^{n+2}(F, \mu_\ell) & \xrightarrow{\text{res}} & H^{n+2}(F[\rho], \mu_\ell) \end{array}$$

The horizontal restriction maps are injective since $[F[\rho] : F] = [L[\rho] : L]\ell - 1$ is coprime to ℓ. Since ξ splits over $L[\rho]$, it also splits over the degree ℓ extension L of F by injectivity.

We note that we have therefore, in effect, proved the following theorem.

Theorem 4.8 *Let F be a semiglobal field over K a discrete valued field and denote k the residue field of the latter. Let m be any positive integer. Suppose that the characteristic of k is coprime to m and* $\mathrm{cd}(k) = n$. *Then any element $\xi \in H^{n+2}(F, \mu_m)$ has index bounded above by m.*

Proof Same argumentation as in Lemma 3.4, using Theorem 4.7 for the inductive assumption. □

5 Open Questions

In this section, we discuss related results and open questions. We first propose a conjecture which extends our results on bounded index for the top cohomology for semiglobal fields. Secondly, we highlight the relationship between symbol length and the u-invariant for quadratic forms.

5.1 Questions on Top Cohomology

Let F be any field and fix a prime number ℓ that is coprime to the characteristic of F. We assume that $\mathrm{cd}_\ell(F) = n$, so that $H^{n+r}(F, \mu_\ell) = 0$ for any $r > 0$. We call $H^n(F, \mu_\ell)$ the top cohomology of F.

We remark that in this generality, there is no uniform bound for indices of elements in top cohomology $H^n(F, \mu_\ell)$. More precisely, in [Mer91], Merkurjev produces examples of fields F with $\mathrm{cd}(F) = 2$ which admit division algebras that are tensor products of n quaternion algebras for any given n. These algebras have index 2^n in $H^2(F, \mu_2)$. We are interested in uniform bounds for indices in top cohomology. As a first step in this program, we highlight the following conjecture.

Conjecture 5.1 *Let K be a field with* $\mathrm{cd}_\ell(K) = n$ *and fix a prime ℓ coprime to the characteristic of K. Let F be a function field of a curve over K. If the indices in $H^n(K, \mu_\ell)$ are bounded, then the indices in $H^{n+1}(F, \mu_\ell)$ are bounded.*

We prove the conjecture if F is a semiglobal field. The conjecture also has a positive answer if K is a global field of characteristic different from ℓ [Sur20].

We recall that the diophantine dimension $\mathrm{dd}(F)$ of a field F is the smallest integer $r \geq 0$ such that any n-form of degree $d \geq 1$ has a nontrivial zero in F, whenever $n > d^r$. If no such r exists, we say that $\mathrm{dd}(F) = \infty$. The following conjecture was posed by Krashen in [Kra16, Conjecture 1, p. 997].

Conjecture 5.2 *Suppose that* $\mathrm{dd}(F) \leq d$. *Then for elements* $\xi \in H^d\left(F, \mu_\ell^{\otimes d}\right)$, *the index* $\mathrm{ind}(\xi)$ *divides* ℓ.

Our main result may be viewed as an analogue of this conjecture with the assumption $\mathrm{dd}(F) \leq d$ replaced by $\mathrm{cd}(k) \leq d-2$, which implies $\mathrm{cd}(F) \leq d$, for a semiglobal field F.

5.2 *Quadratic Forms and u-Invariant*

It is well-known that classical invariants of quadratic forms take values in the Galois cohomology groups. In 1970, Milnor proposed higher invariants for quadratic forms which could determine the isomorphism class of a quadratic form up to hyperbolic planes.

Let $W(K)$ be the Witt ring of a field K and denote by $I(K)$ the fundamental ideal of $W(K)$ of even dimensional forms. The ideal $I^n(K)$ is generate by n-fold Pfister forms

$$\langle\langle a_1, \ldots, a_n\rangle\rangle = \langle 1, -a_1\rangle \otimes \langle 1, -a_2\rangle \otimes \cdots \otimes \langle 1, -a_n\rangle .$$

For every $n > 0$, Milnor defined a map

$$I^n(K) \to H^n(K, \mu_2)$$

that takes the Pfister form $\langle\langle a_1, \ldots, a_n\rangle\rangle$ to the symbol $(a_1, \ldots, a_n)$. Milnor conjectured that these maps are well-defined and onto with kernel precisely $I^{n+1}(K)$. Milnor's conjecture was proved by Voevodsky in [Voe03].

Another invariant of fields associated with quadratic forms is the u-invariant which is defined as the maximum dimension of an anisotropic quadratic form over K. In the case that K is a complete discretely valued field with residue field k, Springer shows in [Spr55] that $u(K) = 2u(k)$. Here, we focus on the case of a semiglobal field F, which is the function field of a curve C over K. Suppose that $\mathrm{char}(k) \neq 2$. If $u(l) \leq d$ for any finite extension l of the residue field k and $u(k(C)) \leq 2d$, where C is any curve over k, then $u(F) \leq 4d$ by [HHK09]. The case that K is a p-adic field was proved in [PS10]. On the other hand, if k is perfect with $\mathrm{char}(k) = 2$, then Parimala and Suresh show in [PS14] that if F is the function field of a curve over a complete discretely valued field K of characteristic zero, then $u(F) \leq 8$. Further, they conjecture that with the exact same hypothesis, but that of k being perfect, $u(F) \leq 8[k : k^2]$. Equivalently, the question is whether every quadratic form in at least nine variables over the function field of a curve over a local field has a nontrivial zero.

We now recall two main results in the area. The first of Krashen, [Kra16, 4.2], relates the bounds on indices of Galois cohomology classes for any field to bounds on the symbol length in Galois cohomology. The second result relates the bound on symbol length in μ_2-cohomology to bounds on the u-invariant. Recall the notation from Sect. 2.2 on the symbol length of an element in $H^n(F, \mu_m)$ as $\mathrm{symb}_n^m(F)$.

Theorem 5.3 ([Kra16]) *Let K be a field, ℓ a prime not equal to the characteristic of K and $n \geq 1$. Suppose that there exists an integer N such that for every finite extension L of K and for all $\beta \in H^d(L, \mu_\ell^{\otimes d})$, $1 \leq d \leq n, \mathrm{ind}(\beta) \leq N$. Then, for any $\alpha \in H^n(K, \mu_\ell^{\otimes n})$, the symbol length of α is bounded in terms of* $\mathrm{ind}(\alpha)$, *N and n.*

The following theorem is a consequence of a theorem of Orlov et al. [OVV07] on the Milnor conjecture (see also [PS15]).

Theorem 5.4 ([OVV07]) *Let K be a field of characteristic not equal to 2. Suppose that there exist integers $M \geq 1$ and N such that $H^M(K, \mu_2) = 0$ and $symb_d^2(K)(\alpha) \leq N$ for all $\alpha \in H^d(K, \mu_2)$, $1 \leq d < M$. Then the u-invariant is bounded by a function of M and N.*

If we could show that the index is bounded for elements in $H^i(F, \mu_2)$ for all $i \leq n + 2$, then using Theorems 5.3 and 5.4, it would follow that the u-invariant of F is bounded. In this paper, letting $\mathrm{cd}(K) = n + 1$ for a discrete valued field K, we have shown that in the top cohomology (i.e., degree $n + 2$) and in the context of a semiglobal fields F over K, the index is necessarily bounded. If we were to show that the index is bounded for all i-cohomology, $i \leq \mathrm{cd}(K)$, then we would obtain a bound depending on this index for each $\alpha \in H^{n+1}(F, \mu_\ell^{\otimes n+1})$. Then applying Theorem 5.4, with $M = n + 2$, we would obtain a bound on the u-invariant as a function of three variables: the integer $n + 2$, some homogeneous bound on the indices for all finite extensions of F (in all degrees) and the index of the element in $H^{n+1}(F, \mu_\ell^{\otimes n+1})$.

Finally, we remark that the converse also holds: The finiteness of the u-invariant for μ_2-cohomology implies the boundedness of the index in all degrees, as proven in [Sal15]. Further, if the u-invariant of F is bounded by 2^n where $\mathrm{cd}_2(F) = n$, then by [Kra16, Proposition 5.2(ii)], the index of elements in $H^n(F, \mu_2)$ is at most 2.

Acknowledgements First, we thank Suresh Venapally for bringing to our attention the questions on top cohomology and for continued support during the project. We also thank Danny Krashen for helpful comments and suggestions. We thank the referee for their careful reading and suggestions. This project was started during the Women in Numbers (WIN) workshop at the Banff International Research Station in 2023. We thank the workshop organizers Shabnam Akhtari, Alina Bucur, Jennifer Park, Renate Scheidler, and the staff at BIRS for their support. We also thank the following for supporting the workshop, childcare, and travel: NSF DMS-2012061, the Number Theory Foundation, the Clay Mathematics Institute, the Journal of Number Theory, the Pacific Institute for Mathematical Sciences, and the AMS-Simons Travel Grant program. Sujatha Ramdorai is supported by the NSERC Discovery Grant 2019-03987.

References

[ABGV11] Asher Auel, Eric Brussel, Skip Garibaldi, and Uzi Vishne. Open problems on central simple algebras. *Transform. Groups*, 16(1):219–264, 2011.

[BNH32] R. Brauer, E. Noether, and H. Hasse. Beweis eines Hauptsatzes in der Theorie der Algebren. *J. Reine Angew. Math.*, 167:399–404, 1932.

[Cas86] J. W. S. Cassels. *Local Fields*. London Mathematical Society Student Texts. Cambridge University Press, 1986.

[Gos22] Saurabh Gosavi. Generalized period-index problem with an application to quadratic forms, 2022. arXiv:1910.02473.

[GS17] Philippe Gille and Tamás Szamuely. *Central simple algebras and Galois cohomology*, volume 165 of *Camb. Stud. Adv. Math.* Cambridge: Cambridge University Press, 2nd revised and updated edition edition, 2017.

[HHK09] David Harbater, Julia Hartmann, and Daniel Krashen. Applications of patching to quadratic forms and central simple algebras. *Invent. Math.*, 178(2):231–263, 2009.

[HHK14] David Harbater, Julia Hartmann, and Daniel Krashen. Local-global principles for Galois cohomology. *Comment. Math. Helv.*, 89(1):215–253, 2014.

[HHK23] David Harbater, Julia Hartmann, and Daniel Krashen. Bounding cohomology classes over semiglobal fields. *Isr. J. Math.,* 257, 353–387, 2023.

[HKP21] David Harbater, Daniel Krashen, and Alena Pirutka. Local-global principles for curves over semi-global fields. *Bull. Lond. Math. Soc.*, 53(1):177–193, 2021.

[Kra16] Daniel Krashen. Period and index, symbol lengths, and generic splittings in galois cohomology. *Bulletin of the London Mathematical Society*, page bdw060, 2016.

[Lip75] Joseph Lipman. Introduction to resolution of singularities. In *Algebraic geometry (Proc. Sympos. Pure Math., Vol. 29, Humboldt State Univ., Arcata, Calif., 1974)*, pages 187–230. Amer. Math. Soc., Providence, R.I., 1975.

[Lip78] Joseph Lipman. Desingularization of two-dimensional schemes. *Ann. of Math. (2)*, 107(1):151–207, 1978.

[Mer91] A. S. Merkurjev. Simple algebras and quadratic forms. *Izv. Akad. Nauk SSSR Ser. Mat.*, 55(1):218–224, 1991.

[NSW08] Jürgen Neukirch, Alexander Schmidt, and Kay Wingberg. *Cohomology of number fields*, volume 323 of *Grundlehren der mathematischen Wissenschaften [Fundamental Principles of Mathematical Sciences]*. Springer-Verlag, Berlin, second edition, 2008.

[OVV07] D. Orlov, A. Vishik, and V. Voevodsky. An exact sequence for $K_*^M/2$ with applications to quadratic forms. *Ann. of Math. (2)*, 165(1):1–13, 2007.

[Pie82] Richard S. Pierce. *Associative algebras*, volume 9 of *Studies in the History of Modern Science*. Springer-Verlag, New York-Berlin, 1982.

[PS98] R. Parimala and V. Suresh. Isotropy of quadratic forms over function fields of p-adic curves. *Inst. Hautes Études Sci. Publ. Math.*, (88):129–150 (1999), 1998.

[PS10] Raman Parimala and V. Suresh. The u-invariant of the function fields of p-adic curves. *Ann. Math. (2)*, 172(2):1391–1405, 2010.

[PS14] R. Parimala and V. Suresh. Period-index and u-invariant questions for function fields over complete discretely valued fields. *Invent. Math.*, 197(1):215–235, 2014.

[PS15] R. Parimala and V. Suresh. On the u-invariant of function fields of curves over complete discretely valued fields. *Advances in Mathematics*, 280:729–742, 2015.

[Sal97] David J. Saltman. Division algebras over p-adic curves. *J. Ramanujan Math. Soc.*, 12(1):25–47, 1997.

[Sal98] David J. Saltman. Correction to: "Division algebras over p-adic curves" [J. Ramanujan Math. Soc. **12** (1997), no. 1, 25–47; MR1462850 (98d:16032)]. *J. Ramanujan Math. Soc.*, 13(2):125–129, 1998.

[Sal15] David J. Saltman. Finite u invariant and bounds on cohomology symbol lengths. *Doc. Math.*, pages 577–590, 2015.

[Ser73] Jean-Pierre Serre. *Cohomologie Galoisienne*, volume Vol. 5 of *Lecture Notes in Mathematics*. Springer-Verlag, Berlin-New York, 1973. Cours au Collège de France, Paris, 1962–1963, Avec des textes inédits de J. Tate et de Jean-Louis Verdier, Quatrième édition.

[Ser97] Jean-Pierre Serre. *Galois Cohomology*. Springer Monographs in Mathematics. Springer-Verlag, New York-Berlin, 1997.

[Spr55] T. A. Springer. Quadratic forms over fields with a discrete valuation. I. Equivalence classes of definite forms. *Indag. Math.*, 17:352–362, 1955. Nederl. Akad. Wetensch. Proc. Ser. A **58**.

[Sur20] Venapally Suresh. Third Galois cohomology group of function fields of curves over number fields. *Algebra Number Theory*, 14(3):701–729, 2020.

[Tat76] John Tate. Relations between K_2 and Galois cohomology. *Invent. Math.*, 36:257–274, 1976.

[Voe03] Vladimir Voevodsky. Motivic cohomology with $\mathbf{Z}/2$-coefficients. *Publ. Math. Inst. Hautes Études Sci.*, (98):59–104, 2003.

On Invariants of Artin-Schreier Curves

Juanita Duque-Rosero, Heidi Goodson, Elisa Lorenzo García, Beth Malmskog, and Renate Scheidler

Abstract The main goal of this article is to expand the theory of invariants of Artin-Schreier curves by giving a complete classification in genus 3 and 4. To achieve this goal, we first establish standard forms of Artin-Schreier curves and determine all isomorphisms between curves in this form. We then compute reconstructing systems of invariants for curves in each connected component of the strata of the moduli spaces for Artin-Schreier curves of genus 3 and 4 for $p > 2$.

2020 Mathematics Subject Classification 11G20 · 14L24 · 13A50 · 14D20 · 14D22 · 14G15 · 14H10 · 14H37 · 14H45 · 14Q05

J. Duque-Rosero (✉)
Department of Mathematics and Statistics, Boston University, Boston, MA, USA
e-mail: juanita@bu.edu
URL: https://juanitaduquer.github.io

H. Goodson
Department of Mathematics, Brooklyn College, Brooklyn, NY, USA
e-mail: heidi.goodson@brooklyn.cuny.edu
URL: https://sites.google.com/site/heidigoodson

E. Lorenzo García
Institut de Mathématiques, Université de Neuchâtel, Neuchâtel, Switzerland
Laboratoire IRMAR, Université de Rennes 1, Rennes Cedex, France
e-mail: elisa.lorenzo@unine.ch; elisa.lorenzogarcia@univ-rennes1.fr
URL: https://sites.google.com/site/elisalorenzo

B. Malmskog
Department of Mathematics and Computer Science, Colorado College, Colorado Springs, CO, USA
e-mail: bmalmskog@coloradocollege.edu
URL: https://malmskog.wordpress.com

R. Scheidler
Department of Mathematics and Statistics, University of Calgary, Calgary, AB, Canada
e-mail: rscheidl@ucalgary.ca
URL: https://cpsc.ucalgary.ca/ rscheidl

S. Akhtari et al. (eds.), *Research Directions in Number Theory*, Association for Women in Mathematics Series 39, https://doi.org/10.1007/978-3-032-11183-8_6

1 Introduction

Artin-Schreier theory derives from a 1927 paper of Emil Artin and Otto Schreier characterizing degree p Galois extensions of a field of characteristic p [AS27]. Artin and Schreier proved that such an extension $K(y)/K$ is precisely the splitting field of a polynomial in y of the form $y^p - y - \alpha$, where $\alpha \in K$ and $\alpha \neq \beta^p - \beta$ for any $\beta \in K$. Such extensions are now referred to as Artin-Schreier extensions, and the varieties defined by the associated polynomials are called Artin-Schreier curves. In this article, we aim to parameterize moduli spaces of Artin-Schreier curves in characteristic $p > 2$ by computing invariants of the curves.

It is well-known that elliptic curve isomorphism classes are given by the their j-invariants. In genus 2, we can describe isomorphism classes of curves using Igusa invariants [Igu60]. There are also examples of invariant computations in higher genus. For example, there are results for some genus 3 curves in [Shi67, Dix87, Ohn07], hyperelliptic curves in [LR12], Picard curves in [KLS20], Ciani curves in [BCK21], and others. In general, these calculations are done over the complex numbers, but they usually extend well to fields of large characteristic (genus 2 curves in [Liu93], genus 3 curves in [Bas15], [LLLR21, Sect. 4]). The sources [CFA06, Sect. 14.5] and [Liu02, Sect. 7.4.3] specifically cover Artin-Schreier curves in characteristic 2, which are hyperelliptic. However, there are still many cases missing from the literature, including "small cases" such as non-hyperelliptic curves of genus $g = 3$ in characteristic $p = 3$.

Artin-Schreier curves have undergone intense research not only as objects of mathematical interest in their own right, but also for their applications to coding theory (see [vdGvdV91] for example). For certain families of Artin-Scheier curves over a rational function field $K = \mathbb{F}_q(x)$, where q is a power of p, it is possible to provide point counts, find their automorphism groups, and determine their zeta functions [vdGvdV92, BHM16]. Such families frequently contain curves with many points, including maximal curves with respect to the Weil bound. Investigations of Artin-Schreier point counts from the perspective of arithmetic statistics can be found in [Ent12, BDF12, BDFL16].

In this paper, we compute invariants for Artin-Schreier curves of genus 3 and 4 in characteristic $p > 2$. Our work builds on that of Pries and Zhu, who determined the strata and dimensions of irreducible components of the moduli spaces of Artin-Schreier curves [PZ12]. We seek to characterize these moduli spaces concretely by parameterizing them via invariants. Our main ideas are taken from Geometric Invariant Theory (GIT), which was first developed by Mumford in 1965 [Mum65]. This theory provides techniques for forming the "quotient" of an algebraic variety (or scheme) X by a group G and is especially useful for constructing moduli spaces as quotients of schemes parameterizing objects. We provide an overview of invariant theory in Sect. 2.1 and of Artin-Schreier curves, their moduli spaces, and their isomorphisms in Sect. 2.2.

As in the genus 1 and 2 cases, it is useful to work with a standard form for Artin-Schreier curves in order to determine their invariants. Section 3 develops a standard

Table 1 Reconstructing invariants for all Artin-Schreier curves of genus $g = 3, 4$ in characteristic $p > 2$. The curves are classified by their p-rank s

g	p	s	Standard form	Set of reconstructing invariants over $\overline{\mathbb{F}}_p$
3	3	0	$y^3 - y = x^4 + ax^2$	$\{a^4\}$
3	3	2	$y^3 - y = x^2 + ax + \frac{b}{x}$	$\{a^4, ab, b^4\}$
3	7	0	$y^7 - y = x^2$	$\emptyset$
4	3	0	$y^3 - y = x^5 + cx^4 + dx^2$	$\{(c^3 + d)^{10}, (-cd - \epsilon^2)^5, (c^3 + d)^2(-cd - \epsilon^2)\}, \epsilon^3 = c$
4	3	2	$y^3 - y = x^2 + ax + \frac{b}{x} + \frac{c}{x^2}$	$\{c, ab, a^4c^2 - b^4\}$
4	3	4	$y^3 - y = x^2 + ax + \frac{b}{x} + \frac{c}{x-1}$	$\{(abc)^2, (abc)(a - b - c), ab + ac - bc\}$
4	5	0	$y^5 - y = x^3 + ax^2$	$\{a^{12}\}$
4	5	1	$y^5 - y = x + \frac{a}{x}$	$\{a^2\}$

Table 2 Partitions $\vec{E}$ of $D + 2$ corresponding to the irreducible components of $\mathcal{AS}_{g,s}$ and the dimension of each irreducible component $\mathcal{AS}_{g,\vec{E}}$ for $g = 3, 4$ and $p \geq 3$

g	p	D	s	$\vec{E}$	$\dim \mathcal{AS}_{g,\vec{E}}$
3	3	3	0	$\{5\}$	1
			2	$\{3, 2\}$	2
	7	1	0	$\{3\}$	0
4	3	4	0	$\{6\}$	2
			2	$\{3, 3\}$	3
			4	$\{2, 2, 2\}$	3
	5	2	0	$\{4\}$	1
			4	$\{2, 2\}$	1

form for Artin-Schreier curves. We also describe the isomorphisms between standard forms in this section. In Sect. 4, we determine reconstructing systems of invariants for all Artin-Schreier curves of genus 3 and 4 for primes $p > 2$, as listed in Table 2. This allows us to prove our main result, given in the following theorem.

Theorem 1.1 *A system of reconstructing invariants for all Artin-Schreier curves of genus $g = 3, 4$ in characteristic $p > 2$ is given in Table 1.*

For these curves, the strata and dimensions of the irreducible components of their moduli spaces are completely characterized in Table 2. The standard forms (see Theorem 3.3) and invariants listed in Table 1 are derived in the examples in Sect. 4.

The remainder of this paper is devoted to the proof of Theorem 1.1. We start by computing sets of invariants directly. We then show that these sets allow us to reconstruct standard forms of Artin-Schreier curves. Finally, standard invariant theoretic results summarized in Sect. 2 establish that these sets must generate the full invariant ring.

2 Background and Preliminaries

In this section we recall the basic definitions and results on invariant theory and Artin-Schreier curves that we will require for the work herein.

2.1 A Brief Introduction to Invariant Theory

Here we mainly follow [Bas15, DK02, DK08, Dol03, Eis95, Eis05]. The most important concepts in this section are the definitions of primary and secondary invariants. The main result is Corollary 2.13 that will allow us to manually compute generating sets of invariants in the examples we consider in this paper. For completeness we also introduce the concept of Hilbert series and how to compute them with the Molien-Weyl formula (see Theorem 2.6). This produces a second method for computing invariants that can be used to double-check our computations. Moreover, the algorithm implemented in Magma [BCP97] to compute invariants is also based on these ideas. Our computations in Sect. 4 can be corroborated by Magma.

Let K be an algebraically closed field. Let G be a linear algebraic group defined over K, acting on an algebraic variety G also defined over K. This action defines another action on $K[X]$ by $(g \cdot f)(x) = f(g^{-1} \cdot x)$ for all $x \in X$, $f \in K[X]$, and $g \in G$.

Definition 2.1 An element $f \in K[X]$ is an *invariant* for G if $g \cdot f = f$ for all $g \in G$. The algebra of invariants is $K[X]^G := \{f \in K[X] : g \cdot f = f, \forall g \in G\}$.

We especially consider here the case in which $X = V$ is a rational representation of G of finite degree, i.e., a linear representation of finite dimension such that the group morphism $G \to \mathrm{GL}(V)$ is also a morphism of varieties. The K-algebras $K[V]$ and $K[V]^G$ are naturally graded K-algebras.

In our situation, G will be finite. Since all finite groups are reductive, we can use the following result due to Emmy Noether [Noe26], here using the wording of [DK02].

Proposition 2.2 (Noether Normalization Lemma) *Let R be a finitely generated algebra over a Noetherian commutative ring K, and let G be a finite group acting on R by automorphisms fixing K elementwise. Then R^G is finitely generated as a K-algebra.*

Noether's result is a special case of the following.

Theorem 2.3 (Hilbert Finiteness Theorem) *If G is a reductive algebraic linear group and V is a rational representation of G of finite degree, then $K[V]^G$ is of finite type.*

Definition 2.4 Let $A = \oplus_{i \geq 0} A_i$ with $A_0 = K$ be a graded K-algebra. A set of homogeneous elements $\theta_1, \ldots, \theta_r \in A$ is a *homogeneous system of parameters* (HSOP) if the following hold.

- $\theta_1, \ldots, \theta_r$ are algebraically independent over K,
- A is a $K[\theta_1, \ldots, \theta_r]$-module of finite type, i.e., there exist $\eta_1, \ldots, \eta_s \in A$ such that

$$A = \eta_1 K[\theta_1, \ldots, \theta_r] + \cdots + \eta_s K[\theta_1, \ldots, \theta_r].$$

When $A = K[V]^G$ and the η_i are homogeneous, the elements θ_i (resp. η_i) are the *primary (resp. secondary) invariants* of A.

The Noether Normalization Lemma (Proposition 2.2) implies that a graded algebra of finite type always admits an HSOP.

Definition 2.5 The *Hilbert series* of a graded K-algebra of finite type $A = \oplus_{i \geq 0} A_i$ with $A_0 = K$ is the power series

$$H(A, t) = \sum_{i=0}^{\infty} \dim_K(A_i) t^i.$$

If $A = \eta_1 K[\theta_1, \ldots, \theta_r] + \cdots + \eta_s K[\theta_1, \ldots, \theta_r]$, then

$$H(A, t) = \frac{\sum_{i=1}^{s} t^{e_i}}{(1 - t^{d_1})...(1 - t^{d_r})},$$

where d_i (resp. e_i) is the degree of θ_i (resp. η_i).

Theorem 2.6 (Molien-Weyl Formula, [DK02], Sect. 4.6.1) *The Hilbert series of a rational representation* (V, ρ) *of finite degree of a compact group* G *is*

$$H(K[V]^G, t) = \int_G \frac{1}{\det(\mathrm{Id} - \rho(g)t)} d\mu(g),$$

where $d\mu$ *is the Haar measure of* G.

Corollary 2.7 *If* G *is a finite group and* $\mathrm{char}(K)$ *does not divide* $|G|$, *then*

$$H(K[V]^G, t) = \frac{1}{|G|} \sum_{g \in G} \frac{1}{\det(\mathrm{Id} - t \cdot g)}.$$

Lemma 2.8 (Noether, Fleischmann, Benson, Fogarty, [DK02], Corollary 3.8.4) *With the previous notation, if* $\mathrm{char}(K)$ *does not divide* $|G|$, *then there exists a set of generators of* $K[X]^G$, *all of degree smaller or equal than* $|G|$.

These results produce algorithms to compute primary and secondary invariants for actions of finite groups in the non-modular case (i.e., $\mathrm{char}(K) = p$ does not divide $|G|$). More precisely, see Algorithms 3.5.4 and 3.7.2 in [DK02]. Moreover, a version of these algorithms is implemented in Magma. They can be extended to work in the modular case as discussed in [DK02, Sects. 3.3 and 3.4.2]; specifically in Algorithm 3.7.5 in loc. cit. and in [DK08]. A general bound for the degree of the generators is given in [Sym11, Corollary 0.2]. Again, some versions of these algorithms are also implemented in Magma.

Definition 2.9 A subset $S \subseteq K[X]^G$ is said to be *separating* if it satisfies the following property. For any two points $x, y \in X$, if there exists an invariant $f \in K[X]^G$ with $f(x) \neq f(y)$, then there exists an element $g \in S$ with $g(x) \neq g(y)$.

Definition 2.10 Let $A \subseteq K[X]$ be a subalgebra of a polynomial ring of positive characteristic p. Then the algebra

$$\hat{A} = \{f \in K[X] : f^{p^r} \in A \text{ for some } r \in \mathbb{N}\} \subseteq K[X]$$

is the *purely inseparable closure* of A in $K[X]$.

Theorem 2.11 ([DK02], Theorem 2.3.15) *Let X be an affine variety and $G \subseteq \mathrm{Aut}(K[X])$ a subgroup of the automorphisms of the coordinate ring $K[X]$. Then there exists a finite separating set $S \subseteq K[X]^G$.*

Theorem 2.12 ([DK02], Theorem 2.3.12) *Let G be a finite group. Set $K = \overline{\mathbb{F}}_p$ and let V be a K-rational representation of G. Let $A \subseteq K[V]^G$ be a finitely generated, graded, separating subalgebra. Then $K[V]^G = \hat{\tilde{A}}$, the purely inseparable closure of the normalization of A.*

Corollary 2.13 *Let X be an affine variety over $K = \overline{\mathbb{F}}_p$ and let $G \subseteq \mathrm{Aut}(K[X])$ be a finite subgroup. There exists $A \subseteq K[X]^G$ generated by a finite number of invariants from which one can reconstruct a point on X from its values. Moreover, for any such A, one has $K[X]^G = \hat{\tilde{A}}$, the purely inseparable closure of the normalization of A.*

A set of invariants that allow reconstruction of a point as described in Corollary 2.13 is referred to as a *reconstructing system*.

2.2 Artin-Schreier Curves

Throughout, let p be a prime and $\overline{\mathbb{F}}_p$ an algebraically closed field of characteristic p. An *Artin-Schreier curve* is a curve over $\overline{\mathbb{F}}_p$ with an affine model of the form

$$C_f : y^p - y = f(x), \tag{2.1}$$

where $f(x) \in \overline{\mathbb{F}}_p(x)$ and $f(x) \neq z^p - z$ for any $z \in \overline{\mathbb{F}}_p(x)$. Let $r + 1$ (with $r \geq 0$) be the number of distinct poles of $f(x)$, and denote by d_i the order of the i-th pole of $f(x)$. By [Sti09, Lemma 3.7.7 (b)], or as we describe in the proof of Theorem 3.3, we can assume that no d_i is a multiple of p. To simplify formulae, set

$$e_i := d_i + 1 \qquad (1 \leq i \leq r + 1).$$

Then by [Sti09, Proposition 3.7.8 (d)], the genus of C_f is given by

$$g = \frac{p-1}{2}D, \quad \text{where} \quad D = -2 + \sum_{i=1}^{r+1} e_i. \tag{2.2}$$

The possible pole counts and orders for C_f thus correspond to partitions of $D+2$ into components e_i.

The *p-rank* of a smooth, irreducible, projective $\overline{\mathbb{F}}_p$-curve X is the integer s such that the cardinality of the p-torsion of its Jacobian, $\mathrm{Jac}(X)[p](\overline{\mathbb{F}}_p)$, is p^s. Then $0 \leq s \leq g$, and the Deuring-Shafarevich formula (see [Sub75] for a full discussion) implies that for an Artin-Schreier curve $X = C_f$, we have

$$s = r(p-1). \tag{2.3}$$

Thus, the p-rank of C_f depends simply on the number of poles of $f(x)$, not the poles themselves or their orders.

Let $\mathcal{AS}_g$ denote the moduli space of Artin-Schreier $\overline{\mathbb{F}}_p$-curves of genus g and $\mathcal{AS}_{g,s}$ the locus corresponding to Artin-Schreier $\overline{\mathbb{F}}_p$-curves of genus g with p-rank exactly s. The following theorem, due to Pries and Zhu [PZ12], characterizes the stratification of $\mathcal{AS}_g$ by p-rank.

Theorem 2.14 ([PZ12], Theorem 1.1) *Let $g = D(p-1)/2$ with $D \geq 1$ and $s = r(p-1)$ with $r \geq 0$.*

(1) The set of irreducible components of $\mathcal{AS}_{g,s}$ is in bijection with the set of partitions $\{e_1, \ldots, e_{r+1}\}$ of $D+2$ into $r+1$ positive integers such that each $e_j \not\equiv 1 \pmod{p}$.

(2) The irreducible component $\mathcal{AS}_{g,\vec{E}}$ of $\mathcal{AS}_{g,s}$ for the partition $\vec{E} = \{e_1, \ldots, e_{r+1}\}$ has dimension

$$\dim \mathcal{AS}_{g,\vec{E}} = D - 1 - \sum_{j=1}^{r+1} \lfloor (e_j - 1)/p \rfloor. \tag{2.4}$$

So the irreducible components of the p-rank strata of the moduli space $\mathcal{AS}_g$ correspond to the different possibilities for the number of distinct poles of $f(x)$ and their orders, given the constraint in (2.2). Note that an irreducible component $\mathcal{C}$ of $\mathcal{AS}_{g,s}$ is not necessarily an irreducible component of $\mathcal{AS}_g$. It may be that $\mathcal{C}$ is open in some higher-dimensional irreducible component of $\mathcal{AS}_g$.

We compute the partitions corresponding to the irreducible components of $\mathcal{AS}_{g,s}$ for small g in Table 2. We omit $p = 2$, because all Artin-Schreier curves in this characteristic are hyperelliptic; as mentioned earlier, their invariants and moduli are already well understood.

We now consider the isomorphisms of Artin-Schreier curves.

Lemma 2.15 *Let C_f and $\widetilde{C}_{\tilde{f}}$ be isomorphic Artin-Schreier curves as given in* (2.1)*. Assume that the function field isomorphism $\varphi : \overline{\mathbb{F}}_p(\tilde{C}) \to \overline{\mathbb{F}}_p(C)$ fixes the p-extension $\overline{\mathbb{F}}_p(C)/\overline{\mathbb{F}}_p(x)$. Then φ must be of the form*

$$(x, y) \mapsto \left(\frac{\alpha x + \beta}{\gamma x + \delta}, \lambda y + h(x)\right), \tag{2.5}$$

where $\alpha, \beta, \gamma, \delta \in \overline{\mathbb{F}}_p$ with $\alpha\delta - \beta\gamma \in \overline{\mathbb{F}}_p^\times$, $\lambda \in \mathbb{F}_p^\times$, and $h(x) \in \overline{\mathbb{F}}_p(x)$.

Proof The isomorphism φ is completely determined by its respective images $\tilde{x}$ and $\tilde{y}$ of x and y. By assumption, the degree p extension $\overline{\mathbb{F}}_p(C)/\overline{\mathbb{F}}_p(x)$ is fixed and $\tilde{x}$ is a function of x not dependent on y. This implies that φ must descend to an isomorphism of $\overline{\mathbb{F}}_p(x)$. Hence

$$\tilde{x} = \varphi(x) = \frac{\alpha x + \beta}{\gamma x + \delta},$$

where $\alpha, \beta, \gamma, \delta \in k$, and $\alpha\delta - \beta\gamma \in \overline{\mathbb{F}}_p^\times$. To determine the image of y under φ, we note that φ is invertible in $\overline{\mathbb{F}}_p(x)(y)$, so it must be of the form

$$\tilde{y} = \varphi(y) = \frac{A(x)y + B(x)}{C(x)y + D(x)},$$

where $A(x), B(x), C(x), D(x) \in \overline{\mathbb{F}}_p(x)$. By substituting $\tilde{x} = \frac{\alpha x + \beta}{\gamma x + \delta}$ into the equation of $\tilde{C}_{\tilde{f}}$ and comparing coefficients for $\tilde{y}$, we may assume that $A(x) = \lambda \in \overline{\mathbb{F}}_p^\times$, $C(x) = 0$, and $D(x) = 1$. □

Corollary 2.16 *Let C_f be an Artin-Schreier curve as given in* (2.1)*. Assume that C_f is not birational to a curve of the form*

(1) $y^p - y = \frac{a}{x^p - x}$, *with* $a \in \overline{\mathbb{F}}_p$*; or*
(2) $y^p - y = \frac{1}{x^\lambda}$, *with* $\lambda \mid p + 1$*; or*
(3) $y^3 - y = \frac{i}{x(x-1)}$, *with* $i^2 = 2$.

Then all the isomorphisms of C_f are given by isomorphisms as in (2.5).

Proof By [VM80, Theorem 6], for C_f not birational to a curve of the form (1)–(3), the cover of $\overline{\mathbb{F}}_p(x)$ of degree p is unique, so we can apply Lemma 2.15 to derive the desired result. □

Remark 2.17 Henceforth, we will refer to curves that are birational to a curve of the form (1)–(3) in Corollary 2.16 as *exceptional*. These curves have genus $(p-1)^2$, $\frac{(p-1)(\lambda-1)}{2}$, and 2, respectively.

Remark 2.18 The exceptional curve in case (2) of Corollary 2.16 does not appear for the range of parameters considered in this article, so we do not examine it further. In addition, it is hyperelliptic, and invariants for curves of this type are known (see [LR12]).

Lemma 2.19 *Let C_f be an exceptional Artin-Schreier curve as given in* (2.1). *Then the automorphisms of C_f are those of type* (2.5) *preserving the curve model, together with the automorphisms*

(1) $\tau : \begin{cases} x \mapsto y \\ y \mapsto x \end{cases}$ *if* $y^p - y = \frac{a}{x^p - x}$*, with* $a \in \overline{\mathbb{F}}_p$*;*

(2) $\phi : \begin{cases} x \mapsto \epsilon x y^{-(p+1)/\lambda)} \\ y \mapsto 1/y \end{cases}$ *if* $y^p - y = x^\lambda$*, with* $\lambda \mid p+1$ *and* $\lambda < p+1$*; where* ϵ *is a* λ*-root of* -1*;*

(2') ϕ *in (2) and* $\begin{cases} x \mapsto x + a^p \\ y \mapsto y + ax + b \end{cases}$ *if* $y^p - y = x^{p+1}$*, where* $a^{p^2} = -a$ *and* $b^p - b = -a^{p+1}$.

Proof It is straightforward to see that the isomorphisms from Corollary 2.16 are also isomorphisms for exceptional curves. The extra function field automorphisms are described in [VM80, Theorem 7] and [Hen78] or in [HKT08, Proposition 11.30, Exercise A.9]. □

Remark 2.20 The automorphism groups of the different cases in Corollary 2.16 and Lemma 2.19 are a semi-direct product of an abelian p-group of order p^2 with a dihedral group of order $2(p-1)$, so with cardinality $2p^2(p-1)$, in case (1), an extension of a cyclic group of order λ in *PGL*(2,p) with $\lambda < p+1$, so of cardinality $\lambda|PGL(2, p)| = \lambda(p+1)p(p-1)$, in case (2), and $PGU(3, p^2)$ in case (2′) where $|PGU(3, p^2)| = (p^3+1)p^3(p^2-1)$. Finally, in case (3), the automorphism group is a extension of a cyclic group of order 2 by S_4.

2.3 *Fractional Linear Transformations Acting on* $\overline{\mathbb{F}}_p(x)$

Let

$$\varphi(x, y) = \left(\frac{\alpha x + \beta}{\gamma x + \delta}, \lambda y + h(x)\right),$$

with $\alpha, \beta, \gamma, \delta, \lambda, h(x)$ as given in Corollary 2.16. For ease of exposition, define

$$M := \begin{pmatrix} \alpha & \beta \\ \gamma & \delta \end{pmatrix} \in \mathrm{GL}_2(\overline{\mathbb{F}}_p) \text{ and } M(x) := \frac{\alpha x + \beta}{\gamma x + \delta}.$$

Let $f(x) \in \overline{\mathbb{F}}_p(x)$. The map $x \mapsto M(x)$ extends to an action of $\mathrm{GL}_2(\overline{\mathbb{F}}_p)$ on $\overline{\mathbb{F}}_p(x)$ by the extension $M(f(x)) = f(M(x))$, which induces an action on the projective line $\mathbb{P}^1(\overline{\mathbb{F}}_p)$ via the poles of functions in $\overline{\mathbb{F}}_p(x)$. When we consider functions with a single pole, this action is the inverse of the well-known action of Möbius transformations on $\mathbb{P}^1(\overline{\mathbb{F}}_p)$. For example, consider $f(x) = \frac{1}{x}$, a function with a single pole at $x = 0$. Then $f(M(x)) = \frac{\gamma x+\delta}{\alpha x+\beta}$, which has a pole at $x = \frac{-\beta}{\alpha}$ when $\alpha \neq 0$, and a pole at ∞ otherwise. The standard Möbius transformation of $\mathbb{P}^1(\overline{\mathbb{F}}_p)$ associated to M satisfies $M(0) = \frac{\beta}{\delta}$ if $\delta \neq 0$, with $M(0) = \infty$ when $\delta = 0$. However, $M^{-1}(0) = \frac{-\beta}{\alpha}$ when $\alpha \neq 0$, ∞ when $\alpha = 0$.

To avoid confusion with these actions, we set the following notation to view poles of functions as points in $\mathbb{P}^1(\overline{\mathbb{F}}_p)$ for the remainder of the paper. Let $P_\infty \in \mathbb{P}^1(\overline{\mathbb{F}}_p)$ be the pole of the function $x \in \overline{\mathbb{F}}_p(x)$. For $\mu \in \overline{\mathbb{F}}_p$, let $P_\mu \in \mathbb{P}^1(\overline{\mathbb{F}}_p)$ be the pole of the function $\frac{1}{x-\mu} \in \overline{\mathbb{F}}_p(x)$. Throughout, we will describe the action of $\mathrm{GL}_2(\overline{\mathbb{F}}_p)$ on poles using this notation. For later convenience, we describe the action of M on poles explicitly.

Lemma 2.21 *The transformation* $M(x) = \frac{\alpha x+\beta}{\gamma x+\delta}$ *sends the pole* P_μ *to* $P_{\mu'}$, *where* $\mu' = M^{-1}(\mu) = \frac{\delta\mu-\beta}{-\gamma\mu+\alpha}$.

3 Standard Form

Every Artin-Schreier curve C_f as given in (2.1) can be transformed into an isomorphic curve in *standard form*. While this form is not unique, there are a finite number of standard forms for a given isomorphism class, so this form facilitates our invariant computations. Informally, the idea is to find an isomorphism that sends the poles of C_f with the three highest (not necessarily distinct) orders to the pole P_∞ at infinity and the poles P_0 and P_1 at $x = 0$ and $x = 1$, respectively. In case C_f has at most two poles, they are sent to P_∞ and P_0 (for two poles) or simply to P_∞ (for only one pole), and certain normalizations are applied to the polynomial part on the right-hand side of (2.1).

The single pole case is the most complicated. Here, the unique pole is moved to infinity, so the right-hand side of C_f becomes a polynomial $f(x) \in \overline{\mathbb{F}}_p[x]$ of degree d, the order of the pole. We address this case in the following lemma.

Lemma 3.1 *Let* C_f *be an Artin-Schreier* $\overline{\mathbb{F}}_p$*-curve as given in* (2.1) *with a unique pole at infinity, where* $f(x) \in \overline{\mathbb{F}}_p[x]$ *is monic of degree* $d > 1$. *Then* C_f *is isomorphic to a curve of the form* $C_g : y^p - y = g(x)$ *with*

$$g(x) = x^d + \sum_{i=0}^{d-1} b_i x^i \in \overline{\mathbb{F}}_p[x],$$

where $b_i = 0$ *whenever* p *divides* i *and* $b_1 = 0$.

Proof We explicitly determine $\beta \in \overline{\mathbb{F}}_p$ and $h(x) \in \overline{\mathbb{F}}_p[x]$ such that the isomorphism $(x, y) \mapsto (x + \beta, y + h(x))$ transforms C_f into the required form.

Write

$$f(x) = a_d x^d + \sum_{i=0}^{d-1} a_i x^i$$

with $a_i \in \overline{\mathbb{F}}_p$ for $0 \le i \le d-1$ and $a_d = 1$. For any $\beta \in \overline{\mathbb{F}}_p$ we define

$$u_i(\beta) := \sum_{j=i}^{d} a_j \binom{j}{i} \beta^{j-i} \quad (0 \le i \le d),$$

so $f(x+\beta) = \sum_{i=0}^{d} u_i(\beta) x^i$ and $u_i(0) = a_i$ for $0 \le i \le d$. Set $M = \lfloor d/p \rfloor$ and $m = \lfloor M/p \rfloor$, where $\lfloor \cdot \rfloor$ denotes the floor function. Then $Mp < d$, as p does not divide d. If $M > 0$, recursively define elements $b_i(\beta) \in \overline{\mathbb{F}}_p$ via

$$b_i(\beta)^p = \begin{cases} u_{ip}(\beta) & \text{for } M \ge i \ge m+1, \\ u_{ip}(\beta) + b_{ip}(\beta) & \text{for } m \ge i \ge 1. \end{cases}$$

Since $i < ip \le mp \le M$ for $1 \le i \le m$, the quantities $b_i(\beta)$ are well defined. Also define $b_0(\beta) \in \overline{\mathbb{F}}_p$ via the identity

$$b_0(\beta)^p - b_0(\beta) = u_0(\beta).$$

Let $f'(x)$ denote the formal derivative of $f(x)$ with respect to x. Since p does not divide d, the degree of $f'(x)$ is $d-1 > 0$, and we note that $u_1(\beta) = f'(\beta)$.

Now choose $\beta \in \overline{\mathbb{F}}_p$ such that $u_1(\beta) + b_1(\beta) = 0$. For this choice of β, define

$$h(x) = \sum_{i=0}^{M} b_i(\beta) x^i.$$

Then the isomorphism $(x, y) \mapsto (x+\beta, y+h(x))$ maps C_f to a curve of the form $C_g : y^p - y = g(x)$ where

$$g(x) = f(x+\beta) - h(x)^p + h(x) \in \overline{\mathbb{F}}_p[x].$$

Since $h(x)^p$ has degree $Mp < d$ and $f(x+\beta)$ is monic of degree d, $g(x)$ is also monic of degree d. It remains to show that the coefficients of $g(x)$ satisfy the desired conditions.

Every monomial of the form x^{ip} potentially appearing in $g(x)$ satisfies $0 \le i \le M$. For $i \ge 1$, the corresponding coefficient of $g(x)$ is

$$b_{ip}(\beta) = \begin{cases} u_{ip}(\beta) - b_i(\beta)^p & \text{for } M \ge i \ge m+1, \\ u_{ip}(\beta) - b_i(\beta)^p + b_{ip}(\beta) & \text{for } m \ge i \ge 1. \end{cases}$$

Moreover, $b_0(\beta) = u_0(\beta) - b_0(\beta)^p + \beta_0(\beta)$. By the definition of $b_i(\beta)$, we have $b_{ip} = 0$ for $0 \le i \le M$. Finally, the coefficient of x in $g(x)$ is $b_1(\beta) = u_1(\beta) + b_1(\beta)$, which also vanishes by our choice of β. □

Remark 3.2 The normalization conditions proposed in [Far10, Proposition 2.1.1] require the removal of the monomial x^{d-1} as well as all p-th power monomials in $f(x)$ via an iterative sequence of suitable isomorphisms. One difficulty arising with these restrictions is that in the case where $d \equiv 1 \pmod{p}$, removal of p-th powers already eliminates the monomial x^{d-1}, in which case it is desirable to remove a second coefficient in order to further simplify the curve model. It is in fact always possible to eliminate at least one of the next two highest order monomials x^{d-2}, x^{d-3}, but the form of the initial curve determines which of them can be removed. Specifically, all linear transformations on x leave the coefficient of x^{d-2} in $f(x)$ fixed, so this monomial can be eliminated precisely when the corresponding term in the initial curve vanishes. In order to avoid this dependence on the shape of the curve, we opt instead to remove the linear term and all p-th power monomials from $f(x)$ at once, with a single isomorphism. This choice of normalization has the advantage of being more canonical and thus mathematically more satisfying. However, in practice, it may come at the expense of complicating the calculation of the curve invariants compared to a model with fewer high degree monomials; the example in Sect. 4.3.1 illustrates this.

We now have all the ingredients to convert an Artin-Schreier curve to standard form.

Theorem 3.3 *Let p be an odd prime and C_f an Artin-Schreier $\overline{\mathbb{F}}_p$-curve as given in (2.1) with $r+1$ poles of respective orders $d_1 \ge d_2 \ge \ldots \ge d_{r+1}$. Then C_f is isomorphic to an Artin-Schreier curve*

$$C_g : y^p - y = g(x),$$

where $g(x) \in \overline{\mathbb{F}}_p(x)$ takes on one of the following forms:

(1) ***Case** $r = 0$:*

$$g(x) = x^{d_1} + Q(x)$$

where $Q(x) \in \overline{\mathbb{F}}_p[x]$ is a multiple of x^2 and no monomial appearing in $Q(x)$ has an exponent that is divisible by p.

(2) ***Case** $r = 1$:*

$$g(x) = F(x) + G\left(\frac{1}{x}\right),$$

where $F(x), G(x) \in \overline{\mathbb{F}}_p[x]$, $F(x)$ is monic, $\deg(F) = d_1$, $\deg(G) = d_2$, and no monomial appearing in $F(x)$ or $G(x)$ has an exponent that is divisible by p.

(3) ***Case*** *$r \geq 2$:*

$$g(x) = F(x) + G\left(\frac{1}{x}\right) + H\left(\frac{1}{x-1}\right) + S(x),$$

where $F(x), G(x), H(x) \in \overline{\mathbb{F}}_p[x]$, $\deg(F) = d_1$, $\deg(G) = d_2$, $\deg(H) = d_3$, either $S(x) = 0$ or

$$S(x) = \sum_{i=4}^{r+1} \frac{g_i(x-\lambda_i)}{(x-\lambda_i)^{d_i}},$$

with $\lambda_i \in \overline{\mathbb{F}}_p \backslash \{0, 1\}$, $g_i(x) \in \overline{\mathbb{F}}_p[x]$ non-zero, $\deg(g_i) < d_i$, and no monomial appearing in $F(x)$, $G(x)$, $H(x)$, or any of the polynomials $x^{d_i} g_i(x^{-1})$ has an exponent that is divisible by p, for $4 \leq i \leq r+1$.

The curve C_g is said to be in standard form.

Proof We represent fractional linear transformations $M(x) = \dfrac{\alpha x + \beta}{\gamma x + \delta}$ by $\begin{pmatrix} \alpha & \beta \\ \gamma & \delta \end{pmatrix}$ as in Sect. 2.3.

We begin with the case $r = 0$, so C_f has a unique pole P_{μ_1} of order d_1. If $\mu_1 \neq \infty$, then $\mu_1 \in \overline{\mathbb{F}}_p$ and $f(x)$ is of the form

$$f(x) = \frac{f_\infty(x-\mu_1)}{(x-\mu_1)^{d_1}},$$

where $f_\infty(x) \in \overline{\mathbb{F}}_p[x]$ is non-zero of degree less than d_1. The matrices

$$\begin{pmatrix} 1 & 0 \\ 0 & 1 \end{pmatrix} \qquad \text{if } \mu_1 = \infty,$$

$$\begin{pmatrix} \mu_1 & -1 \\ 1 & 0 \end{pmatrix} \qquad \text{if } \mu_1 \in \overline{\mathbb{F}}_p,$$

send P_{μ_1} to P_∞, so we produce a curve of the form $y^p - y = \tilde{f}(x)$, where $\tilde{f}(x) \in \overline{\mathbb{F}}_p[x]$ is a polynomial of degree d_1. Let a be the leading coefficient of $\tilde{f}$. Applying a matrix of the form $\begin{pmatrix} a^{-1/p} & 0 \\ 0 & 1 \end{pmatrix}$ yields the curve $y^p - y = \tilde{g}(x)$ where $\tilde{g}(x) = \tilde{f}(a^{-1/d}x)$ is monic of degree d_1.

If $d_1 = 1$, then define $\gamma \in \overline{\mathbb{F}}_p$ via $\gamma^p - \gamma = \tilde{g}(0)$, the constant coefficient of $\tilde{g}(x)$. Then the isomorphism $(x, y) \mapsto (x, y + \gamma)$ produces the curve $y^p - y = x = x^{d_1}$. If $d_1 > 1$, then the isomorphism of Lemma 3.1 produces a curve C_g of the required form.

For the remaining cases (i.e., $r \geq 1$), the conversion of C_f to standard form is more easily done in two stages. The first stage applies a suitable fractional linear transformation of x to C_f that moves the highest order poles as explained earlier.

This is followed by a second sequence of linear transformations applied to y that eliminates p-th powers of x one-by-one.

Assume first that $r = 1$, so C_f contains two poles P_{μ_1}, P_{μ_2} of respective orders $d_1 \geq d_2$. Then the matrices

$$\begin{pmatrix} 1 & \mu_2 \\ 0 & 1 \end{pmatrix} \quad \text{if } \mu_1 = \infty \text{ and } \mu_2 \in \overline{\mathbb{F}}_p,$$
$$\begin{pmatrix} \mu_1 & 1 \\ 1 & 0 \end{pmatrix} \quad \text{if } \mu_1 \in \overline{\mathbb{F}}_p \text{ and } \mu_2 = \infty,$$
$$\begin{pmatrix} \mu_1 & \mu_2 \\ 1 & 1 \end{pmatrix} \quad \text{if } \mu_1 \in \overline{\mathbb{F}}_p \text{ and } \mu_2 \in \overline{\mathbb{F}}_p,$$

send P_1 and P_2 to P_∞ and P_0, respectively, and thus produce a curve of the form $y_p - y = \tilde{f}(x)$, where $\tilde{f}(x) \in \overline{\mathbb{F}}_p[x, x^{-1}]$ is a Laurent polynomial of degree d_1 in x and degree d_2 in x^{-1}. Let a be the coefficient of x^{d_1} in $\tilde{f}(x)$. As in the case $r = 0$, a scaling matrix of the form

$$\begin{pmatrix} a^{-1/p} & 0 \\ 0 & 1 \end{pmatrix}$$

yields an isomorphic curve $C_{\tilde{g}}$ where $\tilde{g}(x)$ is monic with respect to x.

The case $r = 2$ proceeds similarly. Suppose C_f has three or more poles. Let $P_{\mu_1}, P_{\mu_2}, P_{\mu_3}$ be poles of the three largest respective orders $d_1 \geq d_2 \geq d_3$. Then the matrices

$$\begin{pmatrix} \mu_3 - \mu_2 & \mu_2 \\ 0 & 1 \end{pmatrix} \quad \text{if } \mu_1 = \infty, \mu_2 \in \overline{\mathbb{F}}_p \text{ and } \mu_3 \in \overline{\mathbb{F}}_p,$$
$$\begin{pmatrix} \mu_1 & \mu_1 - \mu_3 \\ 1 & 0 \end{pmatrix} \quad \text{if } \mu_1 \in \overline{\mathbb{F}}_p, \mu_2 = \infty \text{ and } \mu_3 \in \overline{\mathbb{F}}_p,$$
$$\begin{pmatrix} -\mu_1 & \mu_2 \\ -1 & 1 \end{pmatrix} \quad \text{if } \mu_1 \in \overline{\mathbb{F}}_p, \mu_2 \in \overline{\mathbb{F}}_p \text{ and } \mu_3 = \infty,$$
$$\begin{pmatrix} \mu_1(\mu_3 - \mu_2) & \mu_2(\mu_1 - \mu_3) \\ \mu_3 - \mu_2 & \mu_1 - \mu_3 \end{pmatrix} \quad \text{if } \mu_i \in \overline{\mathbb{F}}_p \text{ for } i = 1, 2, 3,$$

send P_1, P_2, and P_3 to P_∞, P_0, and P_1, respectively. The same scaling matrix as in the case $r = 1$ produces a rational function $\tilde{g}(x)$ of the specified form that is monic in x.

Next, in both of the cases $r = 1$ and $r \geq 2$, we remove all the monomials of the form x^{kp} for $k \in \mathbb{Z}\backslash\{0\}$. The constant coefficient of $\tilde{g}(x)$ is handled separately at the end.

As described in [Far10, Proposition 2.1.1] and the subsequent discussion, removal of the monomials x^{kp} with $k > 0$ from $\tilde{g}(x)$ is accomplished iteratively via suitable translations of y, leaving x fixed. Suppose $\tilde{g}(x)$ contains a term of the form bx^{pk}, where $b \in \overline{\mathbb{F}}_p^\times$ and $1 \leq k \leq d_1 - 1$. Then the isomorphism $(x, y) \mapsto (x, y + b^{1/p}x^k)$

produces the curve $y^p - y = \tilde{g}(x) - bx^{kp} - e^{1/p}x^k$, where the term bx^{pk} no longer appears on the right-hand side. Systematically applying a finite number of suitable isomorphisms of this form, looping over the values of k in decreasing order from $k = \lfloor d_1/p \rfloor$ to $k = 1$, eliminates all monomials x^{kp} with $k > 0$. An analogous process, looping over all k from $-\lfloor (d_2 - 1)/p \rfloor$ to -1, eliminates all monomials x^{kp} with $k < 0$.

When $r = 1$, this process removes all monomials that are positive p-powers. So suppose that $r \geq 2$. Then for each pole P_μ with $\mu = 1$ or $\mu = \lambda_i$ $(4 \leq i \leq r + 1)$, we apply the isomorphisms $(x, y) \mapsto (x, y + b(x - \mu)^{-kp})$, where k runs from $-\lfloor (d_i - 1)/p \rfloor$ to -1 for $3 \leq i \leq r + 1$, to remove all terms involving powers $(x - \mu)^{kp}$ with $k < 0$.

We are at last left with a curve of the form $y^p - y = \tilde{g}(x)$, where no terms x^{pk} and $(x - \mu)^{-kp}$ for any pole $P_\mu \neq P_\infty$, with $k > 0$, appear in $\tilde{g}(x)$. It remains to eliminate the constant term $\tilde{g}(0)$. To that end, let $\gamma \in \overline{\mathbb{F}}_p$ such that $\gamma^p - \gamma = \tilde{g}(0)$. Then the isomorphism $(x, y) \mapsto (x, y + \gamma)$ yields the curve $y^p - y = g(x)$, where $g(x) = \tilde{g}(x) - \tilde{g}(0)$ is of the desired form. □

As mentioned above, the standard form of an Artin-Schreier curve is not unique. For example, the two standard form curves $y^3 - y = x^4 - x^2$ and $y^3 - y = x^4 + x^2$ are $\overline{\mathbb{F}}_3$-isomorphic via the isomorphism $(x, y) \mapsto (\sqrt{2}x, y)$; this is a special case of Proposition 4.1. However, for any non-exceptional curve, we see that the number of possible standard forms is finite and that the forms are easy to enumerate. In fact, the number of variable coefficients in each standard form is equal to the dimension of the component of the moduli space with the corresponding partition as described in Theorem 2.4.

In the next section, we discuss isomorphisms between Artin-Schreier curves in standard form.

3.1 Isomorphisms Between Curves in Standard Form

If we consider only Artin-Schreier curves in standard form, outside of the exceptional curves in Corollary 2.16, this narrows the possibilities for isomorphisms between them. Let

$$\varphi(x, y) = (M(x), \lambda y + h(x)), \tag{3.1}$$

with $M, \lambda, h(x)$ as in Corollary 2.16.

Lemma 3.4 *Let C_f and $C_{\tilde{f}}$ be isomorphic Artin-Schreier curves in standard form. Then for every choice of $M \in \mathrm{GL}_2(\overline{\mathbb{F}}_p)$ for an isomorphism $\varphi(x, y)$ as in (3.1) between these curves, there is a unique choice of $h(x)$ up to a constant in $\mathbb{F}_p$; that is, up to (multiple) composition with the isomorphism $\sigma : (x, y) \mapsto (x, y + 1)$.*

Proof We apply φ to the equation $y^p - y = f(x)$ to obtain $f(x) = \frac{1}{\lambda}(f(M(x)) - h^p(x) + h(x))$. This enforces a unique choice of $h(x)$ up to a constant in $\mathbb{F}_p$. □

For $M \in \mathrm{GL}_2(\overline{\mathbb{F}}_p)$, we define $h_M(x) \in \overline{\mathbb{F}}_p(x)$ to be a choice of polynomial from Lemma 3.4, up to composition with powers of σ. An isomorphism between standard forms is then determined, up to composition with powers of σ, by the choice of $M \in GL_2(\overline{\mathbb{F}}_p)$ and $\lambda \in \mathbb{F}_p^\times$. Thus we may without confusion define

$$\varphi_{\lambda,M}(x, y) = (M(x), \lambda y + h_M(x)). \tag{3.2}$$

In our application, we consider rational functions $f(x)$ which appear in the standard form of curves $y^p - y = f(x)$ corresponding to a given prime p and partition $\vec{E} = \{e_1, e_2, \dots, e_{r+1}\}$ of $D + 2$. As before, we assume that $e_{i+1} \le e_i$ for $1 \le i \le r$, where $d_i = e_i - 1$ for each i. Since genus and p-rank are invariant under isomorphism, isomorphic curves must correspond to the same partition $\vec{E}$. The isomorphism from one curve to another preserves the orders of distinct poles, but may change their locations.

The possible isomorphisms between non-exceptional Artin-Schreier curves in standard form can be determined by finding all $M \in \mathrm{GL}_2(\overline{\mathbb{F}}_p)$ and $\lambda \in \mathbb{F}_p^\times$ that preserve all the restrictions on poles and curve coefficients imposed by Theorem 3.3 as well as the partition $\vec{E}$. In particular, subject to these conditions, subsets of poles may be permuted and curve coefficients changed by such an isomorphism. We designed standard forms in such a way that the number of free coefficients of $f(x)$ in the right-hand side of a standard form curve (2.1) is equal to the dimension d of the corresponding irreducible component of $\mathcal{AS}_{g,s}$. Thus, the group G of isomorphisms between standard forms must be finite, and we are able to completely enumerate the images of non-exceptional curves under G. Let $a_1, a_2, \dots, a_d$ be the free coefficients of the standard form of some irreducible component. We can think of $\overline{\mathbb{F}}_p[a_1, a_2, \dots, a_d]$ as the space containing all standard form models of curves in $\mathcal{AS}_{g,s}$. The isomorphisms in G act on $\overline{\mathbb{F}}_p[a_1, a_2, \dots, a_d]$, and we seek a separating, or ideally, a reconstructing subset of the invariant ring $\overline{\mathbb{F}}_p[a_1, a_2, \dots, a_d]^G$. Finding a minimal reconstructing set of invariants allows us to essentially parameterize an irreducible component of $\mathcal{AS}_{g,s}$.

Remark 3.5 As just mentioned, the number of free coefficients of a standard form is equal to the dimension of the corresponding irreducible component of $\mathcal{AS}_{g,s}$. This number is also equal to the number of primary invariants for the action of the finite group G acting on the coefficients of the standard form.

3.2 Standard Forms of Exceptional Curves

In Corollary 2.16 we described the isomorphisms from an Artin-Schreier curve C when the function field $\overline{\mathbb{F}}_p(C)$ has a single index p subextension associated to a genus 0 curve. Such a subextension gives a model of the curve of the form $y^p - y = f(x)$ and we discussed its standard forms in Sect. 3.1. We now investigate what standard forms can arise for the cases in which such an index p subextension is not unique. For each order p automorphism σ of C such that the quotient curve $C/\langle\sigma\rangle$ has genus 0, the degree p extension $\overline{\mathbb{F}}_p(C)/\overline{\mathbb{F}}_p(C/\langle\sigma\rangle)$ produces a way to write C as $y^p - y = f(x)$ where $y \in \overline{\mathbb{F}}_p(C)\backslash\overline{\mathbb{F}}_p(C/\langle\sigma\rangle)$ and $f(x) \in \overline{\mathbb{F}}_p(C/\langle\sigma\rangle)$.

Lemma 3.6 *Let $C : y^p - y = \frac{a}{x^p - x}$ be a curve as given in case* (2.16) *of Corollary 2.16. Then any standard form of C can be obtained by applying an isomorphism of the form in* (2.5) *through the procedure in Theorem 3.3 from the given model of C.*

Proof The automorphism group of C is isomorphic to $((C_p \times C_p) \rtimes C_{p-1}) \rtimes C_2$, generated by

$$\begin{aligned}(1, 0, 0, 0)(x, y) &= (x + 1, y),\\ (0, 1, 0, 0)(x, y) &= (x, y + 1),\\ (0, 0, 1, 0)(x, y) &= (\mu x, y/\mu) \text{ with } \mu \in \overline{\mathbb{F}}_p^\times \text{ and}\\ (0, 0, 0, 1)(x, y) &= (y, x).\end{aligned}$$

Any subgroup of order p consists of elements of the form $(\alpha, \beta, 0, 0)$ and is generated by an automorphism of the form $(0, 1, 0, 0)$ or $(1, \beta, 0, 0)$. The first of these does not change the curve equation. The second one produces a degree p extension $\overline{\mathbb{F}}_p(x, y)/\overline{\mathbb{F}}_p(y - \beta x, y^p - y)$ that is isomorphic to $\overline{\mathbb{F}}_p(\mathbb{P}^1)$ if $\beta = 0$; if $\beta \neq 0$, it defines the hyperelliptic curve of genus $p - 1$ given by $v^p - v = \frac{a}{u} - \beta u$. So once again we obtain the original equation $x^p - x = \frac{a}{y^p - y}$ and the result holds. □

A result from [Kon09] explicitly describes the isomorphism classes of these curves in terms of the non-standard model.

Lemma 3.7 ([Kon09, Lemma 3.1]) *Let $a, a' \in \overline{\mathbb{F}}_p^\times$. Then $y^p - y = \frac{a}{x^p - x}$ is isomorphic to $y^p - y = \frac{a'}{x^p - x}$ if and only if $a = \lambda a'$ for some $\lambda \in \mathbb{F}_p^\times$.*

Lemma 3.8 *Let $C : y^p - y = \frac{1}{x^\lambda}$. with $\lambda \mid p + 1$, be a curve as given in case* (2.16) *of Corollary 2.16. Then any standard form of C can be obtained by applying an isomorphism of the form in* (2.5) *through the procedure of Theorem 3.3 from the given model of C. In particular, the only standard form of C is $y^p - y = x^\lambda$.*

Proof The automorphisms of this curve are described in Lemma 2.19. When $\lambda \neq p + 1$, the order p subgroups are generated by

$$(x, y) \mapsto (x, y + 1) \text{ and } (x, y) \mapsto \left(\frac{x}{(y + 1)^{(p+1)/\lambda}}, \frac{y}{y + 1}\right).$$

They all produce curves of the form $y^p - y = x^\lambda$. When $\lambda = p + 1$, there are extra order p subgroups. Rewriting the equation as $y^p - y = x^{p+1}$, it suffices to consider the automorphisms $(x, y) \mapsto (x + a^p, y + ax + b)$ with $a^{p^2} + a = 0$ and $b^p - b - a^{p+1} = 0$, producing the degree p subextension $\overline{\mathbb{F}}_p(x^p + x, P)$ where P is the product of all the images of y under these automorphisms. This subextention is not isomorphic to $\overline{\mathbb{F}}_p(x)$, and hence neither to $\overline{\mathbb{F}}_p(\mathbb{P}^1)$. If it were, C would be birational to $x^p + x = 0$, but this curve is p copies of $\mathbb{P}^1$ and hence reducible. □

Remark 3.9 The upshot of Lemmas 3.6 and 3.8 is that there are only a finite number of standard forms for each isomorphism class of exceptional curves, and they can be easily enumerated. This confirms again that the group of isomorphisms between exceptional curves in standard form is finite.

4 Invariant Computations

In this section we determine invariants for curves within each stratum of $\mathcal{AS}_g$ for Artin-Schreier curves of genus $g = 3, 4$ in odd characteristic. As we see in Table 2, there is a single irreducible component in each stratum in these cases, so each stratum has a unique system of invariants.

4.1 Genus 3, Characteristic 3

In this case, the quantity D defined in (2.2) takes on the value $D = 3$. There are two partitions of $D + 2 = 5$ (up to reordering) that satisfy the conditions of Theorem 2.14, namely

$$\{5\} \quad \text{and} \quad \{3, 2\}, \tag{4.1}$$

so there are two strata in $\mathcal{AS}_g$. These partitions also determine the standard forms of all possible genus 3 Artin-Schreier curves in characteristic 3. Recall that $d_i = e_i - 1$ is the order of the pole P_{μ_i} of $f(x)$ for the Artin-Schreier curve $C_f : y^3 - y = f(x)$. The partitions in (4.1) show that C_f has either one pole of order 4 or two poles with respective orders 2 and 1. We treat these two cases separately below.

4.1.1 One Pole of Order 4

We use (2.4) in Theorem 2.14 with $D = 3$, $r = 0$, and $e_1 = 5$ to determine that the dimension of this stratum is

$$\dim \mathcal{AS}_{3,\{5\}} = 3 - 1 - \lfloor 4/3 \rfloor = 1.$$

We have $r = 0$ and $d_1 = 4 \equiv 1 \pmod 3$, corresponding to Case (1) of Theorem 3.3 and yielding a standard form

$$C : y^3 - y = x^4 + ax^2, \tag{4.2}$$

with $a \in \overline{\mathbb{F}}_3$. When $a = 0$, this curve is isomorphic to an exceptional curve of type (2.16) in Corollary 2.16 which has standard form $y^3 - y = x^4$. This is the only exceptional curve in the family, see Lemma 3.8.

Proposition 4.1 *Every isomorphism between curves in standard form as in* (4.2) *with $a \neq 0$ is given, up to composition with powers of $\sigma : (x, y) \mapsto (x, y+1)$, by*

$$(x, y) \mapsto (\alpha x, \lambda y), \tag{4.3}$$

where $\lambda \in \mathbb{F}_3^\times$ and $\alpha \in \overline{\mathbb{F}}_3$ with $\alpha^4 = \lambda$.

Proof Let C be as in Eq. (4.2). By Corollary 2.16, disregarding composition with powers of σ, every isomorphism φ of C is given by $\varphi_{\lambda,M}$ as in (3.2) with $M = \begin{pmatrix} \alpha & \beta \\ \gamma & \delta \end{pmatrix} \in \mathrm{GL}_2(\overline{\mathbb{F}}_3)$. Since the image of C must be in standard form, $\varphi_{\lambda,M}$ must not move the pole at ∞, so $\gamma = 0$ and we can assume without loss of generality that $\delta = 1$. Since the coefficient of x in $\varphi_{\lambda,M}(C)$ must vanish, we conclude that $\beta = 0$, and that the unique polynomial $h(x)$ that produces a curve in standard form is $h(x) = 0$. So $\varphi_{\lambda,M}$ is of the form $(x, y) \mapsto (\alpha x, \lambda y)$. and yields the curve

$$\lambda(y^3 - y) = \alpha^4 x^4 + a\alpha^2 x^2.$$

Since the right-hand side must be monic in the standard form, we have $\alpha^4 = \lambda$. □

In the language of group actions, we now have a finite group $G \simeq \mathbb{Z}/4\mathbb{Z}$ acting on $\overline{\mathbb{F}}_3[a]$. This action is via isomorphisms of curve models of the form $y^3 - y = x^4 + ax^2$, sending $a \mapsto \epsilon a$ with $\epsilon^4 = 1$. We see that G acts linearly on $\overline{\mathbb{F}}_3[a]$.

Corollary 4.2 *The element $I_1 = a^4$ is an invariant and generates the ring of invariants for Artin-Schreier curves of genus* 3 *in characteristic* 3 *with* 3*-rank* 0*, i.e.,* $\overline{\mathbb{F}}_3[a]^G = \overline{\mathbb{F}}_3[I_1]$.

Proof It is straightforward to check that I_1 is an invariant for the action of G and forms a reconstructing set for the family (namely, different choices of a such that $a^4 = I_1$ for $y^3 - y = x^4 + ax^2$ produce isomorphic curves in standard form). By Corollary 2.13 we have $\overline{\mathbb{F}}_3[a]^G = \widetilde{\overline{\mathbb{F}}_3[I_1]} = \overline{\mathbb{F}}_3[I_1]$. □

Remark 4.3 This corollary can also be proved by computing a set of generators for $\overline{\mathbb{F}}_3[a]^G$ using Magma; or by manually checking for invariants up to degree 4 (see Lemma 2.8).

Remark 4.4 If we start with a model $y^3 - y = ax^4 + bx^3 + cx^2 + dx + e = p(x)$, the invariant I_1 (of a curve in the family $y^3 - y = x^4 + ax^2$ isomorphic to the starting curve) is given by c^4/a^2. More generally, if we start with $y^3 - y = \frac{ax^4+bx^3+cx^2+dx+e}{(x-\lambda)^4}$, the reconstructing invariant can be chosen to be $I_1 = c^4/(a\lambda^4 + b\lambda^3 + c\lambda^2 + d\lambda + e)^2$.

4.1.2 One Pole of Order 2 and One Pole of Order 1

We use (2.4) in Theorem 2.14 with $D = 3$, $r = 1$, $e_1 = 3$, and $e_2 = 2$ to determine that the dimension of this stratum is

$$\dim \mathcal{AS}_{3,\{3,2\}} = 3 - 1 - \lfloor 2/3 \rfloor - \lfloor 1/3 \rfloor = 2.$$

Based on Theorem 3.3, the standard form of curves with a pole of order 2 and a second pole of order 1 is

$$C\colon y^3 - y = x^2 + ax + \frac{b}{x} \tag{4.4}$$

for $a, b \in \overline{\mathbb{F}}_3$ with $b \neq 0$.

Proposition 4.5 *Every isomorphism between curves in standard form as in* (4.4) *is given, up to composition with powers of* $\sigma : (x, y) \mapsto (x, y + 1)$, *by*

$$(x, y) \mapsto (\alpha x, \lambda y), \tag{4.5}$$

where $\lambda \in \mathbb{F}_3^\times$ *and* $\alpha \in \overline{\mathbb{F}}_3$ *with* $\alpha^2 = \lambda$.

Proof No curve in this family is exceptional, so we can directly use Corollary 2.16 to see that, disregarding composition with powers of σ, every isomorphism is of the form $\varphi_{\lambda,M}$ as in (3.2) with $M = \begin{pmatrix} \alpha & \beta \\ \gamma & \delta \end{pmatrix} \in \mathrm{GL}_2(\overline{\mathbb{F}}_3)$. In this particular case, each isomorphism must fix ∞ and 0, so $\gamma = \beta = 0$, and we can choose δ to be equal to 1. This implies that $h(x) = 0$, so the only isomorphisms of C as in (4.4) are of the form $(x, y) \mapsto (\alpha x, \lambda y)$. Applying the isomorphism to C and using the requirement that the right-hand side of a standard form must be monic yields $\alpha^2 = \lambda$, so there are four possible values for α since $\lambda \in \mathbb{F}_3^\times$. This produces the model

$$\widetilde{C}\colon y^3 - y = x^2 + a'x + \frac{b'}{x}, \tag{4.6}$$

where $a' = a/\alpha$ and $b' = b/\alpha^3$. Since a and b can take on any values in $\overline{\mathbb{F}}_3$, we obtain a two-dimensional family of curves from the isomorphism in (4.5). Since the stratum is two-dimensional, this isomorphism yields all curves with the prescribed poles and orders. □

These isomorphisms give a group action of the finite group $G \simeq \mathbb{Z}/4\mathbb{Z}$ acting linearly on $\overline{\mathbb{F}}_3[a, b]$. The action is generated by $(a, b) \mapsto (ia, -ib)$, where $i^2 = -1$.

Corollary 4.6 *The elements* $I_1 = a^4$, $I_2 = ab$, *and* $I_3 = b^4$ *generate the ring of invariants for Artin-Schreier curves of genus* 3 *in characteristic* 3 *with* 3*-rank* 2, *i.e.,* $\overline{\mathbb{F}}_3[a, b]^G = \overline{\mathbb{F}}_3[I_1, I_2, I_3]$. *These elements satisfy the algebraic relation* $I_1 I_3 - I_2^4 = 0$.

Proof Let a' and b' denote the coefficients of $1/x$ and x in (4.6), respectively. Then $(a')^4 = (a/\alpha)^4 = a^4$ and $(b')^4 = (b/\alpha^3)^4 = b^4$ for all $\alpha \in \overline{\mathbb{F}}_3$. Hence I_1 and I_3 are invariants. Furthermore, $a'b' = ab/\alpha^4 = ab$, so I_2 is also an invariant. On the other hand, given values $\{I_1, I_2, I_3\}$ and choosing $a, b \in \overline{\mathbb{F}}_3$ with $a^4 = I_1$, $b^4 = I_3$ and $ab = I_2$, we can reconstruct a unique standard form model as in (4.4). Different choices of a and b yield isomorphic models. By Corollary 2.13 this implies that $\overline{\mathbb{F}}_3[a, b]^G = \overline{\mathbb{F}}_3[I_1, I_2, I_3]$. □

Remark 4.7 Given a more general model $y^3 - y = \frac{ax+b}{x-\lambda_1} + \frac{cx^2+dx+e}{(x-\lambda_2)^2}$, the set

$$\left\{\frac{(c\lambda_1\lambda_2 - d(\lambda_1+\lambda_2)+e)^4}{(c\lambda_2^2+d\lambda_2+e)^2}, \frac{(a\lambda_1+b)^4(c\lambda_2^2+d\lambda_2+e)^2}{(\lambda_1-\lambda_2)^8}, \frac{(c\lambda_1\lambda_2-d(\lambda_1+\lambda_2)+e)(a\lambda_1+b)}{(\lambda_1-\lambda_2)^2}\right\}$$

is a reconstructing set of invariants.

4.2 Genus 3, Characteristic 7

There is only one component in this stratum since the only partition is $\{3\}$. In this case, there is only one pole of order 2 and a unique standard form:

$$y^7 - y = x^2.$$

Note that this is one of the exceptional curves in case (2) of Corollary 2.16.

4.3 Genus 4, Characteristic 3

This case has three strata, consisting of curves with respective 3-ranks 0, 2, and 4.

4.3.1 A Single Pole of Order 5

In this case the 3-rank is $s = 0$ by (2.3), and we have $\vec{E} = \{6\}$. This corresponds to a curve C_f where $f(x)$ has one pole of order 5. The standard form as given in Theorem 3.3 is:

$$C : y^3 - y = x^5 + cx^4 + dx^2 \tag{4.7}$$

for some $c, d \in \overline{\mathbb{F}}_3$. Note that none of the exceptional curves have a standard form of this type, so Corollary 2.16 applies to this stratum.

Identifying isomorphisms between standard forms in this case is cumbersome. However, it is fairly straightforward to compute invariants from a different model of the curve. We first apply isomorphisms to reach a different distinguished form, compute invariants using this model, and then express these invariants in terms of the coefficients in our standard form.

Lemma 4.8 *Every curve with standard form C given in* (4.7) *is isomorphic to one of the form*

$$C_g : y^3 - y = x^5 + ax^2 + bx \tag{4.8}$$

for some $a, b \in \overline{\mathbb{F}}_3$. Specifically, $a = c^3 + d$ and $b = (-cd - \epsilon^2)$, where $\epsilon^3 = c$.

Proof We apply the isomorphism $\varphi_{\lambda,M}$ from (3.2) to C, where $\lambda = 1$ and $M = \begin{pmatrix} 1 & c \\ 0 & 1 \end{pmatrix}$, to obtain

$$y^3 - y = x^5 - c^2x^3 + (c^3 + d)x^2 - cdx + (-c^5 + c^2d) =: \tilde{f}(x).$$

Next, we determine a polynomial $h(x) = h_1x + h_2 \in \overline{\mathbb{F}}_3[x]$ such that $h(x)^3 - h(x)$ has the same x^3-term and constant term as $\tilde{f}$. Let $\epsilon \in \overline{\mathbb{F}}_3$ with $\epsilon^3 = c$. Comparing coefficients at x^3 yields $h_1^3 = -c^2$, so $h_1 = -\epsilon^2$. Comparing constant coefficients defines h_2 via the relation $h_2^3 - h_2 = -c^5 + c^2d$. Then C_f is isomorphic to

$$\begin{aligned} C_g : y^3 - y &= x^5 - c^2x^3 + (c^3 + d)x^2 - cdx + (-c^5 + c^2d) - \left((h_1x + h_2)^3 - (h_1x + h_2)\right) \\ &= x^5 + (c^3 + d)x^2 + (-cd - \epsilon^2)x. \end{aligned}$$

□

Proposition 4.9 *Every isomorphism between curves defined as in* (4.8) *is given, up to composition with powers of $\sigma : (x, y) \mapsto (x, y + 1)$, by*

$$(x, y) \mapsto (\alpha x, \alpha^5 y)$$

for some $\alpha \in \overline{\mathbb{F}}_3$ with $\alpha^{10} = 1$.

Proof Let $M := \begin{pmatrix} \alpha & \beta \\ \gamma & \delta \end{pmatrix} \in \mathrm{GL}_2(\overline{\mathbb{F}}_3)$. Any isomorphism $\varphi_{\lambda,M}$ between curves of the form (4.8) must preserve the pole at infinity, implying $\gamma = 0$. Further, the requirement of a vanishing x^4 term forces $\beta = 0$, and thus $h(x) = 0$. The fact that the resulting form must be monic in x necessitates $\lambda^{-1}\alpha^5/\delta^5 = 1$. Without loss of generality, assume that $\delta = 1$, so $\lambda = \alpha^5$. Since $\lambda^2 = 1$, we must have $\alpha^{10} = 1$. □

These isomorphisms yield a group action of the finite group $G \simeq \mathbb{Z}/5\mathbb{Z}$ acting linearly on $\overline{\mathbb{F}}_3[a, b]$, generated by $(a, b) \mapsto (a/\alpha^3, b/\alpha^4)$ with $\alpha^{10} = 1$.

Corollary 4.10 *The elements $I_1 = a^{10}$, $I_2 = b^5$, and $I_3 = a^2b$ generate the ring of invariants for Artin-Schreier curves of genus* 4 *in characteristic* 3 *with* 3*-rank* 0*, i.e., $\overline{\mathbb{F}}_3[a, b]^G = \overline{\mathbb{F}}_3[I_1, I_2, I_3]$. These elements satisfy the relation $I_1I_2 = I_3^5$.*

Proof For a curve C_g as in (4.8), we apply an isomorphism as given in Proposition 4.9 to obtain

$$C_{\tilde{g}} : y^3 - y = x^5 + \frac{a}{\alpha^3}x^2 + \frac{b}{\alpha^4}x.$$

It is easy to check that the given functions I_1, I_2, I_3. are invariant under this action. Conversely, given values $\{I_1, I_2, I_3\}$ with $I_1 I_2 = I_3^5$, we can choose $a, b \in \overline{\mathbb{F}}_3$ such that $a^{10} = I_1$, $b^5 = I_2$, in which case $a^2 b = I_3$. Suppose $I_1 = \tilde{a}^{10}$ and $I_2 = \tilde{b}^5$ with $\tilde{a}, \tilde{b} \in \overline{\mathbb{F}}_3$, and let $\alpha \in \overline{\mathbb{F}}_3$ be such that $\alpha^3 = \tilde{a}/a$. The relation $a^{10} = \tilde{a}^{10}$ implies that $\alpha^{10} = 1$, so $\alpha^4 = \tilde{a}^2/a^2$. Since $a^2 b = \tilde{a}^2 \tilde{b}$, we have $b/\alpha^4 = \tilde{b}$. Hence, the curves $y^3 - y = x^5 + ax^2 + bx$ and $y^3 - y = x^5 + \tilde{a}x^2 + \tilde{b}x$ are isomorphic via the isomorphism in Proposition 4.9 given by α with $\alpha^3 = \tilde{a}/a$. Corollary 2.13 now implies that I_1, I_2, and I_3 generate the invariant ring. □

By applying the change of variables from Lemma 4.8, we obtain the following.

Corollary 4.11 *Two curves in standard form as in* (4.7) *are isomorphic if and only if they have the same invariants* $I_1 = (c^3 + d)^{10}$, $I_2 = (-cd - \epsilon^2)^5$, *and* $I_3 = (c^3 + d)^2(-cd - \epsilon^2)$, *where* $\epsilon^3 = c$.

4.3.2 Two Poles of Order 2

This case corresponds to the partition $\vec{E} = \{3, 3\}$. According to Theorem 3.3, the curves in this stratum have standard form

$$C : y^3 - y = x^2 + ax + \frac{b}{x} + \frac{c}{x^2} \tag{4.9}$$

for $a, b, c \in \overline{\mathbb{F}}_3$ with $c \neq 0$.

Proposition 4.12 *Every isomorphism between curves in standard form as in* (4.9) *is given, up to composition with powers of* $\sigma : (x, y) \mapsto (x, y + 1)$, *by*

$$(x, y) \mapsto (\alpha x, \alpha^3 y), \tag{4.10}$$

where $\alpha^4 = 1$, *or by*

$$(x, y) \mapsto \left(\frac{1}{\gamma x}, \lambda y\right), \tag{4.11}$$

where $\lambda \in \mathbb{F}_3^\times$ *and* $c\gamma^2 = \lambda$.

Proof Let C be as in Eq. (4.9). Any isomorphism of C that yields another curve in standard form must either fix P_∞ and P_0 or swap these poles. We first consider the case when both poles are fixed. In this situation, once again disregarding composition by powers of σ, the isomorphism is of the form $\varphi_{\lambda, M_\alpha}$ for $M_\alpha = \begin{pmatrix} \alpha & 0 \\ 0 & 1 \end{pmatrix}$, with $\lambda \in \mathbb{F}_3^\times$ and $\alpha \in \overline{\mathbb{F}}_3^\times$. Applying $\varphi_{M_\alpha, \lambda}$ to C gives

$$y^3 - y = \frac{\alpha^2 x^2}{\lambda} + \frac{a\alpha x}{\lambda} + \frac{b}{\lambda \alpha x} + \frac{c}{\lambda \alpha^2 x^2}.$$

Since the image curve must be in standard form, we have $\alpha^2 = \lambda$, so $\alpha^4 = 1$. This isomorphism acts on the standard form model by $(a, b, c) \mapsto \left(\frac{1}{\alpha}a, \alpha b, c\right)$.

In the second case, when the poles are switched, the isomorphism must be of the form $\varphi_{\lambda, M_\gamma}$ for $M_\gamma = \begin{pmatrix} 0 & 1 \\ \gamma & 0 \end{pmatrix}$, with $\lambda \in \mathbb{F}_3^\times$ and $\gamma \in \overline{\mathbb{F}}_3^\times$. The image of C under $\varphi_{M_\gamma, \lambda}$ is

$$y^3 - y = \frac{c\gamma^2 x^2}{\lambda} + \frac{b\gamma x}{\lambda} + \frac{a}{\lambda \gamma x} + \frac{1}{\lambda \gamma^2 x^2}$$

so $c\gamma^2 = \lambda$. Since $\lambda^2 = 1$, the isomorphism acts on the coefficients of a curve in standard form given by Eq. (4.9) via $(a, b, c) \mapsto \left(\frac{\gamma b}{\lambda}, \frac{\lambda a}{\gamma}, c\right)$. □

The isomorphisms in Proposition 4.12 define a non-linear group action of the dihedral group $G \simeq D_4$ of order 8 on the function field $\overline{\mathbb{F}}_3(a, b, c)$. It cannot a priori be descended to the polynomial ring $\overline{\mathbb{F}}_3[a, b, c]$. We proceed as follows to obtain invariants in this case. With the notation in Eq. (4.11) in Proposition 4.12, we have $\left(\frac{\lambda}{\gamma}\right)^2 = \lambda c = \pm c$. This defines a linear action of the dihedral group D_4 of order 8 on the polynomial ring $\overline{\mathbb{F}}_3(\sqrt{c})[a, b]$ over the field $\overline{\mathbb{F}}_3(\sqrt{c})$. It is generated by $(a, b) \mapsto (ia, -ib)$ where $i^2 = -1$ and $(a, b) \mapsto \left(\frac{b}{\sqrt{c}}, \sqrt{c}a\right)$.

Lemma 4.13 *The elements $I_2 = ab$ and $I_3 = a^4c^2 - b^4$ generate $\overline{\mathbb{F}}_3(\sqrt{c})[a, b]^G$.*

Corollary 4.14 *Two curves in standard form as in Eqs.* (4.9) *are isomorphic if and only if they have the same invariants $I_1 = c$, $I_2 = ab$, and $I_3 = a^4c^2 - b^4$.*

4.3.3 Three Poles of Order 1

We now consider the stratum $\mathcal{AS}_{4,4}$ for $p = 3$. There is a single partition yielding $g = 4$ and $s = 4$, namely $\vec{E} = \{2, 2, 2\}$. From Table 2, we see that this component has dimension $\dim \mathcal{AS}_{4,\{2,2,2\}} = 3$. From Theorem 3.3, the standard form of a curve in this component is

$$C : y^3 - y = ax + \frac{b}{x} + \frac{c}{x - 1}, \tag{4.12}$$

where $a, b, c \in \overline{\mathbb{F}}_3^\times$.

Note that there is a subfamily of exceptional curves (see case (2.16) of Corollary 2.16) within this stratum, of the form $C_a = y^3 - y = \frac{a}{x^3 - x}$ for $a \in \overline{\mathbb{F}}_3^\times$. A standard form for this curve is given by $y^3 - y = -ax + \frac{a}{x} + \frac{a}{x-1}$.

Proposition 4.15 *Let C be as in* (4.12)*. The curves in standard form that are isomorphic to C are the ones in Table 3.*

Proof Due to Corollary 2.16 and Lemma 3.6, isomorphisms, up to composition with powers of $\sigma : (x, y) \mapsto (x, y+1)$, of these standard forms must be of the form

$$\varphi_{\lambda,M} : (x, y) \mapsto (Mx, \lambda y + h_M(x)),$$

where $\lambda \in \mathbb{F}_3^\times = \{\pm 1\}$ and $M \in \mathrm{GL}_2(\overline{\mathbb{F}}_3)$ permutes the poles P_∞, P_0, and P_1 in an S_3 action. This gives 6 choices of M and 2 choices of λ, yielding at most 12 labeled models. These 12 models are all distinct, and the action of the isomorphisms on C is described in Table 3. We compute a sample entry to demonstrate the process. To

Table 3 Action of the isomorphism $\varphi_{\lambda,M}$ on the coefficients of (4.12)

λ	Permutation	Matrix M	$f(x)$
1	(P_∞)	$\begin{pmatrix} 1 & 0 \\ 0 & 1 \end{pmatrix}$	$ax + \frac{b}{x} + \frac{c}{x-1}$
1	$(P_\infty P_0)$	$\begin{pmatrix} 0 & 1 \\ 1 & 0 \end{pmatrix}$	$bx + \frac{a}{x} - \frac{c}{x-1}$
1	$(P_\infty P_1 P_0)$	$\begin{pmatrix} 0 & -1 \\ 1 & -1 \end{pmatrix}$	$-bx + \frac{c}{x} - \frac{a}{x-1}$
1	$(P_\infty P_0 P_1)$	$\begin{pmatrix} -1 & 1 \\ -1 & 0 \end{pmatrix}$	$-cx - \frac{a}{x} + \frac{b}{x-1}$
1	$(P_0 P_1)$	$\begin{pmatrix} 1 & -1 \\ 0 & -1 \end{pmatrix}$	$-ax - \frac{c}{x} - \frac{b}{x-1}$
1	$(P_\infty P_1)$	$\begin{pmatrix} -1 & 0 \\ -1 & 1 \end{pmatrix}$	$cx - \frac{b}{x} + \frac{a}{x-1}$
-1	(P_∞)	$\begin{pmatrix} 1 & 0 \\ 0 & 1 \end{pmatrix}$	$-ax - \frac{b}{x} - \frac{c}{x-1}$
-1	$(P_\infty P_0)$	$\begin{pmatrix} 0 & 1 \\ 1 & 0 \end{pmatrix}$	$-bx - \frac{a}{x} + \frac{c}{x-1}$
-1	$(P_\infty P_1 P_0)$	$\begin{pmatrix} 0 & -1 \\ 1 & -1 \end{pmatrix}$	$bx - \frac{c}{x} + \frac{a}{x-1}$
-1	$(P_\infty P_0 P_1)$	$\begin{pmatrix} -1 & 1 \\ -1 & 0 \end{pmatrix}$	$cx + \frac{a}{x} - \frac{b}{x-1}$
-1	$(P_0 P_1)$	$\begin{pmatrix} 1 & -1 \\ 0 & -1 \end{pmatrix}$	$ax + \frac{c}{x} + \frac{b}{x-1}$
-1	$(P_\infty P_1)$	$\begin{pmatrix} -1 & 0 \\ -1 & 1 \end{pmatrix}$	$-cx + \frac{b}{x} - \frac{a}{x-1}$

exchange P_∞ and P_0 while fixing P_1, we solve using the reasoning in Sect. 3.1 to find $M = \begin{pmatrix} 0 & 1 \\ 1 & 0 \end{pmatrix}$. The effect of M on $f(x) = ax + \frac{b}{x} + \frac{c}{x-1}$ is

$$\begin{aligned} M(f(x)) &= \frac{a}{x} + bx + \frac{c}{\frac{1}{x} - 1} \\ &= bx + \frac{a}{x} + \frac{-c}{x-1} - c. \end{aligned}$$

We convert the corresponding curve equation to standard form, which is accomplished by sending $y \mapsto y + m$, where $m \in \overline{\mathbb{F}}_3$ such that $m^3 - m = -c$. This yields the standard form

$$\varphi_{\lambda,M}(C) : y^3 - y = bx + \frac{a}{x} + \frac{-c}{x-1}.$$ □

These isomorphisms define a linear action of a group $G \simeq D_6$, the dihedral group of 12 elements, generated by $(a, b, c) \mapsto (c, a, -b)$ and $(a, b, c) \mapsto (b, a, -c)$ on the polynomial ring $\overline{\mathbb{F}}_3[a, b, c]$.

Corollary 4.16 *The elements $I_1 = (abc)^2$, $I_2 = (abc)(a - b - c)$, $I_3 = ab + ac - bc$, and $I_4 = a^2 + b^2 + c^2$ generate the ring of invariants for Artin-Schreier curves of genus 4 in characteristic 3 with 3-rank 4, i.e., $\overline{\mathbb{F}}_3[a, b, c]^G = \overline{\mathbb{F}}_3[I_1, I_2, I_3, I_4]$. These elements satisfy the relation $I_1(I_3 + I_4) = I_2^2$.*

Proof A model as in (4.12) can be reconstructed from this set of invariants through a naive solve-and-back-substitution procedure that results in a polynomial equation of degree 6. Since all the roots of this equation belong to an (at most degree 6) extension of the (finite) field of definition of the invariants, there is a solution that defines coefficients for (4.12). □

Remark 4.17 The exceptional curves in this family occur exactly when $I_2 = I_3 = I_4 = 0$.

4.4 Genus 4, Characteristic 5

In this case there are two strata.

4.4.1 One Pole of Order 3

Here, the 5-rank is $s = 0$ and we have $\vec{E} = \{4\}$. This corresponds to a curve C_f where f has one pole of order 3, with standard form as given in case (1) of Theorem 3.3:

$$C : y^5 - y = x^3 + ax^2 \tag{4.13}$$

with $a \in \overline{\mathbb{F}}_5$. When $a = 0$, the curve has additional automorphisms as in Lemma 2.19.

Proposition 4.18 *Every isomorphism between curves in standard form as in* (4.13) *is given, up to composition with powers of* $\sigma : (x, y) \mapsto (x, y+1)$, *by* $\varphi_{\lambda,M}$ *with* $\lambda = \alpha^3$ *and* $M = \begin{pmatrix} \alpha & \beta \\ 0 & 1 \end{pmatrix}$, *where* $\alpha \in \overline{\mathbb{F}}_5$ *with* $\alpha^{12} = 1$ *and* $\beta \in \left\{0, -\frac{2a}{3}\right\}$.

Proof Let $M := \begin{pmatrix} \alpha & \beta \\ \gamma & \delta \end{pmatrix} \in \mathrm{GL}_2(\overline{\mathbb{F}}_5)$. Any isomorphism $\varphi_{\lambda,M}$ between standard forms given by (4.13) must preserve the pole at infinity, so $\gamma = 0$. Without loss of generality, assume $\delta = 1$. We then obtain the image curve with model

$$y^5 - y = \frac{1}{\lambda}\alpha^3 x^3 + (3\alpha^2\beta + a\alpha^2)x^2 + (3\alpha\beta^2 + 2a\alpha\beta)x + a\beta^2.$$

Since the right-hand side must be monic, we have $\alpha^3 = \lambda$. Since $\lambda \in \mathbb{F}_5^\times$, $\lambda^4 = 1$, so $\alpha^{12} = 1$. Furthermore, since the standard form requires a vanishing x-term, it follows that $3\alpha\beta^2 + 2a\alpha\beta = 0$, so $\beta = -\frac{2a}{3}$ or $\beta = 0$. Finally, we choose an appropriate $h(x) = h_0 \in \overline{\mathbb{F}}_5$ to eliminate the constant term $a\beta^2$. □

We obtain a linear action of a group $G \simeq \mathbb{Z}/12\mathbb{Z}$ generated by $a \mapsto \alpha a$ with $\alpha^{12} = 1$ on the polynomial ring $\overline{\mathbb{F}}_5[a]$.

Corollary 4.19 *The element* $I_1 = a^{12}$ *generates the ring of invariants for Artin-Schreier curves of genus* 4 *in characteristic* 5 *with* 5*-rank* 0, *i.e.,* $\overline{\mathbb{F}}_5[a]^G = \overline{\mathbb{F}}_5[I_1]$.

Proof It is straightforward to check that $\{I_1\}$ is invariant and a reconstructing set. Now apply Corollary 2.13. □

4.4.2 Two Poles of Order 1

In this case the 5-rank is $s = 4$ and we have $\vec{E} = \{2, 2\}$. This corresponds to a curve C_f where f has two poles of order 1, with standard form as given in case (2) of Theorem 3.3:

$$y^5 - y = x + \frac{a}{x} \tag{4.14}$$

with $a \in \overline{\mathbb{F}}_5^\times$.

Proposition 4.20 *Every isomorphism between curves in standard form as in* (4.14) *is given, up to composition with powers of* $\sigma : (x, y) \mapsto (x, y+1)$, *by*

$$(x, y) \mapsto (\lambda x, \lambda y) \tag{4.15}$$

or by

$$(x, y) \mapsto \left(\frac{a}{\lambda x}, \lambda y\right), \tag{4.16}$$

where $\lambda \in \mathbb{F}_5^\times$.

Proof Let $M := \begin{pmatrix} \alpha & \beta \\ \gamma & \delta \end{pmatrix} \in \mathrm{GL}_2(\overline{\mathbb{F}}_5)$ and $\lambda \in \mathbb{F}_5^{\times}$. Any isomorphism $\varphi_{\lambda,M}$ between standard forms here must either fix the poles P_∞ and P_0 or swap them. So, up to scalar multiple, M must be of the form $M_\alpha = \begin{pmatrix} \alpha & 0 \\ 0 & 1 \end{pmatrix}$ or $M_\gamma = \begin{pmatrix} 0 & 1 \\ \gamma & 0 \end{pmatrix}$. Let $f(x) = x + \frac{a}{x}$. In the first case, we obtain

$$\varphi_{\lambda,M_\alpha}(C_f) : y^5 - y = \frac{\alpha}{\lambda}x + \frac{a}{\alpha\lambda x}.$$

Since the polynomial part on the right-hand side must be monic, we must have $\alpha = \lambda$, leading to the standard form $y^5 - y = x + \frac{a}{\lambda^2 x}$ for $\varphi_{\lambda,M_\alpha}(C_f)$. So the isomorphism acts on the standard form model by sending $a \mapsto \frac{a}{\lambda^2} = \lambda^2 a$. In the second case, the image curve is

$$\varphi_{\lambda,M_\gamma}(C_f) : y^5 - y = \frac{1}{\lambda\gamma x} + \frac{a\gamma x}{\lambda}.$$

Again, since the polynomial part must be monic in standard form, we have $a\gamma = \lambda$, so the standard form of $\varphi_{\lambda,M_\gamma}(C_f)$ can be simplified to $y^5 - y = x + \frac{a}{\lambda^2 x}$. This isomorphism again acts on the standard form model by sending $a \mapsto \frac{a}{\lambda^2} = \lambda^2 a$. $\square$

We obtain a linear action of a group $G \simeq \mathbb{Z}/2\mathbb{Z}$ generated by $a \mapsto -a$ on the polynomial ring $\overline{\mathbb{F}}_5[a]$.

Corollary 4.21 *The element $I_1 = a^2$ generates the ring of invariants for Artin-Schreier curves of genus* 4 *in characteristic* 5 *with* 5*-rank* 4*, i.e.,* $\overline{\mathbb{F}}_5[a]^G = \overline{\mathbb{F}}_5[I_1]$.

Proof The set $\{I_1 = a^2\}$ is a reconstructing set for curves in the stratum with standard form as in (4.14). $\square$

Remark 4.22 The curves in this family are all hyperelliptic, but of characteristic different from 2. They can be written as $v^2 = u^{10} - 2u^6 + u^2 - 4a$ with $v = 2x - y^5 + y$ and $u = y$. Invariants for them inside the family of hyperelliptic curves of genus 4 in characteristic 5 (so inside a larger family) could also be computed with the techniques of [LR12].

5 Conclusion

In this paper, we determined reconstructing invariants for Artin-Schreier curves of genus 3 and 4 in characteristic $p > 2$. Our results on genus 3 represent an important step toward the full characterization of invariants of curves of that genus.

Moving to higher genus curves presents many additional challenges. In particular, curves C_f where f has many poles of the same order exhibit many ismorphisms between standard forms. As the genus grows, the problem becomes more and more complicated. Instead of seeking a general closed form for invariants of all Artin-Schreier curves, an algorithmic method of determining invariants seems appropriate.

Devising the right framework and specializing the results of invariant theory to this setting is the subject of ongoing work.

Acknowledgements The authors thank the Banff International Research Station for Mathematical Innovation and Discovery in Banff (Canada) for sponsoring the Women in Numbers 6 (WIN6) workshop and for providing a productive and enjoyable environment for our initial work on this project. We would especially like to thank the organizers of WIN6, Shabnam Akhtari, Alina Bucur, Jennifer Park, and Renate Scheidler, for making the conference and this collaboration possible. We are indebted to Rachel Pries for helpful conversations around this work. Last but not least, we wish to thank the anonymous referee for useful comments and suggestions. The research of the first author is partially supported by the Simons Foundation grant #550023. The research of the second author is in part supported by National Science Foundation Grant DMS-2201085 and by the School of Natural and Behavioral Sciences at Brooklyn College, City University of New York. The research of the third author is partially funded by the Melodia ANR-20-CE40-0013 project and the 2023-08 Germaine de Staël project. The research of the fourth author is supported by the National Science Foundation under grant DMS-2137661. Lastly, the research of the fifth author is funded by the Natural Sciences and Engineering Research Council of Canada (NSERC) (funding RGPIN-2019-04844).

References

[AS27] Emil Artin and Otto Schreier. Eine Kennzeichnung der reell abgeschlossenen Körper. In *Abh. Math. Sem. Hamburg*, volume 5, pages 225–231. Springer, 1927.

[Bas15] Romain Basson. *Arithmétique des espaces de modules des courbes hyperelliptiques de genre 3 en caractéristique positive*. PhD thesis, Université de Rennes 1, Rennes, 2015.

[BCK21] Irene Bouw, Nirvana Coppola, Pınar Kılıçer, Sabrina Kunzweiler, Elisa Lorenzo García, and Anna Somoza. Reduction types of genus-3 curves in a special stratum of their moduli space. In *Women in Numbers Europe III—Research Directions in Number Theory*, volume 24 of *Assoc. Women Math. Ser.*, pages 115–162. Springer, Cham, 2021.

[BCP97] Wieb Bosma, John Cannon, and Catherine Playoust. The Magma algebra system. I. The user language. *J. Symbolic Comput.*, 24(3-4):235–265, 1997. Computational algebra and number theory (London, 1993).

[BDF12] Alina Bucur, Chantal David, Brooke Feigon, Matilde Lalın, and Kaneenika Sinha. Distribution of zeta zeroes of Artin–Schreier covers. *Math. Res. Lett*, 19(06):1329–1356, 2012.

[BDFL16] Alina Bucur, Chantal David, Brooke Feigon, and Matilde Lalin. Statistics for ordinary Artin–Schreier covers and other p-rank strata. *Trans. AMS*, 368(4):2371–2413, 2016.

[BHM16] Irene Bouw, Wei Ho, Beth Malmskog, Renate Scheidler, Padmavathi Srinivasan, and Christelle Vincent. Zeta functions of a class of Artin-Schreier curves with many automorphisms. In *Directions in Number Theory*, volume 3 of *Assoc. Women Math. Ser.*, pages 87–124. Springer, [Cham], 2016.

[CFA06] Henri Cohen, Gerhard Frey, Roberto Avanzi, Christophe Doche, Tanja Lange, Kim Nguyen, and Frederik Vercauteren, editors. *Handbook of elliptic and hyperelliptic curve cryptography*. Discrete Mathematics and its Applications (Boca Raton). Chapman & Hall/CRC, Boca Raton, FL, 2006.

[Dix87] Jacques Dixmier. On the projective invariants of quartic plane curves. *Adv. in Math.*, 64(3):279–304, 1987.

[DK02] Harm Derksen and Gregor Kemper. *Computational invariant theory*. Invariant Theory and Algebraic Transformation Groups, I. Springer-Verlag, Berlin, 2002. Encyclopaedia of Mathematical Sciences, 130.

[DK08] Harm Derksen and Gregor Kemper. Computing invariants of algebraic groups in arbitrary characteristic. *Adv. Math.*, 217(5):2089–2129, 2008.

[Dol03] Igor Dolgachev. *Lectures on invariant theory*, volume 296 of *London Mathematical Society Lecture Note Series*. Cambridge University Press, Cambridge, 2003.

[Eis95] David Eisenbud. *Commutative algebra*, volume 150 of *Graduate Texts in Mathematics*. Springer-Verlag, New York, 1995. With a view toward algebraic geometry.

[Eis05] David Eisenbud. *The geometry of syzygies*, volume 229 of *Graduate Texts in Mathematics*. Springer-Verlag, New York, 2005. A second course in commutative algebra and algebraic geometry.

[Ent12] Alexei Entin. On the distribution of zeroes of Artin–Schreier L-functions. *Geom. Funct. Anal.*, 22(5):1322–1360, 2012.

[Far10] Shawn Farnell. *Artin-Schreier curves*. PhD thesis, Colorado State University, Fort Collins, Colorado, 2010.

[Hen78] Hans-Wolfgang Henn. Funktionenkörper mit grosser Automorphismengruppe. *J. Reine Angew. Math.*, 302:96–115, 1978.

[HKT08] J. W. P. Hirschfeld, G. Korchmáros, and F. Torres. *Algebraic curves over a finite field*. Princeton Series in Applied Mathematics. Princeton University Press, Princeton, NJ, 2008.

[Igu60] Jun-ichi Igusa. Arithmetic variety of moduli for genus two. *Ann. of Math. (2)*, 72:612–649, 1960.

[KLS20] Pınar Kılıçer, Elisa Lorenzo García, and Marco Streng. Primes dividing invariants of CM Picard curves. *Canad. J. Math.*, 72(2):480–504, 2020.

[Kon09] Aristides Kontogeorgis. Field of moduli versus field of definition for cyclic covers of the projective line. *J. Théor. Nombres Bordeaux*, 21(3):679–692, 2009.

[Liu93] Qing Liu. Courbes stables de genre 2 et leur schéma de modules. *Math. Ann.*, 295(2):201–222, 1993.

[Liu02] Qing Liu. *Algebraic Geometry and Arithmetic Curves*. Oxford University Press, 2002.

[LLLR21] Reynald Lercier, Qing Liu, Elisa Lorenzo García, and Christophe Ritzenthaler. Reduction type of smooth plane quartics. *Algebra Number Theory*, 15(6):1429–1468, 2021.

[LR12] Reynald Lercier and Christophe Ritzenthaler. Hyperelliptic curves and their invariants: geometric, arithmetic and algorithmic aspects. *J. Algebra*, 372:595–636, 2012.

[Mum65] David Mumford. *Geometric invariant theory*. Ergebnisse der Mathematik und ihrer Grenzgebiete, (N.F.), Band 34. Springer-Verlag, Berlin-New York, 1965.

[Noe26] E. Noether. Der Endlichkeitssatz der Invarianten endlicher linearer Gruppen der Charakteristik p. *Nachrichten von der Gesellschaft der Wissenschaften zu Göttingen, Mathematisch-Physikalische Klasse*, 1926:28–35, 1926.

[Ohn07] Toshiaki Ohno. The graded ring of invariants of ternary quartics I, 2007. unpublished.

[PZ12] Rachel Pries and Hui June Zhu. The p-rank stratification of Artin-Schreier curves. *Ann. Inst. Fourier (Grenoble)*, 62(2):707–726, 2012.

[Shi67] Tetsuji Shioda. On the graded ring of invariants of binary octavics. *Amer. J. Math.*, 89:1022–1046, 1967.

[Sti09] Henning Stichtenoth. *Algebraic function fields and codes*, volume 254. Springer Science & Business Media, 2009.

[Sub75] Doré Subrao. The p-rank of Artin–Schreier curves. *Manuscripta Math.*, 16(2):169–193, 1975.

[Sym11] Peter Symonds. On the Castelnuovo-Mumford regularity of rings of polynomial invariants. *Ann. of Math. (2)*, 174(1):499–517, 2011.

[vdGvdV91] Gerard van der Geer and Marcel van der Vlugt. Artin-Schreier curves and codes. *J. Algebra*, 139(1):256–272, 1991.

[vdGvdV92] Gerard van der Geer and Marcel van der Vlugt. Reed-Muller codes and supersingular curves. I. *Compos. Math.*, 84(3):333–367, 1992.

[VM80] Robert C. Valentini and Manohar L. Madan. A hauptsatz of L. E. Dickson and Artin-Schreier extensions. *J. Reine Angew. Math.*, 318:156–177, 1980.

A Note on Large Sums of Divisor-Bounded Multiplicative Functions

Claire Frechette, Mathilde Gerbelli-Gauthier, Alia Hamieh, and Naomi Tanabe

Abstract Given a multiplicative function f, we let $S(x, f) = \sum_{n\le x} f(n)$ be the associated partial sum. In this note, we show that lower bounds on partial sums of divisor-bounded functions result in lower bounds on the partial sums associated to their products. More precisely, we let f_j, $j = 1, 2$ be such that $|f_j(n)| \le \tau(n)^{\kappa}$ for some fixed positive constant κ, and assume their partial sums satisfy $\left|S(x_j, f_j)\right| \ge \eta x_j (\log x_j)^{2^{\kappa}-1}$ for some $x_1, x_2 \gg 1$ and $\eta > \max_j\{(\log x_j)^{-1/100}\}$. We then show that there exists $x \ge \min\{x_1, x_2\}^{\xi^2}$ such that $|S(x, f_1 f_2)| \ge \xi x (\log x)^{2^{2\kappa}-1}$, where $\xi = C\eta^{1+2^{\kappa+3}}$ for some positive constant C that depends only on κ.

2010 Mathematics Subject Classification Primary 11F30; secondary 11F11 · 11F12 · 11M41

The research of Claire Frechette is supported by NSF grant DMS-2203042, and the research of Alia Hamieh is supported by NSERC Discovery grant RGPIN-2018-06313.

C. Frechette (✉)
Department of Mathematics and Mathematical Statistics, Umeå University, Umeå, Sweden
e-mail: claire.frechette@umu.se

M. Gerbelli-Gauthier
Department of Mathematics, University of Toronto, Toronto, ON, Canada
e-mail: m.gerbelli@utoronto.ca

A. Hamieh (✉)
Department of Mathematics and Statistics, University of Northern British Columbia, Prince George, BC, Canada
e-mail: alia.hamieh@unbc.ca

N. Tanabe
Department of Mathematics, Bowdoin College, Brunswick, ME, USA
e-mail: ntanabe@bowdoin.edu

S. Akhtari et al. (eds.), *Research Directions in Number Theory*, Association for Women in Mathematics Series 39, https://doi.org/10.1007/978-3-032-11183-8_7

1 Introduction

A central theme in analytic number theory is studying mean values of multiplicative functions. One enduring quest in this area, which has garnered extensive research over the past century and continues to attract significant interest to date, revolves around understanding the asymptotic behavior of character sums. Let x be a positive real number, and let χ be a primitive Dirichlet character modulo q. We consider the associated character sum $S(x, \chi) = \sum_{n \leq x} \chi(n)$. Trivially, we have the bound

$$|S(x, \chi)| \leq \min(x, q).$$

An important result proven independently by Pólya and Vinogradov [HD80, pages 135–137] asserts that

$$\max_x |S(x, \chi)| \ll \sqrt{q} \log q,$$

which was further improved to

$$\max_x |S(x, \chi)| \ll \sqrt{q} \log \log q,$$

by assuming the Generalized Riemann Hypothesis (GRH) for Dirichlet L-functions thanks to the work of Montgomery and Vaughan [LMCV77]. It follows that

$$S(x, \chi) = o(x) \tag{1.1}$$

when $x \geq q^{\frac{1}{2}+\epsilon}$. In 1957, Burgess established (1.1) in the range $x > q^{\frac{1}{4}+o(1)}$ for any quadratic character χ when q is prime. This was generalized by Burgess himself to any non-principal character provided that q is cubefree, and in the smaller range $x > q^{\frac{3}{8}+o(1)}$ otherwise. In [AH86], Hilderbrand used the theory of mean values of multiplicative functions to improve the range of validity of (1.1) to $x > q^{\frac{1}{4}-o(1)}$ for non-principal characters with prime moduli q. Burgess's range has not been substantially improved over the last five decades, although it is widely believed that (1.1) should hold in the wider range $x \gg_\epsilon q^\epsilon$. In 2001, Granville and Soundararajan [GS01] proved that, for a primitive character $\chi \mod q$, (1.1) holds in the range $\log x / \log \log q \to \infty$ assuming the GRH for $L(s, \chi)$. In [GS18], they also showed that this asymptotic holds under the weaker assumption that "100%" of the zeros of $L(s, f)$ up to height $\frac{1}{4}$ lie on the critical line. This work of Granville and Soundararajan is of particular interest to us because it makes strong connections between character sums and zeros of the associated L-functions using general results from multiplicative number theory pertaining to mean values of 1-bounded multiplicative functions in the pretentious framework. In [GS18], the authors record various other interesting observations on large character sums applying variants of Halász's theorem and Lipschitz estimate as developed in [GS03]. For example, they show that if χ_1 and χ_2 have large character sums then so does their product $\chi_1\chi_2$. This result follows as a direct application of [GS18, Theorem 6.2] which asserts that if the partial sums of two completely multiplicative 1-bounded functions f_1 and f_2 are large, then the partial sum of their product $f_1 f_2$ is large as well. More precisely, they show that if $|S(x_i, f_i)| > \eta x_i$ for some positive real numbers η, x_1 and x_2, then $|S(x, f_1 f_2)| > c\eta^6 x$ for some absolute positive constant c and $x \geq (\min(x_1, x_2))^{c\eta^6}$. In this note, we extend this theorem to multiplicative functions valued outside of the unit disk, but whose magnitude is bounded by a power of the divisor function.

Theorem 1.1 *Let κ be a fixed positive constant. Let f_1 and f_2 be multiplicative functions with $|f_j(n)| \leq \tau(n)^\kappa$ for all $n \in \mathbb{N}$. Suppose that, for $x_1, x_2 \gg 1$, there exists $\eta > \max_j\{(\log x_j)^{-1/100}\}$ such that*

$$\left|\sum_{n \leq x_j} f_j(n)\right| \geq \eta x_j (\log x_j)^{2^\kappa - 1}. \tag{1.2}$$

Then, with $\xi = C\eta^{1+2^{\kappa+3}}$ for some positive constant C that depends only on κ, there exists $x \geq \min\{x_1, x_2\}^{\xi^2}$ such that

$$\left|\sum_{n \leq x} f_1(n) f_2(n)\right| \geq \xi x (\log x)^{2^{2\kappa} - 1}.$$

Theorem 1.1 is proved following the general structure of the argument of [GS18, Theorem 6.1], but we extend it to divisor-bounded multiplicative functions. The technical results employed in the proofs of the various theorems in [GS18] have their roots in the work of Halász [GH71, GH75] and the subsequent works of Montgomery [HM], Tenenbaum [GT08], Granville and Soundararajan [GS03], among other mathematicians, on mean values of multiplicative functions that take their values in the complex unit disk. These results have been extended to a broad class of divisor-bounded multiplicative functions by Tenenbaum in [GT17], Granville, Harper and Soundararajan in [GHS19], Mangerel in [APM18, APM23], and Matthiesen in [LM18]. Our work in the present paper and our previous work [FBHT25] hinges upon such generalizations and is motivated by trying to study a GL_2-analogue of (1.1). In fact, for a primitive cusp form $g(z) = \sum_{n \geq 1} \lambda_g(n) n^{\frac{k-1}{2}} e^{2\pi i n z}$ in $S_k(1)$, Lamzouri [YL19, Corollary 1.2] proved that

$$\sum_{n \leq x} \lambda_g(n) = o(x \log x) \tag{1.3}$$

holds in the range $\log x / \log\log k \to \infty$ assuming the GRH for $L(s, g)$. He also proved unconditionally that this range in x is best possible [YL19, Corollary 1.4]. In [FBHT25], we extend the methods of [GS18] to show that (1.3) holds in the range $x \geq k^\epsilon$ under a weaker assumption than GRH. In connection with the main result of this paper, we derive the following immediate consequence for large sums of Fourier coefficients of primitive cusp forms, which can be viewed as a GL_2-analogue of [GS18, Corollary 1.7].

Corollary 1.2 Let $g \in S_k(r)$ and $h \in S_\ell(q)$ be primitive cusp forms whose Fourier coefficients are denoted by $\lambda_g(n)$ and $\lambda_h(n)$, respectively. Suppose that

$$\left|\sum_{n\leq x_1} \lambda_g(n)\right| \geq \eta x_1 \log x_1 \quad \text{and} \quad \left|\sum_{n\leq x_2} \lambda_h(n)\right| \geq \eta x_2 \log x_2,$$

for some $x_1, x_2 \gg 1$ and $\eta > \max_j\{(\log x_j)^{-1/100}\}$. Then there exists $x \geq (\min(x_1, x_2))^{\xi^2}$ such that

$$\left|\sum_{n\leq x_1} \lambda_g(n)\lambda_h(n)\right| \geq \xi x(\log x)^3,$$

where $\xi = C\eta^{17}$ for some absolute positive constant C.

The paper is organized as follows. In Sect. 2, we record versions of Halász's theorem and Lipschitz formula that are suitable for our setup. We also prove some consequences of these results that are crucial for the proof of our main theorem. In Sect. 3, we introduce a distance function for divisor-bounded multiplicative functions, for which we prove a triangle inequality-type result. We also derive a lower bound on sums of multiplicative functions in terms of the distance function. These results form the main ingredients of the proof of Theorem 1.1, which appears in Sect. 4.

Conventions and Notation. In this work, we adopt the following conventions and notation. Given two functions $f(x)$ and $g(x)$ we write $f(x) \ll g(x)$, $g(x) \gg f(x)$, or $f(x) = O(g(x))$ to mean there exists some positive constant M such that $|f(x)| \leq M|g(x)|$ for x large enough. The notation $f(x) \asymp g(x)$ is used when both estimates $f(x) \ll g(x)$ and $f(x) \gg g(x)$ hold simultaneously. We write $f(x) = o(g(x))$ when $g(x) \neq 0$ for sufficiently large x and $\lim\limits_{x\to\infty} \dfrac{f(x)}{g(x)} = 0$. The letter p will be exclusively used to represent a prime number.

2 Halász's Theorem and Consequences

Let $f(n)$ be a multiplicative function satisfying

$$|f(n)| \leq \tau(n)^\kappa \tag{2.1}$$

for all $n \in \mathbb{N}$, and let $L(s, f)$ be its associated Dirichlet L-series given by

$$L(s, f) = \sum_{n\geq 1} \frac{f(n)}{n^s}, \quad \mathrm{Re}(s) > 1.$$

We begin with presenting slightly modified versions of Halász's theorem and its corollary, following the ideas of [GHS19, Theorem 1.1, Corollary 1.2].

Theorem 2.1 *Let f be a multiplicative function such that $|f(n)| \le \tau(n)^\kappa$ for all $n \in \mathbb{N}$. Let x be sufficiently large. Then for any $1 \ll T \le x^{\frac{9}{10}}$, we have*

$$\frac{1}{x}\sum_{n\le x} f(n) \ll \frac{1}{\log x}\int_{\frac{1}{\log x}}^{1}\left(\max_{|t|\le T}\left|\frac{L(1+\sigma+it,f)}{1+\sigma+it}\right|\right)\frac{d\sigma}{\sigma} + O\left(\frac{(\log T)^{2^\kappa}}{\log x} + \frac{(\log x)^{2^\kappa}}{T}\right).$$

Proof The result follows from [GHS19, Eqs. 2.1, 2.2, and 2.9] with $y = T^2$. □

To further proceed, let $1 \le T_0 \ll (\log x)^{2^\kappa}$, and let $M_f(x, T_0)$ be a real number satisfying

$$\max_{|t|\le T_0}\left|\frac{L\left(1+\frac{1}{\log x}+it, f\right)}{1+\frac{1}{\log x}+it}\right| = e^{-M_f(x,T_0)}(\log x)^{2^\kappa}. \tag{2.2}$$

Then, we obtain the following corollary.

Corollary 2.2 Suppose f is a multiplicative function such that $|f(n)| \le \tau(n)^\kappa$ for all $n \in \mathbb{N}$. Let $T_0 \asymp (\log x)^{2^\kappa}$ for some sufficiently large x, and let $M_f(x, T_0)$ be the quantity defined in (2.2). Then, we have

$$\frac{1}{x}\sum_{n\le x} f(n) \ll (M_f(x,T_0)+1)e^{-M_f(x,T_0)}(\log x)^{2^\kappa-1} + \frac{(\log\log x)^{2^\kappa}}{\log x}.$$

Proof We apply Theorem 2.1 with $T = T_0 \log x$ to get

$$\frac{1}{x}\sum_{n\le x} f(n) \ll \frac{1}{\log x}\int_{\frac{1}{\log x}}^{1}\left(\max_{|t|\le T}\left|\frac{L(1+\sigma+it,f)}{1+\sigma+it}\right|\right)\frac{d\sigma}{\sigma} + O\left(\frac{(\log x)^{2^\kappa-1}}{T_0} + \frac{(\log\log x)^{2^\kappa}}{\log x}\right).$$

Noting that $|f(n)| \le \tau(n)^\kappa = \tau_2(n)^\kappa \le \tau_{2^\kappa}(n)$, where τ_ℓ is the ℓ-fold divisor function (i.e., $\tau_\ell(n)$ is the coefficient of n^{-s} in the Dirichlet series representation of $\zeta(s)^\ell$), and that $\sigma \le 1$, we have

$$|L(1+\sigma+it, f)| \le \zeta(1+\sigma)^{2^\kappa} \le \left(\frac{1}{\sigma} + O(1)\right)^{2^\kappa}, \tag{2.3}$$

and therefore

$$\max_{|t|\le T}\left|\frac{L(1+\sigma+it,f)}{1+\sigma+it}\right| = \max_{|t|\le T_0}\left|\frac{L(1+\sigma+it,f)}{1+\sigma+it}\right| + O\left(\frac{1}{\sigma^{2^\kappa}T_0}\right).$$

It follows that

$$\frac{1}{\log x}\int_{\frac{1}{\log x}}^{1}\left(\max_{|t|\le T}\left|\frac{L(1+\sigma+it,f)}{1+\sigma+it}\right|\right)\frac{d\sigma}{\sigma} = \frac{1}{\log x}\int_{\frac{1}{\log x}}^{1}\left(\max_{|t|\le T_0}\left|\frac{L(1+\sigma+it,f)}{1+\sigma+it}\right|\right)\frac{d\sigma}{\sigma} + O\left(\frac{(\log x)^{2^\kappa-1}}{T_0}\right),$$

and so

$$\frac{1}{x}\sum_{n\leq x} f(n) \ll \frac{1}{\log x}\int_{\frac{1}{\log x}}^{1}\left(\max_{|t|\leq T_0}\left|\frac{L(1+\sigma+it, f)}{1+\sigma+it}\right|\right)\frac{d\sigma}{\sigma} + O\left(\frac{(\log x)^{2^\kappa-1}}{T_0}+\frac{(\log\log x)^{2^\kappa}}{\log x}\right). \tag{2.4}$$

From this point on, the proof follows by applying the same argument employed in the proof of [GHS19, Corollary 1.2]. We include some details for the convenience of the reader. By the definition of $M_f(x, T_0)$, the maximum of $\left|\frac{L(1+\sigma+it,f)}{1+\sigma+it}\right|$ on the vertical line segment $\sigma = \frac{1}{\log x}$ for $|t| \leq T_0$ is $e^{-M_f(x,T_0)}(\log x)^{2^\kappa}$ whereas its maximum on the horizontal line segment $|t| = T_0$ with $\frac{1}{\log x} \leq \sigma \leq 1$ is bounded above by $\frac{\zeta\left(1+\frac{1}{\log x}\right)^{2^\kappa}}{T_0} \ll \frac{(\log x)^{2^\kappa}}{T_0} = O(1)$. A similar bound holds on the vertical line $\sigma = 1$ with $|t| \leq T_0$. Using the maximum modulus principle on the region $\{\sigma + it : \frac{1}{\log x} \leq \sigma \leq 1,\ |t| \leq T_0\}$, we get

$$\max_{|t|\leq T_0}\left|\frac{L(1+\sigma+it, f)}{1+\sigma+it}\right| \leq e^{-M_f(x,T_0)}(\log x)^{2^\kappa} + O(1) \ll e^{-M_f(x,T_0)}(\log x)^{2^\kappa},$$

where the last asymptotic holds since $e^{-M_f(x,T_0)}(\log x)^{2^\kappa} \gg 1$ by [GHS19, Lemma2.7]. By (2.3), we also have the trivial bound

$$\max_{|t|\leq T_0}\left|\frac{L(1+\sigma+it, f)}{1+\sigma+it}\right| \leq \left(\tfrac{1}{\sigma}+O(1)\right)^{2^\kappa}.$$

Using these bounds in the integral appearing in (2.4), we get

$$\begin{aligned}\int_{\frac{1}{\log x}}^{1}\left(\max_{|t|\leq T_0}\left|\frac{L(1+\sigma+it, f)}{1+\sigma+it}\right|\right)\frac{d\sigma}{\sigma} &\ll \int_{\frac{1}{\log x}}^{\frac{e^{M/2^\kappa}}{\log x}} e^{-M_f(x,T_0)}(\log x)^{2^\kappa}\frac{d\sigma}{\sigma} + \int_{\frac{e^{M/2^\kappa}}{\log x}}^{1}\frac{d\sigma}{\sigma^{2^\kappa+1}}\\ &\ll \frac{M}{2^\kappa}e^{-M_f(x,T_0)}(\log x)^{2^\kappa} + e^{-M_f(x,T_0)}(\log x)^{2^\kappa}.\end{aligned}$$

□

Let $\phi_f(x, T_0)$ be a real number in the range $|t| \leq T_0$ for which the function

$$t \mapsto \left|L\left(1+\frac{1}{\log x}+it, f\right)\right|$$

attains its maximum, and let $N_f(x, T_0)$ be a real number satisfying

$$\left|L\left(1+\frac{1}{\log x}+i\phi_f(x, T_0), f\right)\right| = e^{-N_f(x,T_0)}(\log x)^{2^\kappa}. \tag{2.5}$$

As indicated in Corollary 2.2, we will be interested in the case $T_0 \asymp (\log x)^{2^\kappa}$. We also note that

$$e^{-M_f(x,T_0)}(\log x)^{2^\kappa} = \max_{|t|\le T_0} \left| \frac{L(1+\frac{1}{\log x}+it, f)}{1+\frac{1}{\log x}+it} \right| \le \max_{|t|\le T_0} \left| L(1+\tfrac{1}{\log x}+it, f) \right| = e^{-N_f(x,T_0)}(\log x)^{2^\kappa}.$$

It then follows that $M_f(x, T_0) \ge N_f(x, T_0)$, and so $(M_f(x, T_0)+1)e^{-M_f(x,T_0)} \le (N_f(x, T_0)+1)e^{-N_f(x,T_0)}$. Thus, we arrive at the following result.

Corollary 2.3 Suppose f is a multiplicative function such that $|f(n)| \le \tau(n)^\kappa$ for all $n \in \mathbb{N}$. Let $T_0 \asymp (\log x)^{2^\kappa}$ for some sufficiently large x, and let $N_f(x, T_0)$ be the quantity defined in (2.5). Then, we have

$$\frac{1}{x}\sum_{n\le x} f(n) \ll (N_f(x, T_0)+1)e^{-N_f(x,T_0)}(\log x)^{2^\kappa-1} + \frac{(\log\log x)^{2^\kappa}}{\log x}.$$

The following result is inspired by [GS03, Theorem 2a] and will become useful in the proof of Proposition 2.7 at the end of this section.

Proposition 2.4 *Suppose f is a multiplicative function such that $|f(n)| \le \tau(n)^\kappa$ for all $n \in \mathbb{N}$. Let $\phi = \phi_f(x, (\log x)^{2^\kappa})$ as in (2.5) and assume that $|\phi| \asymp (\log x)^{2^\kappa}$. Then*

$$\frac{1}{x}\sum_{n\le x} f(n) \ll \frac{(\log\log x)^{2^{\kappa+1}(1-\frac{2}{\pi})+1}}{(\log x)^{1-\frac{2^{\kappa+1}}{\pi}}} + \frac{(\log\log x)^{2^\kappa}}{\log x}.$$

Proof We apply Corollary 2.3 with $T_0 = |\phi| - 2$ to get

$$\begin{aligned}\frac{1}{x}\sum_{n\le x} f(n) &\ll (N_f(x, |\phi|-2)+1)e^{-N_f(x,|\phi|-2)}(\log x)^{2^\kappa-1} + \frac{(\log\log x)^{2^\kappa}}{\log x} \\ &\ll \frac{1}{\log x}\max_{|t|\le|\phi|-2}\left|L(1+\frac{1}{\log x}+it, f)\right| \log\left(\frac{(\log x)^{2^\kappa}}{\max_{|t|\le|\phi|-2}\left|L(1+\frac{1}{\log x}+it, f)\right|}\right) \\ &\quad + \frac{(\log\log x)^{2^\kappa}}{\log x}.\end{aligned}$$

By [GHS19, Lemma 4.1], we have

$$\begin{aligned}L\left(1+\frac{1}{\log x}+it, f\right) &\ll (\log x)^{\frac{2^{\kappa+1}}{\pi}}\left(\frac{\log x}{1+|t-\phi|\log x} + (\log\log x)^2\right)^{2^\kappa(1-\frac{2}{\pi})} \\ &\ll (\log x)^{\frac{2^{\kappa+1}}{\pi}}(\log\log x)^{2^{\kappa+1}(1-\frac{2}{\pi})},\end{aligned}$$

for all $|t| \le |\phi| - 2$. It follows that

$$\frac{1}{x}\sum_{n\le x} f(n) \ll \frac{1}{\log x}(\log x)^{\frac{2^{\kappa+1}}{\pi}}(\log\log x)^{2^{\kappa+1}(1-\frac{2}{\pi})}\log\left(\frac{(\log x)^{2^\kappa}}{\max_{|t|\le|\phi|-2}\left|L(1+\frac{1}{\log x}+it, f)\right|}\right)$$

$$+\frac{(\log\log x)^{2^\kappa}}{\log x}$$

$$\ll \frac{(\log\log x)^{2^{\kappa+1}(1-\frac{2}{\pi})+1}}{(\log x)^{1-\frac{2^{\kappa+1}}{\pi}}} + \frac{(\log\log x)^{2^\kappa}}{\log x},$$

as required. □

Next, we require the following version of Lipschitz theorem.

Theorem 2.5 (Lipschitz Theorem) *Suppose f is a multiplicative function such that $|f(n)| \le \tau(n)^\kappa$. Let $\phi = \phi_f(x, (\log x)^{2^\kappa})$, as given in* (2.5)*. Then for all $1 \le w \le x^{1/3}$ we have*

$$\left|\frac{1}{x}\sum_{n\le x} f(n)n^{-i\phi} - \frac{1}{x/w}\sum_{n\le x/w} f(n)n^{-i\phi}\right|$$
$$\ll \log\left(\frac{\log x}{\log ew}\right)\left(\frac{\log w + (\log\log x)^2}{\log x}\right)^{\min\{2^\kappa(1-\frac{2}{\pi}),1\}} (\log x)^{2^\kappa - 1}.$$

Proof The proof is very similar to that of [GHS19, Theorem 1.5], and the reader is referred to their detailed exposition. □

Corollary 2.6 Let f and ϕ be as in Theorem 2.5. Then

$$\frac{1}{x}\sum_{n\le x} f(n) = \frac{x^{i\phi}}{1+i\phi}\cdot\frac{1}{x}\sum_{n\le x} f(n)n^{-i\phi} + O(E_\kappa(x)),$$

where

$$E_\kappa(x) = \begin{cases} (\log x)^{-1+4/\pi}(\log\log x)^{5-8/\pi} & \text{if } \kappa = 1 \\ (\log x)^{2^\kappa-2}(\log\log x)^3 & \text{if } \kappa > 1. \end{cases} \tag{2.6}$$

Proof We will show the equivalent statement that

$$\frac{1}{x}\sum_{n\le x} f(n)n^{-i\phi} = \frac{1+i\phi}{x^{1+i\phi}}\sum_{n\le x} f(n) + O(|\phi| E_\kappa(x)).$$

By partial summation, we have

$$\frac{1}{x}\sum_{n\le x} f(n)n^{-i\phi} = \frac{1}{x}\int_1^x u^{-i\phi} d\left\{\sum_{n\le x} f(n)\right\} = \frac{1}{x^{1+i\phi}}\sum_{n\le x} f(n) + \frac{i\phi}{x}\int_1^x \frac{1}{u^{1+i\phi}}\sum_{n\le u} f(n)du.$$

We split the integral into two pieces as follows:

$$\int_1^x \frac{1}{u^{1+i\phi}} \sum_{n\le u} f(n)du = \int_1^{x/(\log x)^2} \frac{1}{u^{1+i\phi}} \sum_{n\le u} f(n)du + \int_{x/(\log x)^2}^x \frac{1}{u^{1+i\phi}} \sum_{n\le u} f(n)du.$$

For the first integral, we apply the trivial bound

$$\sum_{n\le x} |f(n)| \le \sum_{n\le x} \tau_{2^\kappa}(n) \ll x(\log x)^{2^\kappa-1} \tag{2.7}$$

(see [FLLT17] for example) to get

$$\begin{aligned}\frac{i\phi}{x}\int_1^{x/(\log x)^2} \frac{1}{u^{1+i\phi}} \sum_{n\le u} f(n)du &\ll \frac{|\phi|}{x}\int_1^{x/(\log x)^2} \frac{1}{u}\sum_{n\le u}|f(n)|du \\ &\le \frac{|\phi|}{x}\int_1^{x/(\log x)^2} (\log u)^{2^\kappa-1}du \ll |\phi|E_\kappa(x).\end{aligned}$$

Since $w = x/u$ is in the range of Theorem 2.5, the second integral is equal to

$$\begin{aligned}&\frac{i\phi}{x}\int_{x/(\log x)^2}^x \left(\frac{1}{x^{1+i\phi}}\sum_{n\le x} f(n) + O\left(\log\left(\frac{\log x}{\log ew}\right)\left(\frac{\log w + (\log\log x)^2}{\log x}\right)^{\min\{2^\kappa(1-\frac{2}{\pi}),1\}}(\log x)^{2^\kappa-1}\right)\right)du \\ =&\frac{i\phi}{x}\left(\frac{1}{x^{1+i\phi}}\sum_{n\le x} f(n)\right)\int_{x/(\log x)^2}^x du + \frac{i\phi}{x}\cdot O\left(\log\log x\left(\frac{(\log\log x)^2}{\log x}\right)^{\min\{2^\kappa(1-\frac{2}{\pi}),1\}}(\log x)^{2^\kappa-1}\right)\int_{x/(\log x)^2}^x du \\ =&\frac{i\phi}{x^{1+i\phi}}\sum_{n\le x} f(n) + O(|\phi|E_\kappa(x)).\end{aligned}$$

Combining the two integrals gives the desired result. □

We now utilize Proposition 2.4 (when $|\phi|$ is large) and Corollary 2.6 (when $|\phi|$ is small) to get the proposition that will be used in our argument.

Proposition 2.7 *Let f be a multiplicative function such that $|f(n)| \le \tau(n)^\kappa$, and let $N = N_f(x, (\log x)^{2^\kappa})$ and $\phi = \phi_f(x, (\log x)^{2^\kappa})$ for sufficiently large x. Then,*

$$\frac{1}{x}\sum_{n\le x} f(n) \ll (\log x)^{2^\kappa-1}\left(\frac{(N+1)e^{-N}}{1+|\phi|} + \frac{E_\kappa(x)}{(\log x)^{2^\kappa-1}}\right),$$

where $E_\kappa(x)$ is given in (2.6).

Proof Suppose that $|\phi| \ge \frac{1}{2}(\log x)^{2^\kappa}$. By Proposition 2.4, we have

$$\frac{1}{x}\sum_{n\le x} f(n) \ll \frac{(\log\log x)^{2^{\kappa+1}(1-\frac{2}{\pi})+1}}{(\log x)^{1-\frac{2^{\kappa+1}}{\pi}}} + \frac{(\log\log x)^{2^\kappa}}{\log x}.$$

It follows that

$$\frac{1}{x}\sum_{n\le x} f(n) - (\log x)^{2^\kappa-1}\frac{(N+1)e^{-N}}{1+|\phi|} \ll \frac{(\log\log x)^{2^{\kappa+1}(1-\frac{2}{\pi})+1}}{(\log x)^{1-\frac{2^{\kappa+1}}{\pi}}} + \frac{(\log\log x)^{2^\kappa}}{\log x} \ll E_\kappa(x).$$

Hence, we may assume that $|\phi| \le \frac{1}{2}(\log x)^{2^\kappa}$. Applying Corollary 2.3 to the function $g(n) = f(n)n^{-i\phi}$ with $T_0 = \frac{1}{2}(\log x)^{2^\kappa}$ gives

$$\begin{aligned}\frac{1}{x}\sum_{n\le x} f(n)n^{-i\phi} = \frac{1}{x}\sum_{n\le x} g(n) &\ll \left(N_g\left(x, \frac{(\log x)^{2^\kappa}}{2}\right)+1\right)e^{-N_g(x,\frac{1}{2}(\log x)^{2^\kappa})}(\log x)^{2^\kappa-1} + \frac{(\log\log x)^{2^\kappa}}{\log x}\\ &= (N+1)e^{-N}(\log x)^{2^\kappa-1} + \frac{(\log\log x)^{2^\kappa}}{\log x}. \end{aligned} \tag{2.8}$$

The last line follows from the fact that $N_g(x, \frac{1}{2}(\log x)^{2^\kappa}) = N_f(x, (\log x)^{2^\kappa}) = N$. Indeed, this can be observed as follows: Since $L(s, g) = L(s + i\phi, f)$, we have

$$\begin{aligned}\max_{|t|<\frac{1}{2}(\log x)^{2^\kappa}}\left|L\left(1+\frac{1}{\log x}+it, g\right)\right| &= \max_{|t|<\frac{1}{2}(\log x)^{2^\kappa}}\left|L\left(1+\frac{1}{\log x}+i(t+\phi), f\right)\right|\\ &= \max_{|t|<(\log x)^{2^\kappa}}\left|L\left(1+\frac{1}{\log x}+it, f\right)\right|.\end{aligned}$$

This is where we use the assumption that $\phi = \phi_f(x, (\log x)^{2^\kappa})$ satisfies $|\phi| \le \frac{1}{2}(\log x)^{2^\kappa}$.

Applying Corollary 2.6 and (2.8), we get

$$\begin{aligned}\frac{1}{x}\sum_{n\le x} f(n) &\ll \left|\frac{x^{i\phi}}{1+i\phi}\right|\left((N+1)e^{-N}(\log x)^{2^\kappa-1} + \frac{(\log\log x)^{2^\kappa}}{\log x}\right) + O(E_\kappa(x))\\ &\ll \frac{(N+1)e^{-N}}{1+|\phi|}(\log x)^{2^\kappa-1} + O\left(\frac{(\log\log x)^{2^\kappa}}{\log x}\right) + O\left(E_\kappa(x)\right).\end{aligned}$$

The first error term is subsumed into the second one since it is smaller. □

Before concluding this section, we establish the following lemma, which will play a pivotal role in Sect. 4.

Lemma 2.8 *Let f be a multiplicative function such that $|f(n)| \le \tau(n)^\kappa$ for all $n \in \mathbb{N}$. Let $N = N_f(x, (\log x)^{2^\kappa})$ and $\phi = \phi_f(x, (\log x)^{2^\kappa})$ as in (2.5). Suppose that there exists $x \gg 1$ and $\eta > (\log x)^{-1/100}$ such that*

$$\left|\sum_{n\le x} f(n)\right| \ge \eta x(\log x)^{2^\kappa-1}. \tag{2.9}$$

Then,

$$|\phi| \ll \frac{1}{\eta} \quad \textit{and} \quad N \leq 2\log\left(\frac{1}{\eta}\right) + O(1).$$

Proof By Proposition 2.7, we know that

$$\left|\sum_{n\leq x} f(n)\right| \ll x(\log x)^{2^\kappa-1}\left(\frac{(N+1)e^{-N}}{1+|\phi|} + \frac{E_\kappa(x)}{(\log x)^{2^\kappa-1}}\right).$$

Therefore, together with our Assumption (2.9), we have

$$\eta \leq \frac{C(N+1)e^{-N}}{1+|\phi|} + \frac{CE_\kappa(x)}{(\log x)^{2^\kappa-1}}, \tag{2.10}$$

for some absolute constant $C > 0$. Writing the second term on the right-hand side as $A_\kappa(x)$, the inequality (2.10) can be written as

$$(1+|\phi|)(\eta - A_\kappa(x)) \leq C(N+1)e^{-N}.$$

Since $\eta \geq 2A_\kappa(x)$ for sufficiently large x, we see that

$$1+|\phi| \leq \frac{C(N+1)e^{-N}}{\eta - A_\kappa(x)} \leq \frac{2C(N+1)e^{-N}}{\eta}. \tag{2.11}$$

In particular, $|\phi| \ll \frac{1}{\eta}$ and $\eta \ll e^{-\frac{N}{2}}$. It follows that $N \leq 2\log\left(\frac{1}{\eta}\right) + O(1)$ as desired. □

3 A Distance Function for Divisor-Bounded Functions

While the notion of a distance between multiplicative functions makes most sense in the context of functions taking values in the complex unit disk, we adapt the standard definition (see, for example, [GS18, Sect. 2]) and set the following notation for a distance function associated with multiplicative functions f satisfying (2.1):

$$\mathbb{D}^2(f, n^{it}; x) = \sum_{p\leq x} \frac{2^\kappa - \mathrm{Re}(f(p)p^{-it})}{p}. \tag{3.1}$$

This is a special case of distance functions introduced by Tenenbaum [GT17] and later, independently, by Mangerel [APM18].

The distance function $\mathbb{D}^2(f, n^{it}; x)$ is related to the Dirichlet series $L(s, f)$ via

$$\left|L\left(1+\frac{1}{\log x}+it, f\right)\right| \asymp (\log x)^{2^\kappa} \exp(-\mathbb{D}^2(f, n^{it}; x)),$$

which can be observed in the following way. We have

$$\begin{aligned}\left|L\left(1+\frac{1}{\log x}+it, f\right)\right| &= \exp\left(\sum_{p\le x}\frac{\mathrm{Re}(f(p)p^{-it})}{p}+O(1)\right) \\ &= \exp\left(\sum_{p\le x}\frac{2^\kappa}{p}-\sum_{p\le x}\frac{2^\kappa-\mathrm{Re}(f(p)p^{-it})}{p}+O(1)\right) \\ &\asymp (\log x)^{2^\kappa}\exp\left(-\sum_{p\le x}\frac{2^\kappa-\mathrm{Re}(f(p)p^{-it})}{p}\right) \\ &= (\log x)^{2^\kappa}\exp(-\mathbb{D}^2(f, n^{it}; x)),\end{aligned} \tag{3.2}$$

where the first equality follows from [APM18, Lemma 2.2.15].

We note that this distance function is related to $N = N_f(x, (\log x)^{2^\kappa})$, defined in (2.5), in the following way:

$$e^{-N} = \frac{1}{(\log x)^{2^\kappa}}\left|L(1+\frac{1}{\log x}+i\phi, f)\right| \asymp \exp(-\mathbb{D}^2(f, n^{i\phi}; x)). \tag{3.3}$$

It is known that the standard distance function between 1-bounded multiplicative functions satisfies a triangle inequality. While such a result does not hold in our setting, we next establish a triangle-type inequality satisfied by the function $\mathbb{D}(f, n^{it}; x)$ introduced in (3.1).

Lemma 3.1 *For $j = 1, 2$, let f_j be a multiplicative function such that $|f_j(n)| \le \tau(n)^\kappa$, and let t_j be a real number. We have*

$$\mathbb{D}(f_1, n^{it_1}; x_1) + \mathbb{D}(f_2, n^{it_2}; x_2) \ge \frac{1}{\sqrt{2^\kappa}}\mathbb{D}(f_1 f_2, n^{i(t_1+t_2)}; \min\{x_1, x_2\}),$$

where the distance function on the right-hand side is defined with respect to the condition $|f_1(n)f_2(n)| \le \tau(n)^{2\kappa}$.

Proof For simplicity, let us denote $\mathbb{D}_j = \mathbb{D}(f_j, n^{it_j}; x_j)$ for $j = 1, 2$, $x = \min\{x_1, x_2\}$, and $\mathbb{D} = \mathbb{D}(f_1 f_2, n^{i(t_1+t_2)}; x)$. Then, we have

$$\begin{aligned}(\mathbb{D}_1+\mathbb{D}_2)^2 &= \sum_{p\le x_1}\frac{2^\kappa-\mathrm{Re}(f_1(p)p^{-it_1})}{p}+\sum_{p\le x_2}\frac{2^\kappa-\mathrm{Re}(f_2(p)p^{-it_2})}{p} \\ &\quad +2\left(\sum_{p\le x_1}\frac{2^\kappa-\mathrm{Re}(f_1(p)p^{-it_1})}{p}\right)^{1/2}\left(\sum_{p\le x_2}\frac{2^\kappa-\mathrm{Re}(f_2(p)p^{-it_2})}{p}\right)^{1/2} \\ &\ge \sum_{p\le x}\frac{2^{\kappa+1}-\mathrm{Re}(f_1(p)p^{-it_1})-\mathrm{Re}(f_2(p)p^{-it_2})}{p}\end{aligned}$$

$$+2\left(\sum_{p\le x}\frac{2^\kappa-\mathrm{Re}(f_1(p)p^{-it_1})}{p}\right)^{1/2}\left(\sum_{p\le x}\frac{2^\kappa-\mathrm{Re}(f_2(p)p^{-it_2})}{p}\right)^{1/2}$$

$$\ge\sum_{p\le x}\frac{2^\kappa}{p}\left[2-\mathrm{Re}\left(\frac{f_1(p)}{2^\kappa}p^{-it_1}\right)-\mathrm{Re}\left(\frac{f_2(p)}{2^\kappa}p^{-it_2}\right)\right]$$

$$+\sum_{p\le x}\frac{2^\kappa}{p}\sqrt{2\left(1-\mathrm{Re}\left(\frac{f_1(p)}{2^\kappa}p^{-it_1}\right)\right)}\sqrt{2\left(1-\mathrm{Re}\left(\frac{f_2(p)}{2^\kappa}p^{-it_2}\right)\right)}$$

$$\ge\sum_{p\le x}\frac{2^\kappa}{p}\left[2-\mathrm{Re}\left(\frac{f_1(p)}{2^\kappa}p^{-it_1}\right)-\mathrm{Re}\left(\frac{f_2(p)}{2^\kappa}p^{-it_2}\right)+\left|\mathrm{Im}\left(\frac{f_1(p)}{2^\kappa}p^{-it_1}\right)\mathrm{Im}\left(\frac{f_2(p)}{2^\kappa}p^{-it_2}\right)\right|\right],$$

where the last inequality follows from $\left|\frac{f_j(p)}{2^\kappa}p^{-it_j}\right|\le 1$ for $j=1,2$, giving the inequalities

$$\mathrm{Im}\left(\frac{f_j(p)}{2^\kappa}p^{-it_j}\right)^2\le 1-\mathrm{Re}\left(\frac{f_j(p)}{2^\kappa}p^{-it_j}\right)^2\le 2\left(1-\mathrm{Re}\left(\frac{f_j(p)}{2^\kappa}p^{-it_j}\right)\right).$$

The condition $\left|\frac{f_j(p)}{2^\kappa}p^{-it_j}\right|\le 1$ also gives us

$$2-\mathrm{Re}\left(\frac{f_1(p)}{2^\kappa}p^{-it_1}\right)-\mathrm{Re}\left(\frac{f_2(p)}{2^\kappa}p^{-it_2}\right)+\left|\mathrm{Im}\left(\frac{f_1(p)}{2^\kappa}p^{-it_1}\right)\mathrm{Im}\left(\frac{f_2(p)}{2^\kappa}p^{-i\phi_2}\right)\right|$$

$$\ge 1-\mathrm{Re}\left(\frac{f_1(p)}{2^\kappa}p^{-it_1}\right)\mathrm{Re}\left(\frac{f_2(p)}{2^\kappa}p^{-it_2}\right)+\mathrm{Im}\left(\frac{f_1(p)}{2^\kappa}p^{-it_1}\right)\mathrm{Im}\left(\frac{f_2(p)}{2^\kappa}p^{-it_2}\right)$$

$$=1-\mathrm{Re}\left(\frac{f_1(p)f_2(p))}{2^{2\kappa}}p^{-i(t_1+t_2)}\right).$$

It then follows that

$$(\mathbb{D}_1+\mathbb{D}_2)^2\ge\sum_{p\le x}\frac{2^\kappa}{p}\left(1-\mathrm{Re}\left(\frac{f_1(p)f_2(p))}{2^{2\kappa}}p^{-i(\phi_1+\phi_2)}\right)\right)$$

$$=\frac{1}{2^\kappa}\sum_{p\le x}\frac{2^{2\kappa}-\mathrm{Re}(f_1(p)f_2(p)p^{-i(\phi_1+\phi_2)})}{p}=\frac{1}{2^\kappa}\mathbb{D}^2,$$

as desired. □

To conclude this section, we exhibit lower bounds of sums of multiplicative functions in terms of the distance function. More precisely, we show that if the distance $\mathbb{D}(f,n^{i\psi};x)$ is small then the partial sums of f get large in suitable ranges.

Proposition 3.2 *Let f be a multiplicative function satisfying $|f(n)|\le\tau(n)^\kappa$ for all $n\in\mathbb{N}$. Let $\psi=\psi_f(x,(\log x)^{2^\kappa})$ be a number in the range $|t|\le(\log x)^{2^\kappa}$ where the maximum in* (2.2) *is attained. Let $x\gg 1$ and set $\lambda=\mathbb{D}^2(f,n^{i\psi};x)+\log(1+|\psi|)+c$ where c is a suitably large constant (depending only on κ). Then, there exists $y\in[x^\gamma,x]$, with $\gamma=1/(\lambda e^\lambda)$, such that*

$$\left|\sum_{n\le y} f(n)\right| \gg \frac{\exp(-\mathbb{D}^2(f,n^{i\psi};x))}{1+|\psi|}\, y(\log y)^{2^\kappa-1}. \tag{3.4}$$

Proof The strategy of the proof is to produce contradictory upper and lower bounds on the quantity

$$\left|L\left(1+\frac{2\lambda}{\log x}+i\psi, f\right)\right|,$$

under the assumption that (3.4) fails.

For the lower bound, we observe that

$$\begin{aligned}
\left|L\left(1+\frac{2\lambda}{\log x}+i\psi, f\right)\right| &= \exp\left(\sum_{p\le x^{\frac{1}{2\lambda}}} \frac{\mathrm{Re}(f(p)p^{-i\psi})}{p} + O(1)\right)\\
&\ge \exp\left(\sum_{p\le x^{\frac{1}{2\lambda}}} \frac{\mathrm{Re}(f(p)p^{-i\psi})}{p} - \sum_{x^{\frac{1}{2\lambda}}<p\le x} \frac{2^\kappa-\mathrm{Re}(f(p)p^{-i\psi})}{p} + O(1)\right)\\
&\ge \exp\left(\sum_{p\le x} \frac{\mathrm{Re}(f(p)p^{-i\psi})}{p} - \sum_{x^{\frac{1}{2\lambda}}\le p\le x} \frac{2^\kappa}{p} + O(1)\right)\\
&= (2\lambda)^{-2^\kappa}\exp\left(\sum_{p\le x} \frac{\mathrm{Re}(f(p)p^{-i\psi})}{p} + O(1)\right)\\
&= (2\lambda)^{-2^\kappa}(\log x)^{2^\kappa}\exp(-\mathbb{D}^2(f,n^{i\psi};x)).
\end{aligned} \tag{3.5}$$

We now move to the upper bound. Set $\delta := \frac{2\lambda}{\log x}$. By partial summation, we get

$$|L(1+\delta+i\psi, f)| \le |1+\delta+i\psi|\left(\int_1^\infty \frac{1}{y^{2+\delta}}\left|\sum_{n\le y} f(n)\right| dy\right). \tag{3.6}$$

For contradiction, we now assume that the upper bound

$$\left|\sum_{n\le y} f(n)\right| \le e^{-\lambda} y(\log y)^{2^\kappa-1} \tag{3.7}$$

holds for all $y \in [x^\gamma, x]$. We have

$$\begin{aligned}
\int_1^\infty \frac{1}{y^{2+\delta}}\left|\sum_{n\le y} f(n)\right| dy &= \int_1^{x^\gamma} \frac{1}{y^{2+\delta}}\left|\sum_{n\le y} f(n)\right| dy + \int_{x^\gamma}^{x} \frac{1}{y^{2+\delta}}\left|\sum_{n\le y} f(n)\right| dy + \int_{x}^{\infty} \frac{1}{y^{2+\delta}}\left|\sum_{n\le y} f(n)\right| dy\\
&\le \int_1^{x^\gamma} y^{-1-\delta}(\log y)^{2^\kappa-1}dy + e^{-\lambda}\int_{x^\gamma}^{x} y^{-1-\delta}(\log y)^{2^\kappa-1}dy + \int_x^\infty y^{-1-\delta}(\log y)^{2^\kappa-1}dy,
\end{aligned} \tag{3.8}$$

where we used (3.7) to bound $\left|\sum_{n\leq y} f(n)\right|$ in the interval $[x^{\gamma}, x]$ and the trivial bound (2.7) outside of this interval. Applying the change of variable $u = y^{\delta}$ in the three integrals appearing in (3.8), we get

$$\begin{aligned}\int_1^{\infty} \frac{1}{y^{2+\delta}} \left|\sum_{n\leq y} f(n)\right| dy &\leq \frac{1}{\delta^{2^\kappa}} \left(\int_1^{x^{\delta\gamma}} u^{-2}(\log u)^{2^\kappa-1} du + e^{-\lambda} \int_{x^{\delta\gamma}}^{x^{\delta}} u^{-2}(\log u)^{2^\kappa-1} du + \int_{x^\delta}^{\infty} u^{-2}(\log u)^{2^\kappa-1} du\right)\\ &\leq \frac{1}{\delta^{2^\kappa}} \left(\int_1^{x^{\delta\gamma}} u^{-2}(\log u)^{\lceil 2^\kappa\rceil-1} du + e^{-\lambda} \int_{x^{\delta\gamma}}^{x^{\delta}} u^{-2}(\log u)^{\lceil 2^\kappa\rceil-1} du + \int_{x^\delta}^{\infty} u^{-2}(\log u)^{\lceil 2^\kappa\rceil-1} du\right)\\ &= \frac{1}{\delta^{2^\kappa}} \left(\left[-\frac{P(\log u)}{u}\right]_1^{x^{\delta\gamma}} + e^{-\lambda}\left[-\frac{P(\log u)}{u}\right]_{x^{\delta\gamma}}^{x^\delta} + \left[-\frac{P(\log u)}{u}\right]_{x^\delta}^{\infty}\right)\\ &= \frac{e^{-\lambda}(\log x)^{2^\kappa}}{(2\lambda)^{2^\kappa}} \left((\lceil 2^\kappa\rceil - 1)! e^{\lambda} - \frac{e^{\lambda}P(2\gamma\lambda)}{e^{2\lambda\gamma}} + \frac{P(2\gamma\lambda)}{e^{2\lambda\gamma}} - \frac{P(2\lambda)}{e^{2\lambda}} + \frac{e^{\lambda}P(2\lambda)}{e^{2\lambda}}\right),\end{aligned}$$

where

$$P(X) := (\lceil 2^\kappa\rceil - 1)! \sum_{k=0}^{\lceil 2^\kappa\rceil-1} \frac{X^k}{k!}.$$

We want to verify that the expression inside of the parenthesis is bounded as $\lambda \to \infty$. Recall that $\gamma = \frac{1}{\lambda e^{\lambda}}$ and $\kappa > 0$, so that

$$\frac{P(2\gamma\lambda)}{e^{2\lambda\gamma}} - \frac{P(2\lambda)}{e^{2\lambda}} + \frac{e^{\lambda}P(2\lambda)}{e^{2\lambda}} \ll 1.$$

For the first two terms in the parenthesis, let $Q(X) := P(X) - (\lceil 2^\kappa\rceil - 1)!$. We have

$$(\lceil 2^\kappa\rceil - 1)! e^{\lambda} - \frac{e^{\lambda}P(2\gamma\lambda)}{e^{2\lambda\gamma}} = (\lceil 2^\kappa\rceil - 1)! e^{\lambda}(1 - e^{-2\lambda\gamma}) + \frac{e^{\lambda}Q(2e^{-\lambda})}{e^{2e^{-\lambda}}} = (\lceil 2^\kappa\rceil - 1)! e^{\lambda}(1 - e^{-2\lambda\gamma}) + O(1),$$

since $Q(X)$ is a polynomial with vanishing constant term. It remains to bound $(\lceil 2^\kappa\rceil - 1)! e^{\lambda}(1 - e^{-2\lambda\gamma})$. The Taylor expansion of e^X for $X = o(1)$ gives

$$(\lceil 2^\kappa\rceil - 1)! e^{\lambda}(1 - e^{-2\lambda\gamma}) = (\lceil 2^\kappa\rceil - 1)! e^{\lambda}(2\lambda\gamma + O((2\lambda\gamma)^2)) = (\lceil 2^\kappa\rceil - 1)! e^{\lambda}(2e^{-\lambda} + O(e^{-2\lambda})) = O(1),$$

where the implicit constant depends on κ. We conclude that

$$\left|L\left(1 + \frac{2\lambda}{\log x} + i\psi, f\right)\right| \ll \left|1 + \frac{2\lambda}{\log x} + i\psi\right| \frac{e^{-\lambda}(\log x)^{2^\kappa}}{(2\lambda)^{2^\kappa}}.$$

From the definition of λ, we have

$$\lambda = \mathbb{D}^2(f, n^{i\psi}; x) + \log(1 + |\psi|) + c = O(\log\log x),$$

so that

$$\left|L\left(1 + \frac{2\lambda}{\log x} + i\psi, f\right)\right| \ll |1 + i\psi| \frac{\exp(-\mathbb{D}^2(f, n^{i\psi}; x))\exp(-\log|1 + i\psi| - c)(\log x)^{2^\kappa}}{(2\lambda)^{2^\kappa}}$$

$$= \frac{\exp(-\mathbb{D}^2(f, n^{i\psi}; x)) \exp(-c)(\log x)^{2^\kappa}}{(2\lambda)^{2^\kappa}}.$$

Comparing the upper bound from this last line with the conclusion of (3.5), we get

$$\frac{\exp(-\mathbb{D}^2(f, n^{i\psi}; x))(\log x)^{2^\kappa}}{(2\lambda)^{2^\kappa}} \ll \frac{\exp(-\mathbb{D}^2(f, n^{i\psi}; x))e^{-c}(\log x)^{2^\kappa}}{(2\lambda)^{2^\kappa}},$$

which can be made false by choosing c large enough (depending on κ). It follows that Assumption (3.7) is false, and that the result holds. □

4 Proof of Theorem 1.1

Now, we will prove the main theorem. For $j = 1, 2$, we set $\phi_j = \phi_{f_j}(x_j, (\log x_j)^{2^\kappa})$ and $N_j = N_{f_j}(x_j, (\log x_j)^{2^\kappa})$ as in (2.5). We also set $f = f_1 f_2$ and $\phi = \phi_f(x, (\log x)^{2^{2\kappa}})$ with $x = \min\{x_1, x_2\}$. By Lemma 3.1, we have

$$\begin{aligned}\mathbb{D}^2(f_1 f_2, n^{i(\phi_1+\phi_2)}; x) &\le 2^\kappa \left[\mathbb{D}(f_1, n^{i\phi_1}; x_1) + \mathbb{D}(f_2, n^{i\phi_2}; x_2)\right]^2 \\ &\le 2^{\kappa+2} \max_j \left\{\mathbb{D}^2(f_j, n^{i\phi_j}; x_j)\right\}.\end{aligned}$$

Together with (3.3), it follows that

$$\exp\left(-\mathbb{D}^2(f_1 f_2, n^{i(\phi_1+\phi_2)}; x)\right) \ge \min_j \left\{e^{-2^{\kappa+2} N_j}\right\}.$$

Applying Lemma 2.8 yields

$$\exp\left(-\mathbb{D}^2(f_1 f_2, n^{i(\phi_1+\phi_2)}; x)\right) \ge \left(\eta^2\right)^{2^{\kappa+2}} = \eta^{2^{\kappa+3}}. \tag{4.1}$$

On the other hand, by Proposition 3.2, there exists $y \in [x^{1/(\lambda e^\lambda)}, x]$ such that

$$\begin{aligned}\left|\sum_{n\le y} f_1(n) f_2(n)\right| &\gg \frac{\exp(-\mathbb{D}^2(f, n^{i\psi}; x)) y (\log y)^{2^{2\kappa}-1}}{1 + |\psi|} \\ &\gg \frac{\exp(-\mathbb{D}^2(f, n^{i(\phi_1+\phi_2)}; x)) y (\log y)^{2^{2\kappa}-1}}{1 + |\phi_1 + \phi_2|},\end{aligned} \tag{4.2}$$

where we recall that $\psi = \psi_f(x, (\log x)^{2^{2\kappa}})$ is a number in the range $|t| \le (\log x)^{2^{2\kappa}}$, achieving the maximum in (2.2), and $\lambda = \mathbb{D}^2(f, n^{i\psi}; x) + \log(1 + |\psi|) + c$ for some sufficiently large constant c. Observe that the last inequality in (4.2) follows from

$$\begin{aligned}\frac{(\log x)^{2^{2\kappa}}\exp(-\mathbb{D}^2(f,n^{i\psi_f};x))}{\left|1+\frac{1}{\log x}+i\psi_f\right|} &\asymp \frac{\left|L\left(1+\frac{1}{\log x}+i\psi_f,f\right)\right|}{\left|1+\frac{1}{\log x}+i\psi_f\right|} = \max_{|t|\le(\log x)^{2^{2\kappa}}}\frac{\left|L\left(1+\frac{1}{\log x}+it,f\right)\right|}{\left|1+\frac{1}{\log x}+it\right|}\\ &\ge \frac{\left|L\left(1+\frac{1}{\log x}+i(\phi_1+\phi_2),f\right)\right|}{\left|1+\frac{1}{\log x}+i(\phi_1+\phi_2)\right|}\\ &\asymp \frac{(\log x)^{2^{2\kappa}}\exp(-\mathbb{D}^2(f,n^{i(\phi_1+\phi_2)};x))}{\left|1+\frac{1}{\log x}+i(\phi_1+\phi_2)\right|}.\end{aligned}$$

Now, putting (4.1) and (4.2) together gives

$$\left|\sum_{n\le y} f_1(n)f_2(n)\right| \gg \frac{\eta^{2^{\kappa+3}}y(\log y)^{2^{2\kappa}-1}}{1+|\phi_1+\phi_2|}.$$

Finally, since $1+|\phi_1+\phi_2|\le 1+|\phi_1|+|\phi_2|\ll 1/\eta$ by Lemma 2.8, we conclude that

$$\left|\sum_{n\le y} f_1(n)f_2(n)\right| \gg \eta^{1+2^{\kappa+3}}y(\log y)^{2^{2\kappa}-1},$$

which completes the proof by observing that

$$\begin{aligned}\frac{1}{\lambda e^{\lambda}} &\ge e^{-2\lambda}\\ &= \left(\exp(-\mathbb{D}^2(f,n^{i\psi};x)-\log(1+|\psi|)-c)\right)^2\\ &\gg \left(\frac{\exp(-\mathbb{D}^2(f,n^{i(\phi_1+\phi_2)};x))}{\left|1+\frac{1}{\log x}+i(\phi_1+\phi_2)\right|}\right)^2\\ &\gg \left(\eta^{1+2^{\kappa+3}}\right)^2,\end{aligned}$$

and setting $\xi = C\eta^{1+2^{\kappa+3}}$, for some positive constant C depending only on κ. □

Acknowledgements This project originated from Women In Numbers 6 Research Workshop that took place at Banff International Research Station in March 2023. The authors express their utmost gratitude to the organizers for the invaluable opportunity provided by the workshop. They also extend their appreciation to the referee for their suggestions, which have improved the exposition of the paper. The third author would also like to thank Oleksiy Klurman for answering a question related to this work.

References

[HD80] H. Davenport, *Multiplicative number theory*, vol. 74 of Graduate Texts in Mathematics, Springer-Verlag, New York-Berlin, second ed., 1980. Revised by Hugh L. Montgomery.

[FBHT25] C. Frechette, M. Gerbelli- Gauthier, A. Hamieh, and N. Tanabe, *Large sums of fourier coefficients of cusp forms*, J. Théor. Nombres Bordeaux 37 (2025), no. 1, 171–188.

[GHS19] A. Granville, A. J. Harper, and K. Soundararajan, *A new proof of Halász's theorem, and its consequences*, Compos. Math., 155 (2019), pp. 126–163.

[GS01] A. Granville and K. Soundararajan, *Large character sums*, J. Amer. Math. Soc., 14 (2001), pp. 365–397.

[GS03] A. Granville and K. Soundararajan, *Decay of mean values of multiplicative functions*, Canad. J. Math., 55 (2003), pp. 1191–1230.

[GS18] A. Granville and K. Soundararajan, *Large character sums: Burgess's theorem and zeros of L-functions*, J. Eur. Math. Soc. (JEMS), 20 (2018), pp. 1–14.

[GH71] G. Halász, *On the distribution of additive and the mean values of multiplicative arithmetic functions*, Studia Sci. Math. Hungar., 6 (1971), pp. 211–233.

[GH75] G. Halász, *On the distribution of additive arithmetic functions*, Acta Arith., 27 (1975), pp. 143–152.

[AH86] A. Hildebrand, *A note on Burgess' character sum estimate*, C. R. Math. Rep. Acad. Sci. Canada, 8 (1986), pp. 35–37.

[YL19] Y. Lamzouri, *Large sums of Hecke eigenvalues of holomorphic cusp forms*, Forum Math., 31 (2019), pp. 403–417.

[FLLT17] F. Luca and L. Tóth, *The rth moment of the divisor function: an elementary approach*, J. Integer Seq., 20 (2017), pp. Art. 17.7.4, 8.

[APM18] A. P. Mangerel, *Topics in multiplicative and probabilistic number theory*, University of Toronto (Canada), 2018.

[APM23] A. P. Mangerel, *Divisor-bounded multiplicative functions in short intervals*, Research in the Mathematical Sciences, 10 (2023), pp. 1–47.

[LM18] L. Matthiesen, *Generalized Fourier coefficients of multiplicative functions*, Algebra Number Theory, 12 (2018), pp. 1311–1400.

[HM] H. Montgomery, *A note on the mean values of multiplicative functions.*

[LMCV77] H. L. Montgomery and R. C. Vaughan, *Exponential sums with multiplicative coefficients*, Invent. Math., 43 (1977), pp. 69–82.

[GT08] G. Tenenbaum, *Introduction to analytic and probabilistic number theory*, vol. 163 of Graduate Studies in Mathematics, American Mathematical Society, Providence, RI, third ed., 2015. Translated from the 2008 French edition by Patrick D. F. Ion.

[GT17] G. Tenenbaum, *Moyennes effectives de fonctions multiplicatives complexes*, The Ramanujan Journal, 44 (2017), pp. 641–701.

Prime and Thickened Prime Components in Apollonian Circle Packings

Holley Friedlander, Elena Fuchs, Piper Harris, Catherine Hsu, James Rickards, Katherine Sanden, Damaris Schindler, and Katherine E. Stange

Abstract Inspired by a question of Sarnak, we introduce the notion of a *prime component* in an Apollonian circle packing: a maximal tangency-connected subset having all prime curvatures. We also consider *thickened prime components*, which are augmented by all circles immediately tangent to the prime component. In both cases, we ask about the curvatures which appear. We consider the residue classes attained by

This material is based upon work supported by the National Security Agency under Grant No. H98230-19-1-0119, The Lyda Hill Foundation, The McGovern Foundation, and Microsoft Research, while the authors were in residence at the Mathematical Sciences Research Institute in Berkeley, California, during the summer of 2019. Rickards and Stange were supported by NSF-CAREER CNS-1652238 (PI Stange). Stange was supported by NSF DMS-2401580. Schindler was supported by a NWO grant 016.Veni.173.016. Fuchs was supported by NSF Grant DMS-2154624. This work utilized the Alpine high performance computing resource at the University of Colorado Boulder. Alpine is jointly funded by the University of Colorado Boulder, the University of Colorado Anschutz, and Colorado State University.

H. Friedlander
Dickinson College, Carlisle, PA, USA
e-mail: friedlah@dickinson.edu
URL: https://sites.google.com/view/hfriedlander

E. Fuchs
University of California Davis, Davis, CA, USA
e-mail: efuchs@math.ucdavis.edu
URL: https://www.math.ucdavis.edu/ efuchs/

P. Harris
Davis, CA, USA
e-mail: piperhphd@gmail.com
URL: https://www.theliberatedmathematician.com/

C. Hsu
Swarthmore College, Swarthmore, PA, USA
e-mail: chsu2@swarthmore.edu
URL: https://chsu.domains.swarthmore.edu/

J. Rickards
Saint Mary's University, Halifax, NS, Canada
e-mail: james.rickards@smu.ca

S. Akhtari et al. (eds.), *Research Directions in Number Theory*, Association for Women in Mathematics Series 39, https://doi.org/10.1007/978-3-032-11183-8_8

the set of curvatures, the number of circles in such components, the number of distinct integers occurring as curvatures and the number of prime components in a packing. As part of our investigation, we computed and analysed example components up to around curvature 10^{13}; software is available.

2020 Mathematics Subject Classification Primary: 52C26 · 11D09 · 11N32; Secondary: 11E12 · 11N36

1 Introduction

An Apollonian circle packing, or ACP, is a fractal set in the plane, obtained by repeatedly adding circles into an initial constellation of three (Fig. 1). Specifically, starting from a collection of four pairwise tangent circles C_1, C_2, C_3, C_4, one adds all circles that are simultaneously tangent to a subset of three of the original circles: by a theorem of Apollonius, for any set of three pairwise tangent circles, there are exactly two circles simultaneously tangent to that set.[1] One now has eight circles and can continue the process using any new set of three mutually tangent circles in the picture. Figure 1 depicts this process.

Proceeding ad infinitum, one obtains a packing of infinitely many circles. Remarkably, if any four pairwise tangent circles in the packing have integer curvature (the curvature is the reciprocal of the radius), all of the circles in the packing have integer curvature (Fig. 2). Throughout this paper, every integral packing we consider is further assumed to be *primitive*, meaning that the greatest common divisor of all the curvatures of circles in the packing is 1. This simple integrality fact, first observed by Nobel prize laureate in chemistry Frederick Soddy in the early 1900s, inspires many fascinating number theoretic questions about Apollonian packings: for example, which integers appear as curvatures in a given packing? Are there infinitely many primes in any given such packing?

It is known that the number of circles with curvatures $\leq X$ is asymptotic to $cX^{1.3056...}$, for some constant c which depends on the choice of packing [KO11]. In an

URL: https://jamesrickards-canada.github.io/

K. Sanden
Göttingen, Germany
e-mail: katherine.sanden@gmail.com

D. Schindler
Göttingen University, Göttingen, Germany
e-mail: damaris.schindler@mathematik.uni-goettingen.de
URL: https://sites.google.com/site/damarishomepage/

K. E. Stange (✉)
University of Colorado Boulder, Boulder, CO, USA
e-mail: kstange@math.colorado.edu
URL: https://math.katestange.net/

[1] One views a line as a circle with infinite radius.

integral packing, then, the average multiplicity of individual curvatures approaches infinity. It is therefore natural to ask: do all integers appear? In fact they do not, since it is known that any packing may entirely avoid certain residue classes modulo 24 [FS11]. We call a curvature *admissible* if it is allowed by congruence obstructions. Then, according to the philosophy that this *local* (congruence) obstruction is the only obstruction, it was conjectured by [GLM03, FS11] that all sufficiently large admissible integers will appear. This is called the *local-to-global* conjecture. It was only very recently discovered that there is also a type of obstruction arising from quadratic reciprocity, which rules out families of the form cn^2 or cn^4 amongst curvatures [HKRS24], disproving the conjecture. The current conjecture now has the following form.

Conjecture 1 (Apollonian curvature conjecture [HKRS24, Conjecture 1.5], modifying [GLM03, FS11]) *Let $\mathcal{P}^{\mathrm{full}}$ be a primitive integral packing, and let Σ be the union of the residue classes modulo 24 that have representatives in $\mathcal{P}^{\mathrm{full}}$. Let S be the union of the curvature families ruled out by reciprocity obstructions catalogued in [HKRS24, Theorem 2.5]. Then every sufficiently large integer m with $m \in \Sigma \setminus S$ occurs as a curvature in $\mathcal{P}^{\mathrm{full}}$.*

This conjecture remains the central goal in the field. In this paper, we consider certain subsets of the Apollonian circle packing, and ask the same questions.

- A *prime component* is a maximal tangency-connected subset of circles whose curvatures are all odd primes.

Indeed, not a lot is known about circles of prime curvature in an Apollonian packing. Sarnak first defined such components, after observing that, given any circle C in the packing, there is a positive definite binary quadratic form $f_C(x, y)$ such that the curvatures of all circles tangent to C in $\mathcal{P}$ are exactly the integers primitively represented by $f_C(x, y) - a_C$, where a_C is the curvature of C [Sar07]. This implies that any fixed circle has infinitely many tangent circles of prime curvature, and consequently, prime components are infinite (see Proposition 3). Sarnak points out

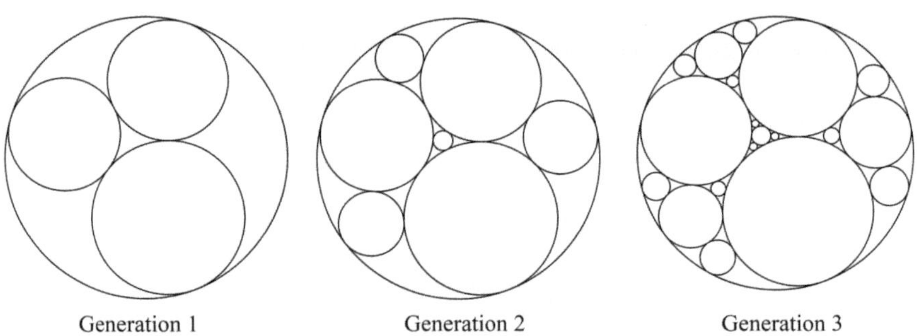

Generation 1 Generation 2 Generation 3

Fig. 1 The construction of an Apollonian packing

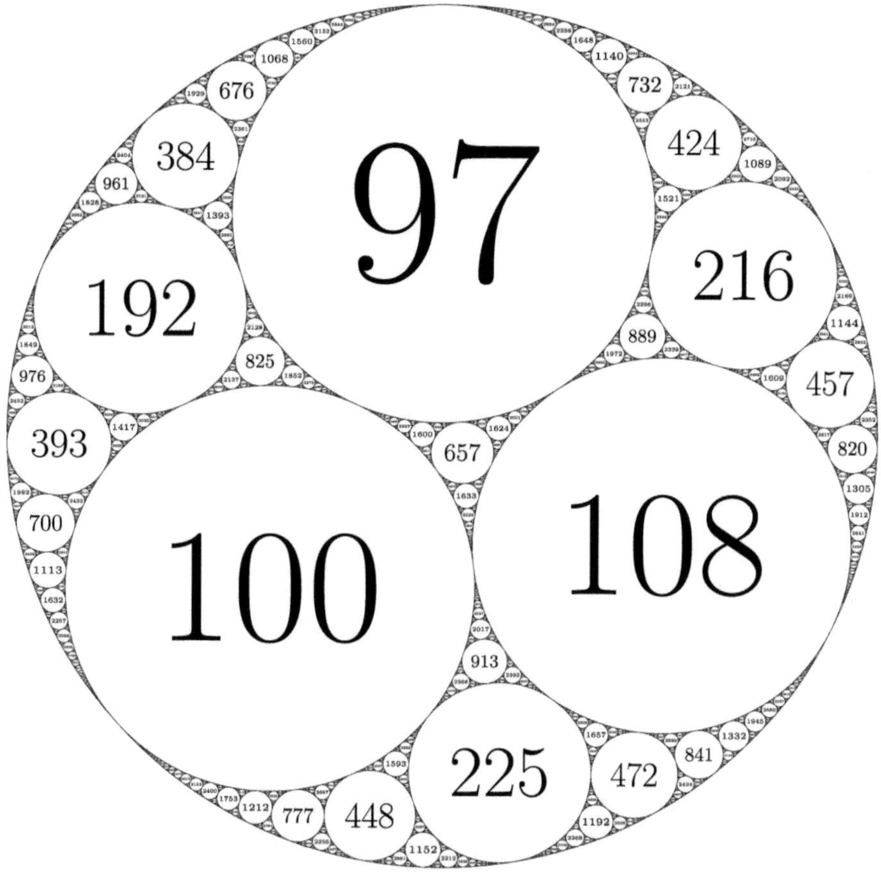

Fig. 2 Circles of curvature ≤ 30000 in the Apollonian circle packing corresponding to $(-47, 97, 100, 108)$

that one gets infinite trees of circles connected by tangencies, all of prime curvature (Fig. 3).

Evidently, prime components cannot satisfy Conjecture 1 as stated, but we might ask, do all sufficiently large primes appear? There is a long list of natural questions to ask about prime components in an ACP:

- *Local Questions*: Modulo a given integer d, do the primes in a given prime component mimic the primes in the whole packing?
- *Growth with Multiplicity*: Given a real number X, how many circles of curvature at most X are there in a fixed prime component?
- *Positive Density or Local-to-Global (growth without multiplicity)*: How many primes less than X appear as curvatures in a fixed prime component? Do a positive density of primes appear? All but finitely many? (Reciprocity obstructions do not apply to primes.)

- *Number of Components*: Can we count prime components in a given ACP?

In an attempt to reach closer to existing tools, it is natural to consider also thickened prime components.

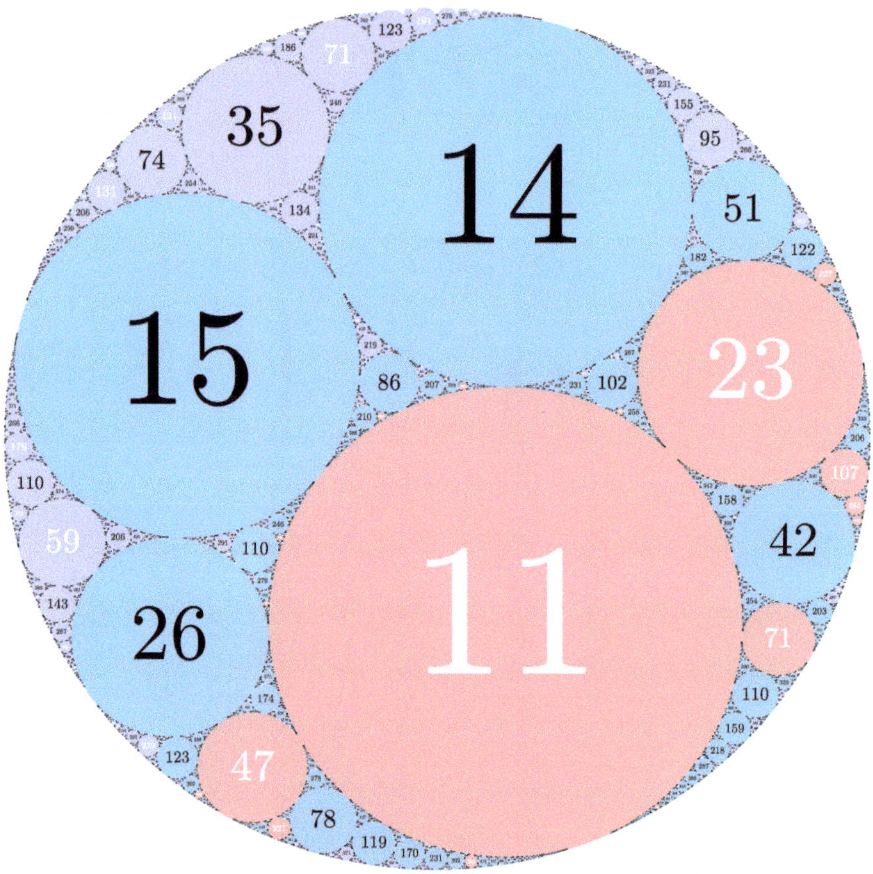

Fig. 3 The packing informally known as the *People's packing*, up to curvature 10,000. In pink, a prime component. In blue, the additional circles in the thickened prime component. All other circles are purple

- A *thickened prime component* is a prime component augmented by adding every circle which is tangent to it.

These might, naïvely, be expected to attain all the integers attained by the ambient packing. Do they? One can hope that, if the arithmetic structure of thickened prime components is rich enough, we might be able to answer some questions:

- *Local Questions*: Modulo a given integer d, do the curvatures in a thickened prime component mimic the curvatures in the whole packing?
- *Growth with Multiplicity*: For a real number X, how many circles of curvature less than X are there in a fixed thickened prime component?
- *Positive Density*: Do a positive density of all integers appear as curvatures in a fixed thickened prime component?
- *Curvature Conjecture*: Do the curvatures in a given thickened prime component satisfy a conjecture similar to Conjecture 1 in the whole packing?

Although we address this wide variety of questions, let us state the analogue to Conjecture 1 as the motivating conjecture for the entire paper:

Conjecture 2 *(1) Let $\mathcal{P}^{\text{pr}}$ be a prime component. Then the set of curvatures appearing in $\mathcal{P}^{\text{pr}}$ is the same as the set of primes appearing in the ambient packing $\mathcal{P}^{\text{full}}$, up to finitely many exceptions.*
(2) Let $\mathcal{P}\text{th}$ be a thickened prime component. Then the set of curvatures appearing in $\mathcal{P}\text{th}$ is the same as the set of curvatures appearing in the ambient packing $\mathcal{P}^{\text{full}}$, up to finitely many exceptions.

This conjecture remains open. Among our varied results, we obtain, for example, the following:

Theorem 1 *Let $\mathcal{P}\text{th}$ be a thickened prime component in a primitive integral Apollonian packing. Then the number $N_{\mathcal{P}\text{th}}(X)$ of distinct integers less than X occuring in $\mathcal{P}\text{th}$ satisfies*

$$N_{\mathcal{P}\text{th}}(X) \gg \frac{X}{(\log\log X)^{1/2}}.$$

Throughout the paper, the notation $g \gg f$ means that g is bounded below by some positive constant multiple of f.

1.1 Local-to-Global for Apollonian Packings

Let us begin with some facts about Apollonian packings as a whole. Progress towards proving Conjecture 1 (in its various forms) dates back to [GLM03], who first examined the local properties of Apollonian packings. They took advantage of certain parameterized sets of circles to show that, if d is relatively prime to 30, every residue class modulo d appears among the curvatures in a primitive ACP. Fuchs gave a stronger condition, showing that the modulus 24 captures all possible congruence obstructions [Fuc11].

Shifting from local to local-to-global, let $N_{\mathcal{P}^{\text{full}}}(X)$ denote the number of distinct integers less than X occurring as curvatures in the packing $\mathcal{P}^{\text{full}}$. Then Conjecture 1 would imply

$$N_{\mathcal{P}^{\text{full}}}(X) = \frac{r_{\mathcal{P}^{\text{full}}}}{24}X + O(\sqrt{X}), \tag{1}$$

where $r_{\mathcal{P}^{\text{full}}} \in \{6, 8\}$ is the number of residue classes modulo 24 which appear in the packing.

A first lower bound on $N_{\mathcal{P}^{\text{full}}}(X)$ that comes relatively close to the expected growth was obtained by Sarnak [Sar07].

Theorem 2 ([Sar07]) *We have*

$$N_{\mathcal{P}^{\text{full}}}(X) \gg \frac{X}{\sqrt{\log X}}.$$

This theorem follows from Sarnak's observation on quadratic forms. It is this observation that led to a series of results, each coming closer to the desired local to global conjecture. Building on this observation, Bourgain-Fuchs showed that a positive proportion of integers occurs in an integral Apollonian circle packing.

Theorem 3 ([BF11]) *We have*

$$N_{\mathcal{P}^{\text{full}}}(X) \gg X.$$

Taking this method further, Bourgain and Kontorovich showed that the local-global conjecture holds for almost all positive integers $m \leq X$ [BK14], a result which has later been generalized to a larger class of Kleinian groups in [FSZ19].

Theorem 4 ([BK14]) *For some $\epsilon > 0$,*

$$N_{\mathcal{P}^{\text{full}}}(X) = \frac{r_{\mathcal{P}^{\text{full}}}}{24}X + O(X^{1-\epsilon}).$$

Though this falls just short of (1), it has a nice consequence concerning *prime* curvatures: since X^{ϵ} exceeds $\log(X)$, it implies positive density of primes as curvatures in a primitive Apollonian packing. This is not apparent from techniques in [BF11] alone, but has already been proved by Bourgain in [Bou12].

The study of prime components presents one novel new difficulty. The curvatures of circles in the whole ACP can be viewed as coordinates of vectors in an orbit of a linear group called the *Apollonian group*. This group acts on quadruples of curvatures of four pairwise tangent circles in the packing. By contrast, neither the prime components nor the thickened prime components constitute a full group orbit. Our results rely heavily on the surviving tool: Sarnak's observation about the relationship of curvatures in the packing and integers represented by shifted binary quadratic forms.

Our main results are as follows.

1.2 Local Results

In Sects. 5 and 6, we investigate local obstructions in prime and thickened prime components (meaning congruence restrictions with respect to a modulus m).

In Sect. 5, the methods we use depend on Sarnak's original observation in [Sar07] that certain families of curvatures appearing in a primitive Apollonian circle packing are represented by shifted binary quadratic forms. In particular, our first result will depend on some statements concerning primes represented primitively by shifted quadratic forms which appear in [FHR25]; we state these results in Theorem 9. Using this theorem, we prove the following:

Theorem 5 (Theorem 3) *Let $m \in \mathbb{Z}_{>1}$ be coprime to 6, and let $\ell \in (\mathbb{Z}/m\mathbb{Z})^\times$. Then, a prime component $\mathcal{P}^{\mathrm{pr}}$ (respectively a thickened prime component $\mathcal{P}\mathrm{th}$) contains infinitely many primes congruent to $\ell \pmod{m}$ among those circles within two tangencies of any sufficiently small fixed circle.*

In Sect. 6, we revisit the question of local obstructions in prime and thickened prime components from the perspective taken in [GLM03, Sect. 6]. In place of binary quadratic forms, we construct special types of matrix words in the Apollonian group, written in terms of the generators S_1, S_2, S_3 and S_4, and show, conditional on a well-known conjecture of Bunyakovsky, that the orbits of a Descartes quadruple under these special matrix words contain circles with curvatures in all residue class for moduli m coprime to 30. One interest of this approach is that instead of bounding the number of tangencies to reach a given residue class, we study the length of the word in the Apollonian group (in other words, the path in the Cayley graph). This is closer in spirit to the way strong approximation is proven for the Apollonian group.

Theorem 6 (Theorem 4) *The two-tangency path of Theorem 3 above corresponds to a walk in the Cayley graph whose length is bounded in terms of Bunyakovsky's Conjecture 4.*

1.3 Size of Prime Components

As far as growth of circles *with multiplicity* in a prime component, we can only show a trivial lower bound (Proposition 4). Let us write

$$C_{\mathcal{P}^{\mathrm{pr}}}(X) := \#\{C \in \mathcal{P}^{\mathrm{pr}} : \mathrm{curv}(C) \leq X\},$$
$$C_{\mathcal{P}\mathrm{th}}(X) := \#\{C \in \mathcal{P}\mathrm{th} : \mathrm{curv}(C) \leq X\}.$$

We make the following conjecture, based on experimental data:

Conjecture 3 *Let $\mathcal{P}^{\mathrm{pr}}$ be a prime component. Then*

$$\lim_{X\to\infty} \frac{C_{\mathcal{P}^{\mathrm{pr}}}(X)}{\pi(X)} = \infty.$$

Let $\mathcal{P}$th be a thickened prime component. Then

$$\lim_{X\to\infty} \frac{C_{\mathcal{P}\mathrm{th}}(X)}{X} = \infty.$$

1.4 Counting Integers in Prime Components

Combining Conjectures 1 and 2, we expect all but finitely many admissible primes to occur in $\mathcal{P}^{\mathrm{pr}}$; and for $\mathcal{P}$th, up to finitely many exceptions, to be subject to only congruence and reciprocity obstructions, as for the whole packing. More precisely, one expects firstly

$$N_{\mathcal{P}^{\mathrm{pr}}}(X) = \sum_{a \text{ admissible}} \#\{p \text{ prime} < X : p \equiv a \bmod 24\} + O(1) \sim \frac{s_{\mathcal{P}^{\mathrm{full}}}}{8}\pi(X),$$

where $s_{\mathcal{P}^{\mathrm{full}}} \in \{1, 2\}$ is the number of invertible admissible residue classes (see [HKRS24, Proposition 1.2] for a classification) and $8 = \varphi(24)$, and secondly,

$$N_{\mathcal{P}\mathrm{th}}(X) = \frac{r_{\mathcal{P}^{\mathrm{full}}}}{24}X + O(\sqrt{X}).$$

In Sect. 7, we show the first lower bound on integers appearing in a thickened prime component: Theorem 1 stated above. This uses the methods of [BF11], and it appears that this is the best one can do with these methods alone. The key idea in [BF11] is to consider all integers that are curvatures of circles two levels of tangencies away from a fixed circle. These are regarded as integers represented by an infinite family of shifted binary forms $f_a(x, y) - a$. A crucial part of this is to balance how many of these forms one considers (one needs enough to get a positive fraction of integers being represented) with how many integers are represented by more than one form (one doesn't want this to overwhelm the positive fraction of integers obtained by summing over all the forms). In our case every form must come from a circle of prime curvature, and there are simply not enough of these to obtain a positive fraction of integers while still keeping the size of the intersections small enough.

Nonetheless, in forthcoming work, we replace primes with r-almost primes—numbers with at most r prime factors—for sufficiently large r, and revisit positive density in thickened r-almost prime components.

1.5 Counting Prime Components

Another question that arises naturally is the number of prime components in a single Apollonian packing. It is not difficult to show there are infinitely many, but we provide a detailed heuristic count in Sect. 4, leading to the following conjecture.

Conjecture (Conjecture 5) *The number of prime components in an Apollonian circle packing $\mathcal{P}^{\text{full}}$ is asymptotic to*

$$c\frac{C_{\mathcal{P}^{\text{full}}}(X)}{\log X},$$

where c is an explicit constant.

1.6 Experimental Data

Finally, the paper includes an extensive computational investigation of the questions mentioned above, and accompanying software is available (see below). In particular, curvatures with multiplicity in prime and thickened prime components were computed up to around 10^{13}, and the results support Conjecture 3; we postulate some estimated growth rates derived naïvely from exploring the data. We also compute the frequency of curvature multiplicities, which have some unexpected and unexplained irregularities. Considering the slow growth rate expected of prime components, it is not surprising that a large proportion of integers are still of multiplicity zero in this range. Finally, we examine Conjecture 5 concerning the number of prime components, finding it plausible. The data, although extensive, is probably best interpreted as not having 'settled down' to its asymptotic behaviour.

1.7 A Note on the Images and Software

The PDF images here, in their original form, are of sufficient quality that an arbitrary-zoom-capable PDF viewer will allow the user to read any curvature up to the stated bound. Images were created using publicly available open source software created by the fifth author [Ric24a] which runs with Pari/GP [The24], with the aid of Sage Mathematics Software [The23] in the case of colour images.

The experimental data was collected using C and PARI/GP code, and methods to replicate the results can be found in [Ric24b].

1.8 Paper Structure

The paper is structured as follows. In Sects. 2–4, we go over the background on Apollonian packings and prime components and provide some heuristics towards counting prime components in an Apollonian circle packing. Sections 5 and 6 are devoted to local properties of prime and thickened prime components. Section 7 contains the proof of Theorem 1. In Sect. 8, we present data illustrating our results and supporting conjectures we make throughout the paper.

2 Background

2.1 Descartes Quadruples and the Apollonian Group

A natural way to think about the structure of the packing is to study its quadruples. A *Descartes quadruple* of circles is a collection of four circles which are pairwise tangent to one another. A *Descartes quadruple* of numbers is the quadruple of curvatures of a Descartes quadruple of circles. In this paper, we use this term both in reference to quadruples of circles and quadruples of curvatures.

Indeed, having determined one Descartes quadruple of a packing, one can find all quadruples in the packing as an orbit of the first. This is a consequence of the following theorem, which is commonly attributed to Descartes, but appeared in correspondence with Princess Elisabeth of Bohemia [Sha07, *The Correspondence*].

Theorem 7 (Descartes, Princess Elisabeth of Bohemia, 1643) *Let a, b, c, d be curvatures of four pairwise tangent circles, where we take the curvature of a circle internally tangent to the other three to have negative curvature. Then*

$$Q(a, b, c, d) := 2(a^2 + b^2 + c^2 + d^2) - (a + b + c + d)^2 = 0. \tag{2}$$

Thus, one can move from one quadruple to another, by fixing three of the variables in the equation above: given a Descartes quadruple (a, b, c, d) in a packing, one has that (a', b, c, d) is another Descartes quadruple in that packing, where

$$a + a' = 2(b + c + d).$$

From a more geometric point of view, according to Apollonius there are exactly two circles tangent to the circles of curvature b, c and d, and they are exactly the ones of curvature a and a'. If the original quadruple lies in a particular Apollonian circle packing, this 'swapping' process (swapping out one circle for its alternate) results in a new quadruple in the same packing.

There are four possible 'swaps' of this type. If $\mathcal{A} = \langle S_1, S_2, S_3, S_4\rangle$ is the subgroup of $\mathrm{O}_Q(\mathbb{Z})$ generated by

$$S_1 = \begin{pmatrix} -1 & 2 & 2 & 2 \\ 0 & 1 & 0 & 0 \\ 0 & 0 & 1 & 0 \\ 0 & 0 & 0 & 1 \end{pmatrix}, \quad S_2 = \begin{pmatrix} 1 & 0 & 0 & 0 \\ 2 & -1 & 2 & 2 \\ 0 & 0 & 1 & 0 \\ 0 & 0 & 0 & 1 \end{pmatrix},$$

$$S_3 = \begin{pmatrix} 1 & 0 & 0 & 0 \\ 0 & 1 & 0 & 0 \\ 2 & 2 & -1 & 2 \\ 0 & 0 & 0 & 1 \end{pmatrix}, \quad S_4 = \begin{pmatrix} 1 & 0 & 0 & 0 \\ 0 & 1 & 0 & 0 \\ 0 & 0 & 1 & 0 \\ 2 & 2 & 2 & -1 \end{pmatrix},$$

then the set of all Descartes quadruples in a packing containing the quadruple $\mathbf{v} = (a, b, c, d)$ is exactly the orbit $\mathcal{A} \cdot \mathbf{v}^T$. This was most likely first discovered by Hirst [Hir67], who called $\mathcal{A}$ the *Apollonian group.*

In [GLM03], it is shown that every packing has a unique *root quadruple* of curvatures, which essentially correspond to the four largest pairwise tangent circles in the packing. There are exactly two integral ACP's where this root quadruple corresponds to more than one Descartes quadruple of circles in the packing: the one generated by the quadruple $(-1, 2, 2, 3)$, where there are two Descartes quadruples with these curvatures, and the one generated by $(0, 0, 1, 1)$, where there are infinitely many quadruples with these curvatures. In these two cases, one simply makes a choice to call just one of these quadruples of circles the root, and all results in [GLM03] hold regardless of the choice.

More generally, we have the following result about how circles in a packing correspond to vectors in $\mathcal{A} \cdot \mathbf{v}^T$.

Lemma 1 ([GLM03], Theorem 3.3) *Let $\mathcal{P}^{\text{full}}$ be an Apollonian circle packing with root $\mathbf{v}$. Also, let $(\mathcal{A} \cdot \mathbf{v}^T)'$ denote the set $\mathcal{A} \cdot \mathbf{v}^T$ with $\mathbf{v}^T$ removed. Then, the set of circles in $\mathcal{P}^{\text{full}}$ that are not in the root quadruple are in one to one correspondence with $(\mathcal{A} \cdot \mathbf{v}^T)'$ through the following relation: the curvatures of the circles in $\mathcal{P}^{\text{full}}$, counted with multiplicity, consist of the four elements of $\mathbf{v}$ plus the largest elements of each vector in $(\mathcal{A} \cdot \mathbf{v}^T)'$.*

We have the following useful corollary:

Lemma 2 *If $C \in \mathcal{P}^{\text{full}}$ is a circle not in the root quadruple, then C appears as the largest curvature in exactly one Descartes quadruple of $\mathcal{P}^{\text{full}}$.*

We call this the *birth quadruple* of C.

Moreover, given a Descartes quadruple $\mathbf{v}$, if $S_{i_1} S_{i_2} \cdots S_{i_k}$ is a reduced word in the generators S_i above, meaning $i_j \neq i_{j+1}$ for any j, it is shown in the proof of [GLM03, Theorem 3.3] that the i_1-th coordinate of $(S_{i_1} S_{i_2} \cdots S_{i_k}) \cdot \mathbf{v}^T$ is the largest of all the coordinates, and in particular is larger than the i_1-th coordinate of $(S_{i_2} S_{i_3} \cdots S_{i_k}) \cdot \mathbf{v}^T$.

2.2 Local Properties

Let S be the set of curvatures of a primitive Apollonian circle packing $\mathcal{P}^{\text{full}}$. Then, when $\gcd(m, 6) = 1$ the set of residues of S modulo m is known to be the full set of residues modulo m. In fact, any existing congruence restriction on S is captured by listing the residue classes attained modulo 24, see [Fuc11]. The possible residue classes modulo 24 are classified in [HKRS24, Proposition 2.1]. The following more precise statement modulo 8 will sometimes be useful.

Proposition 1 *Let $\mathcal{P}^{\text{full}}$ be an Apollonian circle packing. Every Descartes quadruple has two even curvatures and two odd curvatures. Furthermore, exactly one of the following is true:*

(1) all odd curvatures in $\mathcal{P}^{\text{full}}$ are $1 \pmod 8$; or
(2) all odd curvatures in $\mathcal{P}^{\text{full}}$ are $5 \pmod 8$ or
(3) all odd curvatures in $\mathcal{P}^{\text{full}}$ are $3, 7 \pmod 8$, and tangent odd curvatures are not equivalent modulo 8.

Proof This follows from [HKRS24, Lemma 3.3]. □

2.3 Quadratic Families

Much of our paper utilizes the following observation of Sarnak's on quadratic forms related to Apollonian packings. The following was first observed in [Sar07], although we provide a proof here for completeness.

Proposition 2 ([Sar07]) *Let C_a be a circle in $\mathcal{P}^{\text{full}}$, having curvature a and lying within a Descartes quadruple (a, b, c, d). Let S be the set of circles tangent to C_a. Define*

$$f_a(x, y) - a = (b + a)x^2 + (a + b + d - c)xy + (d + a)y^2 - a.$$

Then, the multiset of curvatures of the circles in S is exactly the multiset of values $f_a(x, y) - a$ for (x, y) coprime integers.

An example is given in Fig. 4.

Proof Let $\mathcal{A}_1$ denote the subgroup of $\mathcal{A}$ generated by S_2, S_3 and S_4, i.e. $\mathcal{A}_1 = \langle S_2, S_3, S_4\rangle$. This result is obtained by noting that the circles tangent to C_a correspond to exactly those circles obtained by acting on the quadruple (a, b, c, d) by $\mathcal{A}_1$. More specifically, recall from [Sar07, BF11] that the orbit $\mathcal{A}_1 \cdot (a, b, c, d)^T$ can be expressed as the set of points

$$\begin{pmatrix} a \\ A_0(2k+1)^2 + 2B_0(2k+1)(2m) + C_0(2m)^2 - a \\ A_0(2k+1-2l)^2 + 2B_0(2k+1-2l)(2m-2n-1) + C_0(2m-2n-1)^2 - a \\ A_0(2l)^2 + 2B_0(2l)(2n+1) + C_0(2n+1)^2 - a \end{pmatrix} \tag{3}$$

where $A_0 = b + a,\ 2B_0 = a + b + d - c,\ C_0 = d + a$ and k, l, m, n are such that

$$\begin{pmatrix} 2k+1 & 2l \\ 2m & 2n+1 \end{pmatrix} \in \mathrm{SL}_2(\mathbb{Z}) \tag{4}$$

and thus, correspond to the integer values of $f_a(x, y) - a$ for coprime integers (x, y). □

An important ingredient in our analytic considerations are bounds concerning primes represented by shifted quadratic forms. The following is a special case of a result of Iwaniec.

Theorem 8 ([Iwa72, Theorem 1]) *Let f be a primitive positive definite integral binary quadratic form, with discriminant Δ not a perfect square. Let ω be a non-zero integer. Write $b_{f,\omega}(n)$ for the characteristic function for whether n is represented by $f(x, y) - \omega$. Then*

$$\sum_{\substack{p \leq X, \\ p \text{ prime}}} b_{f,\omega}(p) \gg\ll \frac{X}{(\log X)^{3/2}}.$$

This result uses the notation $g \gg\ll f$, which denotes that g is bounded both above and below by positive constant multiples of f.

Sarnak claimed (without proof) in [Sar07] that Iwaniec's result can be modified to give the same estimate for primitively represented forms. In Sect. 5, where we investigate residue classes of primes appearing in a fixed prime component, we also need an analogous statement for primes in congruence classes. (As an aside, Theorem 14.7 of [FI10] is exactly of this flavour, but for integers represented as a sum of two squares, as opposed to shifted primes represented by a quadratic form.)

Fig. 4 A quadratic family (all circles tangent to the circle of curvature 11) in the People's packing. The quadratic family is shown in grey. For circles not in the quadratic family, colours indicate residue class modulo 7. Font colour distinguishes primes and composites

In a forthcoming paper, a subset of the present authors generalize Iwaniec's results to include primitivity and congruence classes, simultaneously. Here we state only the result in the case of curvature forms like those in Proposition 2.

Theorem 9 ([FHR25, Theorem 1.3]) *Let f_a be a primitive positive definite integral binary quadratic form of the form given in Proposition 2, in particular, having discriminant $\Delta = -4a^2$. Assume also that a is odd. Write $b_{f_a,a}(n)$ for the characteristic function for whether n is represented primitively by $f_a(x, y) - a$.*

(1) Then

$$\sum_{\substack{p \le X, \\ p \text{ prime}}} b_{f_a,a}(p) \gg\ll \frac{X}{(\log X)^{3/2}}.$$

(2) Let ℓ and m be coprime integers satisfying $(m, 2\Delta) = 1$ and $(\ell + a, m) = 1$. Then

$$\sum_{\substack{p \le X, \\ p \text{ prime} \\ p \equiv \ell \bmod m}} b_{f_a,a}(p) \gg \frac{X}{(\log X)^{3/2}},$$

where the implicit constant may depend on m.

The main idea of the proof in [FHR25] is to obtain a result on the number of square-free shifted primes $< X$ represented by the genus of $f_a(x, y)$, which turns out to be still bounded above and below by a positive constant multiple of $X/(\log X)^{3/2}$. Note that all of these are represented primitively. This entails combining a sieving argument with Iwaniec's argument in [Iwa72]. We then use the method of Bourgain–Fuchs in [BF12] to show that of these square-free shifted primes, those which are represented by some but not all forms in the genus (i.e. potentially not represented by f_a) are few enough that one can conclude the lower bound in the first statement of Theorem 9. The second statement of Theorem 9 requires incorporating the additional condition on p into Iwaniec's argument.

2.4 Pinch Families

In our investigation of the local properties of prime components, it is useful to consider smaller subsets of the packing called *pinch families*.

Definition 1 Let C_a and C_b be two tangent circles in $\mathcal{P}^{\text{full}}$. The *pinch family* they govern is the family of circles $C \in \mathcal{P}^{\text{full}}$ which are simultaneously tangent to C_a and C_b.

An example is given in Fig. 5.

The following lemma states that the curvatures of circles appearing in a pinch family are parameterized by a single-variable quadratic polynomial. This is to be

Fig. 5 A pinch family in the People's packing. The pinch family is shown in grey. For circles not in the pinch family, colours indicate residue class modulo 7. Font colour distinguishes primes and composites

expected, since, as compared to the quadratic families parameterized by a binary quadratic form, we are now fixing two circles instead of one.

Lemma 3 ([HKRS24, Lemma 4.5]) *Let (a, b, c, d) be a Descartes quadruple, where curvatures a, b correspond to circles C_a and C_b, respectively. The curvatures of pinch family of circles tangent to both C_a and C_b is parametrized by*

$$f(x) = (a+b)x^2 - (a+b+c-d)x + c, \quad x \in \mathbb{Z}.$$

In fact, $c = f(0)$, $d = f(1)$ and we have

$$(S_4 S_3)^s \cdot (a, b, c, d)^T = (a, b, f(4s), f(4s+1))^T, \quad s \in \mathbb{Z}.$$

For future reference, using Eq. (2), we may compute

$$\mathrm{Disc}(f) = 4ab. \tag{5}$$

Our methods in Sect. 6 require the following lemma that describes the possible residue classes in a pinch family in terms of the residue classes of the governing circles C_a and C_b.

Lemma 4 *Let $m = p^n$ be an odd prime power, and continue the notation of Lemma 3, in particular the function $f(x)$. Consider the set*

$$F_m := \{f(x) : x \in \mathbb{Z}/m\mathbb{Z}\}.$$

Let Q_m denote the set of squares modulo m. Then there are three cases:

(1) *If $(a, b) \equiv (0, 0) \pmod{p}$, then $c \equiv d \pmod{m}$, and $F_m = \{c\}$, and $\#F_m = 1$.*
(2) *Otherwise, if $p \mid a + b$, then $p \nmid a + b + c - d$, $F_m = \mathbb{Z}/m\mathbb{Z}$, and hence $\#F_m = m$.*
(3) *Otherwise, if $a + b$ is invertible,*

$$F_m = (a+b)Q_m - \frac{ab}{(a+b)},$$

and $\#F_m = \#Q_m$.

In particular, the set F_m depends only on a and b (not c or d).

Proof If $a + b \equiv 0 \pmod{p}$, then $c \not\equiv d \pmod{p}$ unless we have a quadruple of the form $(0, 0, c, c) \pmod{p}$. Thus, we are either the first or second case whenever $p \mid a + b$.

For the second case, let $k \in \mathbb{Z}_p$. Then $f(x) - k$ is linear modulo p, and therefore has a root modulo p. Provided that $a + b + c - d \not\equiv 0 \pmod{p}$ (which is the first case, so not this case), then $f'(x) \not\equiv 0$ modulo p. This implies that $f(x) - k$ has a root modulo p^k also.

Otherwise, $a + b \not\equiv 0 \pmod{m}$. By completing the square, since $2(a + b)$ is invertible modulo m,

$$f(X) = (a+b)X^2 + c - \frac{(a+b+c-d)^2}{4(a+b)},$$

where $X = x + \frac{a+b+c-d}{2(a+b)}$. Applying Eq. (2), we simplify to

$$f(X) = (a+b)X^2 - \frac{ab}{(a+b)}.$$

□

The elementary fact that a quadratic polynomial has infinitely many composite values has as a consequence the following.

Lemma 5 *A pinch family contains infinitely many composite circles.*

The question of whether a pinch family contains infinitely many primes (or any primes) is a special case of the following conjecture of Bunyakovsky (also spelled Bouniakowsky); see [AZFG20] for a modern overview.

Conjecture 4 (Bunyakovsky's conjecture) *Suppose that $f(x) \in \mathbb{Z}[x]$ is a quadratic polynomial that satisfies the following three conditions:*

(1) the leading coefficient is positive,
(2) $f(x)$ is irreducible,
(3) the infinite sequence $f(1)$, $f(2)$, $f(3)$, ... has no common factor.

Then $f(n)$ is prime for infinitely many positive integers n.

Note that this conjecture is quite out of reach at this point, at least in the sense that there is no single known polynomial in one variable of degree 2 which represents infinitely many primes.

Remark 1 Our study of residue classes appearing in prime and thickened prime components can be split into two strategies: one is to look at all circles tangent to a fixed one, which is where Theorem 9(2) comes into play, and the other is to travel along chains of circles tangent to two fixed ones, which is where Conjecture 4 is relevant. The latter gives more precise information, in a sense, about where one finds a given residue class, but the former relies on a conjecture which is likely much easier to prove.

3 Prime Components

In this section, we give the basic definitions and geometric properties of prime components. By convention, negative curvatures are not considered prime.

3.1 Basic Properties of Prime Components

Definition 2 Let $\mathcal{P}$ be a primitive integral Apollonian circle packing. Let C_p be a circle of prime curvature.

- The *prime component* containing C_p is the largest tangency-connected subset of $\mathcal{P}$ containing C_p and consisting entirely of circles of odd prime curvature.
- The *thickened prime component* containing C_p is the prime component containing C_p, together with every circle tangent to it.

The birth quadruple of the largest prime curvature of a prime component can be thought of as the root of that component.

Definition 3 A *prime component root* is a Descartes quadruple (a, b, c, d) for which d is prime, a, b and c are composite, and $d \geq a, b, c$.

Such a quadruple is illustrated in Fig. 6.

Proposition 3 *Any prime component contains infinitely many circles.*

Proof This follows from Theorem 9(1), which implies that any circle of odd curvature in the component is tangent to infinitely many prime circles. □

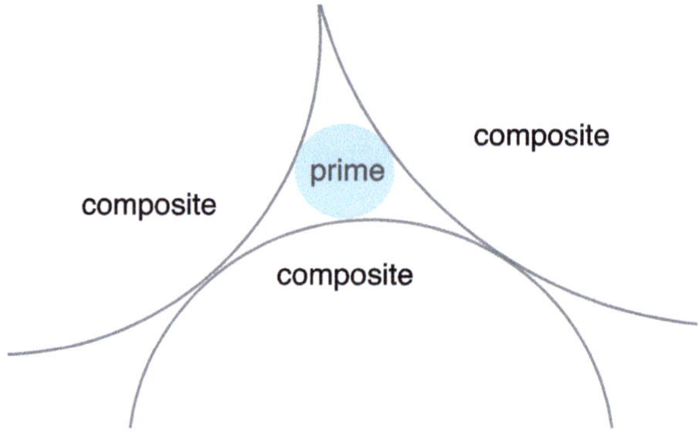

Fig. 6 A typical prime component root in its birth quadruple

Lemma 6 *Fix an Apollonian packing $\mathcal{P}^{\text{full}}$. With at most one exception, any prime component contains a unique prime component root.*

Proof Suppose the component does not include a circle of curvature 2. There are at most two circles of curvature 2 in a packing (this extremal case being achieved by the packing $(-1, 2, 2, 3)$), and they are tangent, so there is at most one component of this type. Suppose also that the component does not intersect the root quadruple of the packing. At most one component can intersect the root quadruple, and in the case that the packing contains a circle of curvature 2, this coincides with the component containing that circle. Thus we have ruled out at most one prime component.

For any other prime component, consider the smallest prime p and a circle C_p of that curvature in the component. Consider all quadruples containing C_p. Among these, since C_p is not in the root quadruple, there exists one for which $a, b, c \leq d = p$ (Lemma 1). By assumption, a, b, c are composite. This is a prime component root.

It remains to consider whether a component can have two such roots, involving distinct circles $C_{p,1}$ and $C_{p,2}$. The circle $C_{p,i}$ and therefore the prime component is contained in the triangle bounded by the other three circles of its prime component root, which are composite. Since the circles are distinct, these triangles are distinct. Hence the circles cannot be part of the same prime component. □

Observe that a prime component root is essentially the same information as the smallest bounding composite triangle for the prime component.

Lemma 7 *Suppose C_a, C_b, C_c are mutually tangent circles in $\mathcal{P}^{\text{full}}$. Then within the triangle they create, there is a circle of prime curvature. Hence, within any mutually tangent triangle of circles of composite curvature, there lies a prime component.*

Proof Choose a circle of odd curvature within the triangle. The family of curvatures tangent to it, which lie within the triangle, have curvatures equal to the values of a translated quadratic form, and as such, includes at least one primitively represented prime by Theorem 9(1). □

Assuming that Conjecture 4, we can strengthen Lemma 7 as follows:

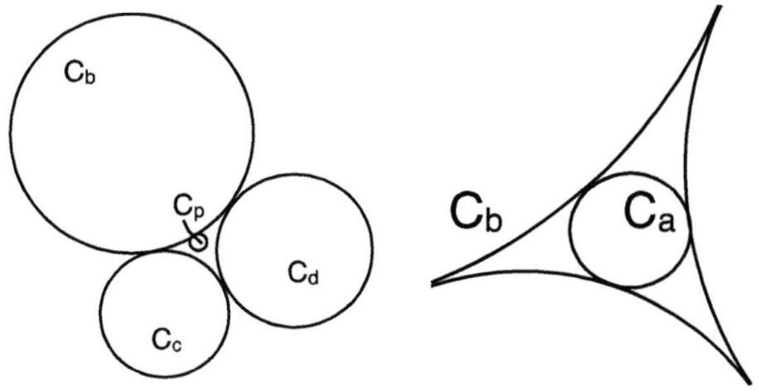

Fig. 7 Locating prime circle between tangent circles

Theorem 10 *Let $\mathcal{P}^{\text{full}}$ be a primitive integer ACP, and consider any circle C_b in the packing of curvature $b \in \mathbb{Z}$. Assume Conjecture 4. If C_c and C_d (of curvatures c and d, respectively) are two touching circles that are both tangent to C_b and satisfy $c \not\equiv d \pmod 2$, then there is a circle C_p of prime curvature p which is tangent to C_b and lies within the triangular interstice bordered by C_b, C_c, and C_d as in the first picture in Fig. 7.*

Proof Let C_a be the unique circle (of curvature a) inside this interstice which is tangent to C_b, C_c, and C_d, and let $\mathbf{v}$ denote the Descartes quadruple $\mathbf{v} = (a, b, c, d)$. With $\mathcal{A}_1$ defined as in Proposition 2, the orbit $\mathcal{A}_1 \cdot \mathbf{v}^T$ is precisely the collection of curvatures of all quadruples of pairwise tangent circles in $\mathcal{P}^{\text{full}}$ which include C_a. Without loss of generality, assume that $b > a, c, d$, so that C_b is the only circle of curvature b tangent to C_a. (If this is not the case, replace C_c and C_d by two smaller tangent circles which are also tangent to C_b in the interstice between C_c and C_d until no circle of curvature b can fit into the resulting interstice.) Hence, to prove this theorem, it suffices to find Descartes quadruples $(a, b, x_3, x_4) \in \mathcal{A}_1 \cdot \mathbf{v}^T$ with at least one of x_3, x_4 prime, since these quadruples will yield a circle of prime curvature in the region depicted in the second picture in Fig. 7, tangent to both C_b and C_a. But, finding quadruples (a, b, x_3, x_4) with at least one of x_3, x_4 prime is equivalent to finding circles of prime curvature in the pinch family governed by C_a and C_b. Assuming Conjecture 4, one can show that the quadratic form that parameterizes the curvatures appearing in this pinch family (defined in Proposition 3) represents infinitely many prime values, and if this is the case, there are infinitely many circles of prime curvature in the region depicted in the second picture in Fig. 7 that are tangent to the circle C_b. □

Remark 2 This theorem can be thought of as a statement about topological density of circles of prime curvature and hence of prime components in a given packing.

As we see in Sect. 6, our ability to locate circles of prime curvature within a fixed pinch family plays an important role when analysing the local structure of prime and thickened prime components.

3.2 The Size of Prime and Thickened Prime Components

We now briefly consider the number of circles in a prime or thickened prime component. Recall that

$$C_{\mathcal{P}}(X) := \#\{C \in \mathcal{P} : \text{curv}(C) \leq X\},$$

where $\mathcal{P}$ is either a prime or thickened prime component. This counts the circles in $\mathcal{P}$ including multiplicity. In order to motivate the study in this paper, we conjecture the following (Conjecture 3): For a prime component,

$$\lim_{X \to \infty} \frac{C_{\mathcal{P}^{\text{pr}}}(X)}{\pi(X)} = \infty,$$

and for a thickened prime component,

$$\lim_{X \to \infty} \frac{C_{\mathcal{P}^{\text{th}}}(X)}{X} = \infty.$$

These statements would imply that the average multiplicity of a curvature in either packing is increasing.

Theorem 9(1) implies an immediate lower bound on the growth of a prime component:

Proposition 4 *Every prime component $\mathcal{P}^{\text{pr}}$ contains infinitely many circles. In fact,*

$$C_{\mathcal{P}^{\text{pr}}}(X) \gg \frac{X}{(\log X)^{3/2}},$$

where the implied constant depends on the prime component.

Proof Let C_p be any odd prime circle of the prime component. Then, by Theorem 9(1), there are $\gg X/(\log X)^{3/2}$ prime curvatures in the family of circles tangent to C_p. □

Conjecture 3 is supported by experimental data; see Sects. 8.2 and 8.3 for estimates on the exact growth rates.

4 Counting Prime Components

In this section, we show that there are infinitely many prime components, prove an upper bound, and give a heuristic estimate and conjecture for the number of such components.

Theorem 1 *There are infinitely many prime components.*

Proof Choose two circles C_a and C_b in the packing $\mathcal{P}^{\text{full}}$ which are composite and tangent. For the existence of such, take a quadruple with sufficiently large curvatures and choose the two even curvatures (see Proposition 1). Consider the pinch family governed by these two circles (Definition 1). It contains infinitely many circles of composite curvature $D_1, D_2, \ldots$ by Lemma 5. Within each triangle C_a, C_b, D_n of composite curvatures, there exists a prime component, by Lemma 7. For any finite list of such components, we can choose n large enough that the triangle C_a, C_b, D_n does not contain it. Therefore there are infinitely many such components. □

Let $\mathcal{P}^{\text{full}}$ be a primitive integral Apollonian packing and let $C_{\mathcal{P}^{\text{full}}}(X)$ be the number of circles of curvature $\leq X$ in the packing. It is known that $C_{\mathcal{P}^{\text{full}}}(X) \sim X^{\delta}$ where $\delta \approx 1.3056\ldots$ is the Hausdorff dimension of an Apollonian circle packing [Boy82, KO11, LO13, Vin12]. Let

$$N^{\text{root}}_{\mathcal{P}^{\text{full}}}(X) := \#\{(a, b, c, p) \text{ a prime component root} : a, b, c \leq p \leq X, p \text{ prime}\}$$

count the number of prime component roots with prime not exceeding X. By Lemma 6, this also counts the number of prime components.

We propose the following conjecture.

Conjecture 5

$$N^{root}_{\mathcal{P}^{\text{full}}}(X) \sim c\frac{C_{\mathcal{P}^{\text{full}}}(X)}{\log X} - c''\frac{C_{\mathcal{P}^{\text{full}}}(X)}{(\log X)^2} \sim c\frac{C_{\mathcal{P}^{\text{full}}}(X)}{\log X},$$

where $c = \prod_{\substack{p \text{ prime} \\ p\equiv 1\,(4)}} \frac{p^2}{p^2-1} \prod_{\substack{p \text{ prime} \\ p\equiv 3\,(4)}} \frac{p^2}{p^2+1} = 0.9159\ldots$, *and* c'' *is a constant between* 0 *and* 2.

The best upper bound we can prove is the following.

Theorem 2

$$N^{root}_{\mathcal{P}^{\text{full}}}(X) \ll \frac{C_{\mathcal{P}^{\text{full}}}(X)}{\log X}.$$

Proof Let $\mathcal{Q}$ be the set of Descartes quadruples of $\mathcal{P}^{\text{full}}$. Consider the two quantities

$$\Sigma_1(X) := \#\{\mathbf{v} \in \mathcal{Q} : \max(\mathbf{v}) \leq X \text{ and prime}\}$$
$$\Sigma_2(X) := \#\{\mathbf{v} \in \Sigma_1(X) : \text{two coordinates of } \mathbf{v} \text{ are prime}\}.$$

As an immediate consequence of Proposition 1, we have

$$N^{\text{root}}_{\mathcal{P}^{\text{full}}}(X) = \Sigma_1(X) - \Sigma_2(X). \tag{6}$$

Therefore, bounding $N^{\text{root}}_{\mathcal{P}^{\text{full}}}(X)$ from above by $\Sigma_1(X)$, we use the upper bound on $\Sigma_1(X)$ from [KO11, Theorem 1.4]. □

We devote the rest of this section to giving a heuristic justification for Conjecture 5. We continue to use the notation of the proof of Theorem 2.

One has an upper bound

$$\Sigma_2(X) \ll \frac{C_{\mathcal{P}^{\text{full}}}(X)}{(\log X)^2}$$

from [KO11, Theorem 1.4]. But the best known lower bound on the first sum is

$$\Sigma_1(X) \gg \frac{X}{\log X}$$

from the work of [BK14], which is quite far from what is needed.

However, heuristically we can do better. In [FS11], the authors consider

$$\psi_{\mathcal{P}^{\text{full}}}(X) = \sum_{\substack{a(C)\le X \\ a(C) \text{ prime}}} \log\big(a(C)\big) \tag{7}$$

where C is a circle in the packing $\mathcal{P}^{\text{full}}$ and $a(C)$ is its curvature. They give the following heuristic in their Conjecture 1.2:

$$\psi_{\mathcal{P}^{\text{full}}}(X) \sim L(2, \chi_4) \cdot C_{\mathcal{P}^{\text{full}}}(X)$$

where $L(2, \chi_4) = 0.9159\ldots$ is the value of the Dirichlet L-series at 2 with character $\chi_4(p) = 1$ for $p \equiv 1\ (4)$ and $\chi_4(p) = -1$ for $p \equiv 3\ (4)$.

Classically, it is known that $\pi(X) \sim \frac{\psi(X)}{\log X}$, where $\psi(X) = \sum_{p \text{ prime}} \log p$, and we expect the same for $\pi_{\mathcal{P}^{\text{full}}}(X)$, the number of circles of prime curvature less than X in a packing $\mathcal{P}^{\text{full}}$. In other words, we expect that

$$\Sigma_1(X) \sim c\frac{C_{\mathcal{P}^{\text{full}}}(X)}{\log X}$$

where $c = 0.9159\ldots$.

Regarding Σ_2, a heuristic in Conjecture 1.3 of [FS11] gives that

$$\sum_{\substack{(C,C')\in S \\ a(C),\, a(C')\le X}} \log(a(C)) \cdot \log(a(C')) \sim c' \cdot C_{\mathcal{P}^{\text{full}}}(X), \tag{8}$$

where S is the set of pairs of tangent circles in $\mathcal{P}^{\text{full}}$ both of which have prime curvature, and

$$c' = 2 \cdot \prod_{p\equiv 1\,(4)} \frac{p^4}{(p^2-1)^2} \prod_{p\equiv 3\,(4)} \frac{p^4}{(p^2+1)^2} \cdot \left(1 - \frac{2}{p(p-1)^2}\right) = 1.3808\ldots.$$

This is the count of pairs of tangent prime circles that we need, except weighted by products of log's of the curvatures. In spirit, this suggests that without the log weights one obtains

$$c'' \frac{C_{\mathcal{P}^{\text{full}}}(X)}{(\log X)^2}$$

for some constant c''. Given this heuristic, the expression in (6) becomes

$$N^{\text{root}}_{\mathcal{P}^{\text{full}}}(X) \sim c \frac{C_{\mathcal{P}^{\text{full}}}(X)}{\log X} - c'' \frac{C_{\mathcal{P}^{\text{full}}}(X)}{(\log X)^2} \sim c \frac{C_{\mathcal{P}^{\text{full}}}(X)}{\log X},$$

as desired.

See Sect. 8 for supporting data, which supports the fact that c'' is between 0 and 2.

5 Local Properties of a Prime Component

In some sense the 'first' question to address about a prime or thickened prime component is the existence of local obstructions. Given a modulus m, which residue classes modulo m are represented in the component? In the case of a full packing $\mathcal{P}^{\text{full}}$, the congruence restrictions can be understood by studying the Cayley graph of the group modulo m. There is no natural way to insert primality into this type of analysis, so our methods rely on studying walks within the component, relying on results concerning the representation of primes by quadratic forms and quadratic functions.

5.1 Primes Represented by Binary Quadratic Forms

The following is a version of a standard fact about quadratic forms. See [Ser73, Chap. 4], for example, for a more thorough discussion.

Lemma 8 *Let $f(x, y)$ be an integral binary quadratic form with discriminant $\Delta \neq 0$, and fix $m \in \mathbb{Z}_{>1}$ coprime to 2Δ. Also, let $\ell \in \mathbb{Z}/m\mathbb{Z}$. Then, the following holds:*

- *if $\Delta \in (\mathbb{Z}/m\mathbb{Z})^\times$ is a square, $f(x, y) \equiv \ell \pmod m$ has a non-zero solution;*
- *otherwise, $f(x, y) \equiv \ell \pmod m$ has a solution only for $\ell \in (\mathbb{Z}/m\mathbb{Z})^\times$.*

Moreover, if $f(x, y) \equiv \ell \pmod m$ has a non-zero solution, it also has a primitive solution.

Proof Working modulo p for any prime dividing m, the first statement (without a primitivity condition) follows from [Ser73, Proposition 5, Chap. IV]. Moreover, since any such p is coprime to 2Δ, a solution to $f(x, y) \equiv \ell \pmod{p}$, for $\ell \in \mathbb{F}_p$, can be lifted via Hensel's lemma to a solution in $\mathbb{Z}_p$. Thus, applying Sun-tzu's theorem, we obtain the first statement for any choice of $m \in \mathbb{Z}_{>1}$ coprime to 2Δ.

It remains to show that given a non-zero solution, we can find a primitive solution. Suppose that $(x_0, y_0) \in \mathbb{Z}^2$ is a solution to $f(x, y) \equiv \ell \pmod{m}$. We first note that $\gcd(m, \gcd(x_0, y_0)) = 1$ since $(\gcd(x_0, y_0))^2 \mid f(x_0, y_0)$. Then, for any $k \in \mathbb{Z}$ satisfying

$$k \not\equiv -x_0 m^{-1} \pmod{p} \quad \text{for } p \;\Bigg|\; \frac{y_0}{\gcd(y_0, m)},$$

one can show that $(x_0 + mk, y_0)$ is a primitive solution to $f(x, y) \equiv \ell \pmod{m}$, as desired. □

For a fixed modulus m, we apply this lemma within the context of an Apollonian circle packing to give a description of the set of curvatures modulo m tangent to a fixed circle:

Lemma 9 *Let C_a be a circle in $\mathcal{P}^{\text{full}}$, having curvature a and lying within a Descartes quadruple (a, b, c, d), and fix $m \in \mathbb{Z}_{>1}$ coprime to $2a$. Let $S_m(a)$ denote the set of curvatures tangent to C_a modulo m. This set depends only on a modulo m and the following holds:*

- *if $-1 \in (\mathbb{Z}/m\mathbb{Z})^\times$ is a square, then $S_m(a) = \mathbb{Z}/m\mathbb{Z}$;*
- *otherwise, $S_m(a) = \{b : a + b \in (\mathbb{Z}/m\mathbb{Z})^\times\}$.*

Proof By Proposition 2, the circle C_a has an associated translated quadratic form

$$f_a(x, y) - a = (b + a)x^2 + (a + b + d - c)xy + (d + a)y^2 - a$$

that parameterizes the curvatures tangent to C_a. In particular, the discriminant of $f_a(x, y)$ is $\Delta = -4a^2$, so the result follows from a direct application of Lemma 8 with $f_a(x, y)$. □

We now use the relationship discussed in Sect. 2.3 between curvatures of circles in a prime component and primes primitively represented by shifted binary quadratic forms to study the residue classes of curvatures appearing in prime components and thickened prime components. The following is an immediate consequence of Theorem 9(2).

Proposition 5 *Continuing with the notation from Lemma 9, and supposing a is odd, for any*

$$\ell \in S_m(a) \cap (\mathbb{Z}/m\mathbb{Z})^\times \cap \left((\mathbb{Z}/m\mathbb{Z})^\times - a\right),$$

the collection of curvatures of circles tangent to C_a contains infinitely many primes satisfying $p \equiv \ell \pmod{m}$. In fact, the number of such primes below X is $\gg \frac{X}{(\log X)^{3/2}}$.

5.2 Congruence Classes in Prime Components

We now combine Lemma 9 and Proposition 5 to show that, using Theorem 9(2), a prime component in an Apollonian circle packing contains infinitely many primes congruent to $\ell \pmod{m}$ for any invertible residue class $\ell \pmod{m}$.

Theorem 3 *Let $m \in \mathbb{Z}_{>1}$ be coprime to 6 and $\ell \in (\mathbb{Z}/m\mathbb{Z})^{\times}$. Then, a prime component $\mathcal{P}^{\mathrm{pr}}$ (respectively, a thickened prime component $\mathcal{P}\mathrm{th}$) contains infinitely many primes congruent to $\ell \pmod{m}$ among those circles within two tangencies of any sufficiently small fixed circle. In fact, the number of such primes below X is $\gg \frac{X}{\log^{3/2} X}$.*

Proof Choose a circle C_a of curvature $a > m$ in $\mathcal{P}^{\mathrm{pr}}$ (throughout the proof we can also take a thickened component $\mathcal{P}\mathrm{th}$ instead).

First, suppose $\ell + a$ is invertible modulo m. In this case, Proposition 5 immediately implies that there are infinitely many primes congruent to ℓ modulo m among those circles tangent to C_a.

It remains to consider the case where $\ell + a$ is not invertible modulo m. Here, it suffices to find any circle C_p of sufficiently large prime curvature p in $\mathcal{P}^{\mathrm{pr}}$ such that $\ell + p$ is invertible modulo m, and then run the same argument again, this time considering circles tangent to C_p. Indeed, we just showed that, tangent to C_a, there are circles of prime curvature p congruent to any invertible residue class k such that $k + a$ is invertible modulo m. Hence, the result follows provided that, whenever $\ell + a$ is not invertible modulo m, there exists an invertible residue class k such that $k + a$ and $\ell + k$ are both invertible modulo m.

Assume $\ell + a$ is not invertible modulo m. If m is prime, then $\ell \equiv -a \pmod{m}$. Let k be a residue such that $k \not\equiv \pm a$ or $0 \pmod{m}$, which exists provided $m \geq 5$. Then $k + a$ and $\ell + k = k - a$ are invertible modulo m. Note that if one can find a suitable k in the case that $m = q$ is prime, then one can find a suitable k modulo q^n as well. Hence by Sun-tzu's theorem, for any m coprime to 6, one can find a suitable k, or, equivalently, a circle tangent to C_a which is tangent to a circle of prime curvature congruent to $\ell \pmod{m}$. □

Remark 3 Theorem 3 does not provide a bound on the length of the word in the Cayley group of the Apollonian group that is required to reach the desired residue class. In the next section we show how the length of this word depends on an effective form of Bunyakovsky's conjecture.

6 Revisiting the Local Structure of Prime and Thickened Prime Components

In the previous section, we showed that, using Theorem 9(2), all invertible residue classes modulo m, for $m \in \mathbb{Z}_{>1}$ coprime to 6, appear in a prime or thickened prime component, within two tangencies of a starting circle. In this section, we revisit this result from the perspective of explicit paths of bounded length through the Cayley graph. This is perhaps in the spirit of a combinatorial spectral gap as studied for the Apollonian group (e.g. [BK14, FSZ19]).

To be more precise, in this section, we work directly with the generating set of swaps S_i for the Apollonian group $\mathcal{A}$ and restrict to matrix words built of units of the form $(S_i S_j)^k$. These types of matrix words lead to single variable quadratic polynomials and allow us to exploit certain algebraic properties of matrices in the Apollonian group. As a result of this shift in perspective, we require Conjecture 4 instead of Theorem 9(2).

The main result of this section can be stated as follows: suppose m is a positive integer coprime to 30, $\mathbf{v} = (a, b, c, p_0)$ is a Descartes quadruple with p_0 prime and $\gcd(c + p_0, m) = 1$. We show that there exist integers s_0 and r_0 such that for

$$\mathcal{W}_{s_0,r_0}(t) := \underbrace{\ldots S_4 S_3 S_4 S_3}_{s_0 \text{ letters}} \underbrace{\ldots S_3 S_2 S_3 S_2}_{r_0 \text{ letters}} \underbrace{\ldots S_2 S_1 S_2 S_1}_{t \text{ letters}},$$

the union of the curvatures appearing in $\{\mathcal{W}_{s_0,r_0}(t) \cdot \mathbf{v}^T\}_{t\in\mathbb{Z}}$ contains all residue classes modulo m. By definition, circles contained in these quadruples are a subset of those within two tangency levels of C_{p_0}, but the subset is rather more restricted. Furthermore, assuming additionally that $c + p_0 > 0$, cp_0 is not a perfect square, a is odd, and Conjecture 4 holds, then for infinitely many t, the quadruple $\mathcal{W}_{s_0,r_0}(t) \cdot \mathbf{v}^T$ is contained in the thickened prime component containing C_{p_0}.

Our method for proving this is heavily inspired by the proof of [GLM03, Theorem 6.2]. The most salient difference is that we work within two tangency levels of a fixed circle (rather than three), and the requirement to stay within prime or thickened prime components requires some new ingredients.

6.1 *Pinch Families in Apollonian Circle Packings*

Our main tool throughout this section is the study of pinch families as in Definition 1. The following notation will be helpful in discussing these collections of circles in more detail.

By a *word*, we mean a matrix word written in the generators S_1, S_2, S_3, S_4 of $\mathcal{A}$. Following notation in [GLM03, Sect. 6], we define length-s words of alternating form:

$$W_{ji}(s) := \underbrace{\ldots S_j S_i S_j S_i}_{s \text{ letters}}.$$

We call these *s-term swap products*. For odd s, $W_{ji}(s)$ is the identity matrix with entries in the ith and jth rows replaced by

$$(i,k)\text{-th entry} = \begin{cases} -s & \text{if } k = i \\ s+1 & \text{if } k = j \\ s(s+1) & \text{else} \end{cases} \quad \text{and} \quad (j,k)\text{-th entry} = \begin{cases} s & \text{if } k = j \\ -(s-1) & \text{if } k = i \\ s(s-1) & \text{else} \end{cases}.$$

For even s, we interchange the ith and jth rows in the above formulæ.

Fixing a Descartes quadruple $\mathbf{v} = (c_1, c_2, c_3, c_4)$, fixing $i, j \in \{1, 2, 3, 4\}$, and applying all words $W_{ji}(s)$, $s \in \mathbb{Z}_{\geq 0}$, the resulting new circles lie within a pinch family, i.e. the family of circles simultaneously tangent to c_k and c_ℓ, where $\{k, \ell\} = \{1, 2, 3, 4\} \setminus \{i, j\}$. When a fixed Descartes quadruple is understood, the term W_{ji}-*pinch family* denotes this pinch family.

Let $\mathbf{v} = (a, b, c, d)$ be a Descartes quadruple, and recall from Lemma 3 that the curvatures in the W_{43}-pinch family of $\mathbf{v}$, i.e. curvatures of circles tangent to both C_a and C_b, are parameterized by

$$f(x) = (a+b)x^2 - (a+b+c-d)x + c, \ \ x \in \mathbb{Z},$$

with

$$W_{43}(2s) \cdot (a, b, c, d)^T = (a, b, f(4s), f(4s+1))^T, \ \ s \in \mathbb{Z}.$$

As s increases, this gives a formula for the mod m residue classes in the W_{43}-pinch family. Note that while

$$f(x + m\ell) \equiv f(x) \pmod{m}, \quad \forall \ell \in \mathbb{Z},$$

the actual curvatures appearing in the W_{43}-pinch family always increase as ℓ increases. Moreover, conditional on Conjecture 4, pinch families will contain infinitely many circles with prime curvature in each residue class appearing in the pinch family:

Lemma 10 *Continuing with the above notation, let m be coprime to 6 and ℓ modulo m be an invertible congruence class represented by $f(x)$. Assume Conjecture 4. If $a + b > 0$, ab is not a perfect square, and c is odd, then the W_{43}-pinch family contains infinitely many circles of prime curvature congruent to $\ell \pmod{m}$.*

Proof We first note that under these assumptions on a, b and c, using Eq. (5), one can check that $f(x)$ satisfies the conditions in Conjecture 4. Next, we fix some $x_0 \in \mathbb{Z}_{\geq 1}$ such that $f(x_0) \equiv \ell \pmod{m}$ and consider the quadratic polynomial $g(k) := f(x_0 + mk)$. Since the leading coefficient of $g(k)$ is $(a+b)m^2 > 0$ and $\{g(k) \mid k \in \mathbb{Z}_{\geq 1}\} \subseteq \{f(x) \mid x \in \mathbb{Z}_{\geq 1}\}$, $g(k)$ satisfies conditions (i) and (iii) in Conjecture 4. Moreover, since

$$\mathrm{Disc}(g(k)) = m^2\, \mathrm{Disc}(f(x)) = 4m^2 ab,$$

$g(k)$ is irreducible and thus also satisfies the conditions in Conjecture 4. We can therefore conclude that $g(k) = f(x_0 + mk)$ is prime for infinitely many values of k, and each such k corresponds to a circle in the W_{43}-pinch family of prime curvature congruent to $\ell \pmod{m}$. □

Remark 4 For $f(x)$ as in Lemma 10, the main result of [Iwa78] implies there are infinitely many 2-almost primes in the W_{43}-pinch family of curvatures of $\mathbf{v}$. So, in the discussion that follows below, we could replace the role of prime components with 2-almost prime components and remove any dependence on Conjecture 4.

Now, by Lemma 4, when $a + b \not\equiv 0 \pmod{m}$, the pinch family governed by C_a and C_b cannot include all residue classes mod m. So, to use s-term swap products to construct an explicit collection of tangent circles that includes all residue classes mod m, we need to allow movement between different pinch families. As we see below, Lemma 10 plays a pivotal role in ensuring we stay within a prime or thickened prime component during switches between pinch families.

6.2 *Walks in Prime and Thickened Prime Components*

For the remainder of this section, our goal is to develop a method for constructing matrix words in the Apollonian group that give paths (of tangent circles) between specific residue classes modulo m. Specifically, we want an explicit matrix word in $\mathcal{A}$ that, for any invertible residue class modulo m, gives a path to any other residue class modulo m satisfying the following properties:

- The initial circle in the path has prime curvature.
- The path is contained within two tangency levels from the initial circle.
- The path is contained within the thickened prime component of the initial circle.

We might try (naïvely) to construct this type of matrix word using just three of the four generators of $\mathcal{A}$, i.e. construct paths contained within the first tangency level of the initial circle, but this is not possible by Lemma 9. As such, we need to use all four generators $S_i \in \mathcal{A}$ and will necessarily swap our initial prime circle at some point in this process. The following generalization of a pinch family will play an important role in our method:

Definition 4 Let $\mathcal{W} = W_{j_1 i_1}(s_1) \cdots W_{j_k i_k}(s_k) \in \mathcal{A}$ be a matrix constructed by concatenating s-term swap products, and let $\mathbf{v}_0 = (a, b, c, d)$ be a Descartes quadruple. Let

$$\mathbf{v}_i^T = S_* \mathbf{v}_{i-1}^T,$$

where $S_* \in \{S_1, S_2, S_3, S_4\}$ runs over the matrices in $\mathcal{W}$ from right to left. Then, we define the *$\mathcal{W}$-family of* $\mathbf{v}_0$ to be the collection of tangent circles that correspond to the curvatures appearing in the Descartes quadruples $\{\mathbf{v}_i : 0 \le i \le s_1 + \cdots + s_k\}$.

Although we use the word *family*, this is a finite set (in contrast to a pinch family, which is infinite). Our approach is somewhat complicated by the fact that we wish to consider both paths in the Cayley graph of $\mathcal{A}$ (i.e. words or sequences of quadruples) and tangency paths of circles in the packing, which are related but not synonymous concepts. Given a word $\mathcal{W}$ as above, we therefore develop some terminology for the various related sets of circles and paths of tangencies.

Definition 5 With $\mathcal{W}$ as above, let C_a and C_ℓ be circles in an ACP $\mathcal{P}^{\text{full}}$.

- We say the circle C_ℓ is $\mathcal{W}$*-tied* to C_a if there exists a Descartes quadruple $\mathbf{v}_a$ containing C_a such that C_ℓ is in the $\mathcal{W}$-family of $\mathbf{v}_a$.
- When C_ℓ is $\mathcal{W}$-tied to C_a, a $\mathcal{W}$*-geodesic* from C_a to C_ℓ is any minimal-length sequence of tangent circles appearing the $\mathcal{W}$-family of $\mathbf{v}_a$ that connects C_a and C_ℓ. Note that a $\mathcal{W}$-geodesic from C_a to C_ℓ is not unique, but its length is. An upper bound for the length of a $\mathcal{W}$-geodesic is k, i.e. the number of changes between pinch families in $\mathcal{W}$.
- Given a prime component $\mathcal{P}^{\text{pr}}$ containing C_a, we say that a $\mathcal{W}$-geodesic from C_a is a *core geodesic* if it is contained in $\mathcal{P}^{\text{pr}}$, with the exception of the geodesic's terminal circle. This means a core $\mathcal{W}$-geodesic from C_a is contained in either the prime component $\mathcal{P}^{\text{pr}}$ or its thickened prime component.

To summarize our goal in this new language, we want to construct a collection of words

$$\mathcal{W}(s_1, s_2, s_3) = W_{j_1 i_1}(s_1) W_{j_2 i_2}(s_2) W_{j_3 i_3}(s_3)$$

such that as s_1, s_2, or s_3 vary, the core $\mathcal{W}(s_1, s_2, s_3)$-geodesics tie circles of all residue classes modulo m to C_a. Since we have taken $k = 3$ in Definition 4, these core geodesics are at most length two. We accomplish this goal by carefully choosing values for all j_k, i_k, as well as s_1 and s_2 in such a way that the curvatures appearing in the third coordinate of $\mathcal{W}(s_1, s_2, s_3) \cdot \mathbf{v}^T$ can be represented as a linear polynomial in s_3. To ensure we stay within the thickened prime component under consideration, we use Lemma 10 to find prime curvature circles when we switch between pinch families. This restriction is the most difficult part of this process (and we only achieve it conditionally).

We start by disregarding the primality aspect. That is, by giving a method for constructing a collection of words $\mathcal{W}$ such that the set of (not necessarily core) $\mathcal{W}$-geodesics terminate in all residue classes, i.e. disregarding whether we stay in the thickened prime component.

Proposition 6 *Let* $m \in \mathbb{Z}_{>1}$ *be coprime to* 30*, and let* $\mathbf{v} = (a, b, c, d)$ *be a Descartes quadruple of curvatures in an ACP such that* $\gcd(c + d, m) = 1$*. There exist integers* s_0 *and* r_0 *such that for*

$$\mathcal{W}_{s_0, r_0}(t) := W_{43}(s_0) W_{32}(r_0) W_{21}(t),$$

the set of $\mathcal{W}_{s_0, r_0}(t)$*-families of* $\mathbf{v}$*, as* t *varies over* $\mathbb{Z}$*, contains all residue classes modulo* m*. In particular, for any residue class* $\ell \pmod{m}$*, we can find an integer* t_0

such that there is a $\mathcal{W}_{s_0,r_0}(t_0)$*-geodesic, of length at most two, that ties a circle of curvature* $\ell \pmod{m}$ *to* C_d*. Furthermore, the word* $\mathcal{W}_{s_0,r_0}(t_0)$ *is length at most* $5m$.

Proof This proof is largely inspired by the proof of [GLM03, Theorem 6.2]. For $r, s \in \mathbb{Z}_{\geq 0}$, we consider $W_{43}(s)W_{32}(r) \in \mathcal{A}_1$. Setting $u := rs + 1$ and assuming that both r and s are odd, the third row of $W_{43}(s)W_{32}(r)$ is

$$(u^2 + s^2 - 1 + su, -su, -rs + (r+1)s(s+1), 1 + s - rs(r-1) + r(r+1)s(s+1)).$$

The sum of the first and second coordinates is $u^2 + s^2 - 1$ and the difference is $2su$. To simplify the presentation, we wish to find u and s such that $u^2 + s^2 \equiv 1 \pmod{m}$ and $2su \not\equiv 0 \pmod{m}$.

By assumption, we suppose $m > 5$. Let p be prime with $p^w \mid\mid m$. Necessarily, $p > 5$, and choosing $u_p = 3 \cdot 5^{-1}$, $s_p = 4 \cdot 5^{-1}$ modulo p^w implies $u_p^2 + s_p^2 \equiv 1 \pmod{p^w}$ and $2s_p u_p \not\equiv 0 \pmod{p^w}$. These choices for u_p, s_p give the relation $r_p \equiv -2 \cdot 4^{-1} \pmod{p^w}$. Using Sun-tzu's theorem, we next find positive odd integers S and R such that $S \equiv s_p \pmod{p^w}$ and $R \equiv r_p \pmod{p^w}$ for all primes p dividing m. Setting $U = RS + 1$, we have $U^2 + S^2 - 1 \equiv 0 \pmod{m}$ and $2SU \not\equiv 0 \pmod{m}$, as desired.

Now, let $A = SU$. Note that $\gcd(A, m) = 1$, and we can rewrite the third row of $W_{43}(S)W_{32}(R)$ in the form $(A, -A, C, D) \pmod{m}$. Applying S_2 and S_1 alternating in succession to this row yields

$$((-1)^t A, (-1)^{t+1} A, -2tA + C, -2tA + D) \pmod{m}$$

for $t \in \mathbb{Z}_{\geq 0}$. In other words, setting $s_0 = S$ and $r_0 = R$, the third row of the word

$$\mathcal{W}(t) = W_{43}(s_0)W_{32}(r_0)W_{21}(t)$$

is of the form

$$(A, -A, -2tA + C, -2tA + D) \pmod{m} \tag{9}$$

for $t \in 2\mathbb{Z}$. Thus, the $\mathcal{W}(t)$-family of $\mathbf{v}$ contains circles of curvatures in the residue class

$$-2A(c+d)t + A(a-b) + Cc + Dd \pmod{m}. \tag{10}$$

Since $\gcd(2A(c+d), m) = 1$, our conclusion follows as t varies over $2\mathbb{Z}$. Note that if t_0 is chosen so that the residue class in Eq. (10) is congruent to some $\ell \pmod{m}$, then a length-two $\mathcal{W}(t_0)$-geodesic from C_d to a circle C_ℓ of curvature $\ell \pmod{m}$ is given by

$$C_d \to C_{a'} \to C_\ell,$$

where $C_{a'}$ is the circle corresponding to the first coordinate in $W_{21}(t_0) \cdot \mathbf{v}$.

The final statement of the proposition is a consequence of the fact that s_0, r_0 are only specified modulo $2m$, and t_0 modulo m. □

We now return to our original goal of studying the residual structure of prime and thickened prime components. To this end, we want to apply Proposition 6 starting from a Descartes quadruple containing a prime curvature. The first step is to check that an appropriate starting quadruple exists, which is guaranteed by the following lemma:

Lemma 11 *Let $m \in \mathbb{Z}_{>1}$ be odd, and let $\mathbf{v} = (a, b, c, p_0)$ be a Descartes quadruple of curvatures such that p_0 is prime and a is odd. There is circle with curvature $c' > 0$ in the W_{32}-pinch family of $\mathbf{v}$ such that* $\gcd(c' + p_0, mp_0) = 1$.

Proof The proof is similar to the proof of the claim in [GLM03, Theorem 6.2], but we need a small modification since we want the fourth coordinate of $\mathbf{v}$ to stay fixed. For even $k \in \mathbb{Z}$, define $q(k)$ to be the sum of the third and fourth coordinates of $W_{32}(k) \cdot \mathbf{v}^T$, that is,

$$q(k) := k^2(a + d) + k(a - b + c + d) + d, \quad \text{for } k \in 2\mathbb{Z}.$$

Then, for any fixed odd prime p, at least one of the values of $q(2)$, $q(4)$ or $q(6)$ must be non-zero modulo p or else the primitivity of $\mathbf{v}$ would be violated. The rest of the proof follows as in [GLM03, Theorem 6.2]. □

Theorem 4 *Assume the conjecture of Bunyakovsky, Conjecture 4. Let $m \in \mathbb{Z}_{>1}$ be coprime to* 30*, and let ℓ represent a residue class modulo m. Let C_p be a circle of prime curvature, and let $\mathcal{P}$*th *be the thickened prime component containing C_p.*

*Then there exists a circle C_ℓ at distance two tangencies from C in $\mathcal{P}$*th *such that*

$$\operatorname{curv}(C_\ell) \equiv \ell \pmod{m}.$$

Furthermore, the two-tangency path is of the restricted form described in Proposition 6.

Proof By Lemma 11, we can chose a starting quadruple $\mathbf{v} = (a, b, c, p)$ containing C_p such that $c + p > 0$, cp is not a perfect square, a is odd and Proposition 6 applies.

We need only re-run the proof of Proposition 6 within the thickened prime component, verifying whether the $\mathcal{W}(t)$-geodesics that we construct are core geodesics. Because of how we have constructed the $\mathcal{W}(t)$-families of $\mathbf{v}$ in Proposition 6, it suffices to examine the initial switch between the W_{21} and W_{32}-pinch families. However, Lemma 10 implies that, assuming Conjecture 4, for any $t_0 \in \mathbb{Z}_{>0}$, we can find some $k \in \mathbb{Z}_{\geq 0}$ such that $W_{21}(t_0 + mk)$ has prime curvature. In particular, since

$$W_{21}(t_0 + mk) \equiv W_{21}(t_0) \pmod{m},$$

this means that we can construct a core $\mathcal{W}(t)$-geodesic, of length at most two, to any residue class modulo m within the thickened prime component. We repeat this process and apply Lemma 10 at the end of $\mathcal{W}(t)$ if we also want the terminal circle of our $\mathcal{W}(t)$-geodesic to have prime curvature. □

We conclude this section with a modification of Theorem 4 that replaces the dependence on Conjecture 4 with Theorem 9(2) under some additional assumptions on the matrix word $W_{43}(s_0)W_{32}(r_0)$ defined in Proposition 6. The key difference in our modification is that instead of searching for a circle of prime curvature within the W_{21}-family of the starting quadruple in Theorem 4, we instead arrange for one of the governing circles in this pinch family to have prime curvature. Under the additional hypotheses on $W_{43}(s_0)W_{32}(r_0)$, this means we only need to find a specific residue class within the W_{21}-pinch family to prove Theorem 4.

The following lemma gives a Descartes quadruple that serves as the starting point for our modification of Theorem 4:

Lemma 12 *Let $m \in \mathbb{Z}_{>1}$ be coprime to 30, and let $k \in \mathbb{Z}$. For a circle $C_d \in \mathcal{P}^{\text{full}}$ of prime curvature d coprime to m, there exists a Descartes quadruple $\mathbf{v} = (a, b, c, d)$ containing C_d such that C_a has prime curvature and $a - b \equiv k \pmod{m}$.*

Proof To emphasize the difference between integers and residue classes, we use $\overline{x}$ to denote a residue class of x modulo m. By Lemma 4, the set of residue classes modulo m which appear in a pinch family with parent circles C_a and C_d is determined only by the residues $\overline{a}$ and $\overline{d}$. We denote this set by $F_m(\overline{a}, \overline{d})$.

The goal is to find circles C_a and C_b tangent to C_d satisfying $a - b \equiv k \pmod{m}$ and a prime. By Proposition 5, it suffices to find a pair of residue classes $(\overline{a}, \overline{b})$ (which we will then lift to integer curvatures a and b) such that

(ia) $\overline{a}$ is invertible modulo m,
(ib) $\overline{a} + \overline{d}$ is invertible modulo m,
(ii) $\overline{a}$ is represented by $f_{C_d}(x, y) - d$ modulo m,
(iii) $\overline{b} \in F_m(\overline{a}, \overline{d})$ and
(iv) $\overline{a} - \overline{b} \equiv k \pmod{m}$.

It suffices to achieve this modulo p^e for each maximal prime power dividing m, and then combine these with the Theorem of Sun-tzu. Thus, we assume m is an odd prime power p^e.

By Lemma 4, a residue class $\overline{b}$ appears in pinch families between circles with curvatures congruent to $\overline{a}$ and $\overline{d}$ if the expression

$$\overline{a}\overline{d} + \overline{b}(\overline{a} + \overline{d}) \tag{11}$$

is in Q_m (the set of squares; notation from Lemma 4). Substituting $\overline{b} = \overline{a} - \overline{k}$ into Eq. (11), we have that conditions (iii) and (iv) hold if the value at $\overline{a}$ of the polynomial

$$g(z) := z^2 + (2\overline{d} - \overline{k})z - \overline{kd}$$

is in Q_m. Thus, we just want to find such an $\overline{a}$ which satisfies (ia), (ib) and (ii).

Observe that the value set G_m of $g(z)$ includes

$$\begin{aligned} g(-d) &= g(k-d) = -d^2, \\ g(k/2) &= g(k/2-2d) = -(k/2)^2, \\ g(0) &= g(k-2d) = -dk, \\ g(k) &= g(-2d) = dk. \end{aligned} \tag{12}$$

Case I: -1 is a quadratic residue modulo p for p odd. Provided the set

$$T := \{-\overline{d}, \overline{k} - \overline{d}, \overline{k}/2, \overline{k}/2 - 2\overline{d}\}$$

has at least three elements, we can choose $\overline{a}$ from this set so that $\overline{a}$ and $\overline{a} + \overline{d}$ are coprime to p. Thus, we may assume this set has at most two elements. In this case, recalling that $(d, p) = 1$, we have $\overline{k} \in \{-2\overline{d}, 2\overline{d}\}$. Therefore, either $T = \{\pm\overline{d}\}$ or $T = \{-\overline{d}, -3\overline{d}\}$. Since d is coprime to p, taking $\overline{a} = \overline{d}$ in the first case and $\overline{a} = -3\overline{d}$ in the second case will guarantee that $\overline{a}$ and $\overline{a} + \overline{d}$ are coprime to p.
Case II: -1 is a quadratic non-residue modulo p. From Eq. (12), there exists $\overline{x} \in \mathbb{Z}/m\mathbb{Z}$ with $g(\overline{x}) \in Q_m$: either $\overline{x} \in \{0, \overline{k} - 2\overline{d}\}$ or $\overline{x} \in \{\overline{k}, -2\overline{d}\}$ will do. By assumption, $\overline{d}$ is non-zero modulo p. Therefore $g(-\overline{d}) = -\overline{d}^2$ is not a square modulo p, which implies $\overline{x} + \overline{d}$ is not divisible by p, hence invertible modulo p. Thus, by Lemma 9, we have satisfied (ii) and (ib). It remains to show that we can find such an $\overline{x}$ that is invertible modulo p. If $\overline{x} = -2\overline{d}$ we are done, so the remaining case is $\overline{x} \in \{0, \overline{k} - 2\overline{d}\}$. This is done unless we are the case $\overline{k} = r\overline{d}$, $r \in \{1, 2\}$. But then, $-\overline{r}\overline{d}^2$ is a square, i.e. $-\overline{r}$ is a square. This is a contradiction for $\overline{r} = 1$, so we have $\overline{k} = 2\overline{d}$, $g(z) = z^2 - 2\overline{d}^2$ and $g(0) = -2\overline{d}^2$ is a square. But then, $g(5 \cdot 3^{-1}\overline{d}) = -2(5^2/3^2 - 1)\overline{d}^2 = -2(4/3)^2\overline{d}^2$ is a square and so we can take $\overline{a} = 5 \cdot 3^{-1}\overline{d}$ (this uses that $\gcd(m, 30) = 1$).

Thus, for the pair $(\overline{a}, \overline{b})$ satisfying conditions (i)–(iv) above, we can find a circle C_a with prime curvature $a \equiv \overline{a} \pmod{m}$ and then find a circle C_b that is in the pinch family between C_a and C_d and satisfies $b \equiv \overline{b} \pmod{m}$. With these choices of a and b and either choice of c that completes a Descartes quadruple, we have the desired quadruple. □

Combining Lemma 12 with Proposition 6 gives a way to construct core geodesics in a thickened prime component using Theorem 9(2) in place of Conjecture 4:

Proposition 7 *Let $m \in \mathbb{Z}_{>1}$ be coprime to* 30, *and suppose*

$$\mathcal{W} := W_{43}(s_0) W_{32}(r_0)$$

has been constructed following Proposition 6. Additionally, suppose that $C \equiv 0 \pmod{m}$ in the third row of $\mathcal{W}$. Then, for any circle C_d of prime curvature and for any residue class $\ell \pmod{m}$, there is a Descartes quadruple $\mathbf{v}$ containing C_d such that the $\mathcal{W}$-family of $\mathbf{v}$ contains a circle of curvature $\ell \pmod{m}$. In particular, every geodesic in this $\mathcal{W}$-family is a core geodesic of length at most 2.

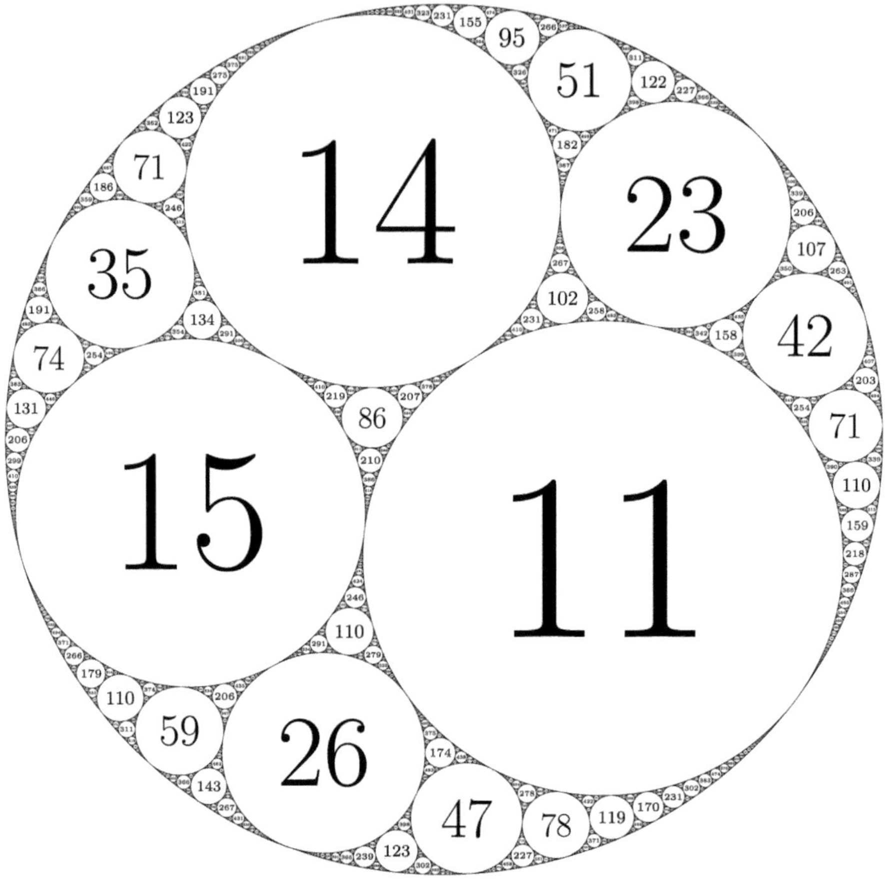

Fig. 8 Circles of curvature ≤ 10000 in the People's packing

Proof This follows immediately by applying Proposition 6 with the starting quadruple obtained from Lemma 12 with $k = A^{-1}(\ell - Dd)$, where A and D are from the third row of $\mathcal{W}$. By design, we can take $t = 0$ in the construction of $\mathcal{W}(t) = W_{43}(s_0)W_{32}(r_0)W_{21}(t)$, and so the proposition follows. □

Remark 5 The condition $C \equiv 0 \pmod{m}$ in Proposition 7 is fundamental to this modified approach. Specifically, we rely on the freedom to choose the residue class of c after we have fixed the residue classes of a and b. If $C \not\equiv 0 \pmod{m}$, we lose this freedom and revert back to requiring Conjecture 4 to find prime curvatures in a fixed pinch family.

6.3 Explicit Examples

We now illustrate a few explicit $\mathcal{W}$-geodesics in the People's packing (Fig. 8) for different choices of moduli m. The first example below gives a straightforward application of Proposition 6 for the People's packing with different choices of moduli m. The second example shows that the choice $m = 7$ satisfies the hypotheses of Proposition 7; we use this method to verify computationally that the thickened prime component of C_{23} in the People's packing contains all residue classes modulo 7 (Fig. 9).

Example 1 For each choice of modulus $m \in \mathbb{Z}_{\geq 7}$ below, we compute the odd integers s_0 and r_0 that are used in the Proposition 6 to construct a row of the form

$$(A, -A, C, D) \pmod{m}$$

in the word $W_{43}(s_0)W_{32}(r_0)$. Note that this computation depends only on the choice of modulus m (and not on the variable t in the Proposition). Then, for the Descartes quadruple $\mathbf{v} = (14, 15, -6, 11)$, which generates the People's packing, we compute the linear polynomial in Eq. (10) that represents residual curvatures appearing in the $\mathcal{W}_{s_0,r_0}(t)$-family of $\mathbf{v}$.

m	s_0	r_0	$(A, -A, C, D)$	$\mathcal{W}_{s_0,r_0}(t)$-family
7	5	3	(3, 4, 0, 0)	$-2t - 3$
11	3	5	(4, 7, 2, 7)	$4t - 5$
13	19	19	(1, 12, 11, 5)	$3t + 1$
17	11	25	(10, 7, 12, 9)	$2t$
19	35	9	(2, 17, 11, 13)	$-t - 1$
23	33	11	(6, 17, 14, 22)	$9t - 9$
29	53	43	(26, 3, 22, 2)	$t + 9$

As t varies over $2\mathbb{Z}$, the $\mathcal{W}_{s_0,r_0}(t)$-families of $\mathbf{v}$ will contain geodesics tying each residue class modulo m to C_d. When we choose t so that the first coordinate of $W_{21}(t) \cdot \mathbf{v}^T$ is prime, we can construct core geodesics in those $\mathcal{W}_{s_0,r_0}(t)$-families of $\mathbf{v}$.

Example 2 Working modulo $m = 7$, we illustrate how Lemma 12 can be used in the People's packing to construct length-two core geodesics without a dependence on Conjecture 4 or Theorem 9(2). Recall from Example 1 that for $m = 7$, the construction in Proposition 6 gives $s_0 = 5$ and $r_0 = 3$, leading the special row (3, 4, 0, 0) (mod 7) in

$$\mathcal{W} = W_{43}(5)W_{32}(3).$$

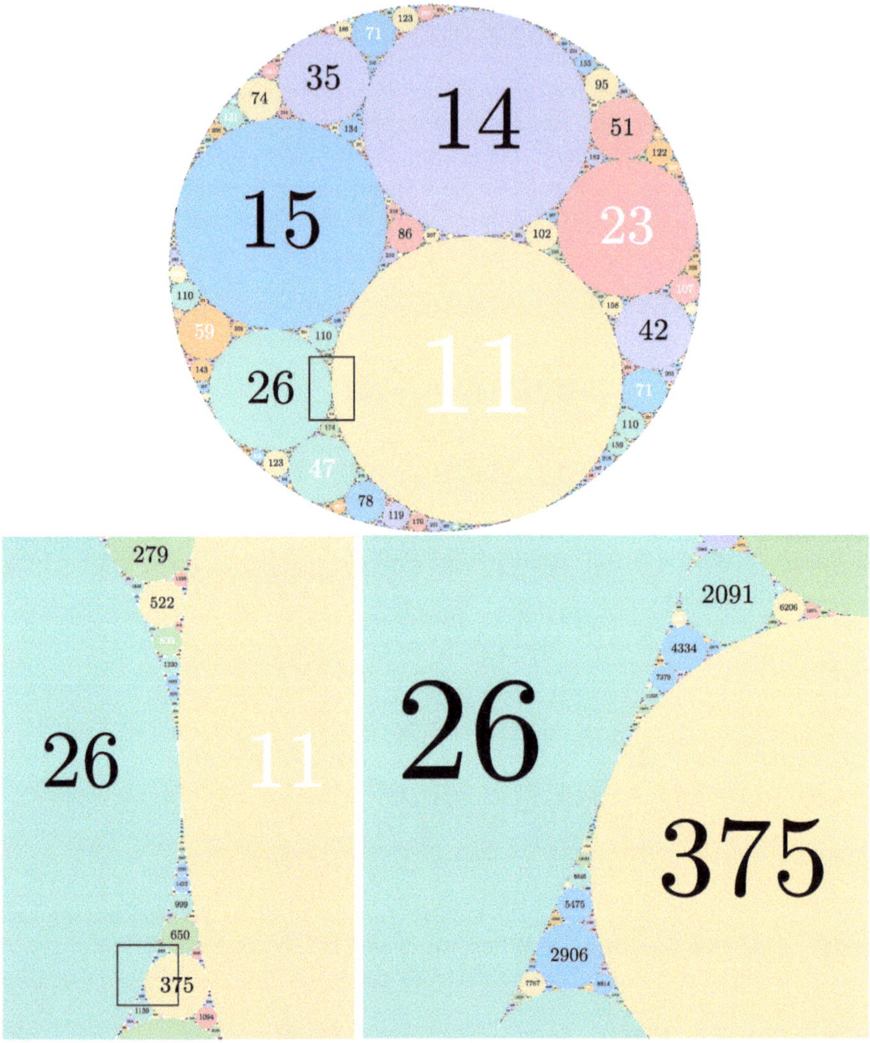

Fig. 9 Three successive zooms following the matrix word $W_{43}(5)W_{32}(3)W_{21}(1)$ applied to the quadruple $(14, 15, -6, 11)$, illustrating the first row in the table of Example 1. Colours indicate residue class modulo 7. Font colour distinguishes primes and composites

Let C_{23} be the circle of curvature 23 in the People's packing. In the table below, for each residue class $\ell \pmod 7$, we give a Descartes quadruple $\mathbf{v} = (a, b, c, 23)$ satisfying the conditions in Lemma 12 so that by Proposition 7, the third coordinate of $\mathcal{W} \cdot \mathbf{v}^T$ is congruent to $\ell \pmod m$.

$\ell \pmod{m}$	$a \pmod{m}$	$b \pmod{m}$	Starting quadruple
0	4	4	(11, 102, 58, 23)
1	4	6	(11, 258, 102, 23)
2	4	1	(11, 1562, 1134, 23)
3	2	1	(107, 918, 350, 23)
4	2	3	(107, 206, 678, 23)
5	4	0	(11, 42, −6, 23)
6	2	0	(107, 350, 42, 23)

By Lemma 9, there are no circles of curvature $\ell \equiv 5 \pmod 7$ tangent to C_{23}; Proposition 7 applied with the starting quadruple in the above table gives an explicit length-two core geodesic path

$$C_{23} \longrightarrow C_{11} \longrightarrow C_{7698}.$$

Moreover, the union of circles in the set of quadruples $\mathcal{W} \cdot \mathbf{v}^T$, where $\mathbf{v}$ runs through the starting quadruples given above, contains all residue classes modulo 7. Thus, the thickened prime component of C_{23} contains all residue classes modulo 7.

7 A Lower Bound for Integers Represented by Thickened Prime Components

In this section, we find a lower bound on the number of distinct curvatures in a thickened prime component by considering circles within two tangencies of the root. We prove the following:

Theorem 5 *Let $\mathcal{P}^{th}$ be a thickened prime component in a primitive integral Apollonian packing. Let $\kappa^{(2)}(\mathcal{P}^{th}, X)$ be the set of distinct integers less than or equal to X appearing as curvatures of a circle at most two tangencies away from the root of $\mathcal{P}^{th}$ (so the circle must either be tangent to the root or tangent to a circle tangent to the root). Then*

$$|\kappa^{(2)}(\mathcal{P}^{th}, X)| \gg \frac{X}{(\log\log X)^{1/2}}.$$

As a consequence, we immediately obtain Theorem 1.

7.1 Outline: Curvatures in Two Layers

Let C_ω be the root of $\mathcal{P}^{\text{th}}$ with curvature ω. As discussed in Sect. 2.3, the curvatures of the circles tangent to it are given by a translated quadratic form $f_{C_\omega}(x, y) - \omega$,

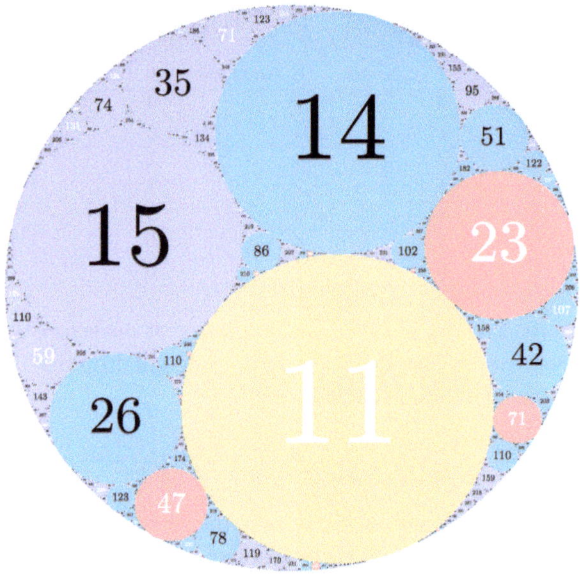

Fig. 10 An illustration of the approach in this section. In the People's packing (shown up to curvature 3000), the initial circle C_ω is of curvature 11, in yellow. The circles of prime curvature tangent to it (the set $\mathcal{B}$) are in red (this is the first 'layer'). The curvatures S_α of circles tangent to red circles are shown in blue (this is the second 'layer'). Taken together, these two layers form a subset of the thickened prime component

which we will more simply denote $f_\omega(x, y) - \omega$. We denote the set of odd prime such values by

$$\mathcal{B} := \{\alpha \in \mathbb{Z} : \alpha \text{ is an odd prime and primitively represented by } f_{C_\omega}(x, y) - \omega\}.$$

These are the odd prime curvatures at the first 'layer' surrounding C_ω.

We will now consider the corresponding collection of circles and their respective translated quadratic forms. To each $\alpha \in \mathcal{B}$, there is at least one positive definite binary quadratic form $f_\alpha(x, y)$ such that $f_\alpha(x, y) - \alpha$ represents the curvatures of all circles tangent to some circle of curvature α which is itself tangent to C_ω. For α which occur multiple times as values of $f_{C_\omega}(x, y) - \omega$, there may be several forms from which to choose; we choose one arbitrarily. We refer to $f_\alpha(x, y) - \alpha$ as a *translated curvature form* and define

$$S_\alpha = \{n \in \mathbb{Z},\ n \leq X : n \text{ is primitively represented by } f_\alpha(x, y) - \alpha\}.$$

See Fig. 10 for an illustration of these two 'layers' in the approach.

The sets S_α, taken together, give a subset of the curvatures that are at most two tangencies away from C_ω in the thickened prime component:

$$\bigcup_{\alpha \in \mathcal{B}} S_\alpha \subseteq \kappa^{(2)}(\mathcal{P}^{\text{th}}, X).$$

The overall strategy is inclusion-exclusion:

$$\left|\kappa^{(2)}(\mathcal{P}^{\text{th}}, X)\right| \geq \left|\bigcup_{\alpha\in\mathcal{B}_X} S_\alpha\right| \geq \sum_{\alpha\in\mathcal{B}_X} |S_\alpha| - \sum_{\alpha_1\neq\alpha_2\in\mathcal{B}_X} |S_{\alpha_1}\cap S_{\alpha_2}|,$$

where we consider a subset $\mathcal{B}_X \subseteq \mathcal{B}$ growing with a parameter X. We roughly follow the proof of positive density of curvatures in Apollonian packings in [BF11].

7.2 Dyadic Subdivision

The set $\mathcal{B}_X$ is defined as follows

$$\mathcal{B}_X = \bigcup_{\substack{2\log\log X\\ \leq k\leq\\ 3\log\log X}} \mathcal{B}^{(k)}, \text{ where } \mathcal{B}^{(k)} = \mathcal{B}\cap[2^k, 2^{k+1}).$$

By Theorem 9(1) on the representation of primes by shifted quadratic forms, we obtain

$$|\mathcal{B}\cap[2^k, 2^{k+1})| \gg \frac{2^k}{k^{3/2}}. \tag{13}$$

The idea will now be to consider the values represented by the translated curvature forms associated to elements of the set in Eq. (13). Using an inclusion-exclusion principle, we will obtain lower bounds on the number of values represented by each such form and sum over $\mathcal{B}_X$, and then obtain upper bounds on the number of values represented simultaneously by two such forms.

7.3 Lemmata of Bourgain-Fuchs

We will need a few results from [BF11].

Lemma 13 *Let f be an integral binary quadratic form of discriminant $D = -4a^2$, and suppose that $(\log X)^2 \leq a \leq (\log X)^3$. Then the number of distinct values $\leq X$ primitively represented by f is $\gg X/a$.*

Proof This follows as in Eq. (3.13) of [BF11], using results of Blomer-Granville [BG06]. For a detailed discussion on this lower bound see Eq. (6) in [Tom20]. □

Lemma 14 (Equation (3.27) of [BF11]) *Let $\omega(n)$ denote the number of distinct prime factors of n. Let $(\log X)^2 \leq \alpha_1, \alpha_2 \leq (\log X)^3$. Then for $\alpha_1 \neq \alpha_2$,*

$$|S_{\alpha_1} \cap S_{\alpha_2}| \ll X \frac{2^{\omega(\gcd(\alpha_1,\alpha_2))}}{\alpha_1\alpha_2} \prod_{\substack{p|\alpha_1\alpha_2(\alpha_1-\alpha_2) \\ p \text{ prime} \\ p\nmid\gcd(\alpha_1,\alpha_2)}} \left(1+\frac{1}{p}\right).$$

7.4 Lower Bound on $\sum |S_\alpha|$

With the notation as above, we have the following.

Proposition 8

$$\sum_{\alpha\in\mathcal{B}_X} |S_\alpha| \gg \frac{X}{(\log\log X)^{1/2}}.$$

Proof From Lemma 13, the number of distinct values less than X primitively represented by f_α for a fixed value α is $\gg \frac{X}{\alpha}$. Therefore we have, using (13),

$$\sum_{\alpha\in\mathcal{B}_X} |S_\alpha| \gg \sum_{\alpha\in\mathcal{B}_X} \frac{X}{\alpha} = X \sum_{\substack{2\log\log X \\ \leq k \leq \\ 3\log\log X}} \sum_{\alpha\in\mathcal{B}^{(k)}} \frac{1}{\alpha} \geq X \sum_{\substack{2\log\log X \\ \leq k \leq \\ 3\log\log X}} \sum_{\alpha\in\mathcal{B}^{(k)}} \frac{1}{2^{k+1}}$$

$$\gg X \sum_{\substack{2\log\log X \\ \leq k \leq \\ 3\log\log X}} \frac{2^k}{k^{3/2}} \frac{1}{2^k} = X \sum_{\substack{2\log\log X \\ \leq k \leq \\ 3\log\log X}} \frac{1}{k^{3/2}} \gg X \frac{\log\log X}{(3\log\log X)^{3/2}}$$

$$\gg \frac{X}{(\log\log X)^{1/2}}.$$

□

7.5 Upper Bound on $\sum |S_{\alpha_1} \cap S_{\alpha_2}|$

We now proceed to compute an upper bound on the number of values simultaneously represented by two curvature forms associated to two values in the set $\mathcal{B}_X$.

Proposition 9

$$\sum_{\alpha_1\neq\alpha_2\in\mathcal{B}_X} |S_{\alpha_1} \cap S_{\alpha_2}| \ll \frac{X\log\log\log X}{\log\log X}.$$

We begin by proving the following lemma.

Lemma 15

$$\sum_{\alpha\in\mathcal{B}_X} \frac{1}{\alpha} \ll \frac{1}{(\log\log X)^{1/2}}.$$

Proof First note that

$$\sum_{\alpha \in \mathcal{B}_X} \frac{1}{\alpha} \leq \sum_{\substack{2 \log\log X \\ \leq k \leq \\ 3 \log\log X}} \frac{1}{2^k} \sum_{\alpha \in \mathcal{B}^{(k)}} 1 \ll \sum_{\substack{2 \log\log X \\ \leq k \leq \\ 3 \log\log X}} \frac{1}{2^k} \frac{2^k}{k^{3/2}} \ll (\log\log X)^{-1/2}.$$

□

By Lemma 14 (dropping the condition that $p \nmid \gcd(\alpha_1, \alpha_2)$),

$$|S_{\alpha_1} \cap S_{\alpha_2}| \ll X \frac{2^{\omega(\gcd(\alpha_1,\alpha_2))}}{\alpha_1 \alpha_2} \prod_{\substack{p | \alpha_1 \alpha_2 (\alpha_1 - \alpha_2) \\ p \text{ prime}}} \left(1 + \frac{1}{p}\right).$$

Therefore,

$$\sum_{\alpha_1 \neq \alpha_2 \in \mathcal{B}_X} |S_{\alpha_1} \cap S_{\alpha_2}| \ll X \sum_{\alpha_1 \neq \alpha_2 \in \mathcal{B}_X} \frac{2^{\omega((\alpha_1,\alpha_2))}}{\alpha_1 \alpha_2} \prod_{\substack{p | \alpha_1 \alpha_2 (\alpha_1 - \alpha_2) \\ p \text{ prime}}} \left(1 + \frac{1}{p}\right). \tag{14}$$

In our case we have that α_1 and α_2 are prime, so we obtain

$$\begin{aligned} (14) &= X \sum_{\alpha_1 \neq \alpha_2 \in \mathcal{B}_X} \frac{1}{\alpha_1 \alpha_2} \left(1 + \frac{1}{\alpha_1}\right)\left(1 + \frac{1}{\alpha_2}\right) \prod_{\substack{p | (\alpha_1 - \alpha_2) \\ p \text{ prime}}} \left(1 + \frac{1}{p}\right) \\ &\ll X \sum_{\alpha_1 \neq \alpha_2 \in \mathcal{B}_X} \frac{1}{\alpha_1 \alpha_2} \sum_{\substack{q | \alpha_1 - \alpha_2 \\ q \text{ squarefree}}} \frac{1}{q} \\ &= X \sum_{\substack{q \text{ squarefree} \\ q \leq 2(\log X)^3}} \frac{1}{q} \sum_{\alpha_2 \in \mathcal{B}_X} \frac{1}{\alpha_2} \sum_{\substack{\alpha_1 \equiv \alpha_2 \pmod{q} \\ \alpha_1 \in \mathcal{B}_X}} \frac{1}{\alpha_1}. \end{aligned}$$

We split this summation into two ranges depending on the size of q. Let Q be a parameter to be chosen later. For small values of q we completely drop the congruence condition modulo q and by Lemma 15 we find that

$$\sum_{\substack{q \leq Q \\ q \text{ squarefree}}} \frac{1}{q} \sum_{\alpha_2 \in \mathcal{B}_X} \frac{1}{\alpha_2} \sum_{\substack{\alpha_1 \equiv \alpha_2 \pmod{q} \\ \alpha_1 \in \mathcal{B}_X}} \frac{1}{\alpha_1} \ll \sum_{q \leq Q} \frac{1}{q} \left(\sum_{\alpha \in \mathcal{B}_X} \frac{1}{\alpha} \right)^2 \ll (\log Q)(\log\log X)^{-1}.$$

For large values of q we forget the condition that α_1 is a value of a fixed shifted binary quadratic form and only keep the size restriction on α_1, and the property that α_1 is in the congruence class α_2 modulo q. We obtain the bound

$$\begin{aligned}
&\sum_{\substack{Q<q\le 2(\log X)^3\\ q \text{ squarefree}}} \frac{1}{q}\sum_{\alpha_2\in\mathcal{B}_X}\frac{1}{\alpha_2}\sum_{\substack{\alpha_1\equiv\alpha_2 \pmod q\\ \alpha_1\in\mathcal{B}_X}}\frac{1}{\alpha_1}\\
&\ll \sum_{\substack{Q<q\le 2(\log X)^3\\ q \text{ squarefree}}} \frac{1}{q}\sum_{\alpha_2\in\mathcal{B}_X}\frac{1}{\alpha_2}\sum_{\substack{2\log\log X\\ \le k\le\\ 3\log\log X}}\frac{1}{2^k}\sum_{\substack{\alpha_1\in\mathcal{B}^{(k)}\\ \alpha_1\equiv\alpha_2 \pmod q}} 1\\
&\ll \sum_{\substack{Q<q\le 2(\log X)^3\\ q \text{ squarefree}}} \frac{1}{q}\sum_{\alpha_2\in\mathcal{B}_X}\frac{1}{\alpha_2}\sum_{\substack{2\log\log X\\ \le k\le\\ 3\log\log X}}\frac{1}{2^k}\left(\frac{2^k}{q}+1\right)\\
&\ll (\log\log X)\sum_{Q<q\le(\log X)^3}\frac{1}{q^2}\sum_{\alpha_2\in\mathcal{B}_X}\frac{1}{\alpha_2}+\sum_{Q<q\le(\log X)^3}\frac{1}{q}\sum_{\alpha_2\in\mathcal{B}_X}\frac{1}{\alpha_2}(\log X)^{-2}\\
&\ll (\log\log X)^{1/2}Q^{-1}+(\log\log X)^{1/2}(\log X)^{-2}.
\end{aligned}$$

We conclude that

$$(14) \ll X((\log Q)(\log\log X)^{-1}+(\log\log X)^{1/2}Q^{-1}+(\log\log X)^{1/2}(\log X)^{-2}).$$

Choosing $Q=(\log\log X)^2$ we deduce that

$$\sum_{\alpha_1\ne\alpha_2\in\mathcal{B}_X}|S_{\alpha_1}\cap S_{\alpha_2}| \ll X\frac{\log\log\log X}{\log\log X}.$$

7.6 Putting the Two Together: Proof of Theorem 5

Propositions 8 and 9 give

$$\sum_{\alpha\in\mathcal{B}_X}|S_\alpha| \gg \frac{X}{(\log\log X)^{1/2}},\qquad \sum_{\alpha_1\ne\alpha_2\in\mathcal{B}_X}|S_{\alpha_1}\cap S_{\alpha_2}| \ll \frac{X\log\log\log X}{\log\log X}.$$

For X sufficiently large this gives

$$\left|\bigcup_{\alpha\in\mathcal{B}_X}S_\alpha\right| \ge \sum_{\alpha\in\mathcal{B}_X}|S_\alpha| - \sum_{\alpha_1\ne\alpha_2\in\mathcal{B}_X}|S_{\alpha_1}\cap S_{\alpha_2}| \gg \frac{X}{(\log\log X)^{1/2}}.$$

□

8 Experimental Data

Special-purpose software was developed to collect and analyse data on prime components in Apollonian circle packings [Ric24a, Ric24b]. In this section we collect results.

8.1 Residue Classes

For all prime components generated from a prime root with largest prime at most 100 (255 such components across all primitive integral packings), we verified that there are no further congruence obstructions (i.e. besides those known for the packing as a whole) for any modulus up to 200, for both the prime and the thickened components. For all prime components with prime root having largest entry at most 40 (29 such components), we checked up to modulus 1000.

8.2 Growth Rate of a Prime Component

Conjecture 3 suggests that the average multiplicity of a curvature in a prime component tends to ∞. We support this conjecture with data from two prime components. For the largest prime components in the packings generated from $(-2, 3, 6, 7)$ and $(-6, 11, 14, 15)$, we compute all curvatures up to 10^{12}, and bin the data into 2000 bins of length 5×10^8. By plotting $\frac{C_{\mathcal{P}^{\mathrm{pr}}}(X)}{\pi(X)}$ against $\log(X)$ in Fig. 11, we observe that this ratio is increasing, possibly to ∞.

We compare to the (somewhat arbitrary) estimate

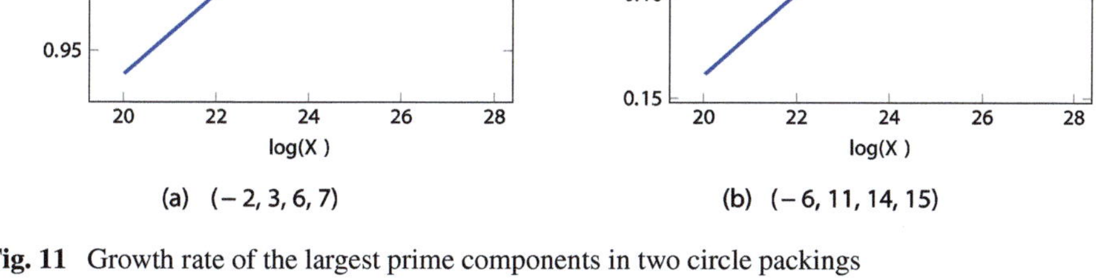

Fig. 11 Growth rate of the largest prime components in two circle packings

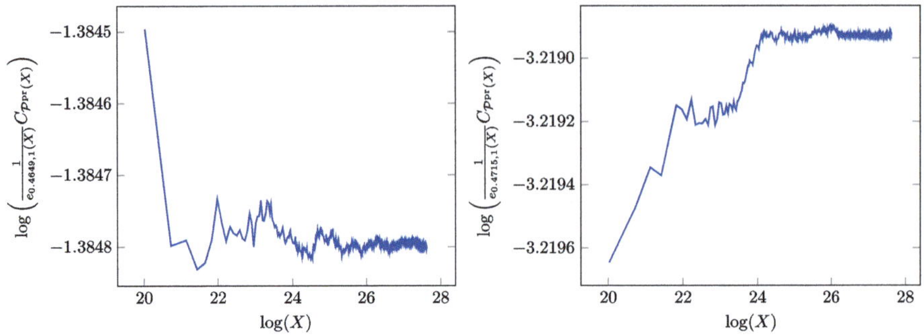

(A) $(-2, 3, 6, 7)$ (B) $(-6, 11, 14, 15)$

Fig. 12 Comparison with estimates for the growth rate of the largest prime components in two circle packings

$$C_{\mathcal{P}^{\mathrm{pr}}}(X) \approx e_{c,\alpha}(X) := \alpha\pi(X)(\log \pi(X))^c \sim \alpha X(\log X)^{c-1},$$

for some constants α and c. To estimate the appropriateness of this approximation, we plot $C_{\mathcal{P}^{\mathrm{pr}}}(X)/e_{c,1}(X)$ versus X on log − log axes in Fig. 12. For the two packings in question, the graph looks 'most' plausible for $c = 0.4649$ and 0.4715, respectively.

8.3 Growth Rate of a Thickened Prime Component

Conjecture 3 suggests that the average multiplicity of a curvature in a thickened prime component also tends to ∞. As in the last section, we consider the largest prime components in the packings generated from $(-2, 3, 6, 7)$ and $(-6, 11, 14, 15)$, we compute all curvatures up to 10^{12}, and bin the data into 2000 bins of length 5×10^8. Plotting $\frac{C_{\mathcal{P}^{\mathrm{th}}}(X)}{X}$ against $\log(X)$ in Fig. 13, we again observe that this ratio is increasing, possibly to ∞.

In the case of $C_{\mathcal{P}^{\mathrm{th}}}(X)$, it is even less clear what potential growth rate to compare to, but we compare again to the same estimate

$$C_{\mathcal{P}^{\mathrm{th}}}(X) \approx e_{c,\alpha}(X) := \alpha\pi(X)(\log \pi(X))^c \sim \alpha X(\log X)^{c-1},$$

for some constants α and c, where we expect the constants for the prime and thickened component to be related by approximately $c(\mathcal{P}\mathrm{th}) \approx 1 + c(\mathcal{P}^{\mathrm{pr}})$. (This might be plausible, for example, if one assumes each circle of curvature p of the prime component contributes around $X/p \log(X)^{1/2}$ circles to the thickening.) To estimate the appropriateness of this approximation, we plot $C_{\mathcal{P}^{\mathrm{th}}}(X)/e_{c,1}(X)$ versus X on log − log axes in Fig. 14.

8.4 Multiplicities of Curvatures

While the data in Sect. 8.3 supports the average multiplicity of a curvature in the thickened prime component tending to ∞, this does not necessarily imply that the multiplicities of admissible curvatures individually tend to ∞. In fact, this statement cannot be true in most packings, as the reciprocity obstructions found in [HKRS24] give infinite families of admissible curvatures with multiplicity 0.

Still, we would like to demonstrate that the multiplicities of curvatures grow in a reasonable fashion, supporting Conjecture 2. To this end, we compute all curvatures in a large range for two packings, and give a histogram of the frequencies of the various multiplicities.

In the packing generated by $(-2, 3, 6, 7)$, we compute the multiplicities of all admissible curvatures c for the largest thickened prime component for c in the range

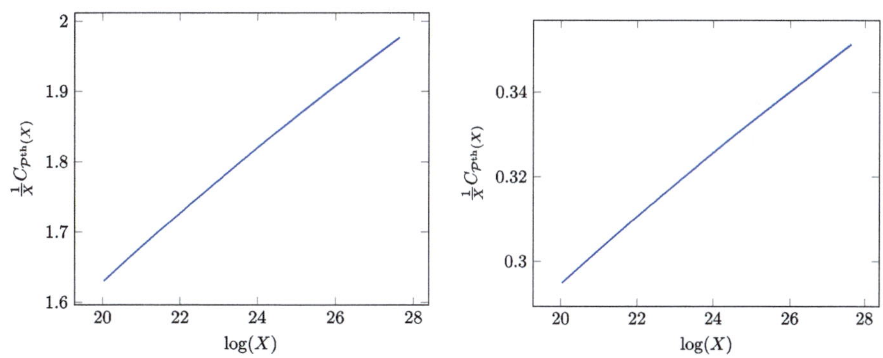

(A) $(-2, 3, 6, 7)$ (B) $(-6, 11, 14, 15)$

Fig. 13 Growth rate of the largest thickened prime components in two circle packings

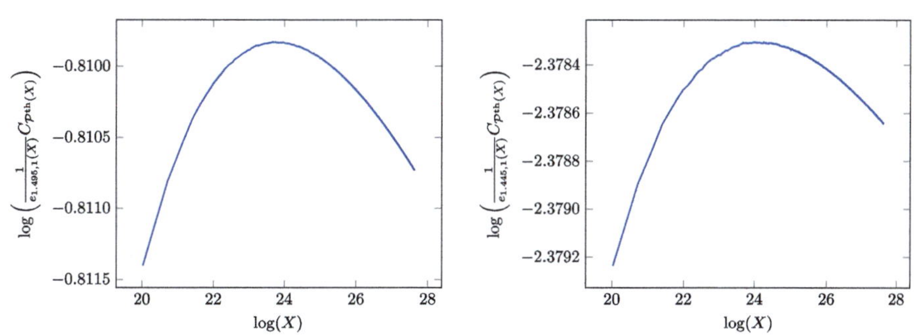

(A) $(-2, 3, 6, 7)$ (B) $(-6, 11, 14, 15)$

Fig. 14 Comparison with estimates for the growth rate of the largest thickened prime components in two circle packings

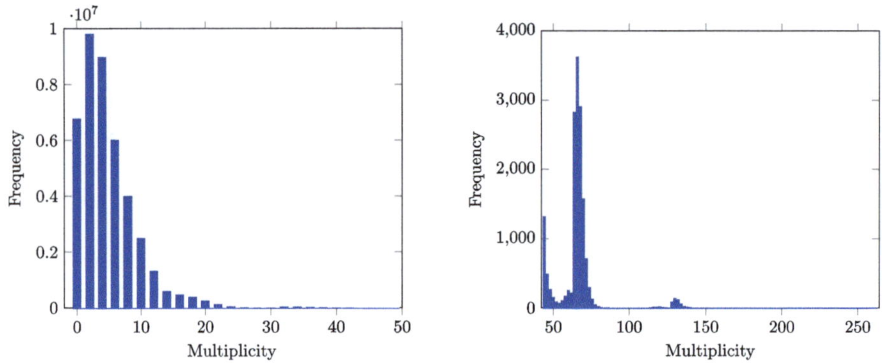

(A) Multiplicity between 0 and 50. (B) Multiplicity between 44 and 262.

Fig. 15 Multiplicity of curvatures $c \equiv 7 \pmod{24}$ for $5 \cdot 10^{13} + 1 \leq c \leq 5 \cdot 10^{13} + 10^9$ in the largest thickened prime component of $(-2, 3, 6, 7)$. The symmetry of the component causes all multiplicities to be even in the given range

$5 \cdot 10^{13} + 1 \leq c \leq 5 \cdot 10^{13} + 10^9$, and generate Fig. 15. This computation took 5.75 days using 64 cores on the Alpine cluster at CU Boulder.

As we choose curvatures c in larger ranges, we expect the average multiplicity to grow, so the 'hump' should move to the right. While the range $[5 \cdot 10^{13} + 1, 5 \cdot 10^{13} + 10^9]$ is not large enough to see the main hump 'depart' from the origin, we can observe that its peak is to the right of the origin, and the average multiplicity has begun to grow. Furthermore, we observe an interesting phenomenon of the right tail: there are multiple further humps, of varying heights (but much smaller than the main one; notice the rescaling of the axes), and unclear genesis.

For example, there are 135 circles of multiplicity between 88 and 126 in this range, yet there are 149 circles of multiplicity 130. Similarly, all multiplicities of 146 and higher occur at most once, save for multiplicity 260, which occurs 3 times. Such a phenomenon does not seem to appear when looking at the entire packing; instead, it appears to be a property of the thickened prime component.

For an example without symmetry, we compute the multiplicities of all admissible curvatures c in the range $10^{14} + 1 \leq c \leq 10^{14} + 10^9$ for the packing generated by $(-6, 11, 14, 15)$. The residue class with the highest multiplicity is $2 \pmod{24}$, and the histogram of multiplicities is Fig. 16.

We observe properties similar to those of the packing $(-2, 3, 6, 7)$: the average multiplicity has started to grow and form a hump, with a few smaller humps of varying heights.

8.5 Counting the Number of Prime Root Components

Conjecture 5 states that

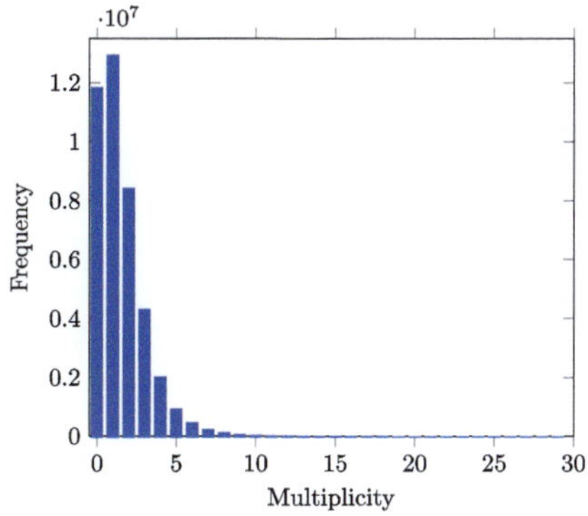

(A) Multiplicity between 0 and 30.

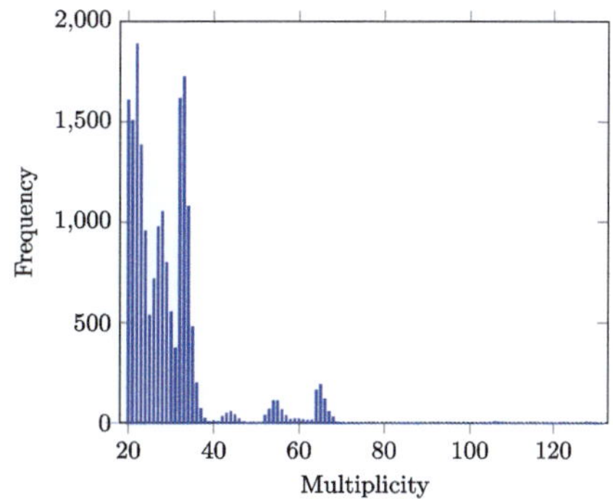

(B) Multiplicity between 18 and 133.

Fig. 16 Multiplicity of curvatures $c \equiv 2 \pmod{24}$ for $10^{14}+1 \le c \le 10^{14}+10^9$ in the largest thickened prime component of $(-6, 11, 14, 15)$

$$N^{\text{root}}_{\mathcal{P}^{\text{full}}}(X) \sim c\frac{C_{\mathcal{P}^{\text{full}}}(X)}{\log X},$$

where $c = 0.9159\ldots$ is an explicit constant. We can collect data, which, at the very least, supports the rough order of magnitude of $\frac{C_{\mathcal{P}^{\text{full}}}(X)}{\log X}$, with a constant plausibly close to c.

The difficulty in obtaining good data is likely due to several factors, including

- The presence of a secondary order term, which is still moderately large at the ranges we can compute to; very roughly, in order for the second term to be less than $1/N$ of the first term, one needs to consider $X > e^{1.5N}$.
- The discrepancy between $\pi(x)$ and $\frac{x}{\log(x)}$ is still quite large for $x \le 10^{10}$.

The size of X required to mitigate these factors sufficiently is likely far beyond any feasibly computable range.

Recalling the heuristic discussion surrounding Conjecture 5, for any $c'' \in \mathbb{R}$ define

$$f_{c''}(X) = N^{\text{root}}_{\mathcal{P}^{\text{full}}}(X) \Bigg/ \left(c\frac{C_{\mathcal{P}^{\text{full}}}(X)}{\log X} - c''\frac{C_{\mathcal{P}^{\text{full}}}(X)}{(\log X)^2}\right).$$

If the main asymptotic is correct, then $\lim_{X\to\infty} f_{c''}(x) = 1$ for any constant c''.

We compute $f_{c''}(X)$ for $X \le 10^{10}$ in the packings generated by $(-2, 3, 6, 7)$ and $(-6, 11, 14, 15)$ for $c'' = 0, 2$, with the results shown in Fig. 17.

When $c'' = 0$, $f_0(X)$ compares the prime root components count to the main term. In both packings, we see that $0.9 < f_0(X) < 1$, and $f_0(X)$ is increasing, possibly to 1. With the correction term coming from $c'' = 2$, we are now overcounting the prime root components, although the graph has started to dip and start to potentially approach 1. In any case, the data supports that Conjecture 5 is plausible.

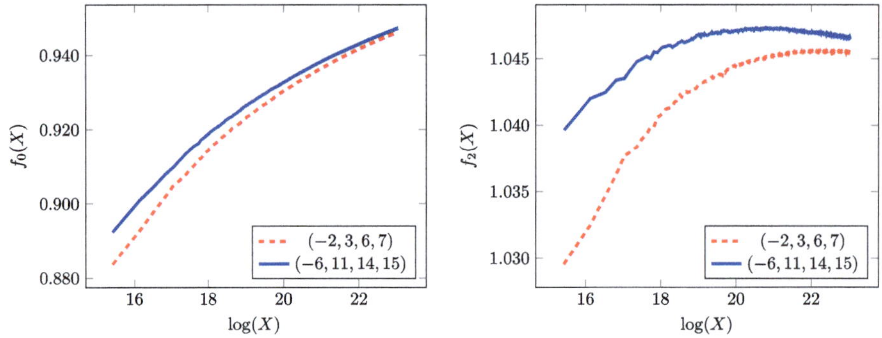

(A) $c'' = 0$ (B) $c'' = 2$

Fig. 17 Testing the asymptotic for $N^{\text{root}}_{\mathcal{P}\text{full}}(X)$ in the packings generated by $(-2, 3, 6, 7)$ and $(-6, 11, 14, 15)$ with $X = i \cdot 10^7$, $1 \le i \le 10^3$

9 Open Questions

We propose some open questions for further investigation:

(1) Is it possible to prove a lower bound on the growth of a prime or thickened prime component (the number of circles with curvatures less than X), or to prove Conjecture 3?
(2) Is it possible to prove bounds approaching the heuristics of Sect. 4 for the number of prime components?
(3) Is it possible to obtain positive-density in prime or thickened prime components? That is, lower bounds on the number of curvatures in the prime component that are $\gg \pi(X)$ (the prime counting function) or $\gg X$ in the case of the thickened prime component? This is work in progress by a subset of the authors here.
(4) What causes the 'humps' observed in Sect. 8.4?
(5) Given the set of primes in a fixed prime component, what bounds can one prove on gaps between these primes (e.g. can one prove that there are infinitely many pairs of primes that are $1,000,000$ apart)? Perhaps Maynard's tools [May15] can be modified to apply in this case.
(6) Do these results hold in other Apollonian-like packings such as those of [FSZ19, KK23, KN19, Sta18]?
(7) What about r-almost prime components? (We consider the count of curvatures in such components in forthcoming work.)
(8) What about thickening twice (by which we mean augmenting a prime component by all circles within two tangencies of the component), or more times? Do multiple thickenings put stronger tools within reach?

Acknowledgements This project originated as part of the Women in Numbers 4 workshop held at the Banff International Research Station. We are grateful to the organizers Jennifer Balakrishnan, Chantal David, Michelle Manes and Bianca Viray for creating a welcoming and stimulating research environment throughout the workshop. We also thank the Mathematical Sciences Research Institute, which hosted us in 2019 as part of their SWiM program. We all thank Peter Sarnak for inspiring this research with his questions about prime components in Apollonian circle packings. Hsu thanks the University of Bristol and the Heilbronn Institute for Mathematical Research for their partial support of this project. Rickards and Stange thank James McVittie and Drew Sutherland for insights into the experimental data analysis. Thank you to Alex Kontorovich for feedback. And finally, we thank the anonymous referee for insightful comments.

References

[AZFG20] Soren Laing Aletheia-Zomlefer, Lenny Fukshansky, and Stephan Ramon Garcia. The Bateman-Horn conjecture: heuristic, history, and applications. *Expo. Math.*, 38(4):430–479, 2020.

[BF11] Jean Bourgain and Elena Fuchs. A proof of the positive density conjecture for integer Apollonian circle packings. *J. Amer. Math. Soc.*, 24(4):945–967, 2011.

[BF12] Jean Bourgain and Elena Fuchs. On representation of integers by binary quadratic forms. *International Mathematics Research Notices*, 2012(24):5505–5553, 2012.

[BG06] Valentin Blomer and Andrew Granville. Estimates for representation numbers of quadratic forms. *Duke Math. J.*, 135(2):261–302, 2006.

[BK14] Jean Bourgain and Alex Kontorovich. On the local-global conjecture for integral Apollonian gaskets. *Invent. Math.*, 196(3):589–650, 2014. With an appendix by Péter P. Varjú.

[Bou12] J. Bourgain. Integral Apollonian circle packings and prime curvatures. *J. Anal. Math.*, 118(1):221–249, 2012.

[Boy82] David W. Boyd. The sequence of radii of the Apollonian packing. *Math. Comp.*, 39(159):249–254, 1982.

[FHR25] Elena Fuchs, Catherine Hsu, James Rickards, Damaris Schindler, and Katherine E. Stange. Primes represented by shifted quadratic forms: on primitivity and congruence classes. *Acta Arithmetica.* https://doi.org/10.4064/aa250502-24-10. https://www.impan.pl/en/publishing-house/journals-and-series/acta-arithmetica/online/116332/primes-represented-by-shifted-quadratic-forms-on-primitivity-and-congruence-classes, 26 February 2026.

[FI10] John Friedlander and Henryk Iwaniec. *Opera de cribro*, volume 57 of *American Mathematical Society Colloquium Publications*. American Mathematical Society, Providence, RI, 2010.

[FS11] Elena Fuchs and Katherine Sanden. Some experiments with integral Apollonian circle packings. *Exp. Math.*, 20(4):380–399, 2011.

[FSZ19] Elena Fuchs, Katherine E. Stange, and Xin Zhang. Local-global principles in circle packings. *Compos. Math.*, 155(6):1118–1170, 2019.

[Fuc11] E. Fuchs. Strong approximation in the Apollonian group. *J. Number Theory*, 131(12):2282–2302, 2011.

[GLM03] R. L. Graham, J. C. Lagarias, C. L. Mallows, A. R. Wilks, and C. H. Yan. Apollonian circle packings: number theory. *J. Number Theory*, 100(1):1–45, 2003.

[Hir67] K. E. Hirst. The Apollonian packing of circles. *J. London Math. Soc.*, 42:281–291, 1967.

[HKRS24] Summer Haag, Clyde Kertzer, James Rickards, and Katherine Stange. The local-global conjecture for Apollonian circle packings is false. *Ann. of Math. (2)*, 200(2):749–770, 2024.

[Iwa72] H. Iwaniec. Primes of the type $\phi(x, y) + A$ where ϕ is a quadratic form. *Acta Arith.*, 21:203–234, 1972.

[Iwa78] Henryk Iwaniec. Almost-primes represented by quadratic polynomials. *Invent. Math.*, 47(2):171–188, 1978.

[KK23] Michael Kapovich and Alex Kontorovich. On superintegral Kleinian sphere packings, bugs, and arithmetic groups. *J. Reine Angew. Math.*, 798:105–142, 2023.

[KN19] Alex Kontorovich and Kei Nakamura. Geometry and arithmetic of crystallographic sphere packings. *Proc. Natl. Acad. Sci. USA*, 116(2):436–441, 2019.

[KO11] Alex Kontorovich and Hee Oh. Apollonian circle packings and closed horospheres on hyperbolic 3-manifolds. *J. Amer. Math. Soc.*, 24(3):603–648, 2011. With an appendix by Oh and Nimish Shah.

[LO13] Min Lee and Hee Oh. Effective circle count for Apollonian packings and closed horospheres. *Geom. Funct. Anal.*, 23(2):580–621, 2013.

[May15] James Maynard. Small gaps between primes. *Ann. of Math. (2)*, 181(1):383–413, 2015.

[Ric24a] James Rickards. Apollonian. https://github.com/JamesRickards-Canada/Apollonian, 2024.

[Ric24b] James Rickards. Apollonian-Prime. https://github.com/JamesRickards-Canada/Apollonian-Prime, 2024.

[Sar07] Peter Sarnak. Letter to J. Lagarias. https://web.math.princeton.edu/sarnak/AppolonianPackings.pdf, 2007.

[Ser73] J.-P. Serre. *A course in arithmetic*. Graduate Texts in Mathematics, No. 7. Springer-Verlag, New York-Heidelberg, 1973. Translated from the French.

[Sha07] Lisa Shapiro, editor. *The Correspondence Between Princess Elisabeth of Bohemia and René Descartes*. University of Chicago Press, 2007.

[Sta18] Katherine E. Stange. The Apollonian structure of Bianchi groups. *Trans. Amer. Math. Soc.*, 370(9):6169–6219, 2018.

[The23] The Sage Developers. *SageMath, the Sage Mathematics Software System (Version 10.2.beta2)*, 2023. https://www.sagemath.org.

[The24] The PARI Group, Univ. Bordeaux. *PARI/GP version* `2.16.2`, 2024. available from http://pari.math.u-bordeaux.fr/.

[Tom20] Radu Toma. Estimates for representation numbers of binary quadratic forms and Apollonian circle packings. *J. of Number Theory*, 214:399–413, 2020.

[Vin12] Ilya Vinogradov. Effective Bisector Estimate with Application to Apollonian Circle packings. *International Mathematics Research Notices*, 2014:3217–3262, 2012.

Machine Learning for Modular Multiplication

Kristin Lauter, Cathy Yuanchen Li, Krystal Maughan, Rachel Newton, and Megha Srivastava

Abstract Motivated by cryptographic applications, we investigate two machine learning approaches to modular multiplication: namely circular regression and a sequence-to-sequence transformer model. The limited success of both methods demonstrated in our results gives evidence for the hardness of tasks involving modular multiplication upon which cryptosystems are based.

1 Introduction

Machine learning approaches to modular arithmetic have recently entered the limelight, motivated at least in part by the reliance on modular addition and multiplication of several cryptographic schemes, including RSA, Learning With Errors (LWE) and Diffie–Hellman, to name but a few.

In the SALSA series of papers [WCCL22, LSWMGCL23, LWAZCL23], transformers are used to develop novel machine learning attacks on LWE-based cryptographic schemes. Such schemes are based on the following problem, which is

K. Lauter
Meta AI, Seattle, WA, USA
e-mail: klauter@meta.com

C. Y. Li
University of Chicago, Chicago, IL, USA
e-mail: yuanchen@cs.uchicago.edu

K. Maughan
University of Vermont, Burlington, VT, USA
e-mail: Krystal.Maughan@uvm.edu

R. Newton (✉)
King's College London, London, UK
e-mail: rachel.newton@kcl.ac.uk

M. Srivastava
Stanford University, Stanford, CA, USA
e-mail: megha@cs.stanford.edu

S. Akhtari et al. (eds.), *Research Directions in Number Theory*, Association for Women in Mathematics Series 39, https://doi.org/10.1007/978-3-032-11183-8_9

assumed to be hard. Given a dimension n, an integer modulus q, and a secret vector $s \in \mathbb{Z}_q^n$, solving the Learning With Errors problem [Reg05] entails finding the secret s given a data set consisting of pairs of vectors (a_i, b_i) where $a_i \in \mathbb{Z}_q^n$ and $b_i \equiv a_i \cdot s + e_i \pmod{q}$ with e_i a small "error" or "noise" sampled from a narrow centered Gaussian distribution. We call b_i a "noisy inner product". In [WCCL22, LSWMGCL23, LWAZCL23], transformer models are (partially) trained on noisy inner product data and used along with statistical cryptanalysis techniques to recover the secret s, under certain conditions on the dimension, modulus, and secret size/sparseness.

In dimension $n = 1$, recovering the secret s is akin to finding a modular inverse for a_i. Therefore, it is natural to ask whether the SALSA architecture has in fact learnt modular arithmetic operations such as multiplication and division. However, the latest paper in the series, SALSA VERDE [LWAZCL23], calls this into question. The authors of that paper observe that secret recovery is harder when the modulus q is smaller and suggest that this may be a modular arithmetic issue. When q is large compared to a_i, s, and e_i, computing b does not involve any modular arithmetic (modular multiplication is replaced by ordinary multiplication). This reopens the question of whether transformers or other machine learning techniques can tackle modular multiplication and provides motivation for our investigations in the present paper.

We experiment with two different machine learning approaches to modular multiplication. In Sect. 2 we describe an approach to solving the 1-dimensional Learning With Errors problem using circular regression. We show that the method, as implemented, does not consistently give a significant improvement on exhaustive search. Indeed, some recent theoretical results suggest that in general a large number of iterations are needed in order for methods based on gradient descent to successfully tackle modular multiplication [TTPBA23]. In Sect. 3 we explore an alternative approach using a sequence-to-sequence transformer, and show poor generalization performance of our method for several different values of a secret s and base representations of numbers, likewise demonstrating the difficult nature of this problem, in line with evidence presented in [Pal17, JdADEWLC23]. We end with a discussion of the relevance of modular arithmetic to cryptographic schemes such as Diffie–Hellman in Sect. 4.

2 Circular Regression for Modular Multiplication

2.1 The Task

We focus on the 1-dimensional version of LWE. Let $p \in \mathbb{Z}_{>0}$ be a prime number and let $s \in \mathbb{Z}/p\mathbb{Z}$. We will refer to s as the *secret*. Given a data set consisting of pairs of integers $\{(a_i, b_i)\}_{1 \le i \le m}$, where $b_i \equiv a_i s + e_i \pmod{p}$ and the "noise" or "error" values e_i are sampled from a centered discrete Gaussian distribution with standard deviation σ, the task is to find the unknown secret s. This problem was studied in the case of binary secrets in [GMPLCL22], where the authors explored using circular regression to solve Learning With Errors (LWE) in small dimension up to 28.

2.2 Transforming to a Circular Regression Problem

Following [GMPLCL22], we rescale to view our integers modulo p as points on the unit circle: define $y_i = \frac{2\pi}{p} b_i$ so that the congruence $b_i \equiv a_i s + e_i \pmod{p}$ becomes

$$y_i \equiv \frac{2\pi}{p} a_i s + \frac{2\pi}{p} e_i \pmod{2\pi}.$$

As in [GMPLCL22], we assume the target variables y_i follow a discrete von Mises distribution. The von Mises distribution is also known as circular normal distribution or Tikhonov distribution and it closely approximates a wrapped normal distribution.

Definition 1 (*von Mises distribution*) The von Mises distribution is a continuous probability distribution on the circle. The von Mises probability density function for the angle $\theta \in [-\pi, \pi]$ is

$$f(\theta \mid \mu, \kappa) = \frac{\exp(\kappa \cos(\theta - \mu))}{2\pi I_0(\kappa)}$$

where the parameters μ and $1/\kappa$ are analogous to μ and σ^2 (the mean and variance) in the normal distribution:

- μ is a measure of location (the distribution is clustered around μ), and
- κ is a measure of concentration (a reciprocal measure of dispersion, so $1/\kappa$ is analogous to σ^2).

$I_0(\kappa)$ is a scaling constant chosen so that the distribution is a probability distribution, i.e., it sums to 1. I_0 denotes the modified Bessel function of the first kind of order 0, which satisfies

$$\int_{-\pi}^{\pi} \exp(\kappa \cos x) dx = 2\pi I_0(\kappa).$$

Since our samples y_i correspond to the integers b_i, they can only take certain discrete values, namely those angles of the form $2\pi n/p$ for integers $n \in [\frac{-p+1}{2}, \frac{p-1}{2}]$. To build a discrete version of the von Mises distribution, take a parameter c and define a modified probability distribution function $\tilde{f}_c(\theta \mid \mu, \kappa)$ as follows:

$$\tilde{f}_c(\theta \mid \mu, \kappa) = \begin{cases} c \cdot f(\theta \mid \mu, \kappa) & \text{if} \theta = 2\pi n/p \text{ for some } n \in \mathbb{Z} \cap [\frac{-p+1}{2}, \frac{p-1}{2}]; \\ 0 & \text{otherwise.} \end{cases}$$

To make this a probability distribution, we need the distribution to sum to 1; this is achieved by choosing an appropriate value for c. Assume from now on that such a value c has been fixed.

We consider each y_i as being sampled from a discrete von Mises distribution $\tilde{f}_c(\theta \mid \mu_i, \kappa)$ where $\mu_i = \frac{2\pi}{p} a_i s$ and κ is the same for each b_i since it corresponds to the variance σ^2 of the distribution of the errors e_i.

The events y_i are independent so the probability density function for a collection of samples $y_1, \ldots, y_m$ is

$$\prod_{i=1}^{m} \tilde{f}_c(y_i \mid \mu_i, \kappa) = c^m \prod_{i=1}^{m} f(y_i \mid \mu_i, \kappa).$$

Therefore, the likelihood function for the parameters $(\mu_1, \ldots, \mu_m; \kappa) = (\frac{2\pi}{p} a_1 s, \ldots, \frac{2\pi}{p} a_m s; \kappa)$, given the collection of samples, is

$$\mathcal{L}((\mu_1, \ldots, \mu_m; \kappa) \mid (y_1, \ldots, y_m)) = c^m \prod_{i=1}^{m} f(y_i \mid \mu_i, \kappa).$$

Since c is fixed, maximizing the (log-)likelihood function is equivalent to maximizing

$$\sum_{i=1}^{m} \log f(y_i \mid \mu_i, \kappa) = \kappa \sum_{i=1}^{m} \cos\left(y_i - \frac{2\pi}{p} a_i s\right) - m \log(2\pi I_0(\kappa)).$$

Thus, we seek to maximize $\sum_{i=1}^{m} \cos\left(y_i - \frac{2\pi}{p} a_i s\right)$ over all choices of s. Equivalently, we minimize the *circular regression loss* $-\sum_{i=1}^{m} \cos\left(y_i - \frac{2\pi}{p} a_i s\right)$. In pursuit of this goal, we will treat s as if it were a continuous real variable, use gradient descent to find a real value of s that minimizes $-\sum_{i=1}^{m} \cos\left(y_i - \frac{2\pi}{p} a_i s\right)$, and our final output will be the integer closest to this real value of s.

Differentiating $-\sum_{i=1}^{m} \cos\left(y_i - \frac{2\pi}{p} a_i s\right)$ with respect to s gives the *gradient*

$$-\frac{2\pi}{p} \sum_{i=1}^{m} a_i \sin\left(y_i - \frac{2\pi}{p} a_i s\right).$$

So we start with an initial guess for s, call it s_0, and at each time step t we define

$$s_{t+1} = s_t + \eta \frac{2\pi}{p} \sum_{i=1}^{m} a_i \sin\left(y_i - \frac{2\pi}{p} a_i s_t\right),$$

where $\eta \in \mathbb{R}_{>0}$ is the learning rate. The theoretical minimum of the circular regression loss $-\sum_{i=1}^{m} \cos\left(y_i - \frac{2\pi}{p} a_i s\right)$ is $-m$. We choose a tolerance $\epsilon \in \mathbb{R}_{>0}$ and halt our process once we have

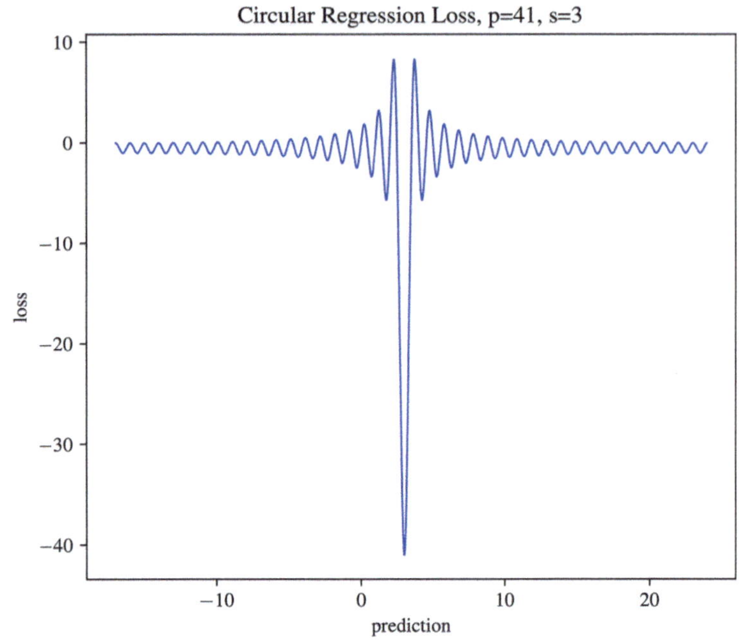

Fig. 1 Circular regression loss for $p = 41$, $s = 3$, plotted using the data set $\{(a_i, b_i = a_i s \pmod{p}) \mid 0 \leq a_i < p, a_i \in \mathbb{Z}\}$, which does not include errors in b_i, and has size $m = p$

$$-\sum_{i=1}^{m} \cos\left(y_i - \frac{2\pi}{p} a_i s_t\right) \leq -m + \epsilon.$$

Our guess for s is then the unique integer in the interval $(s_t - \frac{1}{2}, s_t + \frac{1}{2}]$.

2.3 Analysis of the Algorithm

First, we visualize the circular regression loss. As shown in Fig. 1, the loss reaches the lowest value $-m$ at s, and oscillates with decreasing magnitude as the prediction deviates from s. The loss is periodic, hence we only show one interval of length p for simplicity. The minimum within an interval of length p is the global minimum. For various values of the prime p and secret s, the loss exhibits a shape similar to the plot in Fig. 1 (empirically, we observed the loss for $p = 23, 41, 71, 113, 251, 367, 967, 1471$). It is highly non-convex, making the search for optima with gradient descent quite challenging.

Since the loss consists of a deep valley at s but is much closer to 0 everywhere else, we observe that it is reasonable to relax the condition for halting the process.

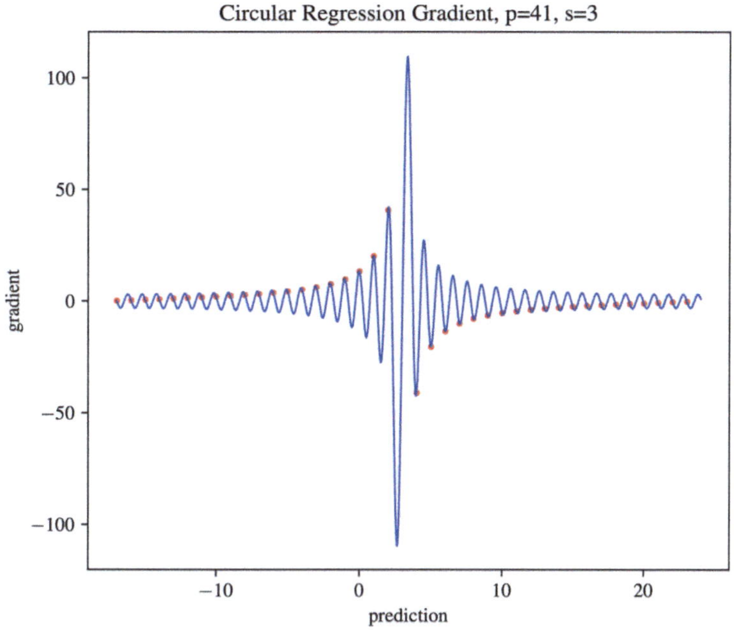

Fig. 2 Circular regression gradient for $p = 41, s = 3$, data set $\{(a_i, b_i = a_i s \pmod{p}) \mid 0 \le a_i < p, a_i \in \mathbb{Z}\}$. The red dots mark the gradient values when the predictions are at integer points

For example, instead of using a tolerance ϵ, we can check the closest integer to the prediction s_t when $-\sum_{i=1}^{m} \cos\left(y_i - \frac{2\pi}{p} a_i s_t\right) \le -\frac{m}{2}$.

For gradient descent to be successful, ideally the direction of the gradient should point toward the closest optimum, and the magnitude of the gradient should be smaller when the prediction is closer to an optimum. Then, as we optimize the prediction using the gradient, we would rapidly approach an optimum and stay close once we reach proximity of an optimum. An example of the circular regression gradient is shown in Fig. 2. It oscillates with higher magnitude around s, which means the magnitude of the gradient displays the opposite behavior to what we would want in an ideal gradient for gradient descent. However, the direction of the gradient is good, at least when the predictions are at integer points. When the predictions are at integer points, the gradient at these points (marked with red dots on the plot) always has the sign that points to the closest correct answer.

Since we would like a function with the same sign as the gradient and a magnitude with better behavior, we are led to consider a version of gradient descent in which we replace the gradient by its reciprocal. We denote the reciprocal of the gradient by *grad_r*. It has the same sign as the gradient and its magnitude has nicer properties, see Fig. 3 as an example, where grad_r has opposite signs on different sides separated by s and has smaller magnitudes when the prediction is closer to s. Grad_r may explode

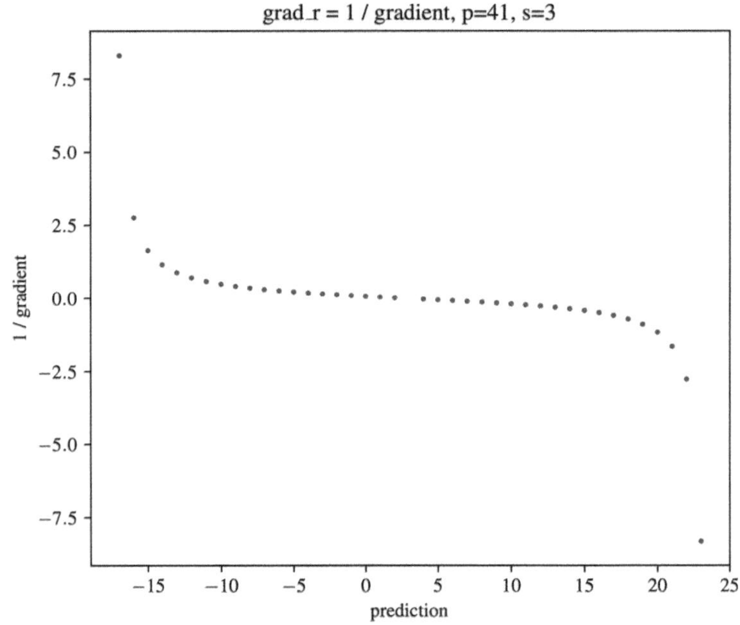

Fig. 3 Reciprocal of the circular regression gradient for $p = 41$, $s = 3$, data set $\{(a_i, b_i = a_i s \pmod{p}) \mid 0 \le a_i < p, a_i \in \mathbb{Z}\}$, when the predictions are at integer points

at various points where the predictions are not integers, where the original gradient is 0. The gradient is also 0 when the prediction is precisely s.

Note that in practice, we would use a small subset of $\{(a_i, a_i s \pmod{p}) \mid 0 \le a_i < p, a_i \in \mathbb{Z}\}$ of size k to compute the gradient for efficiency reasons, and the b_i will have errors, which makes the gradient less accurate. Therefore, the grad_r computed in our implementation does not always have the sign that points to the closest correct answer, nor does it always have smaller magnitude when the prediction is closer to s.

2.4 Experiment Setup

We build the data set with vectors a being integers from 1 to $p - 1$, and $b = as + e \pmod{p}$, where e has standard deviation $\sigma = 3$. For each prime number p, we run 20 integer values of s, randomly chosen from 1 to $p - 1$ without replacement.

For regression, we calculate the gradient on batches (with *batch size* k) of data randomly chosen from the whole data set, and adjust it by scaling with $\frac{1}{k}$. Essentially, that means instead of taking the summation, we are taking the mean, so that the batch size does not directly affect how much s_t is updated each step. Starting with a random integer s_0 as the initial guess, we update the prediction with the reciprocal of the adjusted gradient, scaled by the learning rate η, as follows:

$$s_{t+1} = s_t + \eta \left(\frac{2\pi}{p} \frac{1}{k} \sum_{i=1}^{k} a_i \sin \left(y_i - \frac{2\pi}{p} a_i s_t \right) \right)^{-1}.$$

Although we could use the circular regression loss (Sect. 2.3) to evaluate whether our prediction is likely to be correct, in our implementation we instead verify whether s_t matches s by rounding s_t to the nearest integer and checking the magnitude of $as_t - b \pmod p$, since it is quite cheap and more reliable. A run terminates if s_t matches s, or if the number of steps reaches p. The run is successful at step t if it terminates at step t and the prediction s_t, rounded to the nearest integer, matches s. Henceforth, we will refer to this whole process, starting with a random integer s_0 and terminating with an output $s_t \in \mathbb{Z}$, as circular regression. Our implementation and code to reproduce visualizations in Sect. 2.3 and results in Sect. 2.5 are available at: https://github.com/meghabyte/mod-math/circ_reg.ipynb.

2.5 Empirical Results

To choose the parameters, we ran experiments with various learning rates η and batch sizes k and counted the number of successes. While $\eta = 1$ had more successful trials when $k \in \{256, 512\}$, the performance varied less with p when $\eta = 2$, which is desirable (see Table 1). Hence, for the following experiments, we use $\eta = 2$. And we set the batch size to $k = 256$ because it has similar performance with $k = 512$, but smaller batch size costs less compute.

Figure 4 shows the distribution of number of steps for successful trials with batch size $k = 256$, for $p \in \{251, 1471, 11197, 20663, 42899, 115301, 222553\}$. As p increases, the success rate remains roughly the same, but the number of steps increases. Unfortunately, with batch size $k = 256$, the number of steps needed for circular regression to succeed does not consistently give a significant improvement

Table 1 Number of successes in 20 trials for learning rate $\eta = 0.5, 1, 2$ and batch size $k = 64, 128, 256, 512$, ran for p of different sizes and s randomly selected from 1 to $p - 1$ (see Sect. 2.4). We upper-bound the batch size with the size of the data set, i.e., for $p = 251$ and $k = 256, 512$, each batch is the entire data set

η	0.5			1			2		
p	251	1471	11197	251	1471	11197	251	1471	11197
$k = 64$	6/20	0/20	0/20	16/20	12/20	4/20	15/20	17/20	11/20
$k = 128$	14/20	4/20	0/20	15/20	19/20	6/20	14/20	8/20	11/20
$k = 256$	18/20	11/20	4/20	19/20	17/20	15/20	16/20	13/20	14/20
$k = 512$	17/20	15/20	11/20	20/20	15/20	18/20	14/20	14/20	15/20

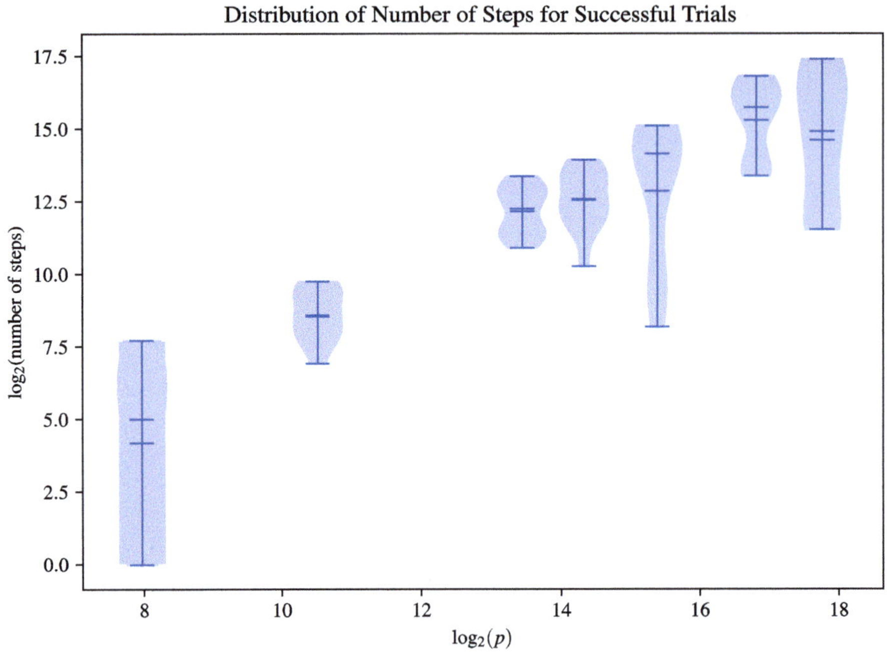

Fig. 4 Distribution of number of steps. For each value of p, we run circular regression on 20 random values of s, with $\eta = 2$ and $k = 256$

on exhaustive search. Possible directions for future work could include scaling η with p, and learning rate decay.

3 Transformers for Modular Multiplication

We now move on to an alternative machine learning-based approach to modular multiplication, namely the use of *transformers*, which are a class of deep learning models designed for "sequence-to-sequence" tasks: transforming one sequence of elements (e.g., words) to another.

3.1 The Task

We consider the noiseless version of the task described in Sect. 2.1—namely, given a data set consisting of m pairs of integers $\{(a_i, b_i)\}_{1 \leq i \leq m}$, where $b_i \equiv a_i s \pmod{p}$, the task is to find the unknown secret s. Knowledge of s together with the ability to perform multiplication modulo p would allow one to take some a_j as an input and generate a valid sample (a_j, b_j) where $b_j \equiv a_j s \pmod{p}$. Moreover, being able to reliably predict b_j given a_j would imply knowledge of s (take $a_j = 1$).

We train a model $\mathcal{M}$ on the dataset $\{(a_i, b_i)\}_{1 \leq i \leq m}$, and determine successful task performance as the model $\mathcal{M}$'s ability to *generalize* to a held-out *test* set of n unknown samples $\{(a_j, b_j)\}_{1 \leq j \leq n}$ not seen during training. Truly learning the secret s and modular multiplication would imply perfect accuracy on a held-out test set – as we demonstrate, we are currently unable to observe this for modular multiplication, suggesting the difficulty of the task.

Recent works have demonstrated success in training *transformers*, powerful encoder-decoder deep neural networks that are behind some of the best-performing language models (e.g., ChatGPT), for modular addition [Gro23, NCLSS23]. These works have demonstrated a surprising phenomenon called *grokking* [PBEBM21], where training a model for a large number of steps (with 0 training loss) leads to a surprising "jump" in generalization capabilities. We therefore consider the transformer architecture as the model class for $\mathcal{M}$, and specifically frame our task as a sequence-to-sequence task: we represent the integer a_i as an input sequence of t tokens $x_{i,1} \ldots x_{i,t}$ in a given base $\mathcal{B}$, and train a transformer-based $\mathcal{M}$ to output b_i represented as an output sequence of t tokens $y_{i,1} \ldots y_{i,t}$ in the same base $\mathcal{B}$. For example, if we use base 10 then the tokens are the usual digits of an integer written in its decimal representation. We note that in the previously mentioned works, models trained to perform modular addition output a *single* token, and therefore the size of the transformer's vocabulary $|\mathcal{V}|$, or number of possible tokens, is equivalent to the modulus p. This is different from our setting, where $\mathcal{M}$ outputs a *sequence* of tokens. In our case, $|\mathcal{V}|$ is equivalent to the base $\mathcal{B}$, and therefore $\mathcal{B}$ influences the overall sequence length that $\mathcal{M}$ needs to generate. Furthermore, the value of the modulus p dictates the total number of input/outputs we can train and evaluate on, so smaller values are more likely to lead to memorization. Indeed, this is what we observe—see Sect. 3.3.

3.2 Representation and Model

Following [WCCL22], we train a sequence-to-sequence transformer, varying the number of encoder-decoder layers, but with a fixed model dimension of 512 and 8 attention heads. The vocabulary size is equivalent to the base ($|\mathcal{V}| = \mathcal{B}$) as described above. *Positional encoding* is used to describe the relative positions of tokens in a sequence. Since the order of the digits is of the utmost importance when representing a number, this should be accounted for in our model. We experiment with two kinds of positional encodings: fixed sinusoidal, as is standard in language models such as GPT-2, and randomly initialized encodings that are learned over the course of the task, and view optimal representation of position for arithmetic tasks with transformers as an interesting direction for future work. We optimize our model by minimizing the KL-divergence [KL51] between the predicted distribution across all tokens in the vocabulary and the ground truth for each token in the output sequence $y_{i,1} \ldots y_{i,t}$. We also experiment with a *weighted loss* objective that places a higher penalty on divergence in the most significant bits (e.g., $y_{i,1}$) than in the least significant bits (e.g., $y_{i,t}$). We specifically use the weight 1.25 for the first 1/3 most significant bits, 1 for the middle 1/3 significant bits, and 0.75 for the 1/3 least significant bits. Finally, we implement *early-stopping* by ending training either after 5000 (the maximum) number of epochs, or when the loss on a held-out valid set has monotonically increased for 5 epochs. Our model implementation and code are publicly available at: https://github.com/meghabyte/mod-math.

3.3 Memorization

We generally observe that for a small prime p and a small secret s, the task of memorization results in a high ($\approx$100%) model accuracy. This is not entirely surprising as a smaller prime means there are fewer possible inputs and outputs, and therefore there is less to learn. On the other hand, when the prime p is large and the secret s is small, modular multiplication often coincides with ordinary multiplication. Previous work shows that transformers are capable of learning ordinary multiplication, see e.g., [Cha22, SBELLZ23] and the references therein. In our experiments we do not observe increased memorization accuracy when the secret is small compared to the prime, but time constraints mean we have not run experiments with very large primes, so this could be a direction for further investigation. Moreover, memorization

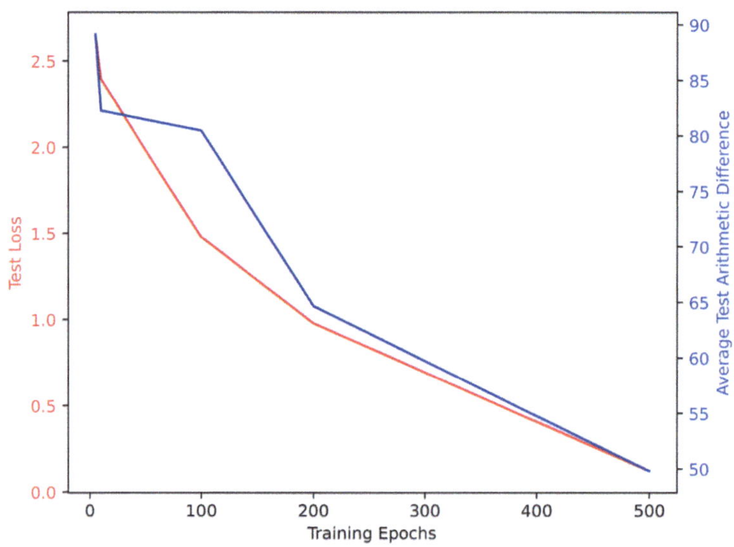

Fig. 5 Training curve for modular multiplication task with $p = 251, s = 3$, and base $\mathcal{B} = 7$ shows that optimizing sequence-to-sequence accuracy also helps improve arithmetic accuracy, as both test loss and arithmetic difference between generated outputs $\hat{y}_i$ and true values y_i decrease during training

may well improve with increased training time. In the case of $p = 83$, memorization accuracy was 100% using our current model for base 8, 9 and 11. For $p = 97$, memorization accuracy is 100% for base 9, 94.12% for base 8, and varying memorization accuracy for base 11. This accuracy quickly decreases for larger primes p. For a selection of secrets $3 \leq s \leq 83$, we evaluated primes p from 83 to 293 and found that accuracy decreased from 100% to ≈40–60% as shown for a selection of primes in Figs. 6, 7, 8, 9, 10, and 11. We evaluated bases $\mathcal{B} \in \{8, 9, 11\}$ for 5000 epochs with Beam = 6 (see Sect. 3.4 below for an explanation of beam search). We note that not all bases are equal: in Fig. 7 we see that for bases 8 and 9, memorization is stable and high across all secrets, but for base 11 we start to see mixed results. This shows that base representation matters when training the model, and this would then influence the model's ability to generalize, cf. [WCCL22, Sect. 5.5].

3.4 Evaluation

In order to evaluate our model's ability to generalize, and therefore successfully learn a given secret s, we must first consider the *decoding* method. We experiment with two such methods: *greedy decoding*, where the most likely token $\hat{y}_{i,t}$ is selected conditioned on the input x_i and previously generated tokens, and *beam search*, where k possible candidates are retained at each step, and the output sequence $\hat{y}$ with the highest likelihood is selected. We then compare the predicted output sequence $\hat{y}_i$ with the true output of modular multiplication for each instance of our test data, y_i.

We evaluate model outputs in two ways: accuracy and arithmetic difference. Consider the task instance $p = 251$, $s = 3$, and base $\mathcal{B} = 7$. For an input $x = 426$, the output of modular multiplication would be $y = 266$ in base $\mathcal{B} = 7$. Perfect accuracy from our model would require $\mathcal{M}$, under a given decoding method, to generate the sequence $\hat{y}_1 = 2$, $\hat{y}_2 = 6$, $\hat{y}_3 = 6$. Unfortunately, we largely observe 0% accuracy on the test set (inputs not observed during training) across different choices of prime modulus, base and secret, and therefore choose to also measure the arithmetic difference (in base 10) between the predicted output $\hat{y}_i$ and ground truth y_i. A predicted output sequence of $\hat{y}_1 = 2$, $\hat{y}_2 = 6$, $\hat{y}_3 = 3$ would be considered closer in arithmetic difference than $\hat{y}_1 = 3$, $\hat{y}_2 = 6$, $\hat{y}_3 = 6$, even though the overall sequence "loss" is equivalent (due to a mismatch in one token). A model that can perform modular multiplication under a certain error range in the least significant bits could still be useful for cryptographic attacks. It is not immediately intuitive that optimizing for the sequence-based loss (generating correct tokens) helps decrease the arithmetic difference. However, in Fig. 5 we show that this generally holds true over a large scale, where both test loss and arithmetic difference decrease as we train $\mathcal{M}$.

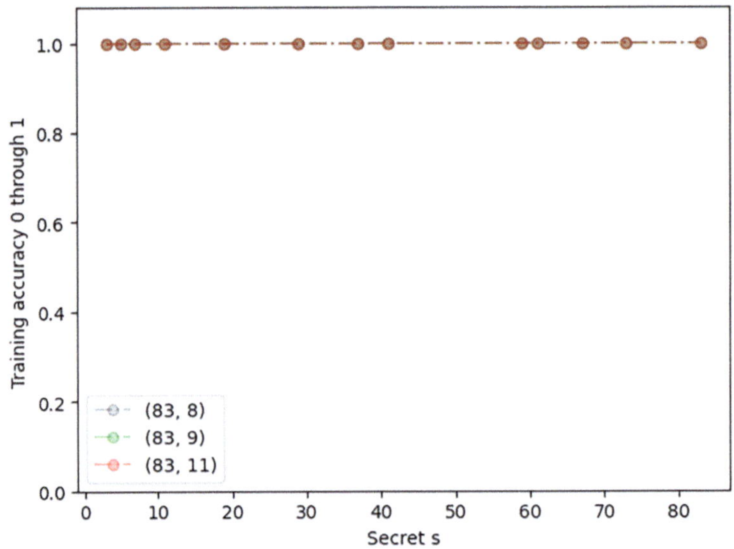

Fig. 6 Train accuracy for $p = 83$, $3 \leq s \leq 83$, and $\mathcal{B} \in \{8, 9, 11\}$, after training for 5000 epochs with learning rate 0.007

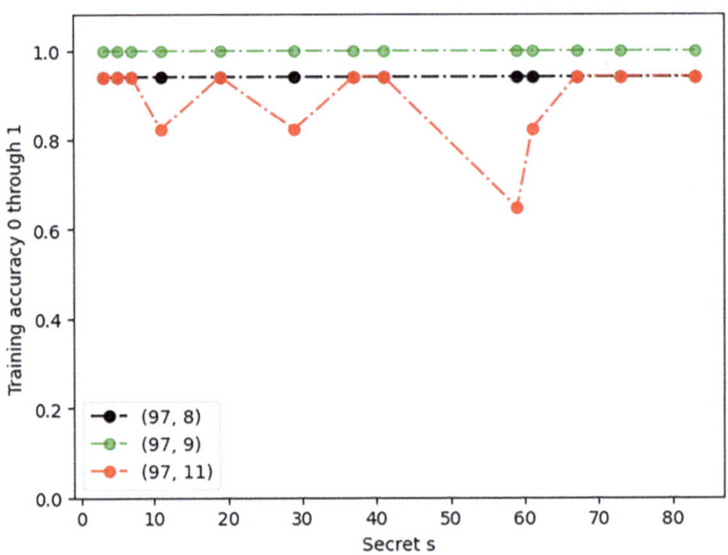

Fig. 7 Train accuracy for $p = 97$, $3 \leq s \leq 83$, and $\mathcal{B} \in \{8, 9, 11\}$, after training for 5000 epochs with learning rate 0.007

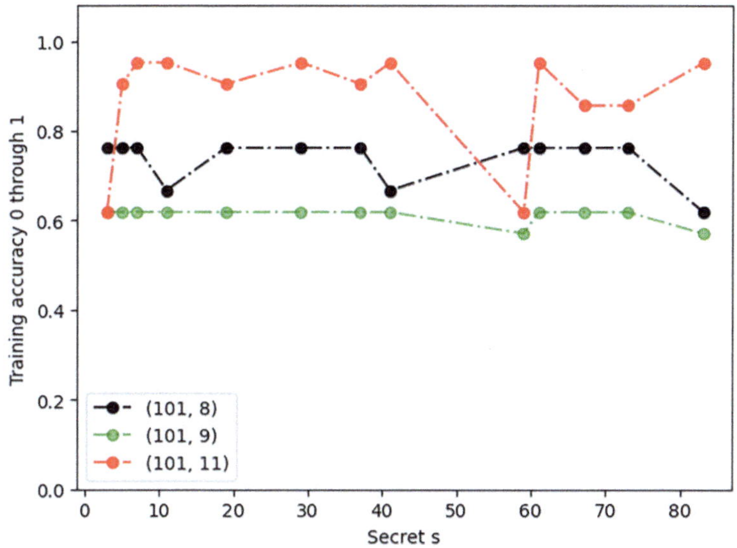

Fig. 8 Train accuracy for $p = 101$, $3 \leq s \leq 83$, and $\mathcal{B} \in \{8, 9, 11\}$, after training for 5000 epochs with learning rate 0.007

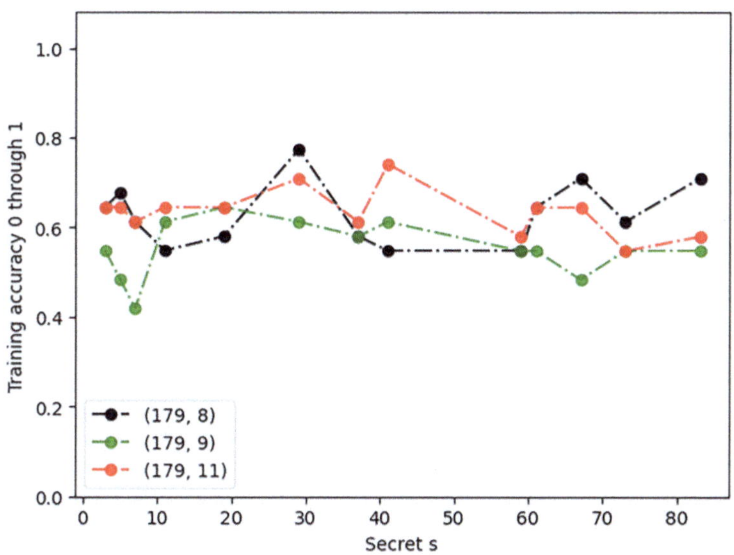

Fig. 9 Train accuracy for $p = 179$, $3 \leq s \leq 83$, and $\mathcal{B} \in \{8, 9, 11\}$, after training for 5000 epochs with learning rate 0.007

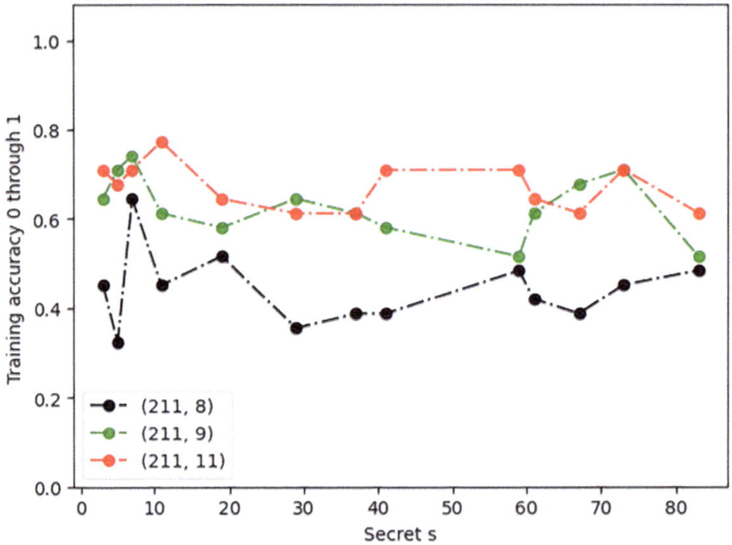

Fig. 10 Train accuracy for $p = 211, 3 \leq s \leq 83$, and $\mathcal{B} \in \{8, 9, 11\}$, after training for 5000 epochs with learning rate 0.007

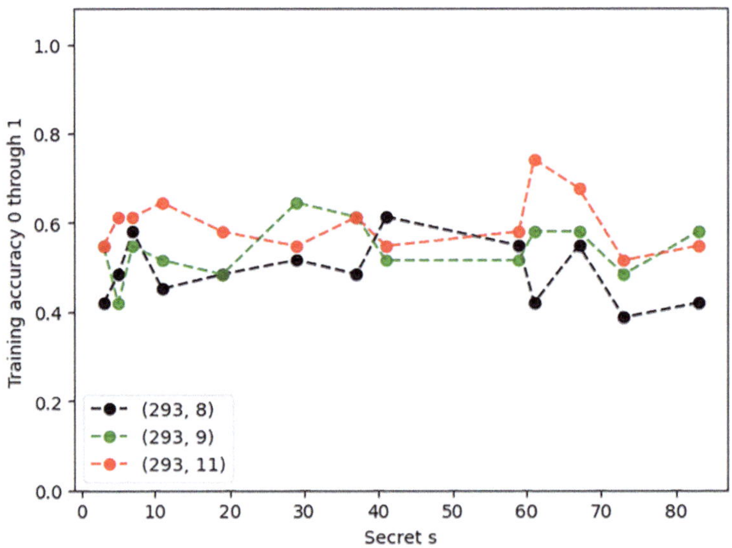

Fig. 11 Train accuracy for $p = 293, 3 \leq s \leq 83$, and $\mathcal{B} \in \{8, 9, 11\}$, after training for 5000 epochs with learning rate 0.007

3.5 Generalization Results

We generally observe that beam search outperforms greedy decoding for all three instances of $(p, s, \mathcal{B})$, that increasing beam size beyond 3 makes no difference in performance, that sinusoidal position encoding only improves performance over random positional encoding for smaller values of the p modulus, and that weighting the loss generally leads to stronger generalization, see Table 2.

Given the low test accuracy (see Sect. 3.4), a natural question to ask is whether this can be improved by having larger models. However, as shown in Fig. 12, increasing the number of layers only improves performance for smaller modulus p, suggesting that expressiveness of the model is not a sufficient reason for low generalization performance.

Finally, in Fig. 13 we show generalization results (average arithmetic distance on held-out test examples) for 5 different values of p, 5 different values of s, and 5 different values of $\mathcal{B}$, in order to explore the effects of prime, secret, and base. All results are from training models identically, with early-stopping, as described earlier. As expected, smaller values of p result in lower arithmetic difference between the secret s and the generated output, as the space of possible differences is smaller (once normalized by the largest possible difference, which is the value of p, all differences lie between 0.3 and 0.33). Meanwhile, we observe no trend in performance related to the base $\mathcal{B}$ or the secret s, though it is possible that some trend emerges at higher values of p.

4 Discussion

One motivation for studying whether modular multiplication is easily tackled by machine learning algorithms, and whether models can learn the representation of a fixed unknown factor (the secret s) in modular multiplication, comes from cryptography. In particular, learning modular multiplication can potentially yield solutions to more advanced problems. For example, consider the problem of recovering a secret s from a data set $\{(a_i, y_i)\}_{1 \le i \le m}$ where $a_i \in \mathbb{Z}$ and $y_i \equiv g^{a_i s} \pmod{p}$ for a public choice of primitive root g modulo the prime p. This is related to a Diffie–Hellman scheme, where Alice picks a random number a and sends $g^a \pmod{p}$ to Bob, Bob picks a random number s and sends $g^s \pmod{p}$ to Alice, and they both compute $y \equiv g^{as} \pmod{p}$ as the shared secret. Note that Alice does not have access to Bob's random number s. However, Alice has control of a and access to the values of y to build a data set for training to predict y from a. (Note that this assumes that the secret s belonging to Bob remains fixed, while Alice's a is changing, cf. semi-static or ephemeral/static Diffie–Hellman encryption schemes such as ElGamal [ElG85].) If an algorithm learns to predict $y_i \equiv g^{a_i s} \pmod{p}$ from a_i, it has implicit knowledge of s and one can potentially extract s.

Table 2 Average test arithmetic difference (lower is better) for different ablations of our transformer modeling approach applied to 3 settings $(p, s, \mathcal{B})$ of modular multiplication. **Std** refers to training a 2-layer transformer with our weighted loss, random positional encodings, and greedy decoding, with early-stopping implemented up until 5000 epochs

$(p, s, \mathcal{B})$	Std.	Unweighted Loss	Sinusoidal Enc.	Beam = 3	Beam = 5
(97, 11, 9)	36.5	44.475	24.613	34.975	34.975
(101, 3, 7)	37.45	36.775	32.213	36.138	36.138
(109, 29, 8)	27.288	33.913	35.425	26.963	26.963
(179, 29, 8)	68.88	75.725	83.7	57.075	57.075

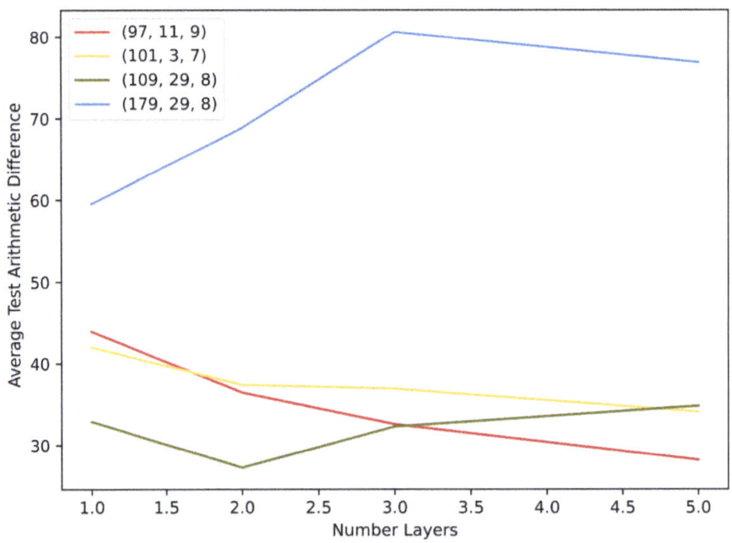

Fig. 12 Performance versus number of encoder-decoder layers of the model. Legend = $(p, s, \mathcal{B})$. Lower average test arithmetic difference means better performance

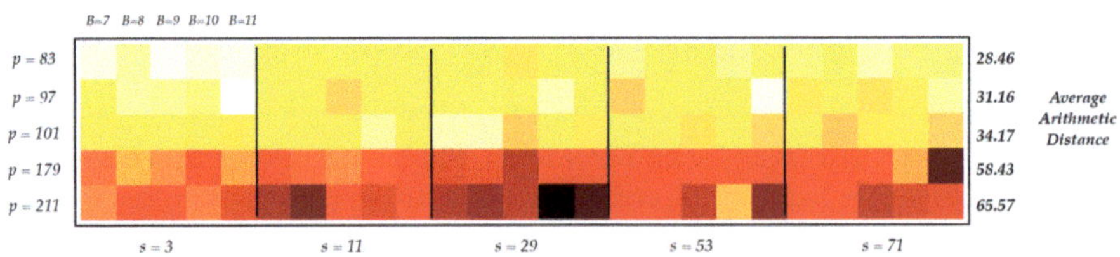

Fig. 13 Smaller values of p result in lower average arithmetic difference (error) on held out test examples for our transformers-based approach. Meanwhile, different values of s and the encoding base $\mathcal{B}$ do not show a strong effect on generalization performance. Heatmap colors represent average arithmetic difference with lighter colors meaning smaller values

In order to have $(g^a)^s \equiv g^b \pmod p$, we only need $a \cdot s \equiv b \pmod{(p-1)}$, which is a modular multiplication problem. Hence, a gradient-based algorithm can try to predict $pred = a \cdot s \pmod{(p-1)}$. However, in the problem described above, the data set has $g^b \pmod p$ accessible but b unknown. In fact, solving for b from $y = g^b \pmod p$ is itself a famous hard problem known as the discrete logarithm problem.

Therefore, the loss function would need to involve a comparison with $y = g^b \pmod p$. For example, one might try using $g^{pred} \pmod p - y$ as the loss function. But this function involves modular arithmetic and raising g to the power of $pred$. Both of these features are challenging for current gradient-based methods, for the following reasons:

(1) the reduction modulo p function is not differentiable;
(2) $pred$ is an integer of the same scale as p so for large p, g^{pred} is a huge number, and the usual ways of handling this (e.g., binary exponentiation) involve modular arithmetic as part of the calculation, which is not differentiable as remarked above.

To circumvent issue (1), one could replace reduction modulo p by a smooth function that gives a close approximation to reduction modulo p. For issue (2), writing the numbers in some base $\mathcal{B}$ (see Sect. 3.1) could be helpful, illustrated in Example 2.

Example 2 Suppose we would like to train a transformer to predict $pred = a \cdot s \pmod{(p-1)}$ from a. The data set consists of pairs (a, y), where $y \equiv (g^a)^s \pmod p$. The transformer could be set up to output a sequence that writes $pred$ in base $\mathcal{B}$. Let us denote that sequence as $[y_k, ..., y_1, y_0]$, so $pred = y_k \cdot \mathcal{B}^k + ... + y_1 \cdot \mathcal{B} + y_0$, where $0 \le y_i < \mathcal{B}$ for all i. Let $g_i = g^{\mathcal{B}^i} \pmod p$. All the g_i are just constants $< p$. Let mod_p denote a smooth function approximating reduction modulo p. A possible loss function to be minimized could be

$$g^{pred} \pmod p - y \approx mod_p(g^{pred}) - y.$$

With $pred$ written in base $\mathcal{B}$, this is

$$\begin{aligned} mod_p\left(g^{\sum_{i=0}^{k} y_i \cdot \mathcal{B}^i}\right) - y &= mod_p\left(\prod_{i=0}^{k} g_i^{y_i} \cdot \ldots \cdot g_1^{y_1} \cdot g^{y_0}\right) - y \\ &\approx mod_p\left(\prod_{i=0}^{k} mod_p(g_i^{y_i})\right) - y. \end{aligned}$$

If we choose $\mathcal{B}$ to be relatively small, which means all the y_i are relatively small, then the $g_i^{y_i}$ should be reasonable to compute. The terms $mod_p(g_i^{y_i})$ are in the interval $(0, p)$, so their product is memory efficient to compute too. Furthermore, the result is differentiable, and therefore may be amenable to gradient-based methods.

In addition to modular arithmetic remaining hard to learn for machine learning algorithms, the discreteness and the scale of the numbers used in cryptographic applications also bring engineering challenges. Example 2 above illustrates one possible way forward to mitigate the difficulties with an algorithmic approach.

Acknowledgements We are grateful to Julio Brau and Kolyan Ray for useful discussions. We thank the anonymous referee for their careful reading of our paper and helpful comments. Rachel Newton was supported by UKRI Future Leaders Fellowship MR/T041609/1 and MR/T041609/2. Megha Srivastava was supported by an IBM PhD Fellowship and the NSF Graduate Research Fellowship Program under Grant No. DGE-1656518. This project began at the WIN6: Women in Numbers 6 workshop in 2023. We are grateful for the hospitality and support of the Banff International Research Station during the workshop and we would especially like to thank Shabnam Akhtari, Alina Bucur, Jennifer Park, and Renate Scheidler for organizing WIN6 and making this collaboration possible.

References

[Cha22] François Charton. Linear algebra with transformers. *Transactions on Machine Learning Research*, 2835–8856 (2022).

[ElG85] Taher ElGamal. A Public-Key Cryptosystem and a Signature Scheme Based on Discrete Logarithms. *IEEE Transactions on Information Theory* **31**(4), (1985), 469–472. (conference version appeared in CRYPTO'84, pp. 10–18).

[GMPLCL22] Evrard Garcelon, Mohamed Malhou, Matteo Pirotta, Cathy Yuanchen Li, François Charton and Kristin Lauter. Solving Learning with Errors with Circular Regression. *Preprint, Meta AI Papers, October 2022.*

[Gro23] Andrey Gromov. Grokking modular arithmetic. Preprint (2023). Available at arxiv:2301.02679

[JdADEWLC23] Samy Jelassi, Stéphane d'Ascoli, Carles Domingo-Enrich, Yuhuai Wu, Yuanzhi Li and François Charton. Length Generalization in Arithmetic Transformers. Preprint (2023). Available at arxiv:2306.15400

[KL51] Solomon Kullback and Richard A. Leibler. On information and sufficiency. *Annals of Mathematical Statistics*. **22** (1) (1951), 79–86.

[LSWMGCL23] Cathy Yuanchen Li, Jana Sotáková, Emily Wenger, Mohamed Malhou, Evrard Garcelon, François Charton and Kristin Lauter. SALSA PICANTE: a machine learning attack on LWE with binary secrets. *Proceedings of the ACM Conference on Computer and Communications Security (CCS)*, November 2023.

[LWAZCL23] Cathy Yuanchen Li, Emily Wenger, Zeyuan Allen-Zhu, François Charton and Kristin Lauter. SALSA VERDE - A machine learning attack on LWE with small sparse secrets. *Proceedings of the 37th Conference on Neural Information Processing Systems (NeurIPS)*, November 2023.

[NCLSS23] Neel Nanda, Lawrence Chan, Tom Lieberum, Jess Smith and Jacob Steinhardt. Progress measures for grokking via mechanistic interpretability. *The Eleventh International Conference on Learning Representations*, 2023.

[Pal17] Theodoros Palamas. Investigating the Ability of Neural Networks to Learn Simple Modular Arithmetic. MSc Thesis, Edinburgh, 2017. Available at https://project-archive.inf.ed.ac.uk/msc/20172390/msc_proj.pdf

[PBEBM21] Alethea Power, Yuri Burda, Harri Edwards, Igor Babuschkin and Vedant Misra. Grokking: Generalization beyond overfitting on small algorithmic datasets. *1st Mathematical Reasoning in General Artificial Intelligence Workshop, ICLR 2021.*

[Reg05] Oded Regev. On lattices, learning with errors, random linear codes, and cryptography. *Proceedings of the thirty-seventh annual ACM symposium on Theory of computing* (Baltimore, MD, USA: ACM, 2005), 84–93.

[SBELLZ23] Ruoqi Shen, Sébastien Bubeck, Ronen Eldan, Yin Tat Lee, Yuanzhi Li and Yi Zhang. Positional Description Matters for Transformers Arithmetic. Preprint (2023), available at arxiv:2311.14737

[TTPBA23] Rustem Takhanov, Maxat Tezekbayev, Artur Pak, Arman Bolatov and Zhenisbek Assylbekov. Gradient Descent Fails To Learn High-frequency Functions and Modular Arithmetic. Preprint (2023), available at arxiv:2310.12660

[WCCL22] Emily Wenger, Mingjie Chen, François Charton and Kristin Lauter. SALSA: Attacking Lattice Cryptography with Transformers. *Proceedings of the 36th Conference on Neural Information Processing Systems (NeurIPS)*, November 2022.

An Explicit Theta Lift to Siegel Paramodular Forms

Jennifer Johnson-Leung and Nina Rupert

Abstract Let E/L be a real quadratic extension of number fields. We construct an explicit map from an irreducible cuspidal automorphic representation of $\mathrm{GL}(2, E)$ which contains a Hilbert modular form with Γ_0 level to an irreducible automorphic representation of $\mathrm{GSp}(4, L)$ which contains a Siegel paramodular form and exhibit local data which produces a paramodular invariant vector for the local theta lift at every finite place, except when the local extension has wild ramification.

1 Introduction

The theta correspondence, introduced by Roger Howe in 1979 [How79], provides a robust and explicit method of relating automorphic forms on pairs of reductive groups. In general, the theta correspondence is a tool for relating automorphic adelic representations (or admissible local representations) of certain pairs of subgroups (S_1, S_2) of a symplectic group $\mathrm{Sp}(X)$ defined over a number field (or a local field of characteristic zero). Such a map is generally called a theta lift, and this theory has been carefully developed over the past forty-five years by the work of many people. See, for example, [Gan23] for a modern introduction.

The global theta lift from S_1 to S_2 takes as input a cusp form f on the adeles of S_1 and defines an automorphic form on the adeles of S_2 by integrating f against a theta kernel; this theta kernel depends on a choice of a certain local Schwartz function φ. In the Archimedean case Schwartz functions are rapidly decreasing away from 0, and the in non-Archimedean case Schwartz functions are locally constant and compactly supported. Determining a choice for φ, which gives automorphic forms

J. Johnson-Leung (✉)
Department of Mathematics and Statistical Science, University of Idaho, Moscow, ID, USA
e-mail: jenfns@uidaho.edu

N. Rupert
Department of Sciences and Mathematics, University of Washington - Tacoma, Tacoma, WA, US
e-mail: ninaru@uw.edu

S. Akhtari et al. (eds.), *Research Directions in Number Theory*, Association for Women in Mathematics Series 39, https://doi.org/10.1007/978-3-032-11183-8_10

with desirable qualities, requires a theory of local theta lifts that are commensurable with the global theory.

While the global theta lift has a natural construction as an integral operator derived from the work of Weil [Wei64], there is nothing so ubiquitous in the local theory. In [WAL80], the author studies the local and global theta lift when W is a 2-dimensional symplectic space, V is a rank 3 quadratic space, and $X = W \otimes V$. In [Wal81], the author investigated the correspondence between modular forms of half integral weight and those with integral weight and computed special values of L-functions in terms of Fourier coefficients of half integral weight modular forms.

In this work, we consider an analogous problem when W is a 4-dimensional symplectic space, V is a rank 4 quadratic space, and $X = W \otimes V$. The key technical result is the construction of an explicit theta lift from $\mathrm{GO}(X)$ to $\mathrm{GSp}(4)$. In this setting, the local theta correspondence was studied by Roberts in a series of papers, culminating in [Rob01]. An integral map for the local theta lift was studied in [JLR12], and the authors proved that for every Hilbert cusp form of level $\Gamma_0(\mathfrak{N})$ there exists a Siegel paramodular newform with weight, level, Hecke eigenvalues, epsilon factor and L-function determined explicitly by the data of the Hilbert modular form.

We continue this work by calculating explicit input data for the local theta lifts. In particular, we study an integral realization for the local theta lifts and give an explicit recipe for Schwartz functions at all places that are not wildly ramified. This allows for explicit construction of the Siegel paramodular forms appearing in this correspondence as lifts of Hilbert modular forms over any real quadratic field, except $\mathbb{Q}(\sqrt{2})$. These paramodular forms correspond to abelian surfaces which are Weil restrictions of elliptic curves over real quadratic fields as a consequence of the results of [FLHS15]. This provides further evidence for the paramodular conjecture of Brumer and Kramer [BK14]. Our result opens up a new approach to calculate Fourier coefficients of these paramodular lifts.

This paper is organized as follows: in Sect. 2 we set up the language of theta correspondence with similitudes which is required to state our main results. Section 3 contains known results about Weil representations and Whittaker models which are required for the proof. In Sect. 4 we study the intertwining maps which give the local integral map of the theta lift, and finally in Sect. 5 we complete the proof of the Main Theorem by specifying specific Schwartz functions to produce the prescribed paramodular vector at every prime.

2 Statement of the Main Theorem

We define the algebraic group $\mathrm{GSp}(4)$ with respect to the symplectic form $J = \left[\begin{smallmatrix} & I \\ -I & \end{smallmatrix}\right]$ to be the subgroup of $g \in \mathrm{GL}(4)$ such that ${}^t g J g = \lambda(g) J$, for some $\lambda(g) \in \mathrm{GL}(1)$. The Siegel upper half space $\mathcal{H}$ is the space of elements of $\left[\begin{smallmatrix} \tau & z \\ z & \tau' \end{smallmatrix}\right] \in \mathrm{M}(2, \mathbb{C})$ whose imaginary part is positive definite. For natural numbers N and n and a rational prime p, we define global and local paramodular groups of level N and p^n, respectively, to be

$$\mathrm{K}(N) = \begin{bmatrix} \mathbb{Z} & \mathbb{Z} & N^{-1}\mathbb{Z} & \mathbb{Z} \\ N\mathbb{Z} & \mathbb{Z} & \mathbb{Z} & \mathbb{Z} \\ N\mathbb{Z} & N\mathbb{Z} & \mathbb{Z} & N\mathbb{Z} \\ N\mathbb{Z} & \mathbb{Z} & \mathbb{Z} & \mathbb{Z} \end{bmatrix} \cap \mathrm{Sp}(\mathbb{Q})$$

$$\mathrm{K}(p^n) = \begin{bmatrix} \mathbb{Z}_p & \mathbb{Z}_p & p^{-n}\mathbb{Z}_p & \mathbb{Z}_p \\ p^n\mathbb{Z}_p & \mathbb{Z}_p & \mathbb{Z}_p & \mathbb{Z}_p \\ p^n\mathbb{Z}_p & p^n\mathbb{Z}_p & \mathbb{Z}_p & p^n\mathbb{Z}_p \\ p^n\mathbb{Z}_p & \mathbb{Z}_p & \mathbb{Z}_p & \mathbb{Z}_p \end{bmatrix} \cap \mathrm{GSp}(\mathbb{Q}_p).$$

A Siegel paramodular form is a holomorphic function on $\mathcal{H}$ such that $F(\gamma\langle Z\rangle) = j(\gamma, Z)^{-k}F(Z)$ for all $\gamma \in \mathrm{K}(N)$ and $Z \in \mathcal{H}$, where $\gamma\langle Z\rangle$ is the action by fractional linear transformation. By Theorem 1.4.1 of [JLRS23] there is a bijection between cuspidal automorphic representations of $\mathrm{GSp}(4, \mathbb{A})$ of weight k which are paramodular with conductor N and cuspidal Siegel eigenforms with respect to $\mathrm{K}(N)$ that are new of level N.

For any even-dimensional non-degenerate symmetric bilinear space X over a field of characteristic not equal to 2, we let $\mathrm{GO}(X)$ be the group of $h \in \mathrm{GL}(X)$ such that there exists a $\lambda \in \mathrm{GL}(1)$ such that $\langle hx, hy\rangle = \lambda\langle x, y\rangle$ for all $x, y \in X$. The scalar λ is unique, and will be denoted by $\lambda(h)$. We let $\mathrm{O}(X)$ be the subgroup of $h \in \mathrm{GO}(X)$ such that $\lambda(h) = 1$, and we let $\mathrm{SO}(X)$ be the subgroup of $h \in \mathrm{O}(X)$ such that $\det(h) = 1$. We see that for $h \in \mathrm{GO}(X)$ we have $\det(h)^2 = \lambda(h)^{\dim X}$. We set $\mathrm{GSO}(X)$ to be the subgroup of $h \in \mathrm{GO}(X)$ such that $\det(h) = \lambda(h)^{\dim X/2}$ and $\mathrm{SO}(X) = \mathrm{O}(X) \cap \mathrm{GSO}(X)$.

Let L be a number field, let E be a real quadratic extension of L, let $\mathbb{A}_L$ be the adeles of L, and let $\mathbb{A}_E$ be the adeles of E. Let π be a cuspidal automorphic representation of $\mathrm{GL}(2, \mathbb{A}_E)$ that has trivial central character and is not Galois invariant. The Jacquet-Langlands correspondence [JL70a] produces a cuspidal automorphic representation π' of a quaternion algebra $B^\times(\mathbb{A}_E)$, from the data of π. Let $X(\mathbb{A}_L)$ be a certain symmetric bilinear 4-dimensional $\mathbb{A}_L$-subspace of $B(\mathbb{A}_E)$. For $x, y \in X$, the symmetric product on X is given by $\langle x, y\rangle = \frac{1}{2}\mathrm{Tr}(xy^*)$. The following sequence is exact [Knu91]:

$$1 \to \mathbb{A}_E^\times \to \mathbb{A}_L^\times \times B(\mathbb{A}_E)^\times \xrightarrow{\rho} \mathrm{GSO}(X(\mathbb{A}_E)) \to 1. \tag{2.1}$$

So, from the data of π', it is simple to produce a cuspidal automorphic representation of $\mathrm{GSO}(X(\mathbb{A}_L))$, also denoted by π'. Let σ be the associated cuspidal automorphic representation of $\mathrm{GO}(X)$, as in Theorem 7 of [Rob01]. Since $\mathrm{GO}(X)$ and $\mathrm{GSp}(4)$ form a dual reductive pair, we have the existence of the Weil representation ω on the group $R = \{(g, h) \in \mathrm{GSp}(4, \mathbb{A}_L) \times \mathrm{GO}(X(\mathbb{A}_L)) \mid \lambda(g) = \lambda(h)\}$ and hence the existence of the theta lift, which is the focus of this paper.

The space of ω is the space of Schwartz functions on $X(\mathbb{A}_L)^2$, which we denote by $\mathcal{S}(X(\mathbb{A}_L)^2)$. The key to determining an explicit theta lift is choosing a $\varphi \in \mathcal{S}(X(\mathbb{A}_L)^2)$ wisely, but we can define the theta lift for any choice of φ. Let f be a cusp form on $\mathrm{GO}(X, \mathbb{A}_L)$ and $\varphi \in \mathcal{S}(X(\mathbb{A}_L)^2)$. Let $\mathrm{GSp}(4, \mathbb{A}_L)^+$ be the subgroup of $g \in \mathrm{GSp}(4, \mathbb{A}_L)$ such that $\lambda(g) \in \lambda(\mathrm{GO}(X(\mathbb{A}_L)))$. For $g \in \mathrm{GSp}(4, \mathbb{A}_L)^+$ define:

$$\theta(f, \varphi)(g) = \int_{\mathrm{O}(X,\mathbb{Q})\backslash \mathrm{O}(X,\mathbb{A}_L)} \vartheta(g, h_1 h; \varphi) f(h_1 h) dh_1$$

where $h \in \mathrm{GO}(X, \mathbb{A}_L)$ is any element such that $(g, h) \in R(\mathbb{A}_L)$ and ϑ is the global theta kernel given by

$$\vartheta(g, h; \varphi) = \sum_{x \in X(L)^2} (\omega(g, h) \cdot \varphi)(x).$$

Then $\theta(f, \varphi)$ can be extended uniquely to all of $\mathrm{GSp}(4, \mathbb{A}_L)$ which is left invariant under $\mathrm{GSp}(4, L)$ and is, in fact, an automorphic form of $\mathrm{GSp}(4, \mathbb{A}_L)$.

In order to make a good choice for $\varphi \in \mathcal{S}(X(\mathbb{A}_L)^2)$ we need to examine the problem locally. Now let L_v be a local field and let E_w be a quadratic extension of L_v. Using the path outlined in the global case one may determine a $\mathrm{GO}(X, L_v)$ representation $(\sigma_v, \mathcal{W})$ from the data of an infinite-dimensional irreducible admissible representation π_v of $\mathrm{GL}(2, E_w)$. Let $R_v = \{(g, h) \in \mathrm{GSp}(4, L_v) \times \mathrm{GO}(X(L_v)) \mid \lambda(g) = \lambda(h)\}$ where the similitude factor of each coordinate matches and let $\mathcal{S}(X(L_v)^2)$ be the Schwartz functions of $X(L_v)^2$. When σ_v is irreducible as a $\mathrm{GO}(X)$-representation, it is the direct sum of two isomorphic $\mathrm{GSO}(X)$-representations. Throughout this work, we choose a distinguished $\mathrm{GSO}(X)$-representation, which we denote with a $+$. We provide data so that the local theta correspondence can be explicitly realized. Define the intertwining map

$$B(g, \varphi, W, s) = \int_{H\backslash \mathrm{SO}(X)} \big(\omega(g, hh')\varphi\big)(x_1, x_2) Z(s, \pi(hh')W)\, dh \tag{2.2}$$

where $g \in \mathrm{GSp}(4, L_v)$, $\varphi \in \mathcal{S}(X(L_v)^2)$, $W \in \mathcal{W}$, $\Re(s) \gg 0$, $Z(s, W)$ is the zeta integral as defined in (3.3) or (3.6), and $x_1, x_2 \in X$ are certain specified elements and H is their stabilizer in $\mathrm{SO}(X)$. Indeed, the space of functions $\{B(\cdot, \varphi, W, s) \mid \varphi \in \mathcal{S}(X^2), W \in V\}$ is an irreducible admissible representation $\Theta(\pi_v^+) = \Pi_v$ of $\mathrm{GSp}(4, L_v)$. The representation Π_v has a canonical paramodular level $\mathfrak{p}_L^N$. Furthermore, by [Rob01] the intertwining map is unique. The main result of this work is summarized below.

Local Main Theorem *Let L be a non-Archimedean local field of characteristic* 0 *and let E be either a real quadratic field extension of L or let $E = L \times L$. If E is a field let $\mathfrak{P}$ be the unique maximal ideal of the ring of integers of E, $\mathfrak{o}_E$, and let ϖ_E be the uniformizer of $\mathfrak{P}$. Assume that if the residual characteristic of L is even then E/L is unramified.*

If E is a field let τ_0 be an infinite-dimensional, irreducible, admissible representation of $\mathrm{GL}(2, E)$ *with trivial central character. If $E = L \times L$ let τ_1 and τ_2 be infinite-dimensional, irreducible, admissible representations of* $\mathrm{GL}(2, L)$ *with trivial* *central character. For* $i \in \{0, 1, 2\}$, *we assume that the space of* τ_i *is its unique Whittaker model* $\mathcal{W}(\tau_i)$. *If* $E = L \times L$ *let* $(\pi(\tau_1, \tau_2), V)$, *where* $V = \mathcal{W}(\tau_1) \times \mathcal{W}(\tau_2)$ *is the associated* $+$*-representation of* $\mathrm{GSO}(X)$.

If E *is a field let* $(\pi(1, \tau_0), V)$, *where* $V = \mathcal{W}(\tau_0)$ *is the associated* $+$*-representation of* $\mathrm{GSO}(X)$, *and let* $W \in V$ *be a local* $\mathrm{GL}(2, E)$*-newform with* $\Gamma_0(\mathfrak{p}^n)$*-invariance. If* $E = L \times L$ *then let* $W_i \in V$ *be local* $\mathrm{GL}(2, L)$*-newforms with* $\Gamma_0(\mathfrak{p}^{n_i})$*-invariance, for* $i \in \{1, 2\}$ *and set* $W = (W_1, W_2)$.

If $E = L \times L$ *then set* $N = n_1 + n_2$ *and let* $\varphi \in \mathcal{S}(X^2)$ *be given by as in Theorem 5.4, if* E/L *is inert then set* $N = 2n$ *and let* $\varphi \in \mathcal{S}(X^2)$ *be given as in Theorem 5.8, and if* E/L *is tamely ramified then set* $N = n + 2$, *let* χ *be the non-trivial quadratic character of* E/L, *and let* $\varphi \in \mathcal{S}(X^2)$ *be given as in Theorem 5.17. Finally, let* $s \in \mathbb{C}$ *be such that* $\Re(s) \gg 0$, *and let* $B(\cdot, \varphi, W, s) : \mathrm{GSp}(4, L) \to \mathbb{C}$ *be the local intertwining map as in* (2.2).

Then $B(\cdot, \varphi, W, s) : \mathrm{GSp}(4, L) \to \mathbb{C}$ *is non-zero and is invariant under right translation by elements of* $\mathrm{K}(\mathfrak{p}^N)$.

The compatibility of this local map with the global theta lift follows from Corollary 4.16 and [Rob01]. The remainder of the paper is devoted to the proof of these results.

3 Preliminaries

In this section, we collect some results on local representations that will be needed in the sequel.

3.1 Whittaker Models

Let ψ be a non-trivial additive character of L. If L is non-Archimedean we let $\mathcal{W}(\mathrm{GL}(2, L), \psi)$ be the $\mathbb{C}$-vector space of functions $W : \mathrm{GL}(2, L) \to \mathbb{C}$ such that

$$W([\begin{smallmatrix} 1 & x \\ & 1 \end{smallmatrix}]g) = \psi(x)W(g) \tag{3.1}$$

for $x \in L$ and $g \in \mathrm{GL}(2, L)$, and there exists a compact, open subgroup K of $\mathrm{GL}(2, L)$ such that $W(gk) = W(g)$ for $g \in \mathrm{GL}(2, L)$ and $k \in K$. If $L = \mathbb{R}$ we let $\mathcal{W}(\mathrm{GL}(2, L), \psi)$ be the space of smooth functions with rapid decay away from 0 that satisfy (3.1). Evidently, $\mathcal{W}(\mathrm{GL}(2, L), \psi)$ is a smooth $\mathrm{GL}(2, L)$-representation under right translation. We say that a representation (τ, V) of $\mathrm{GL}(2, L)$ admits a *Whittaker model*, denoted by $\mathcal{W}(\tau, \psi)$, if it is isomorphic to a subrepresentation of $\mathcal{W}(\mathrm{GL}(2, L), \psi)$.

Theorem 3.1 *Let L be a non-Archimedean local field and let ψ be a non-trivial additive character of L. Let (τ, V) be an irreducible admissible representation of* $\mathrm{GL}(2, L)$. *Then τ is infinite-dimensional if and only if τ admits a unique Whittaker model.*

Proof See Theorem 3.5.3 of [Bum97], for example. □

We note that if (τ, V) is a finite-dimensional irreducible admissible representation of $\mathrm{GL}(2, L)$, then $\tau \cong \beta \circ \det$ for some character $\beta : L^\times \to \mathbb{C}^\times$. In the case that $L = \mathbb{R}$ we have a similar result.

Theorem 3.2 *Let (π, V) be an irreducible admissible $(\mathfrak{g}, K)$-module for* $\mathrm{GL}(2, \mathbb{R})$. *Then there exists at most one space $\mathcal{W}(\pi, \psi) \subset \mathcal{W}(\mathrm{GL}(2, \mathbb{R}), \psi)$ that is invariant under the actions of $U(\mathfrak{g})$ and K on $C^\infty(\mathrm{GL}(2, \mathbb{R}))$ such that $\mathcal{W}(\pi, \psi)$ is isomorphic to (π, V) as a $(\mathfrak{g}, K)$-module.*

Proof See Theorem 2.8.1 of [Bum97], for example. □

3.2 The Split Case

Let L be a local field. Define the following symmetric bilinear space of L:

$$X_M = \mathrm{M}(2, L).$$

Define an action of $\mathrm{GL}(2, L) \times \mathrm{GL}(2, L)$ on X_M by $\rho(d_1, d_2)x = d_1 x d_2^*$. Then $\rho(d_1, d_2) \in \mathrm{GSO}(X_M)$ for $d_1, d_2 \in \mathrm{GL}(2, L)$, and the sequence

$$1 \to L^\times \to \mathrm{GL}(2, L) \times \mathrm{GL}(2, L) \xrightarrow{\rho} \mathrm{GSO}(X_M) \to 1 \tag{3.2}$$

is exact. Let $(\tau_i, \mathcal{W}(\tau_i, \psi))$ be the Whittaker model for an infinite-dimensional, irreducible, admissible representation of $\mathrm{GL}(2, L)$ for $i \in \{1, 2\}$. Assume that τ_1 and τ_2 have equal central character. Then $\pi = \rho(\tau_1, \tau_2)$ is an infinite-dimensional, irreducible, admissible representation of $\mathrm{GSO}(X_M)$, let σ be the induced representation of $\mathrm{GO}(X_M)$, and let π^+ be a particular irreducible subprepresentation of σ. Let $W_i \in \mathcal{W}(\tau_i, \psi)$ for $i \in \{1, 2\}$. Let $s \in \mathbb{C}$ and set

$$Z(s, W_i) = \int\limits_{L^\times} W_i([\begin{smallmatrix} a & \\ & 1 \end{smallmatrix}])|a|^{s-\frac{1}{2}}\, d^\times a.$$

We define $Z(s, W)$ by setting

$$Z(s, W) = Z(s, W_1) \cdot Z(s, W_2) \tag{3.3}$$

for $W = W_1 \otimes W_2 \in V$.

Lemma 3.3 *There exists a positive real number M such that $Z(s, W_i)$ converges absolutely, for all $W_i \in W(\tau_i, \psi)$, for $\Re(s) > M$. If L is non-Archimedean then $Z(s, W_i)$ converges to a rational function in q^{-s}.*

Proof See Theorem 6.12 and Remark 6.13 of [Gel75], for example. □

3.3 The Non-split Case

Let $\delta \in L$ be square free and set $E = L(\sqrt{\delta})$. Define the symmetric bilinear space of L:

$$X_{ns} = \left\{ \left[\begin{smallmatrix} a & b\sqrt{\delta} \\ c\sqrt{\delta} & \alpha(a) \end{smallmatrix} \right] \mid a \in E,\ (b, c) \in L \times L \right\}. \tag{3.4}$$

We define an action ρ of $L^\times \times \mathrm{GL}(2, E)$ on X by $\rho(t, b) \cdot x = t^{-1} b x \alpha(b)^*$ for $t \in L^\times, b \in B^\times$ and $x \in X$. The sequence

$$1 \to E^\times \to L^\times \times \mathrm{GL}(2, E) \xrightarrow{\rho} \mathrm{GSO}(X_{ns}) \to 1 \tag{3.5}$$

is exact. Let $(\tau, \mathcal{W}(\tau, \psi_E))$ be the Whittaker model for an infinite-dimensional, irreducible, admissible representation of $\mathrm{GL}(2, E)$ with trivial central character. Then $\pi = \rho(1, \tau)$ is an infinite-dimensional, irreducible, admissible representation of $\mathrm{GSO}(X_{ns})$ and let σ be the induced representation of $\mathrm{GO}(X_{ns})$, or a particular irreducible subprepresentation. We let

$$Z(s, W) = \int\limits_{E^\times} W(\left[\begin{smallmatrix} a & \\ & 1 \end{smallmatrix} \right]) |a|_E^{s-\frac{1}{2}} \, d^\times a \tag{3.6}$$

for $W \in \mathcal{W}(\tau, \psi_E)$ and $s \in \mathbb{C}$.

Lemma 3.4 *There exists a positive real number M such that $Z(s, W)$ converges absolutely, for all $W \in W(\tau_0, \psi_E)$, for $\Re(s) > M$ to a rational function in q_E^{-s}.*

Proof See Theorem 6.12 and Remark 6.13 of [Gel75], for example. □

We specify the following elements of X

$$x_1 = \begin{cases} \left[\begin{smallmatrix} 0 & 1 \\ 0 & 0 \end{smallmatrix} \right] & X \text{ split} \\ \left[\begin{smallmatrix} 0 & \sqrt{\delta} \\ 0 & 0 \end{smallmatrix} \right] & X \text{ non-split} \end{cases} \quad \text{and,} \quad x_2 = \begin{cases} \left[\begin{smallmatrix} 0 & 0 \\ -2 & 0 \end{smallmatrix} \right] & X \text{ split} \\ \left[\begin{smallmatrix} 0 & 0 \\ \frac{-2\sqrt{\delta}}{\delta} & 0 \end{smallmatrix} \right] & X \text{ non-split.} \end{cases} \tag{3.7}$$

Let H be the stabilizer in $\mathrm{SO}(X)$ of $(x_1, x_2) \in X^2$. We make the following observation.

Lemma 3.5 *Let H be as above. In the non-split case assume that L is non-archimedean. Then, $H = \{\rho(1, \left[\begin{smallmatrix} 1 & \\ & u \end{smallmatrix} \right]) \mid u \in E^1\}$ and in the split case $H = \{\rho(\left[\begin{smallmatrix} a & \\ & 1 \end{smallmatrix} \right], \left[\begin{smallmatrix} a^{-1} & \\ & 1 \end{smallmatrix} \right]) \mid a \in L^\times\}$.*

Lemma 3.6 *Let H be as in Sect. 3, let $X = X_M$ or $X = X_{ns}$, and let τ_1, τ_2, and τ_0 be as above.*

(a) Let $s \in \mathbb{C}$ with $\Re(s) \gg M$. Assume that $X = X_M$. If $W \in V$ and $t_1, t_2 \in L^\times$, then $Z(s, \pi(\rho([{}^{t_1}{}_{t_2}], 1))W) = |t_2/t_1|^{s-\frac{1}{2}} Z(s, W)$.
(b) Let $s \in \mathbb{C}$ with $\Re(s) \gg M$. Assume that $X = X_{ns}$. If $W \in V$ and $t_1, t_2 \in E^\times$, then $Z(s, \pi(\rho(1, [{}^{t_1}{}_{t_2}]))W) = |t_2/t_1|_E^{s-\frac{1}{2}} Z(s, W)$.
(c) Let $s \in \mathbb{C}$ with $\Re(s) \gg M$, $W \in V$, and $h \in H$. Then,

$$Z(s, \pi(h)W) = Z(s, W). \tag{3.8}$$

Proof (a) Let $t_1, t_2 \in L^\times$, let $\tau = \tau_i$ for $i \in \{1, 2\}$ and let $W = W_1 \otimes W_2 \in V$. Then,

$$\begin{aligned} Z(s, \tau_2([{}^{t_1}{}_{t_2}])W_2) &= \int_{L^\times} W_2([{}^{a}{}_{1}][{}^{t_1}{}_{t_2}])|a|^{s-\frac{1}{2}}\, d^\times a \\ &= \int_{L^\times} W_2([{}^{a}{}_{t_2}])|a/t_1|^{s-\frac{1}{2}}\, d^\times a \\ &= \int_{L^\times} W_2(t_2[{}^{a}{}_{1}])|a|^{s-\frac{1}{2}}|t_2/t_1|^{s-\frac{1}{2}}\, d^\times a \\ &= |t_2/t_1|^{s-\frac{1}{2}} Z(s, W_2). \end{aligned}$$

Therefore,
$Z(s, \pi(\rho([{}^{t_1}{}_{t_2}], 1))W) = Z(s, W_1)Z(s, \tau_2([{}^{t_1}{}_{t_2}])W_2) = |t_2/t_1|^{s-\frac{1}{2}} Z(s, W)$.
The result follows for a general $W \in V$.

(b) The calculation is the same as the calculation in part (a), except the factor is $|t_2/t_1|_E^{s-\frac{1}{2}}$.
(c) The proof of (c) follows simply from part (a) and (b) and Lemma 3.5. □

Corollary 3.7 *Assume that L is non-Archimedean and fix a $W \in V$. The function $H\backslash \mathrm{SO}(X) \to \mathbb{C}$ defined by $Hh \mapsto Z(s, \pi(h)W)$ is well defined and locally constant.*

Proof Part (b) of Lemma 3.6 proves the map is well defined. Evidently, the smoothness of π guarantees that the map is locally constant. □

3.4 Additional Results

Lemma 3.8 *Let ψ be a character of $L^\times$ with conductor $\mathfrak{o}_L$. Let dx be the Haar measure on L so that the volume of $\mathfrak{o}_L$ is 1. Let $c \in L^\times$. Then*

$$\int\limits_{\varpi^n \mathfrak{o}_L} \psi(c\alpha)\, d\alpha = \begin{cases} q^{-n} & \nu(c) \geq -n \\ 0 & \nu(c) < -n, \end{cases}$$

and

$$\int\limits_{\varpi^n \mathfrak{o}_L^\times} \psi(c\alpha)\, d\alpha = \begin{cases} q^{-n} - q^{-n-1} & \nu(c) > -n-1 \\ -q^{-n-1} & \nu(c) = -n-1 \\ 0 & \nu(c) < -n-1. \end{cases}$$

Lemma 3.9 *Let N be a non-negative integer. Then $\mathrm{K}(\mathfrak{p}^N)$ is generated by the following:*

(a) $\left[\begin{smallmatrix} A & 0 \\ 0 & \lambda\,{}^t\!A^{-1} \end{smallmatrix}\right]$ *for all $A \in \Gamma_0(\mathfrak{p}^N)$ and $\lambda \in \mathfrak{o}_L^\times$,*

(b) $\begin{bmatrix} 1 & & \varpi^{-N}b_1 & b_2 \\ & 1 & b_2 & b_3 \\ & & 1 & \\ & & & 1 \end{bmatrix}$ *for $b_1, b_2, b_3 \in \mathfrak{o}_L$,*

(c) $s_2 = \begin{bmatrix} 1 & & & \\ & & & 1 \\ & & 1 & \\ & -1 & & \end{bmatrix}$, *an element of the Weyl group, and*

(d) $t_N = \begin{bmatrix} & & \varpi^{-N} & \\ & 1 & & \\ -\varpi^N & & & \\ & & & 1 \end{bmatrix}.$

Proof This follows from the Iwahori factorization and the decomposition of $\mathrm{K}(\mathfrak{p}^N)/\mathrm{Kl}(\mathfrak{p})$. These results can be found in [RS07] in (2.7) and Lemma 3.3.1, respectively. We reproduce the later in this text for easy reference.

$$\mathrm{K}(\mathfrak{p}^N) = \bigsqcup_{u \in \mathfrak{o}_L/\mathfrak{p}^N} \begin{bmatrix} 1 & & u\varpi^{-N} & \\ & 1 & & \\ & & 1 & \\ & & & 1 \end{bmatrix} \mathrm{Kl}(\mathfrak{p}^N) \sqcup \bigsqcup_{o \in \mathfrak{o}_L/\mathfrak{p}^{N-1}} t_N \begin{bmatrix} 1 & & u\varpi^{-N+1} & \\ & 1 & & \\ & & 1 & \\ & & & 1 \end{bmatrix} \mathrm{Kl}(\mathfrak{p}^N) \quad (3.9)$$

for N a positive integer. Notice that if $N = 0$ that both sides are equal to $\mathrm{GSp}(4, \mathfrak{o})$. □

We now demonstrate a more user friendly, but equivalent, Haar measure in the split and non-split cases. In this we will need the following simple observations

$$2\langle \left[\begin{smallmatrix} x_1 & x_2 \\ x_3 & x_4 \end{smallmatrix}\right], \left[\begin{smallmatrix} y_1 & y_2 \\ y_3 & y_4 \end{smallmatrix}\right] \rangle = y_1x_4 - y_2x_3 - y_3x_2 + y_4x_1. \quad (3.10)$$

$$2\langle \left[\begin{smallmatrix} x_1+x_2\sqrt{\delta} & x_3 \\ x_4 & x_1-x_2\sqrt{\delta} \end{smallmatrix}\right], \left[\begin{smallmatrix} y_1+y_2\sqrt{\delta} & y_3 \\ y_4 & y_1-y_2\sqrt{\delta} \end{smallmatrix}\right] \rangle = 2x_1y_1 - 2x_2y_2\delta - x_3y_4 - x_4y_3. \quad (3.11)$$

Lemma 3.10 *If* $y = \begin{bmatrix} y_1 & y_2 \\ y_3 & y_4 \end{bmatrix} \in X_M = \mathrm{M}(2, L)$ *then* $dy_1\, dy_2\, dy_3\, dy_4$ *is a Haar measure on* X_M*. Since* X_M *is locally compact there must exist a non-zero constant* c *such that* $dy = c \cdot dy_1\, dy_2\, dy_3\, dy_4$*. The claim is that* $c = 1$*.*

Assume that $E = L(\sqrt{\delta})$ *where* $\delta \in L^\times$ *is square-free and let* $X = X_{ns}$*. We denote the maximal ideal of* $\mathfrak{o}_E$ *by* $\mathfrak{P}$*. If* $y = \begin{bmatrix} y_1+y_2\sqrt{\delta} & y_3 \\ y_4 & y_1-y_2\sqrt{\delta} \end{bmatrix}$ *then* $dy_1\, dy_2\, dy_3\, dy_4$ *is a Haar measure on* X_{ns}*. Since* X_{ns} *is locally compact there must exist a positive constant* c *such that* $dy = c \cdot dy_1\, dy_2\, dy_3\, dy_4$*. The claim is that* $c = |\delta|_{\mathfrak{p}} = q_E^{-\nu_L(\delta)}$*.*

Proof Assume that the residual characteristic of L is odd.

To align with some future notation chose $\varphi_2 \in \mathcal{S}(X)$ to be

$$\varphi_2 = \begin{cases} f_{M(2,\mathfrak{o}_L)}, & E/L \text{ is split} \\ f_{\begin{bmatrix} \mathfrak{o}_L & \mathfrak{P}^{-r} \\ \mathfrak{P}^r & \mathfrak{o}_L \end{bmatrix} \cap X}, & E/L \text{ is non-split, and } \nu_L(\delta) = \nu_E(\sqrt{\delta}) = r. \end{cases}$$

We choose dy so that $\mathcal{F}_1(\mathcal{F}_1(\varphi_2))(x) = \varphi_2(-x) = \varphi_2(x)$. On the other hand, we can compute this directly using Lemma 3.8 and (3.10) or (3.11):

$$\begin{aligned} \mathcal{F}_1(\varphi_2) &= \int_X \varphi_2(y)\psi(2\langle x, y\rangle)\, dy = \int_{\begin{bmatrix} \mathfrak{o}_L & \mathfrak{o}_L \\ \mathfrak{o}_L & \mathfrak{o}_L \end{bmatrix}} \psi(2\langle x, y\rangle)\, dy \\ &= c \int_{y_1 \in \mathfrak{o}_L} \psi(y_1x_4)\, dy_1 \int_{y_2 \in \mathfrak{o}_L} \psi(y_2x_3)\, dy_2 \int_{y_3 \in \mathfrak{o}_L} \psi(y_3x_2)\, dy_3 \int_{y_4 \in \mathfrak{o}_L} \psi(y_4x_1)\, dy_4 \\ &= c\varphi_2(x), \end{aligned}$$

$$\begin{aligned} \mathcal{F}_1(\varphi_2)(x) &= \int_X \varphi_2(y)\psi(2\langle x, y\rangle)\, dy = \int_{\begin{bmatrix} \mathfrak{o}_E & \mathfrak{P}^{-r} \\ \mathfrak{P}^r & \mathfrak{o}_E \end{bmatrix} \cap X} \psi(2\langle x, y\rangle)\, dy \\ &= c \int_{\mathfrak{o}_E} \psi(2x_1y_1)\, dy_1 \int_{\mathfrak{P}^{-r}} \psi(-2x_2y_2)\delta\, dy_2 \int_{\mathfrak{P}^{-r}} \psi(-x_4y_3)\, dy_3 \int_{\mathfrak{P}^r} \psi(-x_3y_4)\, dy_4 \\ &= cq_E^r \cdot \varphi_2(x). \end{aligned}$$

Therefore $\varphi_2(x) = \mathcal{F}_1(\mathcal{F}_1\varphi_2)(x) = c^2 q_E^{2r} \varphi_2(x)$ and so $c = q_E^{-r}$, in particular c is equal to 1 when E/L is unramified. □

3.5 Weil Representations

Let $\psi : L \to \mathbb{C}^\times$ be a non-trivial continuous unitary character. For $c \in L$ set the notation $\psi^c(x) = \psi(cx)$ for $x \in L$. Assume L is a local field. If L is non-Archimedean we define $\mathcal{S}(X)$ the space of Schwartz functions on X to be locally constant and compactly supported. If L is Archimedean then $\mathcal{S}(X)$ will be functions that decay rapidly away from 0. The following formulas determine a unique Weil representation $(\omega_1, \mathcal{S}(X))$ of $\mathrm{GL}(2, L) \times \mathrm{O}(X)$ with respect to ψ:

$$\begin{aligned}
\big(\omega_1(1,h)\varphi\big)(x) &= \varphi(h^{-1}x),\\
\big(\omega_1([\begin{smallmatrix} a & \\ & a^{-1}\end{smallmatrix}],1)\varphi\big)(x) &= \chi_{E/L}(a)|a|^{\dim X/2}\varphi(ax),\\
\big(\omega_1([\begin{smallmatrix} 1 & b\\ & 1\end{smallmatrix}],1)\varphi\big)(x) &= \psi(b\langle x,x\rangle)\varphi(x),\\
\big(\omega_1([\begin{smallmatrix} & 1\\ -1 & \end{smallmatrix}],1)\varphi\big)(x_1,x_2) &= \gamma_1(X)\big(\mathcal{F}_1(\varphi)\big)(x).
\end{aligned}$$

Here $\gamma_1(X)$ is the fourth root of unity discussed in [JL70b] and $\mathcal{F}_1(\varphi)$ is the Fourier transform of φ given by

$$\mathcal{F}_1(\varphi)(x) = \int_X \varphi(y)\psi(2\langle x,y\rangle)\,dy, \tag{3.12}$$

where the Haar measure on X is the unique Haar measure such that $\mathcal{F}_1(\mathcal{F}_1(\varphi))(x) = \varphi(-x)$ for $x \in X$ and all $\varphi \in \mathcal{S}(X)$. We can extend ω_1 to all of $R = \{(g,h) \in \mathrm{GL}(2,L) \otimes \mathrm{GO}(X) \mid \det(g) = \lambda(h)\}$ via the formula

$$\omega_1(g,h)\varphi = |\lambda(h)|^{-1}\omega_1\big(g[\begin{smallmatrix} 1 & \\ & \det(g)^{-1}\end{smallmatrix}],1\big)(\varphi\circ h^{-1}). \tag{3.13}$$

The following formulas determine a unique *Weil representation* ω of $\mathrm{Sp}(4,L) \times \mathrm{O}(X)$ on $\mathcal{S}^2(X^2) = \mathcal{S}^2(X\times X)$ with respect to ψ:

$$\big(\omega(1,h)\varphi\big)(x_1,x_2) = \varphi(h^{-1}x_1, h^{-1}x_2), \tag{3.14}$$

$$\Big(\omega\Big(\begin{bmatrix} [\begin{smallmatrix} a_1 & a_2\\ a_3 & a_4\end{smallmatrix}] & \\ & [\begin{smallmatrix} a_1 & a_2\\ a_3 & a_4\end{smallmatrix}]^{t-1}\end{bmatrix},1\Big)\varphi\Big)(x_1,x_2) \tag{3.15}$$

$$= \chi_{E/L}(a_1a_4 - a_2a_3)|a_1a_4 - a_2a_3|^{\dim X/2}\varphi(a_1x_1 + a_3x_2, a_2x_1 + a_4x_2),$$

$$\Big(\omega\Big(\begin{bmatrix} 1 & & b_1 & b_2\\ & 1 & b_2 & b_3\\ & & 1 & \\ & & & 1\end{bmatrix},1\Big)\varphi\Big)(x_1,x_2) \tag{3.16}$$

$$= \psi(b_1\langle x_1,x_1\rangle + 2b_2\langle x_1,x_2\rangle + b_3\langle x_2,x_2\rangle)\varphi(x_1,x_2),$$

$$\Big(\omega\Big(\begin{bmatrix} & & 1 & \\ & & & 1\\ -1 & & & \\ & -1 & & \end{bmatrix},1\Big)\varphi\Big)(x_1,x_2) = \gamma(X)\mathcal{F}(\varphi)(x_1,x_2) \tag{3.17}$$

where $g \in \mathrm{GSp}(4)$, $h \in \mathrm{SO}(X)$, $x_1, x_2 \in X$, and $\varphi \in \mathcal{L}^2(X^2)$. Here $\chi_{E/L} : L^\times \to \mathbb{C}^\times$ is the quadratic character associated to the discriminant of X. That is, $\chi_{E/L}$ is the unique quadratic character on L that is trivial on $\mathrm{N}_L^E(E^\times)$. In particular if $\det(X) = 1 \in L^\times/L^{\times 2}$ then $\chi_{E/L}$ is trivial. Also, $\gamma(X)$ is a particular fourth root of unity as in remark 1 in [Yos79]. In (3.17) the Fourier transform $\mathcal{F}(\varphi)$ is defined by $\mathcal{F}(\varphi)(x_1,x_2) = \int_{X^2}\varphi(y_1,y_2)\psi(2\langle x_1,y_1\rangle + 2\langle x_2,y_2\rangle)\,dy_1\,dy_2$, where $dy_1\,dy_2$ is the unique Haar measure so that $\mathcal{F}(\mathcal{F}(\varphi))(x) = \varphi(-x)$. Let $R(L)$ be the subgroup of $\mathrm{GSp}(4, L) \times \mathrm{GO}(X)$ given by

$$R(L) = \{(g, h) \in \mathrm{GSp}(4, L) \times \mathrm{GO}(X) : \lambda(g) = \lambda(h)\}. \tag{3.18}$$

Then ω extends to $R(L)$ via the formula

$$\omega(g, h)\varphi = |\lambda(h)|^{-\frac{\dim X}{2}} \omega\big(g \begin{bmatrix} 1 & & & \\ & 1 & & \\ & & \lambda(g)^{-1} & \\ & & & \lambda(g)^{-1} \end{bmatrix}, 1\big)(\varphi \circ h^{-1}) \tag{3.19}$$

for $\varphi \in \mathcal{S}^2(X^2)$ and $(g, h) \in R$.

Lemma 3.11 *The map*

$$T : \mathcal{S}(X) \otimes \mathcal{S}(X) \to \mathcal{S}(X^2),$$

determined by the formula

$$T(\varphi_1 \otimes \varphi_2)(x_1, x_2) = \varphi_1(x_1)\varphi_2(x_2)$$

for φ_1 and φ_2 in $\mathcal{S}(X)$ and x_1 and x_2 in X, is a well defined complex linear isomorphism such that

$$T \circ (\omega_1\big([\begin{smallmatrix} a_1 & b_1 \\ c_1 & d_1 \end{smallmatrix}], h) \otimes \omega_1([\begin{smallmatrix} a_2 & b_2 \\ c_2 & d_2 \end{smallmatrix}], h)\big) = \omega(\begin{bmatrix} a_1 & & b_1 & \\ & a_2 & & b_2 \\ c_1 & & d_1 & \\ & c_2 & & d_2 \end{bmatrix}, h) \circ T \tag{3.20}$$

for $g_1 = [\begin{smallmatrix} a_1 & b_1 \\ c_1 & d_1 \end{smallmatrix}]$ and $g_2 = [\begin{smallmatrix} a_2 & b_2 \\ c_2 & d_2 \end{smallmatrix}]$ in $\mathrm{GL}(2, L)$ and $h \in \mathrm{GO}(X)$ such that

$$\det(g_1) = \det(g_2) = \lambda(h).$$

Proof It is not hard to demonstrate that T is an isomorphism. It suffices to prove (3.20) holds for $(g_1, g_2) = (g, g) \in \mathrm{SL}(2, L) \times \mathrm{SL}(2, L)$, $(g_1, g_2) = (1, g) \in \mathrm{SL}(2, L) \times \mathrm{SL}(2, L)$ for g a generator for $\mathrm{SL}(2, L)$, and $(g_1, g_2) = ([\begin{smallmatrix} 1 & \\ & \lambda \end{smallmatrix}], [\begin{smallmatrix} 1 & \\ & \lambda \end{smallmatrix}])$ and $h \in \mathrm{GO}(X)$ with $\lambda(h) = \lambda$. □

4 Intertwining Maps

In this section, we study the intertwining maps which define an explicit integral model of the local theta lift.

4.1 The Non-archimedean Case

Let the notation be as in Sect. 3. For this section, make the further assumption that L is non-archimedean, though we will occasionally note when results and their proofs are identical in the case that $L = \mathbb{R}$. Let R be as in Sect. 2. Let $(\omega, \mathcal{S}(X^2))$ be the Weil representation of R associated to X and ψ as in Sect. 3.5 and let V be the space of the Whittaker models, as in Sect. 3.1.

The space $\mathcal{S}(X^2) \otimes V$ is an $R' = R \cap \big(\mathrm{GSp}(4, L) \times \mathrm{GSO}(X)\big)$-space with the action being determined by $(g, h) \cdot (\varphi \otimes v) = (\omega(g, h) \cdot \varphi) \otimes (\pi(h) \cdot v)$ for $(g, h) \in R'$, $\varphi \in \mathcal{S}(X^2)$ and $v \in V$. Similarly we find that $\mathcal{S}(X^2) \otimes (V \times V)$ is also an R-space with the actions determined by $(g, h) \cdot (\varphi \otimes (v_1, v_2)) = (\omega(g, h)\varphi) \otimes (\sigma(h) \cdot (v_1, v_2))$ with $(g, h) \in R$, $\varphi \in \mathcal{S}(X^2)$ and $v_1, v_2 \in V$.

Let H, x_1, x_2 be as in Sect. 3. Let $s \in \mathbb{C}$ be such that $\Re(s) > M$. Assume that X is split. Let $g \in \mathrm{GSp}(4, L)^+ = \mathrm{GSp}(4, L)$ and $\varphi \in \mathcal{S}(X^2)$. Let $W \in V$ and let $Z(s, W)$ be as in (3.3). We define

$$B(g, \varphi, W, s) = \int\limits_{H\backslash \mathrm{SO}(X)} \big(\omega(g, hh')\varphi\big)(x_1, x_2) Z(s, \pi(hh')W)\, dh \tag{4.1}$$

where $h' \in \mathrm{GSO}(X)$ is chosen so that the similitude factor $\lambda(h') = \lambda(g)$. We use the Haar measures on $\mathrm{SO}(X)$ and H for which $\mathrm{SO}(X) \cap \rho(\mathrm{GL}(2, \mathfrak{o}_L) \times \mathrm{GL}(2, \mathfrak{o}_L))$ and $H \cap \rho(\mathrm{GL}(2, \mathfrak{o}_L) \times \mathrm{GL}(2, \mathfrak{o}_L))$ both have measure 1, and the integral in (4.1) uses their quotient measure.

Assume that X is non-split. Let $W \in V$ and let $Z(s, W)$ be as in (3.6). For $g \in \mathrm{GSp}(4, L)^+$ and $\varphi \in \mathcal{S}(X^2)$ we define

$$B(g, \varphi, W, s) = \int\limits_{H\backslash \mathrm{SO}(X)} \big(\omega(g, hh')\varphi\big)(x_1, x_2) Z(s, \pi(hh')W)\, dh \tag{4.2}$$

where $h = \rho(t, h_0)$, $h' \in \mathrm{GSO}(X)$ is chosen so that similitude factor $\lambda(h') = \lambda(g)$. Furthermore we use the Haar measures on $\mathrm{SO}(X)$ and H for which $\mathrm{SO}(X) \cap \rho(\mathfrak{o}_E^\times \times \mathrm{GL}(2, \mathfrak{o}_E))$ and $H \cap \rho(\mathfrak{o}_E^\times \times \mathrm{GL}(2, \mathfrak{o}_E))$ both have measure 1 and the integral in (4.2) uses their quotient measure. We will occasionally refer to $B(g, \varphi, W, s)$ as the *Bessel integral.*

Lemma 4.1 *Let $s \in \mathbb{C}$ be such that $\Re(s) > M$. For $g \in \mathrm{GSp}(4, L)^+$, $\varphi \in \mathcal{S}(X^2)$, and $W \in V$ the integral defining $B(g, \varphi, W, s)$ is well defined and converges absolutely.*

Proof Fix a $(g, h') \in R$, $\varphi \in \mathcal{S}(X^2)$, and $s \in \mathbb{C}$ so that $\Re(s) > M$. We consider the function $f : H\backslash \mathrm{SO}(X) \to \mathbb{C}$ defined by $f(Hh) = (\omega(g, hh')\varphi)(x_1, x_2)$ for $h \in \mathrm{SO}(X)$. We claim that f is compactly supported and locally constant. To see this, we note that $f = f_1 \circ f_2$, where $f_2 : H\backslash \mathrm{SO}(X) \to \mathrm{SO}(X)(x_1, x_2)$ is defined by $f_2(Hh) = (h^{-1}x_1, h^{-1}x_2)$ for $h \in \mathrm{SO}(X)$, and $f_1 : \mathrm{SO}(X)(x_1, x_2) \to \mathbb{C}$ is defined by $f_1(x) = \omega(g, h')\varphi(x)$ for $x \in \mathrm{SO}(X)(x_1, x_2) \subset X^2$. The map f_2 is a homeomorphism by 5.14 of [BZ76]. The map f_1 is compactly supported and locally constant because the map $X^2 \to \mathbb{C}$ defined by $x \mapsto (\omega(g, h')\varphi)(x)$ is compactly supported and $\mathrm{SO}(X)(x_1, x_2)$ is a closed subset of X^2. Indeed, $\mathrm{SO}(X)(x_1, x_2)$ is the inverse image of $[\begin{smallmatrix} 0 & 1/2 \\ 1/2 & 0 \end{smallmatrix}]$ under the map $(z_1, z_2) \mapsto [\begin{smallmatrix} (z_1,z_1) & (z_1,z_2) \\ (z_2,z_1) & (z_2,z_2) \end{smallmatrix}]$ for $(z_1, z_2) \in \mathrm{SO}(X)(x_1, x_2)$. In particular the function $|f_1 \circ f_2| : h \mapsto |\omega(g, hh')\varphi(x_1, x_2)|$ is locally constant and compactly supported on $H\backslash\mathrm{SO}(X)$. From Lemma 3.7 we have that $h \mapsto Z(\Re(s), |\pi(h)W|)$ is also locally constant. Therefore, for $(g, h') \in R$, $h \in H$, $\varphi \in \mathcal{S}(X^2)$, $W \in V$ and $s \in \mathbb{C}$ so that $\Re(s) > M$ we have that

$$\int\limits_{H\backslash\mathrm{SO}(X)} |(\omega(g, hh')\varphi)(x_1, x_2)| \cdot Z(\Re(s), |\pi(hh')W|)\, dh$$

is finite. By the triangle inequality we have that integral defining $B(g, \varphi, W, s)$ converges absolutely. □

Lemma 4.2 *Let* $\varphi \in \mathcal{S}(X^2)$, $W \in V$ *and suppose that* $t = \begin{bmatrix} t_1 & & & \\ & t_2 & & \\ & & t_2 & \\ & & & t_1 \end{bmatrix} \in \mathrm{GSp}(4, L)^+$. *Let* $s \in \mathbb{C}$ *such that* $\Re(s) > M$ *and set* $B(\cdot) = B(\cdot, \varphi, W, s)$. *Then we have the following transformation property for the Bessel integral*

$$B(tg) = |t_1/t_2|^{\frac{1}{2}-s} B(g)$$

for $g \in \mathrm{GSp}(4, L)^+$.

Proof First let us handle the case when $t = z \cdot 1_4$. Then $\lambda(t) = z^2$. Set

$$h_z = \begin{cases} \rho(z^{-1}, 1), & X = X_{ns} \\ \rho(z, 1), & X = X_M, \end{cases}$$

which acts as multiplication by z and has similitude factor $\lambda(h_z) = z^2$. Using the formulas (3.15) and (3.19) we calculate that for $(y_1, y_2) \in X^2$,

$$\begin{aligned} \omega(z, h_z)\varphi'(y_1, y_2) &= |\lambda(h_z)|^{-2}\omega(z \cdot \begin{bmatrix} 1 & & & \\ & 1 & & \\ & & z^{-2} & \\ & & & z^{-2} \end{bmatrix}, 1)\varphi'(\frac{1}{z}y_1, \frac{1}{z}y_2) \\ &= |\lambda(h_z)|^{-2}\omega(\begin{bmatrix} z & & & \\ & z & & \\ & & z^{-1} & \\ & & & z^{-1} \end{bmatrix}, 1)\varphi'(\frac{1}{z}y_1, \frac{1}{z}y_2) \\ &= |\lambda(h_z)|^{-2}\chi_{E/L}(z^2)|z^2|^2\varphi'(y_1, y_2) \\ &= \varphi'(y_1, y_2) \end{aligned}$$

for all $\varphi' \in \mathcal{S}(X^2)$. We have $\pi(z)W = W$, for all $W \in V$, so that

$$\begin{aligned} B(zg) &= \int\limits_{H\backslash \mathrm{SO}(X)} (\omega(zg, hh_zh')\varphi)(x_1, x_2)Z(s, \pi(hh')\big(\pi(z)W\big))\,dh \\ &= \int\limits_{H\backslash \mathrm{SO}(X)} \omega(g, hh')\big(\omega(z, h_z)\varphi\big)(x_1, x_2)Z(s, \pi(hh')\big(\pi(z)W\big))\,dh \\ &= B(g) \end{aligned}$$

for all $g \in \mathrm{GSp}(4)^+$. For $g \in \mathrm{GSp}(4, L)^+$ we have

$$B(\begin{bmatrix} t_1 &&& \\ & t_2 && \\ && t_2 & \\ &&& t_1 \end{bmatrix} g) = B(\begin{bmatrix} t_2 &&& \\ & t_2 && \\ && t_2 & \\ &&& t_2 \end{bmatrix}\begin{bmatrix} t_1t_2^{-1} &&& \\ & 1 && \\ && 1 & \\ &&& t_1t_2^{-1} \end{bmatrix} g) = B(\begin{bmatrix} t_1t_2^{-1} &&& \\ & 1 && \\ && 1 & \\ &&& t_1t_2^{-1} \end{bmatrix} g).$$

Thus, we may assume that $t_2 = 1$.

We determine how t acts on φ and $Z(s, W)$, individually. We will need to treat the case when X is split and when X is non-split separately. Assume that X is split. Let $g \in \mathrm{GSp}(4, L) = \mathrm{GSp}(4, L)^+$ and choose $h' \in \mathrm{GSO}(X)$ such that $\lambda(h') = \lambda(g)$. Set $h_t = \rho([\begin{smallmatrix} t_1 & \\ & 1 \end{smallmatrix}], 1)$. Then $\lambda(h_t) = t_1 = \lambda(t)$. Also, $\omega(t, h_t)\varphi'(x_1, x_2) = |t_1|^{-2}\omega(\begin{bmatrix} t_1 &&& \\ & 1 && \\ && t_1^{-1} & \\ &&& 1 \end{bmatrix}, 1)\varphi'(\frac{1}{t_1}x_1, x_2)\varphi'(x_1, x_2)$, for any $\varphi' \in \mathcal{S}(X^2)$. By Lemma 3.6 we have that $Z(s, \pi(h_t)W') = |1/t_1|^{s-\frac{1}{2}}Z(s, W')$, for any $W' \in V$.

Assume that X is non-split. Let $g \in \mathrm{GSp}(4, L)^+$ and let $h' \in \mathrm{GSO}(X)$ with $\lambda(h') = \lambda(g)$. Suppose that $t = \begin{bmatrix} t_1 &&& \\ & 1 && \\ && 1 & \\ &&& t_1 \end{bmatrix} \in \mathrm{GSp}(4, L)^+$. Then there exists some $a \in E$ such that $\mathrm{N}_L^E(a) = \lambda(t) = t_1$. Set $h_t = \rho(1, [\begin{smallmatrix} a & \\ & 1 \end{smallmatrix}])$ and notice that $\lambda(h_t) = \mathrm{N}_L^E(a) = \lambda(t)$. We calculate that

$$\begin{aligned} \omega(\begin{bmatrix} t_1 &&& \\ & 1 && \\ && 1 & \\ &&& t_1 \end{bmatrix}, h_t)\varphi'(x_1, x_2) &= |t_1|^{-2}\omega(\begin{bmatrix} t_1 &&& \\ & 1 && \\ && t_1^{-1} & \\ &&& 1 \end{bmatrix}, 1)\varphi'(\mathrm{N}_L^E(a^{-1})x_1, x_2) \\ &= \chi_{E/L}(t_1)|t_1|^2|t_1|^{-2}\varphi'(t_1\mathrm{N}_E^L(a^{-1})x_1, x_2) \\ &= \varphi'(x_1, x_2), \end{aligned}$$

for any $\varphi' \in \mathcal{S}(X^2)$. From Lemma 3.6 we know that $Z(s, \pi(h_t)W') = |1/t_1|^{\frac{1}{2}-s}Z(s, W')$, for any $W' \in V$.

We have determined that $\omega(t, h_t)\varphi(x_1, x_2) = \varphi(x_1, x_2)$ and that $Z(s, \pi(h_t)W) = |1/t_1|^{s-\frac{1}{2}}Z(s, W)$ when X is split or non-split. Hence, we find that for any X

$$
\begin{aligned}
B(tg) &= \int\limits_{H\backslash \mathrm{SO}(X)} (\omega(tg, hh'h_t)\varphi)(x_1, x_2) Z(s, \pi(hh'h_t)W)\, dh \\
&= \int\limits_{H\backslash \mathrm{SO}(X)} \omega(t, h_t)(\omega(g, h_t^{-1}hh'h_t)\varphi)(x_1, x_2) Z(s, \pi(hh'h_t)W)\, dh \\
&= \int\limits_{H\backslash \mathrm{SO}(X)} \omega(t, h_t)(\omega(g, h_t^{-1}hh_th_t^{-1}h'h_t)\varphi)(x_1, x_2) Z(s, \pi(hh'h_t)W)\, dh \\
&= \int\limits_{H\backslash \mathrm{SO}(X)} (\omega(g, hh_t^{-1}h'h_t)\varphi)(x_1, x_2) Z(s, \pi(h_thh_t^{-1}h'h_t)W)\, dh \\
&= |1/t_1|^{\frac{1}{2}-s} \int\limits_{H\backslash \mathrm{SO}(X)} \omega(g, hh_t^{-1}h'h_t)\varphi(x_1, x_2) Z(s, \pi(hh_t^{-1}h'h_t)W)\, dh \\
&= |1/t_1|^{\frac{1}{2}-s} B(g)
\end{aligned}
$$

since $\lambda(h') = \lambda(h_t^{-1}h'h_t)$. Here we have used that $\int_{H\backslash \mathrm{SO}(X)} f(h_t^{-1}xh_t)\, dx = \int_{H\backslash \mathrm{SO}(X)} f(x)\, dx$, for all measurable function on $H\backslash \mathrm{SO}(X)$. This completes the proof. □

Corollary 4.3 *Let X be non-split. Let $\varphi \in \mathcal{S}(X^2)$, $W \in V$, $g \in \mathrm{GSp}(4, L)^+$ and let g_1 be as in* (4.4). *Let $s \in \mathbb{C}$ be such that $\Re(s) > M$. Then $B(g, \varphi, W, s) = |\lambda(g)|^{-s+\frac{1}{2}} B(g_1, \varphi, W, s)$.*

We extend $B(\cdot) = B(\cdot, \varphi, W, s)$ to all of $\mathrm{GSp}(4, L)$ via the formula

$$B(g) = |\lambda(g)|^{-s+\frac{1}{2}} B(g_1) \tag{4.3}$$

for

$$g_1 = \begin{bmatrix} \lambda(g)^{-1} & & & \\ & 1 & & \\ & & 1 & \\ & & & \lambda(g)^{-1} \end{bmatrix} g \tag{4.4}$$

and $g \in \mathrm{GSp}(4, L)$. Clearly, Corollary 4.3 indicates that this extension is well defined.

Part *(a)* of the following lemma uses the choice of the canonical $\mathrm{GO}(X)$-representation π^+ made in Sect. 3. Let $s \in \mathrm{O}(X)$ be defined by $s(x) = x^*$ and $T : V \to V$ be defined by $T(v_1, v_2) = (v_2, v_1)$ in the split case and by $T(W)(g) = W(\alpha(g))$ for $g \in \mathrm{GL}(2, E)$, $W \in \mathcal{W}(\tau, \psi_E)$, and α is the non-trivial element of the Galois group of E/L in the non-split case.

Lemma 4.4 *Let (π, V) be the infinite-dimensional, irreducible, admissible representation of $\mathrm{GSO}(X)$ obtained from either τ_0 or the pair τ_1, τ_2, as in Sect. 3. Assume that the space V of π is either $\mathcal{W}(\tau_0, \psi_E)$ or $\mathcal{W}(\tau_1, \psi) \otimes \mathcal{W}(\tau_2, \psi)$, respectively. Let $z \in \mathbb{C}$ be such that $\Re(z) > M$. Then, for all $g \in \mathrm{GSp}(4, L)$, $\varphi' \in \mathcal{S}(X^2)$, $W' \in V$ we have*

(a) $B(g, \omega(1,s)\varphi', T(W'), z) = B(g, \varphi', W', z)$,
(b) *if* $b = \left[\begin{smallmatrix} 1 & B \\ & 1 \end{smallmatrix}\right]$ *where* $B = \left[\begin{smallmatrix} b_1 & b_2 \\ b_2 & b_3 \end{smallmatrix}\right] \in \mathrm{M}(2, L)$, *we have* $B(bg, \varphi', W', z) = \psi(b_2)B(g, \varphi', W', z)$, *and*
(c) *if* $t = \begin{bmatrix} t_1 & & & \\ & t_2 & & \\ & & t_2 & \\ & & & t_1 \end{bmatrix}$ *for* $t_1, t_2 \in L^\times$ *we have* $B(tg, \varphi', W', z) = |t_2/t_1|^{z-\frac{1}{2}} B(g, \varphi', W', z)$.

Remark 4.5 In Sect. 4.2 we will define a similar intertwining map in the case that $L = \mathbb{R}$. In order to refrain from reproducing the same work multiple times we note that Lemma 4.4 will have an identical proof to the analogous result.

Proof To prove (a), let $\varphi' \in \mathcal{S}(X^2)$, $W' \in V$. First see that $Z(z, (s \cdot \pi)(h), W') = Z(z, \pi(h)T(W'))$ for all $h \in \mathrm{GSO}(X)$ and $W' \in V$. This is because in the split case $\pi(s)\rho(h_1, h_2) = \rho(h_2, h_1)$, for $h_1, h_2 \in \mathrm{GL}(2, L)$ and in the non-split case $\pi(s)\rho(t, h) = \rho(t, h^*)$ for $t \in L^\times$ and $h \in \mathrm{GL}(2, E)$. Suppose that $g \in \mathrm{GSp}(4, L)^+$. We calculate that

$$\begin{aligned} B(g, \omega(1,s)\varphi', W', z) &= \int\limits_{H\backslash \mathrm{SO}(X)} (\omega(g, hh')\omega(1,s)\varphi')(x_1, x_2) Z(z, \pi(hh')W')\, dh \\ &= \int\limits_{H\backslash \mathrm{SO}(X)} (\omega(g, ss^{-1}hss^{-1}h's)\varphi')(x_1, x_2) Z(z, \pi(hh')W')\, dh \\ &= \int\limits_{H\backslash \mathrm{SO}(X)} (\omega(1,s)\omega(g, s^{-1}hss^{-1}h's)\varphi')(x_1, x_2) Z(z, \pi(hh')W')\, dh \\ &= \int\limits_{H\backslash \mathrm{SO}(X)} (\omega(g, s^{-1}hss^{-1}h's)\varphi')(sx_1, sx_2) Z(z, \pi(hh')W')\, dh \\ &= \int\limits_{H\backslash \mathrm{SO}(X)} (\omega(g, s^{-1}hss^{-1}h's)\varphi')(-x_1, -x_2) Z(z, \pi(hh')W')\, dh. \end{aligned}$$

As in the proof of Lemma 4.2 we change variables $h \mapsto shs^{-1}$. Also, we replace h' with $h'' = sh's^{-1}$. We may write that the above is

$$= \int\limits_{H\backslash \mathrm{SO}(X)} (\omega(g, hh'')\varphi'(-x_1, -x_2) Z(z, s \cdot \pi(hh'')W')\, dh.$$

Denote by $-1 \in \mathrm{SO}(X)$ the action which sends x to $-x$ so that we have $-1 = \rho(-1, 1)$. Therefore the above is

$$= \int\limits_{H\backslash \mathrm{SO}(X)} (\omega(g, hh''(-1))\varphi'(x_1, x_2) Z(z, s \cdot \pi(hh'')W')\, dh$$
$$= \int\limits_{H\backslash \mathrm{SO}(X)} (\omega(g, hh'')\varphi'(x_1, x_2) Z(z, s \cdot \pi(-hh'')W')\, dh$$
$$= \int\limits_{H\backslash \mathrm{SO}(X)} (\omega(g, hh'')\varphi'(x_1, x_2) Z(z, s \cdot \pi(hh'')W')\, dh$$
$$= B(g, \varphi', T(W'), z).$$

Hence we have shown that

$$B(g, \omega(1, s)\varphi', W', z) = B(g, \varphi', T(W'), z)$$

so that

$$B(g, \omega(1, s)\varphi', T(W'), z) = B(g, \varphi', W', z).$$

Now for all $g \in \mathrm{GSp}(4, L)$, $\varphi' \in \mathcal{S}(X^2)$, and $W' \in V$ we use (4.4) to see that

$$\begin{aligned} B(g, \omega(1, s)\varphi', T(W), z) &= |\lambda(g)|^{\frac{1}{2}-z} B(g_1, \omega(1, s)\varphi', T(W), z) \\ &= |\lambda(g)|^{\frac{1}{2}-z} B(g_1, \varphi', W', z) \\ &= B(g, \varphi', W', z) \end{aligned}$$

which proves *(a)*.

Proof of (b). Set $b = \left[\begin{smallmatrix} 1 & B \\ & 1 \end{smallmatrix}\right]$ where $B = \left[\begin{smallmatrix} b_1 & b_2 \\ b_2 & b_3 \end{smallmatrix}\right] \in \mathrm{M}(2, L)$ is symmetric. First notice that $\lambda(b) = 1$ so that $B(bg, \varphi', W', z)$ only differs from $B(g, \varphi', W', z)$ in the Weil representation factor of the integrand. Recall formula (3.16) to see that

$$\begin{aligned} (\omega(b, 1)\varphi')(x_1, x_2) &= \psi(b_1\langle x_1, x_1\rangle) + 2b_2\langle x_1, x_2\rangle + b_3\langle x_2, x_2\rangle)\, \varphi'(x_1, x_2) \\ &= \psi(b_2)\varphi'(x_1, x_2) \end{aligned}$$

for all $\varphi' \in \mathcal{S}(X^2)$. Therefore it is evident that if $g \in \mathrm{GSp}(4, L)^+$ then by (4.4)

$$B(bg, \varphi', W', z) = \psi(b_2) B(g, \varphi', W', z).$$

If $g \in \mathrm{GSp}(4, L)$ then

$$\begin{aligned} B(bg) &= |\lambda(g)|^{\frac{1}{2}-s} B(bg_1) \\ &= |\lambda(g)|^{\frac{1}{2}-s} \psi(b_2) B(g_1) \\ &= \psi(b_2) B(g). \end{aligned}$$

This finishes the proof of *(b)*.

Proof of (c) The case when $g \in \mathrm{GSp}(4, L)^+$ is proved in Lemma 4.2. Assume that $g \in \mathrm{GSp}(4, L)$ and set

$$t = \begin{bmatrix} t_1 & & & \\ & t_2 & & \\ & & t_2 & \\ & & & t_1 \end{bmatrix}.$$

Then $\lambda(t) = t_1 t_2$. Similar to (4.4) set

$$(tg)_1 = \begin{bmatrix} \lambda(tg)^{-1} & & & \\ & 1 & & \\ & & 1 & \\ & & & \lambda(tg)^{-1} \end{bmatrix} tg = \begin{bmatrix} t_2^{-1} & & & \\ & t_2 & & \\ & & t_2 & \\ & & & t_2^{-1} \end{bmatrix} g_1.$$

Therefore,

$$\begin{aligned} B(tg) &= |\lambda(tg)|^{\frac{1}{2}-s} B((tg)_1) \\ &= |\lambda(g)|^{\frac{1}{2}-s} |t_1 t_2|^{\frac{1}{2}-s} B(\begin{bmatrix} t_2^{-1} & & & \\ & t_2 & & \\ & & t_2 & \\ & & & t_2^{-1} \end{bmatrix} g_1). \end{aligned}$$

Finally, since $\begin{bmatrix} t_2^{-1} & & & \\ & t_2 & & \\ & & t_2 & \\ & & & t_2^{-1} \end{bmatrix} g_1 \in \mathrm{Sp}(4, L)$ we get that $B(tg) = |t_1/t_2|^{\frac{1}{2}-s} |\lambda(g)|^{\frac{1}{2}-s} B(g_1)$, which is just $|t_1/t_2|^{\frac{1}{2}-s} B(g)$, completing the proof. □

Proposition 4.6 *Assume that L is non-archimedean. Let $s \in \mathbb{C}$ be such that $\Re(s) \gg 1$. Assume that X is split. Suppose that $W_i \in \mathcal{W}(\tau_i, \psi)$ are such that $Z(s, W_i)$ are not zero for $i \in \{1, 2\}$. Set $W = W_1 \otimes W_2$. Then there exists $\varphi \in \mathcal{S}(X^2)$ such that $B(1, \varphi, W, s) \neq 0$. Assume that X is non-split. Suppose that $W \in \mathcal{W}(\tau_0, \psi_E)$ is such that $Z(s, W)$ is not zero, then there exists $\varphi \in \mathcal{S}(X^2)$ such that $B(1, \varphi, W, s) \neq 0$.*

Proof Assume that X is non-split. Denote $h = \rho(t, h_0) \in \mathrm{GSO}(X)$, for $t \in L^\times$ and $h_0 \in \mathrm{GL}(2, E)$. By Lemma 3.7 the function $h \mapsto Z(s, \pi(h)W)$ is locally constant on $H\backslash \mathrm{SO}(X)$ and, furthermore, is non-zero at the point $H \cdot 1$. Recall that $f_2 : H\backslash \mathrm{SO}(X) \to \mathrm{SO}(X)(x_1, x_2)$, defined by $f_2(Hh) = h^{-1}(x_1, x_2)$, is a homeomorphism and $\mathrm{SO}(X)(x_1, x_2)$ is closed. Let $A \subset H\backslash \mathrm{SO}(X)$ be a compact open neighborhood of $H \cdot 1$ so that function $h \mapsto Z(s, \pi(h)W)$ is constant on A. Now, $f_2(A)$ is a compact open subset of $\mathrm{SO}(X)(x_1, x_2)$ and there exists some open $U \subset X^2$ such that $f_2(A) = U \cap \mathrm{SO}(X)(x_1, x_2)$. For each $u \in U$ choose an open compact neighborhood Y_u of u such that $Y_u \subset U$. Clearly $\{Y_u \cap \mathrm{SO}(X)(x_1, x_2)\}_{u \in U}$ is an open cover of $f_2(A)$ so there exists a finite subset $I \subset U$ such that $\{Y_u \cap \mathrm{SO}(X)(x_1, x_2)\}_{u \in I}$ is a finite cover of $f_2(A)$. Therefore $f_2(A) = \cup_{u \in I}(Y_u \cap \mathrm{SO}(X)(x_1, x_2))$. Set $Y = \bigcup_{u \in I} Y_u$. Then the characteristic function $\varphi = \mathbf{1}_Y$ is in $\mathcal{S}(X^2)$ and fulfills the requirements. Indeed, we calculate that

$$B(1, \mathbf{1}_Y, W, s) = \int_{H\backslash \mathrm{SO}(X)} \omega(1, h)\mathbf{1}_Y(x_1, x_2)Z(s, \tau_0(h_0)W)\,dh$$
$$= \mathrm{vol}(A)Z(s, W).$$

Assume that X is split. The argument is similar. Let $h = \rho(h_1, h_2)$ for $h_1, h_2 \in \mathrm{GL}(2, L)$. Choose compact neighborhood A, of 1 so that the functions $h \mapsto Z(s, \tau_1(h_1)W_1)$ and $h \mapsto Z(s, \tau_1(h_2)W_2)$ are both constant on A, so that $h \mapsto Z(s, \pi(h)W)$ is also constant on A, and the argument runs verbatim. □

Now that we have proved that $B(\cdot, \varphi, W, s)$ is a well defined and non-zero map, for some choices of φ, W and s we define $\Theta(V)$ to be the space generated by all such functions for all choices of $\varphi \in \mathcal{S}(X^2)$ and $W \in V$. Fix $s \in \mathbb{C}$ such that $\Re(s) \gg M$. We define the map

$$\vartheta : \mathcal{S}(X^2) \otimes V \to \Theta(V) \qquad (4.5)$$
$$\varphi \otimes W \mapsto B(\cdot, \varphi, W, s).$$

It is easy to verify that if $B \in \Theta(V)$, then by Lemma 4.4

$$B(tg) = |t_1/t_2|^{\frac{1}{2}-s} B(g) \quad \text{for } t = \begin{bmatrix} t_1 &&& \\ & t_2 && \\ && t_2 & \\ &&& t_1 \end{bmatrix} \text{ for } t_1, t_2 \in L^\times \text{ and} \qquad (4.6)$$

$$B(bg) = \psi(b_2)B(g) \quad \text{for } b = \left[\begin{smallmatrix} 1 & B \\ & 1 \end{smallmatrix}\right] \text{ where } B = \left[\begin{smallmatrix} b_1 & b_2 \\ b_2 & b_3 \end{smallmatrix}\right] \in \mathrm{M}(2, L). \qquad (4.7)$$

Theorem 4.7 *Let $s \in \mathbb{C}$ with $\Re(s) \gg M$. Let $V = \mathcal{W}(\tau_1, \psi) \otimes \mathcal{W}(\tau_2, \psi)$ or $V = \mathcal{W}(\tau_0, \psi_E)$ in the split case or non-split case, respectively. Then*

(a) the map ϑ from (4.5) *is a non-zero R'-map,*
(b) for each $f \in \Theta(V)$ there is an open and compact $K \subset \mathrm{GSp}(4, L)$ such that, for each $k \in K$ we have $f(g) = f(gk)$ for all $g \in \mathrm{GSp}(4, L)$, and
(c) the image of ϑ lies inside $\mathcal{B}(\mathrm{GSp}(4, L), \psi)$.

It then follows that $\Theta(V)$ is a smooth representation of GSp(4, *L*) *sitting inside* $\mathcal{B}(\mathrm{GSp}(4, L), \psi)$.

Proof *Proof of (a).* Let $(g_0, h_0) \in R'$, let $g \in \mathrm{GSp}(4, L)^+$, and let $h' \in \mathrm{GO}(X)$ so that $(g, h') \in R$. We see that

$$B(g, \omega(g_0, h_0)\varphi, \pi(h_0)W, s) = \int_{H\backslash \mathrm{SO}(X)} \omega(gg_0, h(h'h_0))\varphi(x_1, x_2)Z(s, \pi(h(h'h_0))W)\,dh$$
$$= B(gg_0, \varphi, W, s)$$

since $(gg_0, h'h_0) \in R$. Now suppose that $g \in \mathrm{GSp}(4, L)$ and let $g_1 \in \mathrm{GSp}(4, L)^+$ be as in (4.4). Then using the above result, we find that

$$
\begin{aligned}
B(g, \omega(g_0, h_0)\varphi, \pi(h_0)W, s) &= |\lambda(g)|^{-s+\frac{1}{2}} B(g_1, \omega(g_0, h_0)\varphi, \pi(h_0)W, s) \\
&= |\lambda(g)|^{-s+\frac{1}{2}} B(g_1 g_0, \varphi, W, s) \\
&= B(g g_0, \varphi, W, s).
\end{aligned}
$$

By Lemma 4.6 we have that $B(\cdot)$ is non-zero as long as there is some $W \in V$ such that $Z(s, W)$ is not zero for $s \gg M$. Since τ_0, τ_1, τ_2 have Whittaker models it is clear that $Z(s, W)$ is not zero.

Proof of (b). Let $\varphi \otimes W \in \mathcal{S}(X^2) \otimes V$. It suffices to prove (b) for $B = B(\cdot, \varphi, W, s)$. Given the following exact sequence

$$
1 \longrightarrow L^\times \cdot \mathrm{Sp}(4, L) \hookrightarrow \mathrm{GSp}(4, L) \xrightarrow{\ \lambda\ } L^\times / L^{\times 2} \longrightarrow 1
$$

it suffices to find a compact open subgroup of $\mathrm{Sp}(4, L)$ which stabilizes $B(\cdot)$. Since the Weil representation is smooth there exists some $N \in \mathbb{Z}_{>0}$ so that the open and compact full congruence subgroup

$$
\Gamma(\mathfrak{p}^N) = \{k \in \mathrm{Sp}(4, \mathfrak{o}_L) \mid k \equiv 1_4 \pmod{\mathfrak{p}^N}\}
$$

fixes φ under the action of the Weil representation. Since $\Gamma(\mathfrak{p}^N)$ is also contained in $\mathrm{Sp}(4, L)$ we determine that for all $k \in \Gamma(\mathfrak{p}^N)$ and all $g \in \mathrm{GSp}(4, L)$ that

$$
\begin{aligned}
B(gk) &= |\lambda(g)|^{-s+\frac{1}{2}} \int\limits_{H\backslash \mathrm{SO}(X)} \omega(g_1 k, hh')\varphi(x_1, x_2) Z(s, \pi(hh')W)\, dh \\
&= |\lambda(g)|^{-s+\frac{1}{2}} \int\limits_{H\backslash \mathrm{SO}(X)} \omega(g_1, hh')\big(\omega(k, 1)\varphi(x_1, x_2)\big) Z(s, \pi(hh')W)\, dh \\
&= |\lambda(g)|^{-s+\frac{1}{2}} \int\limits_{H\backslash \mathrm{SO}(X)} \omega(g_1, hh')\varphi(x_1, x_2) Z(s, \pi(hh')W)\, dh \\
&= B(g).
\end{aligned}
$$

Proof of (c) This follows from part (b) of this lemma and from (4.6) and (4.7). □

Lemma 4.8 *Assume that the residual characteristic of L is odd. Let $h \in \mathrm{SO}(X)$. If $X = X_{ns}$ is non-split and E/L is unramified then, $h \in \rho(\mathfrak{o}_L^\times \times \mathrm{GL}(2, \mathfrak{o}_E))H$ if and only if $h(x_1, x_2) \in [\mathrm{M}(2, \mathfrak{o}_E) \cap X_{ns}]^2$. If $X = X_M$ is split then, $h \in \rho(\mathrm{GL}(2, \mathfrak{o}_L), \mathrm{GL}(2, \mathfrak{o}_L))H$ if and only if $h(x_1, x_2) \in \mathrm{M}(2, \mathfrak{o}_L)^2$.*

Proof In both the split and non-split cases the forward direction is clear.

Assume that $X = X_{ns}$ is non-split and that E/L is unramified. For the converse notice that for any $[\begin{smallmatrix} a & b \\ c & d \end{smallmatrix}] \in \mathrm{GL}(2, E)$ we can find $[\begin{smallmatrix} a' & b' \\ c' & d' \end{smallmatrix}] \in \mathrm{GL}(2, \mathfrak{o}_E)$ and $[\begin{smallmatrix} x & y \\ 0 & z \end{smallmatrix}] \in \mathrm{GL}(2, E)$ so that $[\begin{smallmatrix} a & b \\ c & d \end{smallmatrix}] = [\begin{smallmatrix} a' & b' \\ c' & d' \end{smallmatrix}][\begin{smallmatrix} x & y \\ 0 & z \end{smallmatrix}] = [\begin{smallmatrix} a' & b' \\ c' & d' \end{smallmatrix}][\begin{smallmatrix} x/z & y/z \\ 0 & 1 \end{smallmatrix}][\begin{smallmatrix} z & \\ & z \end{smallmatrix}]$. Set $h = \rho(t, [\begin{smallmatrix} a & b \\ c & d \end{smallmatrix}])$, for some $t \in L$ and $[\begin{smallmatrix} a & b \\ c & d \end{smallmatrix}] \in \mathrm{GL}(2, E)$ for which $h(x_i) \in X_{ns} \cap \mathrm{M}(2, \mathfrak{o}_E)$ for $i \in \{1, 2\}$.

Since $\rho(\mathrm{N}_L^E(z), \left[\begin{smallmatrix} z & \\ & z \end{smallmatrix}\right]) = 1$ for any $z \in E^\times$, $h = \rho(t\mathrm{N}_L^E(z)^{-1}, 1)\rho(1, \left[\begin{smallmatrix} a' & b' \\ c' & d' \end{smallmatrix}\right]\left[\begin{smallmatrix} x & y \\ 0 & 1 \end{smallmatrix}\right]) = \rho(u, \left[\begin{smallmatrix} a' & b' \\ c' & d' \end{smallmatrix}\right])\rho(\varpi^k, \left[\begin{smallmatrix} x & y \\ 0 & 1 \end{smallmatrix}\right])$, for some $u \in \mathfrak{o}_E^\times$ and $k \in \mathbb{Z}$. Therefore, $\rho(\varpi^k, \left[\begin{smallmatrix} x & y \\ 0 & 1 \end{smallmatrix}\right])(x_i) \in X_{ns} \cap \mathrm{M}(2, \mathfrak{o}_E)$ for $i = 1, 2$.

It suffices to show that $k = 0$, $x \in \mathfrak{o}_E^\times$ and $y \in \mathfrak{o}_E$. Since $h \in \mathrm{SO}(X_{ns})$ we have that $1 = |\lambda(h)| = |\varpi^{-2k}\mathrm{N}_L^E(x)|$ so that $\mathrm{N}_L^E(x) = |\varpi|^{2k}$. Since E/L is unramified, we know $\sqrt{\delta} \in \mathfrak{o}_E^\times$. Furthermore, we have that

$$\rho(\varpi^k, \left[\begin{smallmatrix} x & y \\ 0 & 1 \end{smallmatrix}\right])(x_1) = \varpi^{-k}\left[\begin{smallmatrix} 0 & \mathrm{N}_L^E(x)\sqrt{\delta} \\ 0 & 0 \end{smallmatrix}\right] \text{ and } \rho(\varpi^k, \left[\begin{smallmatrix} x & y \\ 0 & 1 \end{smallmatrix}\right])(x_2) = 2\varpi^{-k}/\delta\left[\begin{smallmatrix} -y\sqrt{\delta} & \mathrm{N}_L^E(y)\sqrt{\delta} \\ -\sqrt{\delta} & \alpha(y)\sqrt{(\delta)} \end{smallmatrix}\right]$$

are both matrices with integral entries. Therefore $\varpi^{-k}, \varpi^{-k}\mathrm{N}_L^E(x), \varpi^{-k}\mathrm{N}_L^E(y), \varpi^{-k}y \in \mathfrak{o}_E$. Clearly $k \le 0$. Furthermore, since $|\varpi^{-k}\mathrm{N}_L^E(x)| = |\varpi^k|$ it follows that $k \ge 0$, so in fact $k = 0$. It follows that $x \in \mathfrak{o}_E^\times$ and $y \in \mathfrak{o}_E$.

Assume that $X = X_M$ is split and set $h = \rho(\left[\begin{smallmatrix} a_1 & b_1 \\ c_1 & d_1 \end{smallmatrix}\right], \left[\begin{smallmatrix} a_2 & b_2 \\ c_2 & d_2 \end{smallmatrix}\right])$ for some $\left[\begin{smallmatrix} a_1 & b_1 \\ c_1 & d_1 \end{smallmatrix}\right]$ and $\left[\begin{smallmatrix} a_2 & b_2 \\ c_2 & d_2 \end{smallmatrix}\right]$ for which $h(x_i) \in \mathrm{M}(2, \mathfrak{o}_L)$ for $i \in \{1, 2\}$. For $i \in \{1, 2\}$ we can write

$$\left[\begin{smallmatrix} a_i & b_i \\ c_i & d_i \end{smallmatrix}\right] = \left[\begin{smallmatrix} a_i' & b_i' \\ c_i' & d_i' \end{smallmatrix}\right]\left[\begin{smallmatrix} w_i & y_i \\ 0 & z_i \end{smallmatrix}\right] = \left[\begin{smallmatrix} a_i' & b_i' \\ c_i' & d_i' \end{smallmatrix}\right]\left[\begin{smallmatrix} w_i/z_i & y_i/z_i \\ & 1 \end{smallmatrix}\right]\left[\begin{smallmatrix} z_i & \\ & z_i \end{smallmatrix}\right],$$

for some $\left[\begin{smallmatrix} a_i' & b_i' \\ c_i' & d_i' \end{smallmatrix}\right] \in \mathrm{GL}(2, \mathfrak{o}_L)$ and $\left[\begin{smallmatrix} w_i & y_i \\ 0 & z_i \end{smallmatrix}\right] \in \mathrm{GL}(2, L)$. Therefore for some, possibly different, $w_i, y_i, z_i \in L$ we have that

$$\begin{aligned} h &= \rho(\left[\begin{smallmatrix} a_1' & b_1' \\ c_1' & d_1' \end{smallmatrix}\right], \left[\begin{smallmatrix} a_2' & b_2' \\ c_2' & d_2' \end{smallmatrix}\right]) \cdot \rho(\left[\begin{smallmatrix} w_1 & y_1 \\ & 1 \end{smallmatrix}\right], \left[\begin{smallmatrix} w_2 & y_2 \\ & 1 \end{smallmatrix}\right]) \cdot \rho(\left[\begin{smallmatrix} z_1 & \\ & z_1 \end{smallmatrix}\right], \left[\begin{smallmatrix} z_2 & \\ & z_2 \end{smallmatrix}\right]) \\ &= z_1 z_2 \cdot \rho(\left[\begin{smallmatrix} a_1' & b_1' \\ c_1' & d_1' \end{smallmatrix}\right], \left[\begin{smallmatrix} a_2' & b_2' \\ c_2' & d_2' \end{smallmatrix}\right]) \cdot \rho(\left[\begin{smallmatrix} w_1 & y_1 \\ & 1 \end{smallmatrix}\right], \left[\begin{smallmatrix} w_2 & y_2 \\ & 1 \end{smallmatrix}\right]) \\ &= \rho(\left[\begin{smallmatrix} a_1' & b_1' \\ c_1' & d_1' \end{smallmatrix}\right], \left[\begin{smallmatrix} a_2' & b_2' \\ c_2' & d_2' \end{smallmatrix}\right])\, h' \end{aligned}$$

for an appropriate choice of $h' \in \mathrm{GSO}(X)$. Set $z_1 z_2 = u\varpi_L^k$ for some $u \in \mathfrak{o}_L^\times$ and $k \in \mathbb{Z}$.

Since $h \in \mathrm{SO}(X)$ we calculate that $1 = |\lambda(h)| = |w_1 w_2 \varpi^{2k}|$. Furthermore we have that $h'(x_1) = u\varpi^k\left[\begin{smallmatrix} 0 & w_1 w_2 \\ 0 & 0 \end{smallmatrix}\right]$ and $h'(x_2) = u\varpi^k\left[\begin{smallmatrix} -2y_1 & 2y_1 y_2 \\ -2 & 2y_2 \end{smallmatrix}\right]$ are both matrices with integral entries. Since $-2u\varpi^k \in \mathfrak{o}_E$, we know $k \ge 0$. We also have $|w_1 w_2 \varpi^k| = |\varpi^{-k}|$, and since $w_1 w_2 \varpi^k \in \mathfrak{o}_E$, we conclude that $k = 0$. It follows that $w_1 w_2 \in \mathfrak{o}_L$. By Lemma 3.5, the observation that

$$h = \rho(\left[\begin{smallmatrix} a_1' & b_1' \\ c_1' & d_1' \end{smallmatrix}\right], \left[\begin{smallmatrix} a_2' & b_2' \\ c_2' & d_2' \end{smallmatrix}\right])\rho(z_1 z_2\left[\begin{smallmatrix} 1 & u_1 \\ & 1 \end{smallmatrix}\right], \left[\begin{smallmatrix} w_2 w_1 & u_2 \\ & 1 \end{smallmatrix}\right])\rho(\left[\begin{smallmatrix} w_1 & \\ & 1 \end{smallmatrix}\right], \left[\begin{smallmatrix} w_1^{-1} & \\ & 1 \end{smallmatrix}\right])$$

completes the proof. □

The following theorem specifies a choice for $\varphi \in \mathcal{S}(X^2)$ so that when $W \in V$ is unramified, so is the local theta lift $B(\cdot, \varphi, W, s)$.

Theorem 4.9 *If $X = X_{ns}$ is non-split then assume that E/L, τ_0, and ψ_E are all unramified, and set $Z = \mathrm{M}(2, \mathfrak{o}_E) \cap X(E)$. If $X = X_M$ is split then assume that τ_1, τ_2, and ψ are all unramified and set $Z = \mathrm{M}(2, \mathfrak{o}_L)$.*

Let $\varphi = f_{Z^2}$ be the characteristic function of $Z^2 \subset X^2$. Let $W_{un} \in V$ be the standard unramified vector as in Theorem 11 of [God70]. Then $B(\cdot, f_{Z^2}, W_{un}, s)$ satisfies $B(gk, f_{Z^2}, W_{un}, s) = B(g, f_{Z^2}, W_{un}, s)$ for $g \in \mathrm{GSp}(4, L)$ and $k \in \mathrm{GSp}(4, \mathfrak{o}_L)$. Also

$$B(1, f_{Z^2}, W_{un}, s) = \begin{cases} L(s, \tau_0) & X \text{ is non-split,} \\ L(s, \tau_1) \cdot L(s, \tau_2) & X \text{ is split.} \end{cases} \tag{4.8}$$

Proof Let $g \in \mathrm{GSp}(4, L)$ and $k \in \mathrm{GSp}(4, \mathfrak{o}_L)$. Suppose that g_1 is as in (4.4). Assume that $X = X_{ns}$ is non-split. Because E/L is unramified, $\mathrm{N}_L^E(\mathfrak{o}_E^\times) = \mathfrak{o}_L^\times$. This allows one to choose $j \in \mathrm{GSO}(X, L)$ such that $\lambda(j) = \lambda(k)$ and $jM = M$, using Lemma 4.8. We calculate

$$\begin{aligned}
B(gk, \varphi, W_{un}, s) &= |\lambda(gk)|^{-(s-\frac{1}{2})} B(\begin{bmatrix} \lambda(gk)^{-1} & & & \\ & 1 & & \\ & & 1 & \\ & & & \lambda(gk)^{-1} \end{bmatrix} gk, \varphi, W_{un}, s) \\
&= |\lambda(g)|^{-(s-\frac{1}{2})} B(g_1 k, \varphi, W_{un}, s) \\
&= |\lambda(g)|^{-(s-\frac{1}{2})} \int\limits_{H\backslash \mathrm{SO}(X)} \big(\omega(g_1 k, hj)\varphi\big)(x_1, x_2) Z(s, \pi(hj) W_{un})\, dh \\
&= |\lambda(g)|^{-(s-\frac{1}{2})} \int\limits_{H\backslash \mathrm{SO}(X)} \big(\omega(g_1, h)\omega(k, j)\varphi\big)(x_1, x_2) Z(s, \pi(h)\pi(j) W_{un})\, dh \\
&= |\lambda(g)|^{-(s-\frac{1}{2})} \int\limits_{H\backslash \mathrm{SO}(X)} \big(\omega(g_1, h)\varphi\big)(x_1, x_2) Z(s, \pi(h) W_{un})\, dh \\
&= B(g, \varphi, W_{un}, s).
\end{aligned}$$

Let $h \in \mathrm{SO}(X)$. By Lemma 4.8, we have that $h(x_1, x_2) \in Z^2$ if and only if $h \in \rho(\mathfrak{o}_L^\times \times \mathrm{GL}(2, \mathfrak{o}_E))H$. By Lemma 3.6 we have that the map $H \to \mathbb{C}$ given by $h \mapsto Z(s, \pi(h) W_{un})$ is constant. Therefore,

$$\begin{aligned}
B(1, \varphi, W, s) &= \int\limits_{H\backslash \mathrm{SO}(X)} f_{Z^2}\big(h^{-1}(x_1, x_2)\big) Z(s, \pi(h) W_{un})\, dh \\
&= \int\limits_{H\backslash [\mathrm{SO}(X) \cap H\rho(\mathfrak{o}_L^\times \times \mathrm{GL}(2, \mathfrak{o}_E))]} Z(s, \pi(h) W_{un})\, dh \\
&= Z(s, W_{un}),
\end{aligned}$$

since $\mathrm{vol}\big(H\backslash(\mathrm{SO}(X) \cap H\rho(\mathfrak{o}_L \times \mathrm{GL}(2, \mathfrak{o}_E)))\big) = 1$. It is observed in Theorem 11 of [God70] that $Z(s, W_{un}) = L(s, \tau_0)$, which completes the proof in the non-split case.

This proof of the split case is similar. □

Lemma 4.10 *Let* (π, V) *be a representation of* $\mathrm{GSO}(X)$, *let* $(\sigma, V \times V)$ *be the induced representation of* $\mathrm{GO}(X)$ *obtained from* π *as in Sect. 3. Let* (Π, W) *be a representation of* $\mathrm{GSp}(4, L)$. *Then we have the following* $\mathbb{C}$*-linear isomorphism*

$$M : \mathrm{Hom}_{R'}(\mathcal{S}(X^2) \otimes V, W) \xrightarrow{\sim} \mathrm{Hom}_R(\mathcal{S}(X^2) \otimes (V \times V), W)$$

determined by $M(f)(\varphi \otimes (v_1, v_2)) = f(\varphi \otimes v_1) + f(\omega(1, s) \cdot \varphi \otimes v_2)$ *for* $\varphi \in \mathcal{S}(X^2)$, $f \in \mathrm{Hom}_{R'}(\mathcal{S}(X^2) \otimes V, W)$, *and* $v_1, v_2 \in V$, *and extended linearly. Here* $(1, s)$ *is a non-trivial coset representative of* R/R'. *For example we could take* s *the map that takes* x *to* x^*. *Additionally the inverse map*

$$N : \mathrm{Hom}_R(\mathcal{S}(X^2) \otimes (V \times V), W) \xrightarrow{\sim} \mathrm{Hom}_{R'}(\mathcal{S}(X^2) \otimes V, W)$$

is given by $N(f)(\varphi \otimes v) = f(\varphi \otimes (v, 0))$ *for* $f \in \mathrm{Hom}_R(\mathcal{S}(X^2) \otimes (V \times V), W)$, $\varphi \in \mathcal{S}(X^2)$, *and* $v \in V$.

Proof To show that M is well defined, let $f \in \mathrm{Hom}_{R'}(\mathcal{S}(X^2) \otimes V, W)$, $(g, h) \in R'$, $\varphi \in \mathcal{S}(X^2)$, and $v_1, v_2 \in V$. We see that

$$\begin{aligned} M(f)\big((g, h) \cdot (\varphi \otimes (v_1, v_2))\big) &= M(f)\big(\omega(g, h) \cdot \varphi \otimes (\pi(h) \cdot v_1, \pi(shs^{-1}) \cdot v_2)\big) \\ &= f(\omega(g, h) \cdot \varphi \otimes \pi(h) \cdot v_1) + f(\omega(1, s)\omega(g, h)\varphi \otimes \pi(shs^{-1}) \cdot v_2) \\ &= f(\omega(g, h) \cdot \varphi \otimes \pi(h) \cdot v_1) + f(\omega(g, shs^{-1})\omega(1, s)\varphi \otimes \pi(shs^{-1}) \cdot v_2) \\ &= g \cdot f(\varphi \otimes v_1) + g \cdot f(\omega(1, s) \cdot \varphi \otimes v_2) \\ &= g \cdot M(f)(\varphi \otimes (v_1, v_2)). \end{aligned}$$

Hence, we see that $M(f) \in \mathrm{Hom}_{R'}(\mathcal{S}(X^2) \otimes (V \times V), W)$ so it only remains to show that

$$(1, s) \cdot M(f)(\varphi \otimes (v_1, v_2)) = M(f)(\omega(1, s) \cdot \varphi \otimes \sigma(s) \cdot (v_1, v_2)).$$

We calculate

$$\begin{aligned} M(f)\big((1, s) \cdot (\varphi \otimes (v_1, v_2))\big) &= M(f)\big(\omega(1, s) \cdot \varphi \otimes (v_2, v_1)\big) \\ &= f(\omega(1, s) \cdot \varphi \otimes v_1) + f(\varphi \otimes v_2) \\ &= M(f)(\varphi \otimes (v_2, v_1)) \\ &= (1, s) \cdot M(f)(\varphi \otimes (v_1, v_2)). \end{aligned}$$

To prove that M is one-to-one and onto, let $f, f' \in \mathrm{Hom}_{R'}(\mathcal{S}(X^2) \otimes V, W)$ and assume that $M(f) = M(f')$. For all $\varphi \in \mathcal{S}(X^2)$ and for all $v_1 \in V$ we have that

$$\begin{aligned} M(f)(\varphi \otimes (v_1, 0)) &= M(f')(\varphi \otimes (v_1, 0)), \quad \text{so that} \\ f(\varphi \otimes v_1) &= f'(\varphi \otimes v_1), \quad \text{and} \\ f &= f'. \end{aligned}$$

To see N is well defined and one-to-one, let $f \in \mathrm{Hom}_R(\mathcal{S}(X^2) \otimes (V \times V), W)$, $(g, h) \in R'$, $\varphi \in \mathcal{S}(X^2)$ and $v \in V$. Then $N(f) \in \mathrm{Hom}_{R'}(\mathcal{S}(X^2) \otimes V, W)$ since

$$\begin{aligned} N(f)((g, h) \cdot (\varphi \otimes v)) &= N(f)((g, h) \cdot \varphi \otimes \pi(h)v) \\ &= f((g, h)\varphi \otimes (\pi(h)v, 0)) \\ &= f((g, h)\varphi \otimes \sigma(h)(v, 0)) \\ &= g \cdot f(\varphi \otimes (v, 0)) \\ &= g \cdot N(f)(\varphi \otimes v). \end{aligned}$$

Let $f, f' \in \mathrm{Hom}_R(\mathcal{S}(X^2) \otimes (V \times V), W)$ and assume that $N(f) = N(f')$. Then for all $\varphi \in \mathcal{S}(X^2)$ and $v \in V$

$$\begin{aligned} N(f) &= N(f'), \quad \text{so that} \\ f(\varphi \otimes (v, 0)) &= f'(\varphi \otimes (v, 0)), \\ (1, s) \cdot f(\varphi \otimes (v, 0)) &= (1, s) \cdot f'(\varphi \otimes (v, 0)), \quad \text{and} \\ f(\varphi \otimes (0, v)) &= f'(\varphi \otimes (0, v)). \end{aligned}$$

We conclude that $f = f'$, so that N is one-to-one.

Next we verify that $N = M^{-1}$.

$$\begin{aligned} \big[(M \circ N)(f)\big](\varphi \otimes (v_1, v_2)) &= N(f)(\varphi \otimes v_1) + N(f)(\omega(1, s) \cdot \varphi \otimes v_2) \\ &= f(\varphi \otimes v_1, 0) + f(\omega(1, s) \cdot \varphi \otimes (v_2, 0)) \\ &= f(\varphi \otimes v_1, 0) + f(\varphi \otimes (0, v_2)) \\ &= f(\varphi \otimes (v_1, v_2)). \end{aligned}$$

So we have that $M \circ N = \mathrm{Id}$. On the other hand,

$$\begin{aligned} \big[(N \circ M)(f)\big](\varphi \otimes v) &= M(f)(\varphi \otimes (v, 0)) \\ &= f(\varphi \otimes v) + f(\omega(1, s) \cdot \varphi \otimes 0) \\ &= f(\varphi \otimes v). \end{aligned}$$

Since M is an isomorphism, so is N. □

From Theorem 4.7 and Lemma 4.10 we can easily establish the following result.

Corollary 4.11 *Let* (π, V) *be the representation of* $\mathrm{GSO}(X)$ *as in Sect. 3 so that* V *is either equal to* $\mathcal{W}(\tau_1, \psi) \otimes \mathcal{W}(\tau_2, \psi)$ *or equal to* $\mathcal{W}(\tau_0, \psi_E)$. *Let* $(\sigma, V \times V) = \mathrm{Ind}_{\mathrm{GSO}(X)}^{\mathrm{GO}(X)} \pi$ *and let* π^+ *be the canonical irreducible subrepresentation of* σ. *Let* ϑ *be as in* (4.5) *and let* M *be the map in lemma 4.10. The composition*

$$M(\vartheta) : \mathcal{S}(X^2) \otimes (V \times V) \to \Theta(V)$$

is a non-zero R*-map. Furthermore, the restriction of* $M(\vartheta)$ *to* π^+ *is a non-zero* R*-map.*

Proof By part (a) of Theorem 4.7 we have that $\vartheta \in \mathrm{Hom}_{R'}(\mathcal{S}(X^2) \otimes V, \Theta(V))$ so that $M(\vartheta) \in \mathrm{Hom}_R(\mathcal{S}(X^2) \otimes (V \times V), \Theta(V))$. Furthermore $M(\vartheta)$ is non-zero by Lemma 4.4. Indeed, for $\varphi \in \mathcal{S}(X^2)$ and $W, W' \in V$ we have that

$$\begin{aligned} M(\vartheta)(\varphi \otimes (W, W')) &= \vartheta(\varphi \otimes W) + \vartheta(\omega(1, s) \cdot \varphi \otimes W') \\ &= B(\cdot, \varphi, W, s) + B(\cdot, \omega(1, s)\varphi, T(W'), s) \\ &= B(\cdot, \varphi, W, s) + B(\cdot, \varphi, T(W'), s). \end{aligned}$$

Since τ_i has trivial central character we have that π^+ has space $V_{\pi^+} = \{(v, T(v)) \mid v \in V\}$, where T is as in Lemma 4.4. Using part (a) of Lemma 4.4 we calculate that $M(\vartheta)(\mathcal{S}(X^2) \otimes V_{\pi^+}) \neq 0$ since for $\varphi \in \mathcal{S}(X^2)$ and $W \in V$

$$M(\vartheta)(\varphi \otimes (W, \pi(s)W)) = 2B(\cdot, \varphi, W, s),$$

which is not zero by Theorem 4.7. □

4.2 The Archimedean Case

In this section we make the assumption that $L = \mathbb{R}$ and that $X = X_M$ is split, so that $E = \mathbb{R} \times \mathbb{R}$. Fix the Lie algebras $\mathfrak{h}, \mathfrak{g}, \mathfrak{r}$ of $\mathrm{GO}(X(L))$, $\mathrm{GSp}(4, L)$, and $R(L)$, respectively, and maximal compact subgroups $J \subset \mathrm{GO}(X(L))$, $K \subset \mathrm{GSp}(4, L)$, and $F \subset R$. Let $(\omega, \mathcal{S}(X^2))$ be the $(\mathfrak{r}, F)$-module given by the Archimedean Weil representation. Let τ_1, τ_2 be infinite-dimensional, irreducible, admissible $(\mathfrak{g}, K)$-modules for $\mathrm{GL}(2, \mathbb{R})$ that have the same central character, but are not isomorphic. Let $\mathcal{W}(\tau_i, \psi)$ be the Whittaker model for τ_i. Set $V = \mathcal{W}(\tau_1, \psi) \otimes \mathcal{W}(\tau_2, \psi)$. Let $W = W_1 \otimes W_2 \in V$ and let $Z(s, W)$ be defined as in (3.3).

The primary purpose of this section is to define the intertwining map analogous to (4.1), in the Archimedean case, and to verify some necessary results. In particular we want to establish that the integral that defines the intertwining map is absolutely convergent, non-zero, and corresponds to an intertwining map in the theta correspondence. Let $g \in \mathrm{GSp}(4, \mathbb{R})^+$ and $\varphi \in \mathcal{S}(X^2)$. Let $W \in V$. We define

$$B(g, \varphi, W, s) = \int\limits_{H \backslash \mathrm{SO}(X)} (\omega(g, hh')\varphi)(x_1, x_2) Z(s, \pi(hh')W)\, dh \qquad (4.9)$$

where $h' \in \mathrm{GSO}(X)$ is chosen so that $\lambda(h') = \lambda(g)$.

Lemma 4.12 *For $g \in \mathrm{GSp}(4, \mathbb{R})^+$, $\varphi \in \mathcal{S}(X^2)$, $W = W_1 \otimes W_2 \in V$ such that $Z(s, W) \neq 0$, and $s > \frac{3}{2}$ the integral in* (4.9)*, defining $B(g, \varphi, W, s)$, converges absolutely. In fact*

$$\int\limits_{H(\mathbb{R}) \backslash \mathrm{SO}(X,\mathbb{R})} \left|\big(\omega(g, hh')\varphi\big)(x_1, x_2)\right| \cdot Z(\Re(s), |\pi(hh')W|)\, dh \qquad (4.10)$$

is finite.

Proof This proof appears in the unpublished manuscript [Rob03]. Since we are in the split case we have that the integral (4.10) is

$$I = \int_{H(\mathbb{R}) \backslash \mathrm{SO}(X,\mathbb{R})} \left|\big(\omega(1, h)\varphi'\big)(x_1, x_2)\right| \cdot$$
$$\int\limits_{\mathbb{R}^\times} |x|^{\Re(s)-\frac{1}{2}} |W_1([\begin{smallmatrix} x & \\ & 1 \end{smallmatrix}] h_1)]|\, d^\times x \int\limits_{\mathbb{R}^\times} |x|^{\Re(s)-\frac{1}{2}} |W_2([\begin{smallmatrix} x & \\ & 1 \end{smallmatrix}] h_2)|\, d^\times x\, dh$$

where $h = \rho(h_1, h_2)$ and $\varphi' = \omega(g, h')\varphi$. Let T be the group $\{[\begin{smallmatrix} y & \\ & y^{-1} \end{smallmatrix}],\ y \in \mathbb{R}^\times\}$ and let $\Delta T = \{(t, t^{-1}) \mid t \in T\}$. Then ρ produces the following homeomorphism

$$\rho : \Delta T \backslash (\mathrm{SL}(2, \mathbb{R}) \times \mathrm{SL}(2, \mathbb{R})) \xrightarrow{\sim} H(\mathbb{R}) \backslash \mathrm{SO}(X, \mathbb{R}).$$

We have $\Delta \subset T \times T \subset \mathrm{SL}(2, \mathbb{R}) \times \mathrm{SL}(2, \mathbb{R})$. Moreover, ΔT and $T \times T$ are closed unimodular subgroups of $\mathrm{SL}(2, \mathbb{R})$ and $\mathrm{SL}(2, \mathbb{R})$ is unimodular. Therefore, we can write (4.10) as

$$I = \int\limits_{T \backslash \mathrm{SL}(2,\mathbb{R}) \times T \backslash \mathrm{SL}(2,\mathbb{R})} \int\limits_{T} \Bigg(\left|\big(\omega(1, \rho(th_1, h_2))\varphi'\big)(x_1, x_2)\right| \int\limits_{\mathbb{R}^\times} |x|^{\Re(s)-\frac{1}{2}} |W_1([\begin{smallmatrix} x & \\ & 1 \end{smallmatrix}] th_1)|\, d^\times x$$
$$\int\limits_{\mathbb{R}^\times} |x|^{\Re(s)-\frac{1}{2}} |W_2([\begin{smallmatrix} x & \\ & 1 \end{smallmatrix}] h_2)|\, d^\times x \Bigg)\, dt\, dh_1\, dh_2.$$

Let $N = \{[\begin{smallmatrix} 1 & z \\ & 1 \end{smallmatrix}],\ z \in \mathbb{R}\}$. Then, a standard integration formula gives

$$I = \int\limits_{\mathrm{SO}(2,\mathbb{R})\times\mathrm{SO}(2,\mathbb{R})} \int\limits_{N\times N} \int\limits_{T} \Big(|(\omega(1, \rho(tn_1k_1, n_2k_2))\varphi')(x_1, x_2)| \cdot \tag{4.11}$$

$$\int\limits_{\mathbb{R}^\times} |x|^{\Re(s)-\frac{1}{2}} |W_1([\begin{smallmatrix} x & \\ & 1 \end{smallmatrix}]t_1n_1k_1)|\, d^\times x$$

$$\int\limits_{\mathbb{R}^\times} |x|^{\Re(s)-\frac{1}{2}} |W_2([\begin{smallmatrix} x & \\ & 1 \end{smallmatrix}]n_2k_2)|\, d^\times x \Big)\, dt\, dn_1\, dn_2\, dk_1\, dk_2.$$

This is an improvement since N and T are explicitly defined. Since $\mathcal{W}(\tau_i, \psi)$ is $\mathrm{SO}(2, \mathbb{R})$-finite, the subspace spanned by $\mathrm{SO}(2, \mathbb{R})$-translates of W_i, for $i \in \{1, 2\}$, is a finite dimensional vector space. Therefore we can choose a positive integer n and functions $W_{1,i}, W_{2,j} \in \mathcal{W}(\pi, \psi)$, for $1 \le i, j \le n$, so that $W_1 = \sum_{i=1}^n W_{1,i}$ and $W_2 = \sum_{j=1}^n W_{2,j}$, where each $W_{1,i}$ and $W_{2,j}$ transforms according to a character of $\mathrm{SO}(2, \mathbb{R})$. Let the functions $F_{i,j} : \mathrm{SO}(X, \mathbb{R}) \to \mathbb{C}$ be the inner integrals of (4.11) but with $W_{1,i}$ and $W_{2,j}$ replacing W_1 and W_2, respectively. With this (4.10) is

$$\le \sum_{\substack{1\le i\le n \\ 1\le j\le n}} \int\limits_{\mathrm{SO}(2,\mathbb{R})\times\mathrm{SO}(2,\mathbb{R})} \int\limits_{N\times N} \int\limits_{T} F_{i,j}(\rho(tn_1k_1, n_2k_2))\, dt\, dn_1\, dn_2\, dk_1\, dk_2.$$

For any pair (i, j), with $1 \le i \le n$ and $1 \le j \le n$, we have

$$\int\limits_{\mathrm{SO}(2,\mathbb{R})\times\mathrm{SO}(2,\mathbb{R})} \int\limits_{N\times N} \int\limits_{T} F_{i,j}(\rho(tn_1k_1, n_2k_2))\, dt\, dn_1\, dn_2\, dk_1\, dk_2$$

$$= \int\limits_{\mathrm{SO}(2,\mathbb{R})\times\mathrm{SO}(2,\mathbb{R})} \int\limits_{\mathbb{R}\times\mathbb{R}} \int\limits_{\mathbb{R}^\times} F_{i,j}(\rho([\begin{smallmatrix} t & \\ & t^{-1} \end{smallmatrix}][\begin{smallmatrix} 1 & z_1 \\ & 1 \end{smallmatrix}]k_1, [\begin{smallmatrix} 1 & z_2 \\ & 1 \end{smallmatrix}]k_2) \frac{dt}{|t|}\, dz_1\, dz_2\, dk_1\, dk_2$$

$$= \int\limits_{\mathbb{R}\times\mathbb{R}} \int\limits_{\mathbb{R}^\times} \int\limits_{\mathrm{SO}(2,\mathbb{R})\times\mathrm{SO}(2,\mathbb{R})} |(\omega(1, \rho(k_1, k_2))\varphi')([\begin{smallmatrix} & t^{-1} \\ & \end{smallmatrix}], -2[\begin{smallmatrix} z_1t & z_1z_2 \\ t & tz_2 \end{smallmatrix}])|\, dk_1\, dk_2 \cdot$$

$$(\int\limits_{R^\times} |x|^{\Re(s)-\frac{1}{2}} |W_{1,i}([\begin{smallmatrix} x & \\ & 1 \end{smallmatrix}][\begin{smallmatrix} t & \\ & t^{-1} \end{smallmatrix}][\begin{smallmatrix} 1 & z_1 \\ & 1 \end{smallmatrix})]|\, d^\times x)$$

$$(\int\limits_{R^\times} |x|^{\Re(s)-\frac{1}{2}} |W_{2,j}([\begin{smallmatrix} x & \\ & 1 \end{smallmatrix}][\begin{smallmatrix} 1 & z_2 \\ & 1 \end{smallmatrix}])|\, d^\times x) \frac{dt}{|t|}\, dz_1\, dz_2.$$

Since each $W_{u,i}$ is in $\mathcal{W}(\pi, \psi)$ and transforms according to a character of $\mathrm{SO}(2, \mathbb{R})$, it can be shown that there are positive constants $C_{1,i}, C_{2,j}$ so that the above is

$$I \leq C_{1,i} C_{2,j} \int\limits_{\mathrm{SO}(2,\mathbb{R})^2} \int\limits_{\mathbb{R}\times\mathbb{R}} \int\limits_{\mathbb{R}^\times} |(\omega(1, \rho(k_1, k_2))\varphi')([\begin{smallmatrix} & t^{-1} \\ & \end{smallmatrix}], -2[\begin{smallmatrix} z_1 t & z_1 z_2 \\ t & t z_2 \end{smallmatrix}])| \\ \frac{dt}{|t|} dz_1\, dz_2\, dk_1\, dk_2.$$

Define $C = \sum_{1 \leq i, j \leq n} C_{1,i} C_{2,j}$ so that (4.10) is

$$I \leq C \int\limits_{\mathrm{SO}(2,\mathbb{R})^2} \int\limits_{\mathbb{R}\times\mathbb{R}} \int\limits_{\mathbb{R}^\times} |(\omega(1, \rho(k_1, k_2))\varphi')([\begin{smallmatrix} & t^{-1} \\ & \end{smallmatrix}], -2[\begin{smallmatrix} z_1 t & z_1 z_2 \\ t & t z_2 \end{smallmatrix}])| \\ \frac{dt}{|t|} dz_1\, dz_2\, dk_1\, dk_2.$$

Let $L, M,$ and N be positive integers which we will determine momentarily. Define a polynomial on X by

$$Q([\begin{smallmatrix} x_1 & x_2 \\ x_3 & x_4 \end{smallmatrix}], [\begin{smallmatrix} x_1' & x_2' \\ x_3' & x_4' \end{smallmatrix}]) = (1 + (\frac{1}{2}x_3')^{2N} + x_2^{2M})(1 + (\frac{1}{2}x_1' x_2)^{2L} + (\frac{1}{2}x_2 x_4')^{2L})$$

where x_i and x_i' in L for $i \in \{1, 2\}$. Let $|| \cdot ||_Q$ be the semi-norm corresponding to Q, on the space of $\mathcal{S}(X^2)$, given explicitly by

$$||\varphi||_Q = \sup_{x \in X^2} |Q(x)\varphi(x)|.$$

Then we have that the integral from (4.10) is bounded as

$$\begin{aligned} I \leq & C \int\limits_{\mathrm{SO}(2,\mathbb{R})^2} ||\omega(1, \rho(k_1, k_2))\varphi'||_Q \int\limits_{\mathbb{R}\times\mathbb{R}} \int\limits_{\mathbb{R}^\times} \frac{|t|^{-2\Re(s-\frac{1}{2})-1}}{Q([\begin{smallmatrix} & t^{-1} \\ & \end{smallmatrix}], -2[\begin{smallmatrix} -z_1 t & -z_1 z_2 \\ t & t z_2 \end{smallmatrix}])} dt\, dz_1\, dz_2\, dk_1\, dk_2 \\ = & C \int\limits_{\mathrm{SO}(2,\mathbb{R})^2} ||\omega(1, \rho(k_1, k_2))\varphi'||_Q\, dk_1\, dk_2 \int\limits_{\mathbb{R}\times\mathbb{R}} \frac{1}{1 + |z_1|^{2L} + |z_2|^{2L}} dz_1\, dz_2 \\ & \int\limits_{\mathbb{R}^\times} \frac{|t|^{2M-2\Re(s-\frac{1}{2})-1}}{1 + |t|^{2(N+M)} + |t|^{2M}} dt. \end{aligned}$$

We choose L to be large enough so that the above integral over $\mathbb{R} \times \mathbb{R}$ converges. Also, we choose M to be large enough so that $2M - 2(\Re(s - \frac{1}{2})) - 1$ is positive and then choose N large enough so that the above integral over $\mathbb{R}^\times$ converges. Therefore, there is a constant $C' \in \mathbb{R}$ so that $I \leq C' \int_{\mathrm{SO}(2,\mathbb{R})^2} ||\omega(1, \rho(k_1, k_2))\varphi'||_Q\, dk_1\, dk_2$.

Define an action of $\mathrm{SO}(2, \mathbb{R})^2$ on polynomials on X_M by $(k_1, k_2)P(y) = R(\rho(k_1, k_2)^{-1} y)$. There exist an integer m and polynomials $\{Q_i\}_{i=1}^m$ which all transform according to a character of $\mathrm{SO}(2,\mathbb{R})^2$ such that $Q=\sum_{i=1}^m Q_i$. Furthermore, for all $(k_1,k_2)\in\mathrm{SO}(2,\mathbb{R})^2$, we have

$$||\omega(1,\rho(k_1,k_2))\varphi'||_Q\leq\sum_{i=1}^m||\varphi'||_{Q_i}.$$

Therefore we conclude that integral (4.10) is bounded above by $C'\mathrm{vol}(\mathrm{SO}(X,\mathbb{R}))\sum_{i=1}^m||\omega(g,h')\varphi||_{Q_i}$ which is finite. □

Lemma 4.13 *Suppose that $W\in\mathcal{V}$ such that $Z(s,W)\neq0$ and for $s\gg M$. There exists $\varphi\in\mathcal{S}(X^2)$ such that $B(1,\varphi,W,s)\neq0$.*

Proof There certainly exists a smooth rapidly decreasing function $\bar{\varphi}:X^2\to\mathbb{C}$ so that the Bessel integral $B(1,\bar{\varphi},W,s)\neq0$. Since $\mathcal{S}(X^2)$ is dense in the Schwartz space on X^2 there exists a sequence $\{\varphi_i\}_{i\geq1}\subset\mathcal{S}(X^2)$ so that $\varphi_i\to\bar{\varphi}$. As in the proof of Lemma 4.12 we have that

$$|B(1,\bar{\varphi},W,s)-B(1,\varphi_n,W,s)|\leq CC'\mathrm{vol}(2,\mathbb{R})^2\sum_{i=1}^m||\varphi-\varphi_n||_{Q_i}.$$

Since $B(1,\bar{\varphi},W,s)\neq0$ there is some integer n so that $B(1,\varphi_n,W,s)\neq0$. □

Now we are sure that $B(\cdot,\varphi,W,s)$ is well defined and non-zero, for some choice of $\varphi\in\mathcal{S}(X^2)$, $W\in V$, and $s\in\mathbb{C}$. We define $\Theta(V)$ to be the $(\mathfrak{g},K)$-module, for $\mathrm{GSp}(4,L)$, of smooth functions generated by $B(\cdot,\varphi,W,s)$ for all choices of $\varphi\in\mathcal{S}(X^2)$, $W\in V$, and $\Re(s)>\frac{3}{2}$. We define the map

$$\begin{aligned}\vartheta:\mathcal{S}(X^2)\otimes V&\to\Theta(V)\\ \varphi\otimes W&\mapsto B(\cdot,\varphi,W,s).\end{aligned}\tag{4.12}$$

If $B\in\Theta(V)$ then by Remark 4.5 we are justified in using Lemma 4.4 to verify that

$$B(tg)=|t_1/t_2|^{\frac{1}{2}-s}B(g)\quad\text{for } t=\begin{bmatrix}t_1&&&\\&t_2&&\\&&t_2&\\&&&t_1\end{bmatrix}\text{ for } t_1,t_2\in L^\times,\text{ and}\tag{4.13}$$

$$B(bg)=\psi(b_2)B(g)\quad\text{for } b=\left[\begin{smallmatrix}1&B\\&1\end{smallmatrix}\right]\text{ where } B=\left[\begin{smallmatrix}b_1&b_2\\b_2&b_3\end{smallmatrix}\right]\in\mathrm{M}(2,L).\tag{4.14}$$

With this in place we can extend the results of Theorem 4.7 to the Archimedean case.

Theorem 4.14 *Let $s\in\mathbb{C}$ with $\Re(s)>\frac{3}{2}$. Let $V=\mathcal{W}(\tau_1,\psi)\otimes\mathcal{W}(\tau_2,\psi)$. Then*

(a) The map ϑ from (4.12) is a non-zero $(\mathfrak{r},F')$-map.
(b) The image of φ lies inside $\mathcal{B}(\mathrm{GSp}(4,\mathbb{R}),\psi)$.

It follows that $\Theta(V)$ is a $(\mathfrak{r},F')$-module sitting inside $\mathcal{B}(\mathrm{GSp}(4,\mathbb{R}),\psi)$.

Proof It follows that ϑ is an R-map from the simple calculation: if $(g_0, h_0) \in R'$, $(g, h') \in R$, $\varphi \in \mathcal{S}(X^2)$, $W \in V$, and $\Re(S) > \frac{3}{2}$ then

$$
\begin{aligned}
B(g, \omega(g_0, h_0)\varphi, \pi(h_0)W, s) &= \int_{H\backslash \mathrm{SO}(X)} \omega(gg_0, h(h'h_0))\varphi(x_1, x_2)Z(s, \pi(h(h'h_0))W)\, dh \\
&= B(gg_0, \varphi, W, s)
\end{aligned}
$$

since $(gg_0, h'h_0) \in R$. It follows that ϑ induces a $(\mathfrak{r}\,, F')$-equivariant map.

Part (b) follows from the definition of $\Theta(V)$ and from (4.13) and (4.14). □

Lastly, we need to extend the results of Corollary 4.11 to the case when $L = \mathbb{R}$.

Lemma 4.15 *Let (π, V) be a $(\mathfrak{h}, J_1)$-module, let $(\sigma, V \times V)$ be the induced $(\mathfrak{h}, J)$-module obtained from π. Let (Π, W) be a $(\mathfrak{g}, K)$-module. Then we have the following $\mathbb{C}$-linear isomorphism*

$$
M : \mathrm{Hom}_{(\mathfrak{r}, F')}(\mathcal{S}(X^2) \otimes V, W) \xrightarrow{\sim} \mathrm{Hom}_{(\mathfrak{r}, F)}(\mathcal{S}(X^2) \otimes (V \times V), W)
$$

determined by $M(f)(\varphi \otimes (v_1, v_2)) = f(\varphi \otimes v_1) + f(\omega(1, s) \cdot \varphi \otimes v_2)$ for $\varphi \in \mathcal{S}(X^2)$, $f \in \mathrm{Hom}_{(\mathfrak{r}, K')}(\mathcal{S}(X^2) \otimes V, W)$, and $v_1, v_2 \in V$, and extended linearly. Here $(1, s)$ is a non-trivial coset representative of R/R'. For example we could take s the map that takes x to x^*. Additionally the inverse map

$$
N : \mathrm{Hom}_{(\mathfrak{r}, F)}(\mathcal{S}(X^2) \otimes (V \times V), W) \xrightarrow{\sim} \mathrm{Hom}_{(\mathfrak{r}, F')}(\mathcal{S}(X^2) \otimes V, W)
$$

is given by $N(f)(\varphi \otimes v) = f(\varphi \otimes (v, 0))$ for $f \in \mathrm{Hom}_{(\mathfrak{r}, K)}(\mathcal{S}(X^2) \otimes (V \times V), W)$, $\varphi \in \mathcal{S}(X^2)$, and $v \in V$.

Proof This is similar to the proof of Lemma 4.10. □

Corollary 4.16 *Let (π, V) be the $(\mathfrak{h}, J')$-module associated to the representation of $\mathrm{GSO}(X)$ as in Sect. 3 so that V is equal to $\mathcal{W}(\tau_1, \psi) \otimes \mathcal{W}(\tau_2, \psi)$. Let $(\sigma, V \times V)$ be the $(\mathfrak{h}, J)$-module, associated to the representation of $\mathrm{GO}(X)$ that is isomorphic to $\mathrm{Ind}_{\mathrm{GSO}(X)}^{\mathrm{GO}(X)} \pi$. We are justified in applying this result to the real case. Let π^+ be the canonical irreducible subrepresentation of σ. Let ϑ be as in (4.12) and let M be the map in 4.15. The composition*

$$
M(\vartheta) : \mathcal{S}(X^2) \otimes (V \times V) \to \Theta(V)
$$

is a non-zero $(\mathfrak{r}, F)$-map. Furthermore, the restriction of $M(\vartheta)$ to π^+ is a non-zero $(\mathfrak{r}, F)$-map.

Proof It is clear, from what is presented in Corollary 4.11, that $M(\vartheta)|_{\pi^+}$ is a non-zero $(\mathfrak{r}, F)$-equivariant map. □

5 The Explicit Schwartz Functions

In this section, we construct explicit Schwartz functions, $\varphi \in \mathcal{S}(X^2)$, which produce a non-zero paramodular invariant vector. We will work separately in the case that E/L is split, inert, and tamely ramified. Unfortunately, when E/L is wildly ramified we do not find any such Schwartz functions. For our global application, we note that the only real quadratic number field for which we do not attain an explicit local lift at every place is $\mathbb{Q}(\sqrt{2})$.

In the split case, assume that $W \in V$ is $\Gamma_0(\mathfrak{p}^{n_1}) \times \Gamma_0(\mathfrak{p}^{n_2})$-invariant for some non-negative integers n_1, n_2. In the non-split case assume that $W \in V$ is $\Gamma_0(\mathfrak{P}^n)$-invariant for some non-negative integers n. For any $\varphi \in \mathcal{S}(X^2)$, let $B(\cdot, \varphi, W, s)$ be as in Sect. 4.

There are three important considerations. First, we want the support of φ to be simple, with regard to $Z(s, W)$. To be specific, if $h \in H \backslash \mathrm{SO}(X)$ so that $(h^{-1}x_1, h^{-1}x_2) \in \mathrm{supp}(\varphi)$ then we want $Z(s, \pi(h)W)$ to be simple to compute. When L has odd residual characteristic we are able to make choices so that $Z(s, \varphi(h)W) = Z(s, W)$, for all such h. Our second consideration is invariance under the Weil representation for $(g, h) \in R \cap (K(\mathfrak{p}^N) \times \mathrm{GSO}(X))$. When E/L is unramified these first two goals have almost perfect overlap, while if E/L is tamely ramified there is still enough overlap to create good candidate Schwartz functions.

If we make a natural choice of Schwartz function in the ramified case we do not get full paramodular invariance in the Weil representation, but we do get invariance under a rather large subgroup of $K(\mathfrak{p}^N)$. We can sum over coset representatives of this subgroup to produce a fully paramodular invariant Schwartz function. Of course, we must verify that our choice of φ does not make $B(\cdot, \varphi, W, s) = 0$. More specifically, we are able to verify that $B(1, \varphi, W, s)$ is a non-zero multiple of $Z(s, W)$.

5.1 The Split Case

Suppose that $X = X_M$ as in Sect. 3. Let τ_1 and τ_2 be irreducible admissible representations of $\mathrm{GL}(2, L)$ with trivial central character. Suppose the space of τ_i is its Whittaker model $\mathcal{W}_{\tau_i}$ and further suppose that there are non-negative integers n_i such that $W_i \in \mathcal{W}_{\tau_i}$ that is invariant under $\Gamma_0(\mathfrak{p}^{n_i})$ for $i \in \{1, 2\}$. Set $W = W_1 \otimes W_2$. Let φ_1 and φ_2 be in $\mathcal{S}(X)$ and let $\varphi = T(\varphi_1 \otimes \varphi_2) \in \mathcal{S}(X^2)$. Set $N = n_1 + n_2$, which we will show is the paramodular level of the lift. Using the Lemmas 3.11 and 3.9 we can easily calculate the action of the generators of $\mathrm{K}(\mathfrak{p}^N)$ on φ. With these calculations in mind we can specify φ_1 and φ_2 so that $B(\cdot, \varphi, W, s)$ is paramodular invariant and non-zero. Choose

$$\varphi_1 = f_{\begin{bmatrix} \mathfrak{p}^{n_2} & \mathfrak{o}_L \\ \mathfrak{p}^N & \mathfrak{p}^{n_1} \end{bmatrix}} \quad \text{and} \quad \varphi_2 = f_{\mathrm{M}(2, \mathfrak{o}_L)}, \tag{5.1}$$

where f_S is the characteristic function of a set S.

Lemma 5.1 *Let φ be as in* (5.1). *For every $(k, h) \in R$ such that $k \in \mathrm{K}(\mathfrak{p}^N)$ we have that $\omega(k, h)\varphi = \varphi$.*

Proof Let us go through the generators listed in Lemma 3.9.

(a) Let $A = [\begin{smallmatrix} a_1 & a_2 \\ a_3 & a_4 \end{smallmatrix}] \in \Gamma_0(\mathfrak{p}^N) \subset \mathrm{GL}(2, \mathfrak{o}_L)$ so that $a_1, a_4 \in \mathfrak{o}_L^\times$, $a_3 \in \mathfrak{p}^N$ and $a_2 \in \mathfrak{o}_L$. Note that $A^{-1} = \frac{1}{\det(A)} \left[\begin{smallmatrix} a_4 & -a_2 \\ -a_3 & a_1 \end{smallmatrix}\right] \in \Gamma_0(\mathfrak{p}^N)$. If $(x, y) \in \mathrm{supp}(\varphi)$ then

$$a_1 x + a_3 y \in [\begin{smallmatrix} \mathfrak{p}^{n_2} & \mathfrak{o}_L \\ \mathfrak{p}^N & \mathfrak{p}^{n_1} \end{smallmatrix}] \quad \text{and} \quad a_2 x + a_4 y \in \mathrm{M}(2, \mathfrak{o}_L)$$

so that $(a_1 x + a_3 y, a_2 x + a_4 y) \in \mathrm{supp}(\varphi)$. On the other hand, if $(a_1 x + a_3 y, a_2 x + a_4 y) \in \mathrm{supp}(\varphi)$ then

$$\begin{aligned} &A^{-1} \cdot (a_1 x + a_3 y, a_2 x + a_4 y) \\ &= \frac{1}{\det(A)} (a_4(a_1 x + a_3 y) + -a_3(a_2 x + a_4 y), \\ &\quad - a_2(a_1 x + a_3 y) + a_1(a_2 x + a_4 y)) \\ &= (x, y). \end{aligned}$$

Since $A^{-1} \in \Gamma_0(\mathfrak{p}^N)$ we conclude that $(x, y) \in \mathrm{supp}(\varphi)$. So, $(x, y) \in \mathrm{supp}(\varphi)$ if and only if $(a_1 x + a_3 y, a_2 x + a_4 y) \in \mathrm{supp}(\varphi)$. Since $\det(A) = a_1 a_4 - a_2 a_3 \in \mathfrak{o}_L^\times$ and $\chi_{E/L}$ is trivial in the split case we see that

$$\begin{aligned} \omega([\begin{smallmatrix} A & \\ & {}^t A^{-1} \end{smallmatrix}], 1)\varphi(x, y) &= \chi_{E/L}(\det A) |\det A|^2 \varphi(a_1 x + a_3 y, a_2 x + a_4 y) \\ &= \varphi(x, y). \end{aligned}$$

Let $u \in \mathfrak{o}_L^\times$ and set

$$g_u = \begin{bmatrix} 1 & & & \\ & 1 & & \\ & & u & \\ & & & u \end{bmatrix} \quad \text{and} \quad h_u = \rho([\begin{smallmatrix} u & \\ & 1 \end{smallmatrix}], [\begin{smallmatrix} 1 & \\ & 1 \end{smallmatrix}])$$

so that $(g_u, h_u) \in R$. Since $u \in \mathfrak{o}_L^\times$ we have $(h_u^{-1} x, h_u^{-1} y) \in \mathrm{supp}(\varphi)$ if an only if $(x, y) \in \mathrm{supp}(\varphi)$. Hence,

$$\begin{aligned} \omega(g_u, h_u)\varphi(x, y) &= |u|^{-2} \omega(1, 1)\varphi(h_u^{-1} x, h_u^{-1} y) \\ &= \varphi(x, y). \end{aligned}$$

(b) Let $B = [\begin{smallmatrix} b_1 & b_2 \\ b_2 & b_3 \end{smallmatrix}] \in [\begin{smallmatrix} \mathfrak{p}^{-N} & \mathfrak{o}_L \\ \mathfrak{o}_L & \mathfrak{o}_L \end{smallmatrix}]$. If $(x, y) \in \mathrm{supp}(\varphi)$ then $\langle x, x \rangle \in \mathfrak{p}^N$ and $\langle x, y \rangle$, $\langle y, y \rangle \in \mathfrak{o}_L$. Hence,

$$\begin{aligned} \omega([\begin{smallmatrix} 1 & B \\ & 1 \end{smallmatrix}], 1)\varphi(x, y) &= \psi(b_1 \langle x, x \rangle + 2b_2 \langle x, y \rangle + b_3 \langle y, y \rangle)\varphi(x, y) \\ &= \varphi(x, y). \end{aligned}$$

(c) In the proof of Lemma 3.10 we verified that $\mathcal{F}_1(\varphi_2) = \varphi_2$. Now, using Lemma 3.11 we find that

$$\begin{aligned}\omega(s_2, 1)\varphi(x, y) &= \omega_1(1, 1)\varphi_1(x)\omega_1([\,_{-1}{}^{1}], 1)\varphi_2(y)\\ &= \varphi_1(x)\mathcal{F}_1(\varphi_2)(y)\\ &= \varphi(x, y).\end{aligned}$$

(d) To determine $\omega(t_N, 1)$ we need the following preliminary calculation. Let $\varphi_1^a(x) = \varphi_1(ax)$. We claim that $\mathcal{F}_1(\varphi_1^{\varpi^N})(x) = q^{2N}\varphi_1(x)$. The simpliest way to do this is to use (3.10) along with the corresponding considerations for the Haar measure made in Lemma 3.10. With these in mind we calculate that

$$\begin{aligned}\mathcal{F}_1(\varphi_1^{\varpi^N})(x) &= \int_X \varphi_1(\varpi^N y)\psi(2\langle x, y\rangle)\, dy = \int_{\left[\begin{smallmatrix}\mathfrak{p}^{-n_1} & \mathfrak{p}^{-N}\\ \mathfrak{o}_L & \mathfrak{p}^{-n_2}\end{smallmatrix}\right]} \psi(2\langle x, y\rangle)\, dy\\ &= \int_{\mathfrak{p}^{-n_1}} \psi(x_4y_1)\, dy_1 \int_{\mathfrak{p}^{-N}} \psi(-x_3y_2)\, dy_2 \int_{\mathfrak{o}_L} \psi(-x_2y_3)\, dy_3 \int_{\mathfrak{p}^{-n_2}} \psi(x_1y_4)\, dy_4.\end{aligned}$$

By Lemma 3.8, we conclude that the above is $\mathcal{F}_1(\varphi_1^{\varpi^N})(x) = q^{2N}\varphi_1(x)$. Now, again using the seesaw embedding in Lemma 3.11 we have

$$\begin{aligned}\omega(t_n, 1)\varphi(x, y) &= \omega_1([\,_{-\varpi^N}{}^{\varpi^{-N}}], 1)\varphi_1(x)\varphi_2(y)\\ &= q^{-2N}\mathcal{F}_1(\varphi_1^{\varpi^N})(x)\varphi_2(y)\\ &= \varphi(x, y).\end{aligned}$$

This completes the proof. □

Corollary 5.2 *For all* $k \in \mathrm{K}(\mathfrak{p}^N)$ *and all* $g \in \mathrm{GSp}(4, L)$ *we have that* $B(gk, \varphi, W, s) = B(g, \varphi, W, s)$.

Proof When $\lambda(k) = 1$ the result follows from Lemma 5.1, the definition of $B(\cdot, \varphi, W, s)$, and the fact that for every $u \in \mathfrak{o}_L^\times$ there is some $b \in \Gamma_0(\mathfrak{p}^n)$ such that $\det(b) = u$. For a general $k \in K(\mathfrak{p}^N)$ we have to do a bit more work. Suppose that $\lambda(k) = u \in \mathfrak{o}_L^\times$ and let $g \in \mathrm{GSp}(4, L)$ and let $g_1 \in \mathrm{GSp}(4, L)$ be as in (4.4) then

$$B(gk) = |\lambda(g)|^{-s+\frac{1}{2}} B(g_2k_1) = |\lambda(g)|^{-s+\frac{1}{2}} B(g_2)$$

where

$$k_1 = \begin{bmatrix} u^{-1} & & & \\ & 1 & & \\ & & 1 & \\ & & & u^{-1}\end{bmatrix} k \quad\text{and}\quad g_2 = \begin{bmatrix} u^{-1} & & & \\ & 1 & & \\ & & 1 & \\ & & & u^{-1}\end{bmatrix} g_1 \begin{bmatrix} u & & & \\ & 1 & & \\ & & 1 & \\ & & & u\end{bmatrix}.$$

It remains to verify that $B(g_2) = B(g_1)$. To do this we only need to check that $\omega(g_1, 1)\varphi = \omega(g_2, 1)\varphi$ for the φ chosen in (5.1) and for each of the generators with similitude factor equal to 1, namely the elements of $\mathrm{GSp}(4, L)$ found in (3.15), (3.16), and (3.17).

(a) Suppose that

$$g_1 = \begin{bmatrix} \left[\begin{smallmatrix} a_1 & a_2 \\ a_3 & a_4 \end{smallmatrix}\right] & \\ & \left[\begin{smallmatrix} a_1 & a_2 \\ a_3 & a_4 \end{smallmatrix}\right]^{t^{-1}} \end{bmatrix} \quad \text{so that} \quad g_2 = \begin{bmatrix} \left[\begin{smallmatrix} a_1 & a_2u^{-1} \\ a_3u & a_4 \end{smallmatrix}\right] & \\ & \left[\begin{smallmatrix} a_1 & a_2u^{-1} \\ a_3u & a_4 \end{smallmatrix}\right]^{t^{-1}} \end{bmatrix}.$$

With φ chosen as in (5.1) it is clear that $\omega(g_1, 1)\varphi = \omega(g_2, 1)\varphi$ by examining (3.15).

(b) Suppose that

$$g_1 = \begin{bmatrix} 1 & & b_2 & b_2 \\ & 1 & b_2 & b_3 \\ & & 1 & \\ & & & 1 \end{bmatrix} \quad \text{so that} \quad g_2 = \begin{bmatrix} 1 & & b_1u^{-1} & b_2 \\ & 1 & b_2 & b_3u \\ & & 1 & \\ & & & 1 \end{bmatrix}. \tag{5.2}$$

With φ chosen as in (5.1) it is clear that $\omega(g_1, 1)\varphi = \omega(g_2, 1)\varphi$ by examining (3.16)

(c) Suppose that $g_1 = J$ so that $g_2 = J\begin{bmatrix} u & & & \\ & u^{-1} & & \\ & & u^{-1} & \\ & & & u \end{bmatrix}$. For φ as chosen in (5.1) we have that $\varphi(u^{-1}x, uy) = \varphi(x, y)$ so we conclude that

$$\begin{aligned} \omega(g_2)\varphi(x, y) &= \omega(J, 1)\omega(\begin{bmatrix} u & & & \\ & u^{-1} & & \\ & & u^{-1} & \\ & & & u \end{bmatrix})\varphi(x, y) \\ &= \omega(J, 1)\varphi(u^{-1}x, uy) = \omega(J, 1)\varphi(x, y). \end{aligned}$$

This completes the proof. □

Lemma 5.3 *Let x_1, x_2 and H be as in Sect. 3. Let $h = \rho(h_1, h_2) \in \mathrm{SO}(X)$ then $h^{-1}(x_1, x_2) \in \mathrm{supp}(\varphi)$ if and only if there is some $h' = \rho(h'_1, h'_2) \in H$ such that $h'_1h_1 \in \Gamma_0(\mathfrak{p}^{n_1})$ and $h'_2h_2 \in \left[\begin{smallmatrix} 2^{-1} & \\ & 1 \end{smallmatrix}\right]\Gamma_0(\mathfrak{p}^{n_2})$.*

Proof Let $m \in \mathbb{Z}_{\geq 0}$ be such that $2\mathfrak{o}_L = \mathfrak{p}^m$. First, suppose that $h = \rho(h_1, h_2)$ where $h_1 \in \Gamma_0(\mathfrak{p}^{n_1})$ and $h_2 \in \left[\begin{smallmatrix} 2^{-1} & \\ & 1 \end{smallmatrix}\right]\Gamma_0(\mathfrak{p}^{n_2})$ where $h_1 = \left[\begin{smallmatrix} a_1 & b_1 \\ c_1 & d_1 \end{smallmatrix}\right]$, and $h_2 = \left[\begin{smallmatrix} a_2/2 & b_2/2 \\ c_2 & d_2 \end{smallmatrix}\right]$, for some $a_i, b_i, c_i, d_i \in L$, for $i \in \{1, 2\}$. Then we have $h^{-1}(x_1) = \left[\begin{smallmatrix} d_1c_2 & d_1d_2 \\ -c_1c_2 & -c_1d_2 \end{smallmatrix}\right] \in \mathrm{supp}(\varphi_1)$ and $h^{-1}(x_2) = \left[\begin{smallmatrix} b_1a_2 & b_1b_2 \\ -a_1a_2 & -a_1b_2 \end{smallmatrix}\right] \in \mathrm{supp}(\varphi_2)$. On the other hand, assume that $h^{-1}(x_1, x_2) \in \mathrm{supp}(\varphi)$. Then we have the following congruences $c_1c_2 \in \mathfrak{p}^N$, $c_1d_2 \in \mathfrak{p}^{n_1}$, $d_1c_2 \in \mathfrak{p}^{n_2}$, $d_1d_2 \in \mathfrak{o}_L$, $a_1a_2 \in \mathfrak{o}_L$, $a_1b_2 \in \mathfrak{o}_L$, $b_1a_2 \in \mathfrak{o}_L$, and $b_1b_2 \in \mathfrak{o}_L$. Choose $r_1 \in \mathbb{Z}$ maximally so that $c_1 \in \mathfrak{p}^{r_1}$ and $d_1 \in \mathfrak{p}^{r_1-n_1}$. Since at least one of c_1 and d_1 is not equal to 0 we know that $c_2 \in \mathfrak{p}^{N-r_1} = \mathfrak{p}^{n_2-(r_1-n_1)}$ and $d_2 \in \mathfrak{p}^{n_1-r_1} = \mathfrak{p}^{0-(r_1-n_1)}$. Similarly, choose $r_2 \in \mathbb{Z}$ maximally so that $a_1, b_1 \in \mathfrak{p}^{r_2}$. Since at least one of a_1 and b_1 is not equal to 0 we know that $a_2, b_2 \in \mathfrak{p}^{-r_2-m}$. Therefore $h_1 \in \left[\begin{smallmatrix} \mathfrak{p}^{r_2} & \mathfrak{p}^{r_2} \\ \mathfrak{p}^{r_1} & \mathfrak{p}^{r_1-n_1} \end{smallmatrix}\right]$ and

$h_2 \in [\begin{smallmatrix} 2^{-1}\mathfrak{p}^{-r_2} & 2^{-1}\mathfrak{p}^{-r_2} \\ \mathfrak{p}^{N-r_1} & \mathfrak{p}^{n_1-r_1} \end{smallmatrix})] \cap \mathrm{SO}(X)$. We choose $h_1' = [\begin{smallmatrix} \varpi^{-r_2} & \\ & \varpi^{n_1-r_1} \end{smallmatrix}]$ and $h_2' = [\begin{smallmatrix} \varpi^{r_2} & \\ & \varpi^{r_1-n_1} \end{smallmatrix}]$ so that $h' = \rho(h_1', h_2') \in H$ and easily see that $h_1' h_1 \in \Gamma_0(\mathfrak{p}^{n_1})$ and that $h_2' h_2 \in \left[\begin{smallmatrix} 2^{-1} & \\ & 1 \end{smallmatrix}\right] \Gamma_0(\mathfrak{p}^{n_2})$. □

The following theorem establishes our main local result in the split case.

Theorem 5.4 *Suppose that E/L is split and that $W \in \mathcal{W}_{\tau_1} \otimes \mathcal{W}_{\tau_2}$ is $\Gamma_0(\mathfrak{p}^{n_1}) \times \Gamma_0(\mathfrak{p}^{n_2})$-invariant. Let φ be as in* (5.1) *and assume that $\Re(s) > M$, then the intertwining map B defined in* (4.1) *is non-zero and $\mathrm{K}(\mathfrak{p}^N)$-invariant. In particular $B(1, \varphi, W, s) \neq 0$.*

Proof By Corollary 5.2 We have already show that $B(\cdot, \varphi, W, s)$ is paramodular invariant. Lemma 5.3 can be used determine the support of $B(1, \varphi, W, s)$. Indeed, we see that

$$\begin{aligned} B(1, \varphi, W, s) &= \int\limits_{H \backslash \mathrm{SO}(X)} \omega(1, h)\varphi(x_1, x_2) Z(s, \pi(h)W) \\ &= \int\limits_{H \backslash \rho(\Gamma_0(\mathfrak{p}^{n_1}) \times \left[\begin{smallmatrix} 2^{-1} & \\ & 1 \end{smallmatrix}\right] \Gamma_0(\mathfrak{p}^{n_2}))} Z(s, \pi(h)W) \\ &= \mathrm{vol}[H \backslash \rho(\Gamma_0(\mathfrak{p}^{n_1}) \times \left[\begin{smallmatrix} 2^{-1} & \\ & 1 \end{smallmatrix}\right] \Gamma_0(\mathfrak{p}^{n_2}))] \cdot |2|^{s-\frac{1}{2}} \cdot Z(s, W) \neq 0. \end{aligned}$$

Note that the additional constant $|2|^{s-\frac{1}{2}}$ comes from an application of Lemma 3.6 part (a). □

5.2 *The Inert Case*

We now let $X = X_{ns}$ and apply the same method to prove that if E/L is inert that there exists a $\varphi \in \mathcal{S}(X^2)$ such that $B(\cdot, \varphi, W, s) \neq 0$, in particular we will show that $B(1) \neq 0$. Suppose that $1 \neq \delta \in \mathfrak{o}_L$ is square-free and that the field extensions $E = L(\sqrt{\delta})$ is inert. In particular this means that $\delta \in \mathfrak{o}_L^{\times}$. Let τ_0 be an irreducible admissible representation of $\mathrm{GL}(2, E)$ with trivial central character. We assume that the space of τ_0 is its Whittaker model $\mathcal{W}_{\tau_0}$ and that there is some $W \in \mathcal{W}_{\tau_0}$ that is $\Gamma_0(\mathfrak{p}^n)$-invariant, for some non-negative integer n. Set $N = 2n$ which will be the paramodular level of $B(\cdot, \varphi, W, s)$. With this information we choose Schwartz function $\varphi = T(\varphi_1 \otimes \varphi_2)$ where

$$\varphi_1 = f_{\left[\begin{smallmatrix} \mathfrak{p}^n & \mathfrak{o}_L \\ \mathfrak{p}^N & \mathfrak{p}^n \end{smallmatrix}\right] \cap X} \quad \text{and} \quad \varphi_2 = f_{\mathrm{M}(2, \mathfrak{o}_E) \cap X}, \tag{5.3}$$

where f_S is the characteristic function of a set S. Evidently we see that we can also write the support of the φ_i as

$$\operatorname{supp}(\varphi_1) = \{[\begin{smallmatrix} x_1+x_2\sqrt{\delta} & x_3 \\ x_4 & x_1-x_2\sqrt{\delta} \end{smallmatrix}] \mid x_1 \in \mathfrak{p}^n, x_2 \in \mathfrak{p}^n, x_3 \in \mathfrak{o}_L, x_4 \in \mathfrak{p}^N\} \tag{5.4}$$

and

$$\operatorname{supp}(\varphi_2) = \{[\begin{smallmatrix} x_1+x_2\sqrt{\delta} & x_3 \\ x_4 & x_1-x_2\sqrt{\delta} \end{smallmatrix}] \mid x_1, x_2, x_3, x_4 \in \mathfrak{o}_L\}. \tag{5.5}$$

Lemma 5.5 *Let φ be as chosen above. Then, for every $(k, h) \in R$ such that $k \in$ $\mathrm{K}(\mathfrak{p}^N)$ we have that $\omega(k, h)\varphi = \varphi$.*

Proof Just as in the split case it suffices to check this for each of the generators of $\mathrm{K}(\mathfrak{p}^N)$, as listed in Lemma 3.9.

(a) Let $A = [\begin{smallmatrix} a_1 & a_2 \\ a_3 & a_4 \end{smallmatrix}] \in \Gamma_0(\mathfrak{p}^N) \subset \mathrm{GL}(2, \mathfrak{o}_L)$ for some choices of $a_i \in \mathfrak{o}_L$. Note that $A^{-1} = \frac{1}{\det(A)} \left[\begin{smallmatrix} a_4 & -a_2 \\ -a_3 & a_1 \end{smallmatrix}\right] \in \Gamma_0(\mathfrak{p}^N)$. If $(x, y) \in \operatorname{supp}(\varphi)$ then $a_1x + a_3y \in \left[\begin{smallmatrix} \mathfrak{p}^n & \mathfrak{o}_L \\ \mathfrak{p}^N & \mathfrak{p}^n \end{smallmatrix}\right]$ and $aa_2x + a_4y \in \mathrm{M}(2, \mathfrak{o}_L)$ so that $(a_1x + a_3y, a_2x + a_4y) \in \operatorname{supp}(\varphi)$. On the other hand, if $(a_1x + a_3y, a_2x + a_4y) \in \operatorname{supp}(\varphi)$ then

$$\begin{aligned} A^{-1} &\cdot (a_1x + a_3y, a_2x + a_4y) \\ &= \frac{1}{\det(A)}(a_4(a_1x + a_3y) + -a_3(a_2x + a_4y), \\ &\quad - a_2(a_1x + a_3y) + a_1(a_2x + a_4y)) \\ &= (x, y). \end{aligned}$$

Since $A^{-1} \in \Gamma_0(\mathfrak{p}^N)$ we conclude that $(x, y) \in \operatorname{supp}(\varphi)$. So, we have just proved that $(x, y) \in \operatorname{supp}(\varphi)$ if and only if $(a_1x + a_3y, a_2x + a_4y) \in \operatorname{supp}(\varphi)$. Recall that when E/L is unramified that $\mathrm{N}_L^E : \mathfrak{o}_E^\times \to \mathfrak{o}_L^\times$ is surjective. Since $\det(A) \in \mathfrak{o}_E^\times$ we see that

$$\begin{aligned} \omega([\begin{smallmatrix} A & \\ & {}^tA^{-1} \end{smallmatrix}], 1)\varphi(x, y) &= \chi_{E/L}(\det(A))|\det(A)|^2\varphi(a_1x + a_3y, a_2x + a_4y) \\ &= \varphi(x, y). \end{aligned}$$

Let $u \in \mathfrak{o}_L^\times$ such that $g_u = \begin{bmatrix} 1 & & & \\ & 1 & & \\ & & u & \\ & & & u \end{bmatrix} \in \mathrm{GSp}(4)^+$. Choose $u_E \in \mathfrak{o}_E^\times$ such that $\mathrm{N}_L^E(u_E) = u$ and set $h_u = \rho(1, [\begin{smallmatrix} u_E & \\ & 1 \end{smallmatrix}])$. Then $(g_u, h_u) \in R$. Since u_E is a unit in $\mathfrak{o}_E$ we can conclude that

$$\begin{aligned} \omega(g_u, h_u)\varphi(x, y) &= |u|^{-2}\varphi(h_u^{-1}x, h_u^{-1}y) \\ &= \varphi(x, y) \end{aligned}$$

for all $(x, y) \in X^2$.

(b) Suppose that $B = \left[\begin{smallmatrix} b_1 & b_2 \\ b_2 & b_3 \end{smallmatrix}\right] \in \left[\begin{smallmatrix} \mathfrak{p}^{-N} & \mathfrak{o}_L \\ \mathfrak{o}_L & \mathfrak{o}_L \end{smallmatrix}\right]$. Then for all $(x, y) \in \mathrm{supp}(\varphi)$ we can easily calculate that

$$b_1\langle x, x\rangle + 2b_2\langle x, y\rangle + b_3\langle y, y\rangle \in \mathfrak{o}_L, \quad \text{so that}$$

$$\begin{aligned} \omega(\left[\begin{smallmatrix} 1 & B \\ & 1 \end{smallmatrix}\right], 1)\varphi(x, y) &= \psi(b_1\langle x, x\rangle + 2b_2\langle x, y\rangle + b_3\langle y, y\rangle)\varphi(x, y) \\ &= \varphi(x, y). \end{aligned}$$

(c) As in part (*c*) in the proof of the the split case (Lemme 5.1) it suffices to show that $\mathcal{F}_1(\varphi_2) = \varphi_2$, which we did in the proof of Lemma 3.10.

(d) Again, similar to the split case it suffices to show that $\mathcal{F}_1(\varphi_1^{\varpi^N}) = q^{2N}\varphi_1$. For a generic element $y \in X$ we write $y = \begin{bmatrix} y_1+y_2\sqrt{\delta} & y_3 \\ y_4 & y_1-y_2\sqrt{\delta} \end{bmatrix}$ with $y_i \in L$ for $i \in \{1, 2, 3, 4\}$. Using (3.11) and the corresponding considerations for the Haar measure in Lemma 3.10 we can calculate this as follows:

$$\begin{aligned} (\mathcal{F}_1\varphi_1^{\varpi^N})(x) &= \int_X \varphi_1(\varpi^N x)\psi(2\langle x, y\rangle)\, dy = \int_{\left[\begin{smallmatrix} \mathfrak{p}^{-n} & \mathfrak{p}^{-N} \\ \mathfrak{o}_E & \mathfrak{p}^{-n} \end{smallmatrix}\right]\cap X} \psi(2\langle x, y\rangle)\, dy \\ &= \int_{\mathfrak{p}^{-n}} \psi(2x_1y_1)\, dy_1 \int_{\mathfrak{p}^{-n}} \psi(-2x_2y_2\delta)\, dy_2 \int_{\mathfrak{p}^{-N}} \psi(-x_4y_3)\, dy_3 \int_{\mathfrak{o}_L} \psi(-x_3y_4\delta)\, dy_4. \end{aligned}$$

Using Lemma 3.8 we can conclude that the above is $(\mathcal{F}_1\varphi_1^{\varpi^N})(x) = q^{2N}\varphi_2(x)$. □

Corollary 5.6 *For all $k \in \mathrm{K}(\mathfrak{p}^N)$ and all $g \in \mathrm{GSp}(4, L)$ we have that $B(gk, \varphi, W, s) = B(g, \varphi, W, s)$.*

Proof The proof is very similar to the proof of Corollary 5.2. □

Lemma 5.7 *Let x_1, x_2 and H be as in Sect. 3 and let $m \in \mathbb{Z}$ be such that $2\mathfrak{o}_L = \mathfrak{p}^m$ and set $\bar{m} = \lfloor \frac{m}{2} \rfloor$. Let $t \in L^\times$ and $h_0 \in \mathrm{GL}(2, E)$ and set $h = \rho(t, h_0) \in \mathrm{SO}(X)$. Then, $h^{-1}(x_1, x_2) \in \mathrm{supp}(\varphi)$ if and only if there is some $t' \in L^\times$ and $h_0' \in \mathrm{GL}(2, E)$ such that $h' = \rho(t', h_0') \in H$, $t't = 1$, and $h_0'h_0 = \left[\begin{smallmatrix} \varpi^{-\bar{m}} & \\ & 1 \end{smallmatrix}\right] A$, for some $A \in \Gamma_0(\mathfrak{p}^n)$.*

Proof First suppose that there is such an h' and an $A = \left[\begin{smallmatrix} \varpi^{-\bar{m}}a & \varpi^{-\bar{m}}b \\ c & d \end{smallmatrix}\right] \in \left[\begin{smallmatrix} \varpi^{-\bar{m}} & \\ & 1 \end{smallmatrix}\right]\Gamma_0(\mathfrak{p}^n)$ such that $h'h = \rho(1, A)$. We calculate

$$h^{-1}x_1 = h^{-1}(h'^{-1}x_1) = (h'h)^{-1}x_1 = Ax_1\alpha(A)^* = \sqrt{\delta}\left[\begin{smallmatrix} d\alpha(c) & d\alpha(d) \\ -c\alpha(c) & -c\alpha(d) \end{smallmatrix}\right] \tag{5.6}$$

and

$$h^{-1}x_2 = h^{-1}(h'^{-1}x_2) = (h'h)^{-1}x_2 = Ax_2\alpha(A)^* = \frac{\sqrt{\delta}}{\delta}\begin{bmatrix} \varpi^{m-2\bar{m}}b\alpha(a) & \varpi^{m-2\bar{m}}b\alpha(b) \\ -\varpi^{m-2\bar{m}}a\alpha(a) & -\varpi^{m-2\bar{m}}a\alpha(b) \end{bmatrix}. \tag{5.7}$$

Since $m - 2\bar{m} \geq 0$ we conclude that $h^{-1}(x_1, x_2) \in \operatorname{supp}(\varphi)$.

On the other hand, let $h = \rho(1, A)$ with $A = \left[\begin{smallmatrix} \varpi^{-\bar{m}}a & \varpi^{-\bar{m}}b \\ c & d \end{smallmatrix}\right] \in \mathrm{GL}(2, E)$ and assume that $h^{-1}(x_1, x_2) \in \operatorname{supp}(\varphi)$. From (5.6) and (5.7), we immediately find the following congruences

$$\begin{aligned} &d\alpha(c) \in \mathfrak{p}^n, d\alpha(d) \in \mathfrak{o}_L, c\alpha(c) \in \mathfrak{p}^N \\ &b\alpha(a), a\alpha(a), b\alpha(b) \in \mathfrak{p}^{2\bar{m}-m} = \begin{cases} \mathfrak{o}_L; & m \text{ is even} \\ \mathfrak{p}^{-1}; & m \text{ is odd.} \end{cases} \end{aligned}$$

Since $|\alpha(x)| = |x|$ for all $x \in E$ we conclude that $a, b, d \in \mathfrak{o}_L$ and $c \in \mathfrak{p}^n$. Let $t \in L^\times$ and let A be as above. Set $h = \rho(t, A)$ and suppose that $h^{-1}(x_1, x_2) \in \operatorname{supp}(\varphi)$. If $\nu(t)$ is odd then $\nu(c\alpha(c))$ is odd, which is impossible. Then $t = \varpi^{2k}u$ for some $k \in \mathbb{Z}$ and some $u \in \mathfrak{o}_L^\times$. Set $h' = \rho(t^{-1}, \left[\begin{smallmatrix} \varpi^k & \\ & \varpi^k u \end{smallmatrix}\right]) \in H$. A priori, $h'h \in \rho(1, \mathrm{GL}(2, E))$ and $(h'h)^{-1}(x_1, x_2) = h^{-1}(x_1, x_2) \in \operatorname{supp}(\varphi)$. By the above calculation when $t = 1$ we can conclude that $h'h \in \rho(1, \left[\begin{smallmatrix} \varpi^{-\bar{m}} & \\ & 1 \end{smallmatrix}\right]\Gamma_0(\mathfrak{p}^n))$. □

The following theorem establishes our main local result in the inert case.

Theorem 5.8 *Suppose that E/L is inert and that $W \in \mathcal{W}_{\tau_0}$ has $\Gamma_0(\mathfrak{p}^n)$-invariance. Let φ be as in* (5.3) *and assume that $s \gg M$, then the intertwining map B defined in* (4.2) *is non-zero and $K(\mathfrak{p}^N)$-invariant. In particular $B(1, \varphi, W, s) \neq 0$.*

Proof By Corollary 5.6 we have already showed that $B(\cdot, \varphi, W, s)$ is paramodular invariant. Lemma 5.7 can be used to determine the support of $B(1, \varphi, W, s)$. Indeed, we see that

$$\begin{aligned} B(1, \varphi, W, s) &= \int_{H\backslash \mathrm{SO}(X)} \omega(1, h)\varphi(x_1, x_2) Z(s, \pi(h)W)\, dh \\ &= \int_{H\backslash\rho(1,\left[\begin{smallmatrix} \varpi^{-\bar{m}} & \\ & 1 \end{smallmatrix}\right]\Gamma_0(\mathfrak{p}^n))} Z(s, \pi(h)W)\, dh \\ &= \mathrm{vol}[H\backslash\rho(1, \left[\begin{smallmatrix} \varpi^{-\bar{m}} & \\ & 1 \end{smallmatrix}\right]\Gamma_0(\mathfrak{p}^n))] \cdot |\varpi|^{\bar{m}(s-\frac{1}{2})} \cdot Z(s, W) \neq 0. \end{aligned}$$

Note that the additional constant $|\varpi|^{\bar{m}(s-\frac{1}{2})}$ comes from an application of Lemma 3.6 part (a). □

5.3 The Ramified Case

Suppose that $\delta \in \mathfrak{o}_L$ is square-free and that the field extensions $E = L(\sqrt{\delta})$ is ramified, so that $\delta \in \mathfrak{p}$. Let τ_0 be an irreducible admissible representation of $\mathrm{GL}(2, E)$ with trivial central character. We assume that the space of τ_0 is its Whittaker model $\mathcal{W}_{\tau_0}$ and that there is some $W \in \mathcal{W}_{\tau_0}$ that is invariant under $\Gamma_0(\mathfrak{P}^n)$ for some non-negative integer n. Set $N = n + 2$, which we will prove is the paramodular level of $B(\cdot, \varphi, W, s)$. In the previous two sections we were able to find Schwartz functions for which $B(g, \varphi, W, s) \neq 0$ is paramodular invariant by inspection. In the ramified case we take a more systematic approach. Define $\tilde{\varphi}(x, y) = T(\varphi_1(x) \otimes \varphi_2(y))$, where

$$\begin{aligned}\varphi_1(x) &= \chi(x_3) f_{\left[\begin{smallmatrix} \mathfrak{P}^{n+1} & \mathfrak{P} \\ \varpi_E^{2n+1}\mathfrak{o}_E^\times & \mathfrak{P}^{n+1} \end{smallmatrix}\right] \cap X}(x) = \chi(x_3) f_{\mathfrak{P}^{n+1}}(x_1) f_{\mathfrak{o}_L}(x_2) f_{\varpi_L^n \mathfrak{o}_L^\times}(x_3), \qquad (5.8)\\ \varphi_2(y) &= \chi(y_3) f_{\left[\begin{smallmatrix} \mathfrak{P}^{-1} & \mathfrak{P}^{-1} \\ \varpi_E^{-1}\mathfrak{o}_E^\times & \mathfrak{P}^{-1} \end{smallmatrix}\right] \cap X}(y), = \chi(y_3) f_{\mathfrak{P}^{-1}}(y_1) f_{\mathfrak{p}^{-1}}(y_2) f_{\varpi_L^{-1} \mathfrak{o}_L^\times}(y_3),\end{aligned}$$

with $x = \left[\begin{smallmatrix} x_1 & x_2\sqrt{\delta} \\ x_3\sqrt{\delta} & \alpha(x_1) \end{smallmatrix}\right]$ and $y = \left[\begin{smallmatrix} y_1 & y_2\sqrt{\delta} \\ y_3\sqrt{\delta} & \alpha(y_1) \end{smallmatrix}\right]$.

Let T be the subgroup of $\mathrm{GSp}(4, L)$ generated by

$$\begin{aligned}\{\left[\begin{smallmatrix} A & \\ & {}^tA^{-1} \end{smallmatrix}\right], \left[\begin{smallmatrix} 1 & B \\ & 1 \end{smallmatrix}\right], \left[\begin{smallmatrix} 1 & \\ C & 1 \end{smallmatrix}\right] \mid A \in \Gamma_0(\mathfrak{p}^N), B \in \left[\begin{smallmatrix} \mathfrak{p}^{-N+1} & \mathfrak{o}_L \\ \mathfrak{o}_L & \mathfrak{p} \end{smallmatrix}\right] \cap \mathrm{Sym}(2),\\ C \in \left[\begin{smallmatrix} \mathfrak{p}^N & \mathfrak{p}^{N-1} \\ \mathfrak{p}^{N-1} & \mathfrak{o}_L \end{smallmatrix}\right] \cap \mathrm{Sym}(2)\}.\end{aligned}$$

The following lemma computes the Fourier transforms of φ_1 and φ_2.

Lemma 5.9 *Let φ_1 and φ_2 be as above. Then*

$$\mathcal{F}_1(\varphi_1)(\varpi^{-N}x) = q_E^{-N} \mathcal{F}_1(\varphi_1^{\varpi^N})(x) = q_E^{-N} \chi(x_2) f_{\left[\begin{smallmatrix} \mathfrak{P}^{n+2} & \varpi_E \mathfrak{o}_E^\times \\ \mathfrak{P}^{2n+3} & \mathfrak{P}^{n+2} \end{smallmatrix}\right] \cap X}(x)$$

where $\varphi_1^a(x) = \varphi_1(ax)$, and

$$\mathcal{F}_1(\varphi_2)(x) = \chi(x_2) f_{\left[\begin{smallmatrix} \mathfrak{o}_E & \varpi_E^{-1}\mathfrak{o}_E^\times \\ \mathfrak{P} & \mathfrak{o}_E \end{smallmatrix}\right] \cap X}(x).$$

Proof First note that, since E/L is ramified, $\psi_E^{\varpi_E^{-1}}$ has conductor $\mathfrak{o}_E$. Evaluations of character sums and Gauss sums in this proof rely on the relevant formulas proven in Sect. 3. We begin with the easier calculation:

$$
\begin{aligned}
\mathcal{F}_1(\varphi_2)(x) &= \int\limits_X \varphi_2(y)\psi(2\langle x, y\rangle)\,dy \\
&= k \int\limits_{\mathfrak{P}^{-1}} \psi_E(x_1\alpha(y_1))\,dy_1 \int\limits_{\mathfrak{p}^{-1}} \psi(-\delta x_3 y_3)\,dy_2 \int\limits_{\varpi_L^{-1}\mathfrak{o}_L^\times} \chi(y_3)\psi(-\delta x_2 y_3)\,dy_3 \\
&= k \int\limits_{\mathfrak{P}^{-1}} \psi_E^{\varpi_E^{-1}}(\varpi_E x_1\alpha(y_1))\,dy_1 \int\limits_{\mathfrak{p}^{-1}} \psi(-\delta x_3 y_2)\,dy_2 \int\limits_{\varpi_L^{-1}\mathfrak{o}_L^\times} \chi(y_3)\psi(-\delta x_2 y_3)\,dy_3 \\
&= k'\chi(x_2) f_{\mathfrak{o}_E}(x_1) f_{\varpi_L^{-1}\mathfrak{o}_L^\times}(x_2) f_{\mathfrak{o}_L}(x_3) \\
&= k'\chi(x_2) f_{\left[\begin{smallmatrix} \mathfrak{o}_E & \varpi_E^{-1}\mathfrak{o}_E^\times \\ \mathfrak{P} & \mathfrak{o}_E \end{smallmatrix}\right]\cap X}(x).
\end{aligned}
$$

We know that $(k')^4 = 1$ since $\mathcal{F}_1(\mathcal{F}_1\varphi_2)(x) = \varphi_2(-x) = \chi(-1)\varphi_2(x)$. For φ_1 we see that the first equality is true because $\left[\begin{smallmatrix} & \varpi_L^{-N} \\ -\varpi_L^N & \end{smallmatrix}\right] = \left[\begin{smallmatrix} & 1 \\ -1 & \end{smallmatrix}\right]\left[\begin{smallmatrix} \varpi_L^{-N} & \\ & \varpi_L^N \end{smallmatrix}\right] = \left[\begin{smallmatrix} \varpi_L^N & \\ & \varpi_L^{-N} \end{smallmatrix}\right]\left[\begin{smallmatrix} & -1 \\ 1 & \end{smallmatrix}\right]$. For the second equality we calculate that

$$
\begin{aligned}
\mathcal{F}_1(\varphi_1)(x) &= \int\limits_X \varphi_1(y)\psi(2\langle x, y\rangle)\,dy \\
&= k \int\limits_{\mathfrak{P}^{n+1}} \psi_E^{\varpi_E^{-1}}(\varpi_E x_1\alpha(y_1))\,dy_1 \int\limits_{\mathfrak{o}_L} \psi(-\delta x_3 y_2)\,dy_2 \int\limits_{\varpi_L^{n}\mathfrak{o}_L^\times} \chi(y_3)\psi(-\delta x_2 y_3)\,dy_3 \\
&= q_E^{-N} k'\chi(x_2) f_{\mathfrak{P}^{-n-2}}(x_1) f_{\varpi_L^{-n-2}\mathfrak{o}_L^\times}(x_2) f_{\mathfrak{p}^{-1}}(x_3).
\end{aligned}
$$

Therefore we have that

$$
\begin{aligned}
\mathcal{F}_1(\varphi_1)(\varpi_L^{-N}x) &= q_E^{-N} k'\chi(x_2) f_{\mathfrak{P}^{n+2}}(x_1) f_{\mathfrak{o}_L^\times}(x_2) f_{\mathfrak{p}^{n+1}}(x_3). \\
&= q_E^{-N}\chi(x_2) f_{\left[\begin{smallmatrix} \mathfrak{P}^{n+2} & \varpi_E\mathfrak{o}_E^\times \\ \mathfrak{P}^{2n+3} & \mathfrak{P}^{n+2} \end{smallmatrix}\right]\cap X}(x).
\end{aligned}
$$

So, both equalities are verified. □

Lemma 5.10 *Let $\tilde{\varphi}$ be as defined before* (5.8). *For every $g \in T$ we have that $\omega(g, 1)\tilde{\varphi} = \tilde{\varphi}$.*

Proof (a) Let $A = \left[\begin{smallmatrix} a_1 & a_2 \\ a_3 & a_4 \end{smallmatrix}\right] \in \Gamma_0(\mathfrak{p}^N)$. Then $(x, y) \in \mathrm{supp}(\tilde{\varphi})$ if and only if $(a_1x + a_3y, a_2x + a_4y) \in \mathrm{supp}(\tilde{\varphi})$. Furthermore, assuming that $(x, y) \in \mathrm{supp}(\tilde{\varphi})$, we have that

$$
\begin{aligned}
&\chi((a_1x_3 + a_3y_3)(a_2x_3 + a_4y_3)) \\
=&\chi(a_1a_2x_3^2 + 2a_2a_3x_3y_3 + a_3a_4y_3^2 + x_3y_3(\det A)) \\
=&\chi\left[(x_3y_3\det A)\left(1 + (x_3y_3\det A)^{-1}(a_1a_2x_3^2 + 2a_2a_3x_3y_3 + a_3a_4y_3^2)\right)\right] \\
=&\chi(x_3y_3\det A)
\end{aligned}
$$

since $(x_3y_3\det A)^{-1}(a_1a_2x_3^2+2a_2a_3x_3y_3+a_3a_4y_4^2)\in\mathfrak{p}$. Therefore

$$\begin{aligned}\omega([\begin{smallmatrix}A&\\&{}^tA^{-1}\end{smallmatrix}],1)\tilde{\varphi}(x,y)&=\chi(\det A)\tilde{\varphi}(a_1x+a_3y,a_2x+a_4y)\\&=\tilde{\varphi}(x,y).\end{aligned}$$

(b) Let $B=[\begin{smallmatrix}b_1&b_2\\b_2&b_3\end{smallmatrix}]\in[\begin{smallmatrix}\mathfrak{p}^{-N+1}&\mathfrak{o}_L\\\mathfrak{o}_L&\mathfrak{p}\end{smallmatrix}]$. If $(x,y)\in\operatorname{supp}(\tilde{\varphi})$ then we get the following congruences

$$\langle x,x\rangle\in\mathfrak{p}^{n+1}=\mathfrak{p}^{N-1},\quad\langle y,y\rangle\in\mathfrak{p}^{-1},\quad\text{and}\quad\langle x,y\rangle\in\mathfrak{o}_L$$

so that,

$$b_1\langle x,x\rangle+2b_2\langle x,y\rangle+b_3\langle y,y\rangle\in\mathfrak{o}_L.$$

Therefore it is now clear that for all $(x,y)\in X^2$ we have

$$\begin{aligned}\omega([\begin{smallmatrix}1&B\\&1\end{smallmatrix}],1)\tilde{\varphi}(x,y)&=\psi(\langle b_1\langle x,x\rangle+2b_2\langle x,y\rangle+b_3\langle y,y\rangle)\tilde{\varphi}(x,y)\\&=\tilde{\varphi}(x,y)\end{aligned}$$

(c) Let $C=[\begin{smallmatrix}c_1&c_2\\c_2&c_3\end{smallmatrix}]\in[\begin{smallmatrix}\mathfrak{p}^N&\mathfrak{p}^{N-1}\\\mathfrak{p}^{N-1}&\mathfrak{o}_L\end{smallmatrix}]$ and set $C'=[\begin{smallmatrix}-c_1\varpi^{-2N}&-c_2\varpi^{-N}\\-c_2\varpi^{-N}&-c_3\end{smallmatrix}]$. If $(x,y)\in\operatorname{supp}(\omega(t_Ns_2,1)\tilde{\varphi})=\operatorname{supp}(\mathcal{F}_1(\tilde{\varphi}_1)(\varpi^{-N}\cdot)\otimes\mathcal{F}_1(\tilde{\varphi}_2)(\cdot))$ then we have the congruences

$$\langle x,x\rangle\in\mathfrak{p}^{n+2},\quad\langle y,y\rangle\in\mathfrak{o}_L,\quad\text{and}\quad\langle x,y\rangle\in\mathfrak{p}$$

so that,

$$-c_1\varpi^{-2N}\langle x,x\rangle-2c_2\varpi^{-N}\langle x,y\rangle-c_3\langle y,y\rangle\in\mathfrak{o}_L.$$

Therefore for all $(x,y)\in X^2$ we find that

$$\begin{aligned}&\omega([\begin{smallmatrix}1&\\C&1\end{smallmatrix}],1)\tilde{\varphi}(x,y)\\&=\omega((t_Ns_2)^{-1}[\begin{smallmatrix}1&C'\\&1\end{smallmatrix}]t_Ns_2,1)\tilde{\varphi}(x,y)\\&=\omega((t_Ns_2)^{-1},1)\psi(-c_1\varpi^{-2N}\langle x,x\rangle-2c_2\varpi^{-N}\langle x,y\rangle\\&\quad-c_3\langle y,y\rangle)[\omega(t_Ns_2,1)\tilde{\varphi}](x,y)\\&=\omega((t_Ns_2)^{-1},1)[\omega(t_Ns_2,1)\tilde{\varphi}](x,y)\\&=\tilde{\varphi}(x,y).\end{aligned}$$

We have verified the claim for all the generators of T, which means that we are done. □

Lemma 5.11 *The set* $\{s_2, \begin{bmatrix} 1&&&\\ &1&&v\\ &&1&\\ &&&1 \end{bmatrix} \mid v \in \mathfrak{o}_L/\mathfrak{p}\}$ *is a complete list of coset representatives of* $\mathrm{Kl}(\mathfrak{p}^N)/[\mathrm{Kl}(\mathfrak{p}^N) \cap T]$.

Proof Let $k \in \mathrm{Kl}(\mathfrak{p}^N)$. Then, there exists $a, b, c, a', b', c' \in \mathfrak{o}_L$, $u \in \mathfrak{o}_L^\times$ and $s = \left[\begin{smallmatrix} s_1 & s_2\\ s_3 & s_4\end{smallmatrix}\right] \in \mathrm{SL}(2, \mathfrak{o}_L)$ such that

$$k = \begin{bmatrix} 1&a&c&b\\ &1&b&\\ &&1&\\ &&-a&1 \end{bmatrix}\begin{bmatrix} u_1&&&\\ &s_1&&s_2\\ &&u^{-1}&\\ &s_3&&s_4 \end{bmatrix}\begin{bmatrix} 1&&&\\ \varpi^N a'&1&&\\ \varpi^N c'&\varpi^N b'&1&-\varpi^N a'\\ \varpi^N b'&&&1 \end{bmatrix} \tag{5.9}$$

by the Iwahori factorization ([RS07], for example). Clearly the matrix on the right is in $T \cap \mathrm{Kl}(\mathfrak{p}^N)$. First, assume that $s_4 \in \mathfrak{o}_L^\times$. In that case we find that

$$\begin{aligned}
&\begin{bmatrix} 1&a&c&b\\ &1&b&\\ &&1&\\ &&-a&1 \end{bmatrix}\begin{bmatrix} u&&&\\ &s_1&&s_2\\ &&u^{-1}&\\ &s_3&&s_4 \end{bmatrix}(T \cap \mathrm{Kl}(\mathfrak{p}^N))\\
&= \begin{bmatrix} 1&a&c&b\\ &1&b&\\ &&1&\\ &&-a&1 \end{bmatrix}\begin{bmatrix} 1&&&\\ &1&&s_4^{-1}s_2\\ &&1&\\ &&&1 \end{bmatrix}\begin{bmatrix} u&&&\\ &s_1-s_4^{-1}s_2s_3&&\\ &&u^{-1}&\\ &s_3&&s_4 \end{bmatrix}(T \cap \mathrm{Kl}(\mathfrak{p}^N))\\
&= \begin{bmatrix} 1&a&c&b\\ &1&b&\\ &&1&\\ &&-a&1 \end{bmatrix}\begin{bmatrix} 1&&&\\ &1&&s_4^{-1}s_2\\ &&1&\\ &&&1 \end{bmatrix}(T \cap \mathrm{Kl}(\mathfrak{p}^N))\\
&= \begin{bmatrix} 1&&c+ab&b\\ &1&b&\\ &&1&\\ &&&1 \end{bmatrix}\begin{bmatrix} 1&a&&\\ &1&&\\ &&1&\\ &&-a&1 \end{bmatrix}\begin{bmatrix} 1&&&\\ &1&&s_4^{-1}s_2\\ &&1&\\ &&&1 \end{bmatrix}(T \cap \mathrm{Kl}(\mathfrak{p}^N))\\
&= \begin{bmatrix} 1&&c+ab&b\\ &1&b&\\ &&1&\\ &&&1 \end{bmatrix}\begin{bmatrix} 1&&a^2s_4^{-1}s_2&as_4^{-1}s_2\\ &1&as_4^{-1}s_2&s_4^{-1}s_2\\ &&1&\\ &&&1 \end{bmatrix}\begin{bmatrix} 1&a&&\\ &1&&\\ &&1&\\ &&-a&1 \end{bmatrix}(T \cap \mathrm{Kl}(\mathfrak{p}^N))\\
&= \begin{bmatrix} 1&&c+ab+a^2s_4^{-1}s_2&b+as_4^{-1}s_2\\ &1&b+as_4^{-1}s_2&s_4^{-1}s_2\\ &&1&\\ &&&1 \end{bmatrix}(T \cap \mathrm{Kl}(\mathfrak{p}^N))\\
&= \begin{bmatrix} 1&&&\\ &1&&s_4^{-1}s_2\\ &&1&\\ &&&1 \end{bmatrix}(T \cap \mathrm{Kl}(\mathfrak{p}^N)).
\end{aligned}$$

Now assume that $s_4 \in \mathfrak{p}$ so that $s_2 \in \mathfrak{o}_L^\times$. In this case we find that

$$\begin{aligned}
&\begin{bmatrix} 1&a&c&b\\ &1&b&\\ &&1&\\ &&-a&1 \end{bmatrix}\begin{bmatrix} u&&&\\ &s_1&&s_2\\ &&u^{-1}&\\ &s_3&&s_4 \end{bmatrix}(T \cap \mathrm{Kl}(\mathfrak{p}^N))\\
&= \begin{bmatrix} 1&a&c&b\\ &1&b&\\ &&1&\\ &&-a&1 \end{bmatrix}\begin{bmatrix} 1&&&\\ &&&1\\ &&1&\\ &-1&& \end{bmatrix}\begin{bmatrix} 1&&&\\ &1&&-s_2^{-1}s_4\\ &&1&\\ &&&1 \end{bmatrix}\begin{bmatrix} u&&&\\ &-s_3+s_2^{-1}s_1s_4&&\\ &&u^{-1}&\\ &s_1&&s_2 \end{bmatrix}(T \cap \mathrm{Kl}(\mathfrak{p}^N))\\
&= \begin{bmatrix} 1&a&c&b\\ &1&b&\\ &&1&\\ &&-a&1 \end{bmatrix}\begin{bmatrix} 1&&&\\ &&&1\\ &&1&\\ &-1&& \end{bmatrix}(T \cap \mathrm{Kl}(\mathfrak{p}^N))
\end{aligned}$$

$$
\begin{aligned}
&= \begin{bmatrix} 1 & & & \\ & & & 1 \\ & & 1 & \\ & -1 & & \end{bmatrix} \begin{bmatrix} 1 & -b & c & a \\ & 1 & a & \\ & & 1 & \\ & & b & 1 \end{bmatrix} (T \cap \mathrm{Kl}(\mathfrak{p}^N)) \\
&= \begin{bmatrix} 1 & & & \\ & & & 1 \\ & & 1 & \\ & -1 & & \end{bmatrix} \begin{bmatrix} 1 & & c-ba & a \\ & 1 & a & \\ & & 1 & \\ & & & 1 \end{bmatrix} \begin{bmatrix} 1 & -b & & \\ & 1 & & \\ & & 1 & \\ & & b & 1 \end{bmatrix} (T \cap \mathrm{Kl}(\mathfrak{p}^N)) \\
&= \begin{bmatrix} 1 & & & \\ & & & 1 \\ & & 1 & \\ & -1 & & \end{bmatrix} (T \cap \mathrm{Kl}(\mathfrak{p}^N))
\end{aligned}
$$

which gives us exactly the cosets we described. □

Lemma 5.12 *The following is an exact list of coset representatives of* $\mathrm{K}(\mathfrak{p})/[\mathrm{K}(\mathfrak{p}^N) \cap T]$*:*

$$
\{\begin{bmatrix} 1 & & \varpi^{-N}u & \\ & 1 & & v \\ & & 1 & \\ & & & 1 \end{bmatrix}, s_2 \begin{bmatrix} 1 & & \varpi^{-N}u & \\ & 1 & & \\ & & 1 & \\ & & & 1 \end{bmatrix}, t_N \begin{bmatrix} 1 & & & \\ & 1 & & v \\ & & 1 & \\ & & & 1 \end{bmatrix}, t_N s_2 \mid u, v \in \mathfrak{o}_L/\mathfrak{p}\}.
$$

Proof Use the Lemma 5.11 and the decomposition of $\mathrm{K}(\mathfrak{p}^N)/\mathrm{Kl}(\mathfrak{p}^N)$ found in [RS07], Lemma 3.3.1. The later is reproduced in (3.9). □

Define the Schwartz function

$$
\varphi(x, y) = \sum_{g \in \mathrm{K}(\mathfrak{p}^N)/T \cap \mathrm{K}(\mathfrak{p}^N)} \omega(g, 1)\tilde{\varphi}(x, y). \tag{5.10}
$$

It is clear from the above discussion that φ is invariant under all of $K_1(\mathfrak{p}^N)$. The lemma below shows that $\varphi \neq 0$

Lemma 5.13 *We have that*

$$
\varphi = \varphi^{(1)} + \varphi^{(2)} + \varphi^{(3)} + \varphi^{(4)} \tag{5.11}
$$

where

$$
\begin{aligned}
\varphi^{(1)}(x, y) &= q_L^2 f_{\mathfrak{p}^{n+2}}(\langle x, x\rangle)\varphi_1(x)\, f_{\mathfrak{o}_L}(\langle y, y\rangle)(y)\varphi_2(y), \\
\varphi^{(2)}(x, y) &= q_L \chi(y_2) f_{\mathfrak{p}^{n+2}}(\langle x, x\rangle)\varphi_1(x)\, f_{\left[\begin{smallmatrix} \mathfrak{o}_E & \varpi_E^{-1}\mathfrak{o}_E^\times \\ \mathfrak{P} & \mathfrak{o}_E \end{smallmatrix}\right] \cap X}(y), \\
\varphi^{(3)}(x, y) &= q \chi(x_2) f_{\left[\begin{smallmatrix} \mathfrak{P}^{n+2} & \varpi_E \mathfrak{o}_E^\times \\ \mathfrak{P}^{2n+3} & \mathfrak{P}^{n+2} \end{smallmatrix}\right]}(x)\, f_{\mathfrak{o}_L}(\langle y, y\rangle)\varphi_2(y), \quad \textit{and} \\
\varphi^{(4)}(x, y) &= \chi(x_2 y_2) f_{\left[\begin{smallmatrix} \mathfrak{P}^{n+2} & \varpi_E \mathfrak{o}_E^\times \\ \mathfrak{P}^{2n+3} & \mathfrak{P}^{n+2} \end{smallmatrix}\right]}(x) f_{\left[\begin{smallmatrix} \mathfrak{o}_E & \varpi_E^{-1}\mathfrak{o}_E^\times \\ \mathfrak{P} & \mathfrak{o}_E \end{smallmatrix}\right] \cap X}(y).
\end{aligned}
$$

Moreover, the supports of the four summands in this formula are pairwise disjoint.

Proof We calculate

$$\sum_{u\in\mathfrak{o}_L/\mathfrak{p}} \psi(\varpi^{-N}u\langle x,x\rangle)\varphi_1(x) = q_L f_{\mathfrak{p}^N}(\langle x,x\rangle)\varphi_1(x),$$

$$\sum_{v\in\mathfrak{o}_L/\mathfrak{p}} \psi(v\langle y,y\rangle)\varphi_2(y) = q_L f_{\mathfrak{o}_L}(\langle y,y\rangle)\varphi_2(y).$$

Therefore, applying Lemma 5.9, we have

$$\varphi^{(1)} = \sum_{u,v\in(\mathfrak{o}_L/\mathfrak{p})^2} \omega(\begin{bmatrix} 1 & & \varpi^{-N}u & \\ & 1 & & v \\ & & 1 & \\ & & & 1 \end{bmatrix})\varphi(x,y) = q_L^2 f_{\mathfrak{p}^N}(\langle x,x\rangle)\varphi_1(x) f_{\mathfrak{o}_L}(\langle y,y\rangle)\varphi_2(y)$$

$$\varphi^{(2)} = \sum_{u\in\mathfrak{o}_L/\mathfrak{p}} \omega(s_2\begin{bmatrix} 1 & & \varpi^{-N}u & \\ & 1 & & \\ & & 1 & \\ & & & 1 \end{bmatrix})\varphi(x,y) = q_L\chi(y_2) f_{\mathfrak{p}^N}(\langle x,x\rangle)\varphi_1(x)\; f_{\left[\begin{smallmatrix} \mathfrak{o}_E & \varpi^{-1}\mathfrak{o}_E^\times \\ \mathfrak{P} & \mathfrak{o}_E \end{smallmatrix}\right]\cap X}(y)$$

$$\varphi^{(3)} = \sum_{v\in\mathfrak{o}_L/\mathfrak{p}} \omega(t_N\begin{bmatrix} 1 & & & \\ & 1 & & v \\ & & 1 & \\ & & & 1 \end{bmatrix})\varphi(x,y) = q\chi(x_2) f_{\left[\begin{smallmatrix} \mathfrak{P}^N & \varpi_E\mathfrak{o}_E^\times \\ \mathfrak{P}^{2N-1} & \mathfrak{P}^N \end{smallmatrix}\right]}(x)\; f_{\mathfrak{o}_L}(\langle y,y\rangle)\varphi_2(y)$$

$$\varphi^{(4)} = \omega(t_N s_2)\varphi(x,y) = \chi(x_2 y_2) f_{\left[\begin{smallmatrix} \mathfrak{P}^N & \varpi_E\mathfrak{o}_E^\times \\ \mathfrak{P}^{2N-1} & \mathfrak{P}^N \end{smallmatrix}\right]}(x) f_{\left[\begin{smallmatrix} \mathfrak{o}_E & \varpi^{-1}\mathfrak{o}_E^\times \\ \mathfrak{P} & \mathfrak{o}_E \end{smallmatrix}\right]\cap X}(y).$$

Since the cosets of $\mathrm{K}(\mathfrak{p}^N)/(\mathrm{K}(\mathfrak{p}^N)\cap T)$ are represented exactly once in the above sums we have proven that (5.11) is valid. □

Recall the choice of x_1, x_2 and H from Sect. 3. We will now determine for which $h \in H\backslash\mathrm{SO}(X)$ we have $(h^{-1}x_1, h^{-1}x_2) \in \mathrm{supp}(\varphi)$. This will help us evaluate the value of the intertwining map, as in (4.1), evaluated at φ.

Lemma 5.14 *Let $h \in \mathrm{SO}(X)$. Then*

(a) $(h^{-1}x_1, h^{-1}x_2) \in \mathrm{supp}(\varphi^{(1)})$ *if and only if there exists some $h' \in H$ such that $h'h = \rho(t, A)$ for some $t \in \mathfrak{o}_L^\times$ and $A \in [\begin{smallmatrix} \mathfrak{o}_E & \mathfrak{o}_E \\ \varpi^n\mathfrak{o}_E^\times & \mathfrak{o}_E^\times \end{smallmatrix}]$ such that $\mathrm{N}_L^E(\det A) = t^2$,*

(b) $(h^{-1}x_1, h^{-1}x_2) \in \mathrm{supp}(\varphi^{(2)})$ *if and only if $n = 0$ and there exists some $h' \in H$ such that $h'h = \rho(t, A)$ for some $t \in \mathfrak{o}_L^\times$ and $A \in [\begin{smallmatrix} \mathfrak{o}_E & \mathfrak{o}_E^\times \\ \mathfrak{o}_E^\times & \mathfrak{P} \end{smallmatrix}]$ such that $\mathrm{N}_L^E(\det A) = t^2$,*

(c) $(h^{-1}x_1, h^{-1}x_2) \in \mathrm{supp}(\varphi^{(3)})$ *if and only if there exists some $h' \in H$ such that $h'h = \rho(t, A)$ for some $t \in \mathfrak{o}_L^\times$ and $A \in [\begin{smallmatrix} \mathfrak{o}_E^\times & \mathfrak{o}_E \\ \mathfrak{P}^{n+1} & \mathfrak{o}_E^\times \end{smallmatrix}]$ such that $\mathrm{N}_L^E(\det A) = t^2$, and*

(d) $(h^{-1}x_1, h^{-1}x_2) \in \mathrm{supp}(\varphi^{(4)})$ *for no $h \in \mathrm{SO}(X)$.*

In particular we can assume that t is one of the two representatives of $L^\times/\mathrm{N}_L^E(E^\times)$.

Proof Let's start with a preliminary observation. Let $A = [\begin{smallmatrix} a & b \\ c & d \end{smallmatrix}]$ and suppose that there is some $h' \in H$ so that $h'h = \rho(1, A^{-1})$. Then we calculate that

$$h^{-1}x_1 = h^{-1}(h'^{-1}x_1) = (h'h)^{-1}x_1 = t^{-1}Ax_1\alpha(A)^* = t^{-1}\begin{bmatrix} a\alpha(c)\sqrt{\delta} & a\alpha(a)\sqrt{\delta} \\ -c\alpha(c)\sqrt{\delta} & -c\alpha(a)\sqrt{\delta} \end{bmatrix} \tag{5.12}$$

and

$$h^{-1}x_2 = h^{-1}(h'^{-1}x_2) = (h'h)^{-1}x_2 = t^{-1}Ax_2\alpha(A)^* = 2t^{-1}\begin{bmatrix} -b\alpha(d)\frac{\sqrt{\delta}}{\delta} & b\alpha(b)\frac{\sqrt{\delta}}{\delta} \\ -d\alpha(d)\frac{\sqrt{\delta}}{\delta} & d\alpha(b)\frac{\sqrt{\delta}}{\delta} \end{bmatrix}. \tag{5.13}$$

Also notice that for $i = 1, 2$ and $g \in \mathrm{GSO}(X)$ we have that $0 = \langle x_i, x_i \rangle = \langle g^{-1}x_i, g^{-1}x_i \rangle$. Considering the current application, in (5.11) we can ignore the restriction of the domain that $\langle x, x \rangle \in \mathfrak{p}^N$ and that $\langle y, y \rangle \in \mathfrak{o}_L$.

Suppose that $h = \rho(s, A')$ for some $s \in L^\times$ and some $A' \in \mathrm{GL}(2, E)$. We distinguish an element of the stabilizer $h' = \rho(\varpi_L^{-\nu_E(s)}, [\begin{smallmatrix} \varpi_E^{\nu_E(s)} & \\ & \varpi_E^{\nu_E(s)} \end{smallmatrix}]) \in H$. So that, $h'h = \rho(t, A)$ for some $t \in \mathfrak{o}_L^\times$ and $A \in \mathrm{GL}(2, E)$. In particular we may assume that t is a coset representative of the L/L^1. Since $(h'h)^{-1}x_i = h^{-1}x_i$, in the following arguments we can restrict our attention to $h = \rho(t, A)$.

(a) Suppose that $(h^{-1}x_1, h^{-1}x_2) \in \mathrm{supp}(\varphi^{(1)})$. By (5.11) we have that $(h^{-1}x_1, h^{-1}x_2) \in \mathrm{supp}(\varphi^{(1)})$ if and only if $h^{-1}x_1 \in \mathrm{supp}(\varphi_1)$ and $h^{-1}x_2 \in \mathrm{supp}(\varphi_2)$. These two conditions imply that $A \in [\begin{smallmatrix} \mathfrak{o}_E & \mathfrak{o}_E \\ \varpi^n\mathfrak{o}_E^\times & \mathfrak{o}_E^\times \end{smallmatrix}]$. On the other hand, assume that $A \in [\begin{smallmatrix} \mathfrak{o}_E & \mathfrak{o}_E \\ \varpi^n\mathfrak{o}_E^\times & \mathfrak{o}_E^\times \end{smallmatrix}]$, then by (5.12) and (5.13) we can see that $h^{-1}x_1 \in \mathrm{supp}(\varphi_1)$ and $h^{-1}x_2 \in \mathrm{supp}(\varphi_2)$. We conclude that $(h^{-1}x_1, h^{-1}x_2) \in \mathrm{supp}\varphi^{(1)}$.

(b) Suppose that $(h^{-1}x_1, h^{-1}x_2) \in \mathrm{supp}(\varphi^{(2)})$. By (5.11) we have that $(h^{-1}x_1, h^{-1}x_2) \in \mathrm{supp}(\varphi^{(1)})$ if and only if $h^{-1}x_1 \in \mathrm{supp}(\varphi_1)$ and $h^{-1}x_2 \in [\begin{smallmatrix} \mathfrak{o}_E & \varpi_E^{-1}\mathfrak{o}_E^\times \\ \mathfrak{P} & \mathfrak{o}_E \end{smallmatrix}]$. These two conditions imply that $A \in [\begin{smallmatrix} \mathfrak{o}_E & \mathfrak{o}_E^\times \\ \varpi_E^n\mathfrak{o}_E^\times & \mathfrak{P} \end{smallmatrix}]$. Since $\rho(t, A) \in \mathrm{SO}(X)$, this can only happen when $n = 0$. On the other hand, assume that $A \in [\begin{smallmatrix} \mathfrak{o}_E & \mathfrak{o}_E^\times \\ \varpi_E^n\mathfrak{o}_E^\times & \mathfrak{P} \end{smallmatrix}]$, then by (5.12) and (5.13) we can see that $h^{-1}x_1 \in \mathrm{supp}(\varphi_1)$ and $h^{-1}x_2 \in [\begin{smallmatrix} \mathfrak{o}_E & \varpi_E^{-1}\mathfrak{o}_E^\times \\ \mathfrak{P} & \mathfrak{o}_E \end{smallmatrix}] \cap X$. We conclude that $(h^{-1}x_1, h^{-1}x_2) \in \mathrm{supp}\varphi^{(2)}$.

(c) Suppose that $(h^{-1}x_1, h^{-1}x_2) \in \mathrm{supp}(\varphi^{(3)})$. By (5.11), this can happen if and only if $h^{-1}x_1 \in [\begin{smallmatrix} \mathfrak{P}^{n+2} & \varpi_E\mathfrak{o}_E^\times \\ \mathfrak{P}^{2n+3} & \mathfrak{P}^{n+2} \end{smallmatrix}]$ and $h^{-1}x_2 \in \mathrm{supp}(\varphi_2)$. These two conditions imply that $A \in [\begin{smallmatrix} \mathfrak{o}_E^\times & \mathfrak{o}_E \\ \mathfrak{P}^{n+1} & \mathfrak{o}_E^\times \end{smallmatrix}]$ On the other hand, assume that $A \in [\begin{smallmatrix} \mathfrak{o}_E^\times & \mathfrak{o}_E \\ \mathfrak{P}^{n+1} & \mathfrak{o}_E^\times \end{smallmatrix}]$, then by (5.12) and (5.13) we can see that $h^{-1}x_1 \in [\begin{smallmatrix} \mathfrak{P}^{n+2} & \varpi_E\mathfrak{o}_E^\times \\ \mathfrak{P}^{2n+3} & \mathfrak{P}^{n+2} \end{smallmatrix}]$ and $h^{-1}x_2 \in \mathrm{supp}(\varphi_2)$. We conclude that $(h^{-1}x_1, h^{-1}x_2) \in \mathrm{supp}\varphi^{(3)}$.

(d) Suppose that $(h^{-1}x_1, h^{-1}x_2) \in \mathrm{supp}(\varphi^{(4)})$. This can happen if and only if $h^{-1}x_1 \in \left[\begin{smallmatrix} \mathfrak{P}^{n+2} & \varpi_E \mathfrak{o}_E^\times \\ \mathfrak{P}^{2n+3} & \mathfrak{P}^{n+2} \end{smallmatrix}\right]$ and $h^{-1}x_2 \in \left[\begin{smallmatrix} \mathfrak{o}_E & \varpi_E^{-1} \\ \mathfrak{P} & \mathfrak{o}_E \end{smallmatrix}\right]$. These two conditions imply that $A \in \left[\begin{smallmatrix} \mathfrak{o}_E^\times & \mathfrak{o}_E^\times \\ \mathfrak{P}^{n+1} & \mathfrak{P} \end{smallmatrix}\right]$, which is impossible since $\rho(t, A) \in \mathrm{SO}(X)$. □

Corollary 5.15 *Let $h \in \mathrm{SO}(X)$. Then $(h^{-1}x_1, h^{-1}x_2) \in \mathrm{supp}(\varphi)$ if and only if there exists some $h' \in H$ such that $h'h = \rho(t, A)$ for t a representative of $L^\times/\mathrm{N}_L^E(E^\times)$ and some $A \in \Gamma_0(\mathfrak{P}^n)$ such that $\mathrm{N}_L^E(\det A) = t^2$.*

Proof If $n = 0$ then

$$\Gamma_0(\mathfrak{o}_E) = \mathrm{GL}(2, \mathfrak{o}_E) = \left(\left[\begin{smallmatrix} \mathfrak{o}_E^\times & \mathfrak{o}_E^\times \\ \mathfrak{o}_E^\times & \mathfrak{P} \end{smallmatrix}\right] \sqcup \left[\begin{smallmatrix} \mathfrak{P} & \mathfrak{o}_E^\times \\ \mathfrak{o}_E^\times & \mathfrak{o}_E^\times \end{smallmatrix}\right] \sqcup \left[\begin{smallmatrix} \mathfrak{P} & \mathfrak{o}_E^\times \\ \mathfrak{o}_E^\times & \mathfrak{P} \end{smallmatrix}\right] \sqcup \left(\left[\begin{smallmatrix} \mathfrak{o}_E^\times & \mathfrak{o}_E \\ \mathfrak{o}_E^\times & \mathfrak{o}_E^\times \end{smallmatrix}\right] \cap \mathrm{GL}(2, \mathfrak{o}_E)\right) \sqcup \left[\begin{smallmatrix} \mathfrak{o}_E^\times & \mathfrak{o}_E \\ \mathfrak{P} & \mathfrak{o}_E^\times \end{smallmatrix}\right]\right).$$

If $n > 0$ then

$$\Gamma_0(\mathfrak{P}^n) = \left(\left[\begin{smallmatrix} \mathfrak{o}_E^\times & \mathfrak{o}_E \\ \varpi_E^n \mathfrak{o}_E^\times & \mathfrak{o}_E^\times \end{smallmatrix}\right] \sqcup \left[\begin{smallmatrix} \mathfrak{o}_E^\times & \mathfrak{o}_E \\ \mathfrak{P}^{n+1} & \mathfrak{o}_E^\times \end{smallmatrix}\right]\right).$$

Examining Lemma 5.14 gives the result. □

To explicitly calculate the $B(1, \varphi, W, s)$ we will need the volumes of the subsets of $\mathrm{GL}(2, \mathfrak{o}_E)$ that appear in the proof of Corollary 5.15, in terms of the volume of some subgroups of $\mathrm{SO}(X)$. When $n = 0$ set $\Gamma = \rho(\mathfrak{o}_L^\times, \Gamma_0(\mathfrak{P})) \cap \mathrm{SO}(X)$. When $n > 0$ set $\Gamma = \rho(\mathfrak{o}_L^\times, \Gamma_0(\mathfrak{P}^n)) \cap \mathrm{SO}(X)$.

Lemma 5.16 *Table 1 summarizes volumes of certain subsets of $H\backslash\big(\rho(\mathfrak{o}_L^\times, \mathrm{GL}(2, \mathfrak{o}_E)) \cap \mathrm{SO}(X)\big)$, up to a positive constant.*

Proof Recall the Haar measures of $\mathrm{SO}(X)$ and H which were chosen in Sect. 4. Suppose that S is a subgroup of $\mathrm{SO}(X)$ and $f_S(x)$ is the characteristic function of S. We have that

$$\int\limits_{\mathrm{SO}(X)} f_S(g)\, dg = \int\limits_{H\backslash \mathrm{SO}(X)} \Big(\int\limits_H f_S(hg)\, dh\Big)\, dg = \int\limits_{H\backslash \mathrm{SO}(X)} \mathrm{vol}_H(H \cap S) f_{HS}(G)\, dh,$$

which is $\mathrm{vol}_H(H \cap S)\mathrm{vol}_{H\backslash \mathrm{SO}(X)}(H\backslash HS)$. Therefore, we have the formula for the quotient measure:

$$\mathrm{vol}_{H\backslash G}(H\backslash HS) = \frac{\mathrm{vol}_{\mathrm{SO}(X)}(S)}{\mathrm{vol}_H(H \cap S)}. \tag{5.14}$$

We begin with the case that $n > 0$. First, note that the following is an exact list of coset representatives:

$$\big(\rho(\mathfrak{o}_L^\times, \Gamma_0(\mathfrak{P}^{n+1})) \cap \mathrm{SO}(X)\big)\backslash\Gamma = \{\rho(1, \left[\begin{smallmatrix} 1 & \\ x & 1 \end{smallmatrix}\right]) \mid x \in \mathfrak{o}_E/\mathfrak{P}\}.$$

Table 1 Volumes of subsets of $\rho(\mathfrak{o}_L^\times, \mathrm{GL}(2, \mathfrak{o}_E))$

$\rho(\mathfrak{o}_L^\times, \cdot) \cap \mathrm{SO}(X)$	Volume in $\rho(\mathfrak{o}_L^\times, \mathrm{GL}(2, \mathfrak{o}_E)) \cap \mathrm{SO}(X)$	Support in $\varphi^{(i)}$
$\begin{bmatrix} \mathfrak{o}_E^\times & \mathfrak{o}_E^\times \\ \mathfrak{o}_E^\times & \mathfrak{P} \end{bmatrix}$	$(1 - \frac{1}{q})\mathrm{vol}(\Gamma)$	2
$\begin{bmatrix} \mathfrak{P} & \mathfrak{o}_E^\times \\ \mathfrak{o}_E^\times & \mathfrak{o}_E^\times \end{bmatrix}$	$(1 - \frac{1}{q})\mathrm{vol}(\Gamma)$	1
$\begin{bmatrix} \mathfrak{P} & \mathfrak{o}_E^\times \\ \mathfrak{o}_E^\times & \mathfrak{P} \end{bmatrix}$	$\frac{1}{q}\mathrm{vol}(\Gamma)$	2
$\begin{bmatrix} \mathfrak{o}_E^\times & \mathfrak{o}_E \\ \mathfrak{o}_E^\times & \mathfrak{o}_E^\times \end{bmatrix} \cap \mathrm{GL}(2, \mathfrak{o}_E)$	$1 - (3 - \frac{1}{q})\mathrm{vol}(\Gamma)$	1
$\begin{bmatrix} \mathfrak{o}_E^\times & \mathfrak{o}_E \\ \mathfrak{P} & \mathfrak{o}_E^\times \end{bmatrix}$	$\mathrm{vol}(\Gamma)$	3
$\begin{bmatrix} \mathfrak{o}_E^\times & \mathfrak{o}_E \\ \varpi_E^n \mathfrak{o}_E^\times & \mathfrak{o}_E^\times \end{bmatrix}$	$(1 - \frac{1}{q})\mathrm{vol}(\Gamma)$	1
$\begin{bmatrix} \mathfrak{o}_E^\times & \mathfrak{o}_E \\ \mathfrak{P}^{n+1} & \mathfrak{o}_E^\times \end{bmatrix}$	$\frac{1}{q}\mathrm{vol}(\Gamma)$	3

Therefore, the volume of $\rho(\mathfrak{o}_L^\times, \Gamma_0(\mathfrak{P}^{n+1})) \cap \mathrm{SO}(X)$ is $\frac{1}{q}\mathrm{vol}(\Gamma)$. By (5.14) we have that

$$\begin{aligned}\mathrm{vol}(H\backslash\rho(\mathfrak{o}_L^\times, \Gamma_0(\mathfrak{P}^{n+1})) &= \mathrm{vol}(H\backslash H\rho(\mathfrak{o}_L^\times, \Gamma_0(\mathfrak{P}^{n+1})) \\ &= \frac{\mathrm{vol}(\rho(\mathfrak{o}_L^\times, \Gamma_0(\mathfrak{P}^{n+1}))}{\mathrm{vol}(\rho(\mathfrak{o}_L^\times, \Gamma_0(\mathfrak{P}^{n+1}) \cap H)} \\ &= \frac{\mathrm{vol}(\rho(\mathfrak{o}_L^\times, \Gamma_0(\mathfrak{P}^{n+1}))}{\mathrm{vol}(H)} = \frac{1}{q}\mathrm{vol}(\Gamma).\end{aligned}$$

Since $H\backslash\rho(\mathfrak{o}_L^\times, \left[\begin{smallmatrix} \mathfrak{o}_E^\times & \mathfrak{o}_E \\ \varpi_E^n \mathfrak{o}_E^\times & \mathfrak{o}_E^\times \end{smallmatrix}\right]) \cap \mathrm{SO}(X)$ is the complement of $H\backslash\rho(\mathfrak{o}_L^\times, \Gamma_0(\mathfrak{P}^{n+1})) \cap \mathrm{SO}(X)$ inside of $H\backslash\Gamma$, we find that

$$\begin{aligned}\mathrm{vol}(H\backslash\rho(\mathfrak{o}_L^\times, \left[\begin{smallmatrix} \mathfrak{o}_E^\times & \mathfrak{o}_E \\ \varpi_E^n \mathfrak{o}_E^\times & \mathfrak{o}_E^\times \end{smallmatrix}\right]) \cap \mathrm{SO}(X)) &= \mathrm{vol}(H\backslash\Gamma) - \mathrm{vol}(H\backslash\rho(\mathfrak{o}_L^\times, \Gamma_0(\mathfrak{P}^{n+1})) \cap \mathrm{SO}(X)) \\ &= (1 - \frac{1}{q})\mathrm{vol}(\Gamma).\end{aligned}$$

Now assume that $n = 0$. The following is a complete list of coset representatives

$$\rho(\mathfrak{o}_L^\times, \left[\begin{smallmatrix} \mathfrak{o}_E^\times & \mathfrak{P} \\ \mathfrak{P} & \mathfrak{o}_E^\times \end{smallmatrix}\right]) \cap \mathrm{SO}(X)\backslash\Gamma = \{[\begin{smallmatrix} 1 & x \\ & 1 \end{smallmatrix}] \mid x \in \mathfrak{o}_E/\mathfrak{P}\}.$$

So, similarly to the above case we find that

$$\begin{aligned}\mathrm{vol}(H\backslash\rho(\mathfrak{o}_L^\times, \begin{bmatrix} \mathfrak{o}_E^\times & \mathfrak{P} \\ \mathfrak{P} & \mathfrak{o}_E^\times \end{bmatrix}) \cap \mathrm{SO}(X)) &= \mathrm{vol}(H\backslash H\rho(\mathfrak{o}_L^\times, \begin{bmatrix} \mathfrak{o}_E^\times & \mathfrak{P} \\ \mathfrak{P} & \mathfrak{o}_E^\times \end{bmatrix}) \cap \mathrm{SO}(X)) \\ &= \frac{\mathrm{vol}(\rho(\mathfrak{o}_L^\times, \begin{bmatrix} \mathfrak{o}_E^\times & \mathfrak{P} \\ \mathfrak{P} & \mathfrak{o}_E^\times \end{bmatrix}) \cap \mathrm{SO}(X))}{\mathrm{vol}(\rho(\mathfrak{o}_L^\times, \begin{bmatrix} \mathfrak{o}_E^\times & \mathfrak{P} \\ \mathfrak{P} & \mathfrak{o}_E^\times \end{bmatrix}) \cap H)} \\ &= \frac{\mathrm{vol}(\rho(\mathfrak{o}_L^\times, \begin{bmatrix} \mathfrak{o}_E^\times & \mathfrak{P} \\ \mathfrak{P} & \mathfrak{o}_E^\times \end{bmatrix}) \cap \mathrm{SO}(X))}{\mathrm{vol}(H)} = \frac{1}{q}\mathrm{vol}(\Gamma).\end{aligned}$$

Since $H\backslash\rho(\mathfrak{o}_L^\times, \begin{bmatrix} \mathfrak{o}_E^\times & \mathfrak{o}_E^\times \\ \mathfrak{P} & \mathfrak{o}_E^\times \end{bmatrix}) \cap \mathrm{SO}(X))$ is the complement of $H\backslash\rho(\mathfrak{o}_L^\times, \begin{bmatrix} \mathfrak{o}_E^\times & \mathfrak{P} \\ \mathfrak{P} & \mathfrak{o}_E^\times \end{bmatrix}) \cap \mathrm{SO}(X))$ inside of $H\backslash\Gamma$, we find that

$$\begin{aligned}\mathrm{vol}(H\backslash\rho(\mathfrak{o}_L^\times, \begin{bmatrix} \mathfrak{o}_E^\times & \mathfrak{o}_E^\times \\ \mathfrak{P} & \mathfrak{o}_E^\times \end{bmatrix}) \cap \mathrm{SO}(X)) &= \mathrm{vol}(H\backslash\Gamma) - \mathrm{vol}(H\backslash\rho(\mathfrak{o}_L^\times, \begin{bmatrix} \mathfrak{o}_E^\times & \mathfrak{P} \\ \mathfrak{P} & \mathfrak{o}_E^\times \end{bmatrix}) \cap \mathrm{SO}(X)) \\ &= (1 - \frac{1}{q})\mathrm{vol}(\Gamma).\end{aligned}$$

Lastly we have that $H\backslash\rho(\mathfrak{o}_L^\times, \begin{bmatrix} \mathfrak{o}_E^\times & \mathfrak{o}_E \\ \mathfrak{o}_E^\times & \mathfrak{o}_E^\times \end{bmatrix} \cap \mathrm{GL}(2, \mathfrak{o}_E)) \cap \mathrm{SO}(X)$ is the complement of $H\backslash\rho(\mathfrak{o}_L^\times, \begin{bmatrix} \mathfrak{o}_E^\times & \mathfrak{o}_E \\ \mathfrak{P} & \mathfrak{o}_E^\times \end{bmatrix} \sqcup \begin{bmatrix} \mathfrak{P} & \mathfrak{o}_E^\times \\ \mathfrak{o}_E^\times & \mathfrak{P} \end{bmatrix} \sqcup \begin{bmatrix} \mathfrak{P} & \mathfrak{o}_E^\times \\ \mathfrak{o}_E^\times & \mathfrak{o}_E^\times \end{bmatrix} \sqcup \begin{bmatrix} \mathfrak{o}_E^\times & \mathfrak{o}_E^\times \\ \mathfrak{o}_E^\times & \mathfrak{P} \end{bmatrix}) \cap \mathrm{SO}(X))$ inside of $H\backslash\rho(\mathfrak{o}_L^\times, \mathrm{GL}(2, \mathfrak{o}_E)) \cap \mathrm{SO}(X)$. Therefore

$$\begin{aligned}&\mathrm{vol}(H\backslash\rho(\mathfrak{o}_L^\times, \begin{bmatrix} \mathfrak{o}_E^\times & \mathfrak{o}_E \\ \mathfrak{o}_E^\times & \mathfrak{o}_E^\times \end{bmatrix}) \cap \mathrm{SO}(X)) \\ &= 1 - \mathrm{vol}(H\backslash\rho(\mathfrak{o}_L^\times, \begin{bmatrix} \mathfrak{o}_E^\times & \mathfrak{o}_E \\ \mathfrak{P} & \mathfrak{o}_E^\times \end{bmatrix} \sqcup \begin{bmatrix} \mathfrak{P} & \mathfrak{o}_E^\times \\ \mathfrak{o}_E^\times & \mathfrak{P} \end{bmatrix} \sqcup \begin{bmatrix} \mathfrak{P} & \mathfrak{o}_E^\times \\ \mathfrak{o}_E^\times & \mathfrak{o}_E^\times \end{bmatrix} \sqcup \begin{bmatrix} \mathfrak{o}_E^\times & \mathfrak{o}_E^\times \\ \mathfrak{o}_E^\times & \mathfrak{P} \end{bmatrix}) \cap \mathrm{SO}(X)) \\ &= 1 - \mathrm{vol}(H\backslash\Gamma) - \mathrm{vol}(H\backslash\rho(\mathfrak{o}_L^\times, \begin{bmatrix} \mathfrak{o}_E^\times & \mathfrak{P} \\ \mathfrak{P} & \mathfrak{o}_E^\times \end{bmatrix}) \cap \mathrm{SO}(X)) - 2\mathrm{vol}(H\backslash\rho(\mathfrak{o}_L^\times, \begin{bmatrix} \mathfrak{o}_E^\times & \mathfrak{o}_E^\times \\ \mathfrak{P} & \mathfrak{o}_E^\times \end{bmatrix}) \cap \mathrm{SO}(X)) \\ &= 1 - (3 - \frac{1}{q})\mathrm{vol}(\Gamma).\end{aligned}$$

This completes the proof. □

The following result completes the proof of the main theorem for the tamely ramified case.

Theorem 5.17 *Suppose that* L *has odd residual characteristic, that* E/L *is ramified, and that* $W \in \mathcal{W}_{\tau_0}$ *has* $\Gamma_0(\mathfrak{p}^n)$*-invariance. Let* φ *be chosen as in* (5.10) *and assume that* $s \gg M$*. Then, the function* $B(\cdot, \varphi, W, s)$ *as defined in* (4.2) *is non-zero and* $\mathrm{K}(\mathfrak{P}^N)$*-invariant, where* $N = n + 2$*. In particular* $B(1, \varphi, W, s) \neq 0$*.*

Proof By construction we have that $\tilde{\varphi}$ is paramodular invariant under the action of the Weil representation. It follows that $B(\cdot, \tilde{\varphi}, W, s)$ is also paramodular invariant.

We write a generic element of $H\backslash\mathrm{SO}(X)$ as $h = \rho(1, [\begin{smallmatrix} a & b \\ c & d \end{smallmatrix}])$. Using Lemma 5.14 to determine the support of the intertwining map we write

$$\begin{aligned}
B(1, \varphi^{(1)}, W, s) &= \int_{H\backslash\mathrm{SO}(X)} \omega(1, h)\varphi^{(1)}(x_1, x_2)Z(s, \pi(h)W)\, dh \\
&= \int_{H\backslash\rho(\mathfrak{o}_L^\times, \left[\begin{smallmatrix} \mathfrak{o}_E & \mathfrak{o}_E \\ \varpi_E^n\mathfrak{o}_E^\times & \mathfrak{o}_E^\times \end{smallmatrix}\right])\cap\mathrm{SO}(X)} \omega(1, h)\varphi^{(1)}(x_1, x_2)Z(s, \pi(h)W)\, dh \\
&= q_L^2 \int_{H\backslash\rho(\mathfrak{o}_L^\times, \left[\begin{smallmatrix} \mathfrak{o}_E & \mathfrak{o}_E \\ \varpi_E^n\mathfrak{o}_E^\times & \mathfrak{o}_E^\times \end{smallmatrix}\right])\cap\mathrm{SO}(X)} \chi(2c\alpha(c)d\alpha(d)\delta^{-1})Z(s, \pi(h)W)\, dh \\
&= q_L^2 \mathrm{vol}(H\backslash\rho(\mathfrak{o}_L^\times, \left[\begin{smallmatrix} \mathfrak{o}_E & \mathfrak{o}_E \\ \varpi_E^n\mathfrak{o}_E^\times & \mathfrak{o}_E^\times \end{smallmatrix}\right]) \cap \mathrm{SO}(X))\chi(2\delta)Z(s, W)
\end{aligned}$$

also,

$$\begin{aligned}
B(1, \varphi^{(3)}, W, s) &= \int_{H\backslash\mathrm{SO}(X)} \omega(1, h)\varphi^{(3)}(x_1, x_2)Z(s, \pi(h)W)\, dh \\
&= \int_{H\backslash\rho(\mathfrak{o}_L^\times, \left[\begin{smallmatrix} \mathfrak{o}_E^\times & \mathfrak{o}_E \\ \mathfrak{P}^{n+1} & \mathfrak{o}_E^\times \end{smallmatrix}\right])\cap\mathrm{SO}(X)} \omega(1, h)\varphi^{(3)}(x_1, x_2)Z(s, \pi(h)W)\, dh \\
&= q_L \int_{H\backslash\rho(\mathfrak{o}_L^\times, \left[\begin{smallmatrix} \mathfrak{o}_E^\times & \mathfrak{o}_E \\ \mathfrak{P}^{n+1} & \mathfrak{o}_E^\times \end{smallmatrix}\right])\cap\mathrm{SO}(X)} \chi(-2d\alpha(d)a\alpha(a)\delta^{-1})Z(s, \pi(h)W)\, dh \\
&= q_L \mathrm{vol}(H\backslash\rho(\mathfrak{o}_L^\times, \left[\begin{smallmatrix} \mathfrak{o}_E^\times & \mathfrak{o}_E \\ \mathfrak{P}^{n+1} & \mathfrak{o}_E^\times \end{smallmatrix}\right]) \cap \mathrm{SO}(X))\chi(2)Z(s, W).
\end{aligned}$$

If $n = 0$ then we also have a contribution from

$$\begin{aligned}
B(1, \varphi^{(2)}, W, s) &= \int_{H\backslash\mathrm{SO}(X)} \omega(1, h)\varphi^{(2)}(x_1, x_2)Z(s, \pi(h)W)\, dh \\
&= \int_{H\backslash\rho(\mathfrak{o}_L^\times, \left[\begin{smallmatrix} \mathfrak{P} & \mathfrak{o}_E^\times \\ \mathfrak{o}_E^\times & \mathfrak{o}_E \end{smallmatrix}\right])\cap\mathrm{SO}(X)} \omega(1, h)\varphi^{(2)}(x_1, x_2)Z(s, \pi(h)W)\, dh \\
&= q_L \int_{H\backslash\rho(\mathfrak{o}_L^\times, \left[\begin{smallmatrix} \mathfrak{P} & \mathfrak{o}_E^\times \\ \mathfrak{o}_E^\times & \mathfrak{o}_E \end{smallmatrix}\right])\cap\mathrm{SO}(X)} \chi(-2d\alpha(d)b\alpha(b)\delta^{-1})Z(s, \pi(h)W)\, dh \\
&= q_L \mathrm{vol}(H\backslash\rho(\mathfrak{o}_L^\times, \left[\begin{smallmatrix} \mathfrak{P} & \mathfrak{o}_E^\times \\ \mathfrak{o}_E^\times & \mathfrak{o}_E \end{smallmatrix}\right] \cap \mathrm{SO}(X)))\chi(2)Z(s, W).
\end{aligned}$$

Therefore, from Lemma 5.14, we find that

$$\begin{aligned} B(1,\varphi,W,s) =& B(1,\varphi^{(1)}+\varphi^{(2)}+\varphi^{(3)},W,s) \\ =& \chi(2)Z(s,W)\Big[\chi(\delta)q_L \mathrm{vol}(H\backslash\rho(1,\left[\begin{smallmatrix} \mathfrak{o}_E & \mathfrak{o}_E \\ \varpi_E^n\mathfrak{o}_E^\times & \mathfrak{o}_E^\times \end{smallmatrix}\right])\cap \mathrm{SO}(X)) \\ &+ q_L^2\mathrm{vol}(H\backslash\rho(1,\left[\begin{smallmatrix} \mathfrak{o}_E^\times & \mathfrak{o}_E \\ \mathfrak{P}^{n+1} & \mathfrak{o}_E^\times \end{smallmatrix}\right])\cap \mathrm{SO}(X)) \\ &+ \epsilon(n)q_L\mathrm{vol}(H\backslash\rho(1,\left[\begin{smallmatrix} \mathfrak{P} & \mathfrak{o}_E^\times \\ \mathfrak{o}_E^\times & \mathfrak{o}_E \end{smallmatrix}\right])\cap \mathrm{SO}(X))\Big] \end{aligned}$$

where $\epsilon(0)=1$ and is equal to zero elsewhere.

When $n>0$ we can use the volumes in Table 1 to calculate that

$$\begin{aligned} B(1,\varphi,W,s) &= \chi(2)Z(s,W)\Big[\chi(\delta)q_L\mathrm{vol}(H\backslash\rho(1,\left[\begin{smallmatrix} \mathfrak{o}_E^\times & \mathfrak{o}_E \\ \varpi_E^n\mathfrak{o}_E^\times & \mathfrak{o}_E^\times \end{smallmatrix}\right])\cap \mathrm{SO}(X)) \\ &\qquad + q_L^2\mathrm{vol}(H\backslash\rho(1,\left[\begin{smallmatrix} \mathfrak{o}_E^\times & \mathfrak{o}_E \\ \mathfrak{P}^{n+1} & \mathfrak{o}_E^\times \end{smallmatrix}\right])\cap \mathrm{SO}(X))\Big] \\ &= \chi(2)Z(s,W)\mathrm{vol}(\Gamma)\Big[\chi(\delta)(1-\frac{1}{q})+\frac{1}{q}\Big] \neq 0 \end{aligned}$$

since $q\neq 2$. When $n=0$ we have

$$\begin{aligned} B(1,\varphi,W,s) &= \chi(2)Z(s,W)\Big[\chi(\delta)q_L\mathrm{vol}(H\backslash\rho(1,\left[\begin{smallmatrix} \mathfrak{o}_E & \mathfrak{o}_E \\ \mathfrak{o}_E^\times & \mathfrak{o}_E^\times \end{smallmatrix}\right])\,\mathrm{SO}(X)) \\ &\qquad + q_L^2\mathrm{vol}(H\backslash\rho(1,\left[\begin{smallmatrix} \mathfrak{o}_E^\times & \mathfrak{o}_E \\ \mathfrak{P} & \mathfrak{o}_E^\times \end{smallmatrix}\right])\,\mathrm{SO}(X)) \\ &\qquad + q_L\mathrm{vol}(H\backslash\rho(1,\left[\begin{smallmatrix} \mathfrak{P} & \mathfrak{o}_E^\times \\ \mathfrak{o}_E^\times & \mathfrak{o}_E \end{smallmatrix}\right])\cap \mathrm{SO}(X))\Big] \\ &= \chi(2)Z(s,W)\Big[\chi(\delta)q\big(1-(3-\frac{1}{q})\mathrm{vol}(\Gamma)+(1-\frac{1}{q})\mathrm{vol}(\Gamma)\big) \\ &\quad + q^2(\mathrm{vol}(\Gamma))+q(\mathrm{vol}(\Gamma))\Big] \\ &= \chi(2)Z(s,W)\Big[\chi(\delta)q+\mathrm{vol}(\Gamma)(\chi(\delta)(4q-2)+q^2+q)\Big] \\ &\neq 0 \end{aligned}$$

since $\mathrm{vol}(\Gamma)=\frac{1}{q+1}$. □

Acknowledgements This project was initiated as part of the second author's Ph.D. thesis at the University of Idaho. Both authors would like to thank Brooks Roberts for providing key insights through many conversations, sharing unpublished work, and providing detailed feedback on the initial form of these results as a member of the dissertation committee.

References

[BK14] Armand Brumer and Kenneth Kramer, *Paramodular abelian varieties of odd conductor*, Trans. Amer. Math. Soc. **366** (2014), no. 5, 2463–2516. MR 3165645

[Bum97] Daniel Bump, *Automorphic forms and representations*, Cambridge Studies in Advanced Mathematics, vol. 55, Cambridge University Press, Cambridge, 1997. MR 1431508 (97k:11080)

[BZ76] I. N. Bernšteĭn and A. V. Zelevinskiĭ, *Representations of the group $GL(n, F)$, where F is a local non-Archimedean field*, Uspehi Mat. Nauk **31** (1976), no. 3(189), 5–70. MR 0425030 (54 #12988)

[FLHS15] Nuno Freitas, Bao V. Le Hung, and Samir Siksek, *Elliptic curves over real quadratic fields are modular*, Invent. Math. **201** (2015), no. 1, 159–206. MR 3359051

[Gan23] Wee Teck Gan, *Explicit constructions of automorphic forms: theta correspondence and automorphic descent.*

[Gel75] Stephen Gelbart, *Automorphic forms and representations of adele groups*, Department of Mathematics, University of Chicago, Chicago, Ill., 1975, Lecture Notes in Representation Theory. MR 0419361 (54 #7382)

[God70] R Godement, *Notes on jacquet-langlands' theory*, The Institute for Advanced Study, 1970.

[How79] R. Howe, *θ-series and invariant theory*, Automorphic forms, representations and L-functions (Proc. Sympos. Pure Math., Oregon State Univ., Corvallis, Ore., 1977), Part 1, Proc. Sympos. Pure Math., vol. XXXIII, Amer. Math. Soc., Providence, RI, 1979, pp. 275–285. MR 546602

[JL70a] H. Jacquet and R. P. Langlands, *Automorphic forms on gl (2)*, Springer Berlin Heidelberg, 1970.

[JL70b] H. Jacquet and R. P. Langlands, *Automorphic forms on* GL(2), Lecture Notes in Mathematics, Vol. 114, Springer-Verlag, Berlin-New York, 1970. MR 0401654 (53 #5481)

[JLR12] Jennifer Johnson-Leung and Brooks Roberts, *Siegel modular forms of degree two attached to Hilbert modular forms*, J. Number Theory **132** (2012), no. 4, 543–564. MR 2887605

[JLRS23] Jennifer Johnson-Leung, Brooks Roberts, and Ralf Schmidt, *Stable klingen vectors and paramodular newforms*, Springer Nature Switzerland, 2023.

[Knu91] Max-Albert Knus, *Quadratic and Hermitian forms over rings*, Grundlehren der Mathematischen Wissenschaften [Fundamental Principles of Mathematical Sciences], vol. 294, Springer-Verlag, Berlin, 1991, With a foreword by I. Bertuccioni. MR 1096299 (92i:11039)

[Rob01] Brooks Roberts, *Global L-packets for* GSp(2) *and theta lifts*, Doc. Math. **6** (2001), 247–314 (electronic). MR 1871665 (2003a:11059)

[Rob03] Brooks Roberts, *Epsilon factors for some representations of gsp(2) and bessel coefficients*, 2003, Unpublished Manuscript.

[RS07] Brooks Roberts and Ralf Schmidt, *Local newforms for GSp(4)*, Lecture Notes in Mathematics, vol. 1918, Springer, Berlin, 2007. MR 2344630 (2008g:11080)

[WAL80] J.-L. Waldspurger, *Correspondance de Shimura*, J. Math. Pures Appl. (9) **59** (1980), no. 1, 1–132. MR 577010

[Wal81] J.-L. Waldspurger, *Sur les coefficients de Fourier des formes modulaires de poids demi-entier*, J. Math. Pures Appl. (9) **60** (1981), no. 4, 375–484. MR 646366

[Wei64] André Weil, *Sur certains groupes d'opérateurs unitaires*, Acta Math. **111** (1964), 143–211. MR 165033

[Yos79] Hiroyuki Yoshida, *Weil's representations of the symplectic groups over finite fields*, J. Math. Soc. Japan **31** (1979), no. 2, 399–426. MR 527552 (82j:20084)

Comparing Densities for Powers in Artin's Conjecture on Primitive Roots

Antonella Perucca and Pietro Sgobba

Abstract We consider Artin's conjecture on primitive roots and related Artin-type problems, working over a number field K. Such problems deal with the reductions of algebraic numbers $\alpha \in K^\times$ modulo primes $\mathfrak{p}$ of K. The key property concerns the value of the index of $(\alpha \bmod \mathfrak{p})$, but it is also customary to require an additional Frobenius condition. The set of primes $\mathfrak{p}$ satisfying such properties admits a density, conditionally under the generalized Riemann hypothesis. In this work, we compare the density for α to the density for its powers, and also address some related questions.

2010 Mathematics Subject Classification. 11A07 · 11M26 · 11R45

1 Introduction

We investigate questions stemming from Artin's conjecture on primitive roots, and we refer the reader to Moree's survey [Mor12] for an introduction to the research area around this conjecture.

Consider a rational number α which is not $0, 1, -1$ and not a square in $\mathbb{Q}^\times$, and exclude the finitely many primes p such that the reduction $(\alpha \bmod p)$ is zero. We may then ask whether α is a primitive root at p, which means that $(\alpha \bmod p)$ has index 1 in the multiplicative group $(\mathbb{Z}/p\mathbb{Z})^\times$. Conditionally under the Generalized Riemann Hypothesis (GRH), Hooley [Hoo67] proved that the set of prime numbers p for which α is a primitive root at p has a positive Dirichlet density, which we call $\operatorname{dens}(\alpha)$. He also gave explicit formulas for $\operatorname{dens}(\alpha)$ (which we recall in Sect. 2).

A. Perucca (✉)
Department of Mathematics, University of Luxembourg, Esch-sur-Alzette, Luxembourg
e-mail: antonella.perucca@uni.lu

P. Sgobba
Department of Pure Mathematics, Xi'an Jiaotong–Liverpool University, Suzhou, China
e-mail: pietro.sgobba@xjtlu.edu.cn

S. Akhtari et al. (eds.), *Research Directions in Number Theory*, Association for Women in Mathematics Series 39, https://doi.org/10.1007/978-3-032-11183-8_11

Since some power of α is a primitive root at p only if α is a primitive root at p, we thus have that

$$\mathrm{dens}(\alpha^m) \leqslant \mathrm{dens}(\alpha)$$

for every positive integer m. The difference between these two densities is due to the contribution to $\mathrm{dens}(\alpha)$ given by the primes p such that $p \equiv 1 \bmod q$ holds for some prime divisor q of m.

In fact, it does happen that $\mathrm{dens}(\alpha^m) = \mathrm{dens}(\alpha)$. One first reason is the following: if α is an m-th power, then there is no contribution to $\mathrm{dens}(\alpha)$ given by the primes considered above. However, the equality $\mathrm{dens}(\alpha^m) = \mathrm{dens}(\alpha)$ can hold even if for every $n > 1$ the rational number α is not an n-th power. For example, Hooley's formula implies that

$$\mathrm{dens}(-3^3) = \mathrm{dens}(-3)\,.$$

Indeed, there is no contribution to $\mathrm{dens}(-3)$ coming from the primes $p \equiv 1 \bmod 3$ because $\mathbb{Q}(\sqrt{-3}) = \mathbb{Q}(\zeta_3)$ and the index of $(-3 \bmod p)$ is even if $p \neq 3$ splits in $\mathbb{Q}(\sqrt{-3})$.

In Theorem 2.2, we provide a complete description for when we have $\mathrm{dens}(\alpha^m) = \mathrm{dens}(\alpha)$.

We also study the above question replacing $\mathbb{Q}$ by an arbitrary number field K and assuming GRH, by which we mean the Riemann hypothesis for the Dedekind zeta functions of number fields. Some of our results can be made unconditional by assuming that the involved densities do exist.

We also replace α by a finitely generated torsion-free subgroup G of $K^\times$. Finally, we consider more general conditions for the index of the reductions of G, see Theorem 4.1 and the results in Sect. 5. Our investigations naturally lead us to consider how the density for an algebraic number changes if we enlarge the number field that contains it, see, for example, Theorem 4.6.

There are several works in the literature addressing Artin's conjecture on primitive roots and its variants. For example, Lenstra [Len77] and Ziegler [Zie06] provide formulas when the condition of the index being 1 is replaced by the index dividing t or the index being t, for some positive integer t. Moreover, for $\alpha \in \mathbb{Q}$, the problem of restricting the classical Artin condition on $\mathrm{ind}_p(\alpha)$ by only considering the primes p in a given arithmetic progression was investigated by Lenstra [Len77, Theorem 8.3], by Moree [Mor99, Mor08], and later by Lenstra et al. [LMS14, Sect. 5]. Finally, notice that in order to prove our results with general conditions on the index we rely on the framework by Järviniemi and Perucca [JP23], later developed by these two authors and Sgobba [JPS25].

2 Comparing Artin Densities for Rational Numbers

In this section, we work over $\mathbb{Q}$, *assuming GRH.* We use the notation ℓ to denote prime numbers. For ℓ odd we set $\ell^* := (-1)^{(\ell-1)/2}\ell$, so $\mathbb{Q}(\sqrt{\ell^*})$ is the quadratic subextension of the cyclotomic field $\mathbb{Q}(\zeta_\ell)$. We write μ for the Möbius function.

Let $\alpha \in \mathbb{Q}^\times \backslash \{\pm 1\}$ be not a square in $\mathbb{Q}^\times$, and call δ the discriminant of the quadratic field $\mathbb{Q}(\sqrt{\alpha})$. Denote by τ the largest integer for which $\alpha \in \mathbb{Q}^{\times\tau}$ (notice that τ is odd). Artin's density [Hoo67, Sect. 7] can then be written as

$$\operatorname{dens}(\alpha) = A(\tau)\begin{cases} 1 & \text{if } \delta \not\equiv 1 \bmod 4, \\ 1 - \mu(|\delta|) f_\tau(\delta) & \text{otherwise}, \end{cases}$$

where

$$A(\tau) := \prod_{\ell \mid \tau}\left(1 - \frac{1}{\ell - 1}\right)\prod_{\ell \nmid \tau}\left(1 - \frac{1}{\ell(\ell-1)}\right)$$

and

$$f_\tau(\delta) = \prod_{\ell \mid \delta,\ \ell \mid \tau} \frac{1}{\ell - 2} \prod_{\ell \mid \delta,\ \ell \nmid \tau} \frac{1}{\ell^2 - \ell - 1}\,.$$

The above formulas, using that the series $\sum_\ell \ell^{-s}$ is convergent for $s > 1$, imply that $\operatorname{dens}(\alpha) > 0$. Indeed, it is well known that the Artin constant $A(1)$ is non-zero (in fact, $A(1) \approx 0.3739558\ldots$) so its positive rational multiple $A(\tau)$ is non-zero.

Also remark that if t is the squarefree part of τ (namely, t is obtained from τ by removing repeated prime factors), then we have $A(t) = A(\tau)$.

Remark 2.1 Let m be an odd positive integer. We then have

$$\frac{\operatorname{dens}(\alpha^m)}{\operatorname{dens}(\alpha)} = \frac{A(\tau m)}{A(\tau)}\begin{cases} 1 & \text{if } \delta \not\equiv 1 \bmod 4, \\ \frac{1-\mu(|\delta|) f_{\tau m}(\delta)}{1-\mu(|\delta|) f_\tau(\delta)} & \text{otherwise}. \end{cases}$$

This ratio does not change if we replace m by its squarefree part, or if we remove from m prime factors that divide τ. In fact, we have

$$\frac{A(\tau m)}{A(\tau)} = \prod_{\ell \mid m,\ \ell \nmid \tau}\left(1 - \frac{1}{\ell - 1}\right)\left(1 - \frac{1}{\ell(\ell-1)}\right)^{-1} = \prod_{\ell \mid m,\ \ell \nmid \tau} \frac{\ell^2 - 2\ell}{\ell^2 - \ell - 1}\,.$$

In particular, if $\delta \not\equiv 1 \bmod 4$ or if the common prime factors of δ and m divide τ (so that $f_{\tau m} = f_\tau$ in the latter case), we have

$$\frac{\operatorname{dens}(\alpha^m)}{\operatorname{dens}(\alpha)} = \prod_{\ell \mid m,\ \ell \nmid \tau} \frac{\ell^2 - 2\ell}{\ell^2 - \ell - 1}\,.$$

In the following result, we characterize when the ratio $\mathrm{dens}(\alpha^m)/\mathrm{dens}(\alpha)$ equals 1.

Theorem 2.2 *If m is a positive odd squarefree integer, then* $\mathrm{dens}(\alpha) = \mathrm{dens}(\alpha^m)$ *if and only if $\alpha \in \mathbb{Q}^{\times m}$ or there is $\ell \mid m$ such that $\alpha \in \mathbb{Q}^{\times m/\ell}$ and $\alpha/\ell^* \in \mathbb{Q}^{\times 2}$.*

Proof The equality $\mathrm{dens}(\alpha) = \mathrm{dens}(\alpha^m)$ is equivalent to $\mathrm{dens}(\alpha) = \mathrm{dens}(\alpha^q)$ holding for every prime $q \mid m$, and the condition for q holds if α is a q-th power, see Theorem 4.10 and Remark 4.3. Notice that this argument holds also over number fields.

Sufficiency. It remains to prove that if α/ℓ^* is a square, then $\mathrm{dens}(\alpha) = \mathrm{dens}(\alpha^\ell)$. Our assumptions imply that $\delta = \ell^* \equiv 1 \bmod 4$. Then the desired equality follows from Remark 2.1, since we have $\mu(|\delta|) = -1$ and

$$\frac{A(\ell)}{A(1)} = \frac{\ell(\ell-2)}{\ell^2-\ell-1} \qquad f_1(\delta) = \frac{1}{\ell^2-\ell-1} \qquad f_\ell(\delta) = \frac{1}{\ell-2}\,.$$

Necessity. Suppose that α is not an m-th power and let $\ell \mid m$ be a prime such that $\alpha \notin \mathbb{Q}^{\times \ell}$. It suffices to prove that α/ℓ^* is a square (equivalently, $\delta = \ell^*$), because this implies that ℓ is unique. Notice that τ is coprime to 2ℓ. We know $\mathrm{dens}(\alpha^\ell) = \mathrm{dens}(\alpha)$ and we invoke Remark 2.1. We exclude $\delta \not\equiv 1 \bmod 4$ because $A(\tau\ell) \neq A(\tau)$. We also get the identity

$$1 - \mu(|\delta|) f_\tau(\delta) = \frac{\ell(\ell-2)}{\ell^2-\ell-1}\big(1 - \mu(|\delta|) f_{\tau\ell}(\delta)\big). \tag{2.1}$$

Let us write

$$f(\delta, \tau) := \prod_{\substack{q \mid \gcd(\delta,\tau) \\ q \neq \ell}} \frac{1}{q-2} \prod_{\substack{q \mid \delta \\ q \nmid \ell\tau}} \frac{1}{q^2-q-1}\,.$$

If $\ell \nmid \delta$, then (2.1) leads to a contradiction while, supposing $\ell \mid \delta$, (2.1) becomes

$$\ell^2 - \ell - 1 - \mu(|\delta|) f(\delta, \tau) = \ell(\ell-2) - \ell\mu(|\delta|) f(\delta, \tau).$$

We deduce $\mu(|\delta|) f(\delta, \tau) = -1$, namely, $f(\delta, \tau) = 1$ and $\mu(|\delta|) = -1$.

The second product in $f(\delta, \tau)$ must be empty, and in the first product we can at most have $q = 3$. This gives $|\delta| = \ell$ (we exclude $|\delta| = 3\ell$ because $\mu(3\ell) = 1$). Recalling that $\delta \equiv 1 \bmod 4$ we obtain $\delta = \ell^*$. □

3 Preliminaries for the General Case

3.1 Setup

We let K be a number field, and work within a fixed algebraic closure. We write ζ_n for a primitive n-th root of unity. Given a set $\mathcal{S}$ of primes of K, we denote by $\pi_{\mathcal{S}}(x)$ the number of primes $\mathfrak{p} \in \mathcal{S}$ with norm $\mathrm{N}\,\mathfrak{p} \leqslant x$.

We let $\alpha \in K^\times$ and G be a finitely generated torsion-free subgroup of $K^\times$ of positive rank r. While considering the reductions of α or G modulo primes $\mathfrak{p}$ of K we tacitly exclude the finitely many $\mathfrak{p}$ such that the reduction modulo $\mathfrak{p}$ is zero and hence not contained in the multiplicative group of the residue field $k_\mathfrak{p}$. It suffices to discard the primes $\mathfrak{p}$ such that $v_\mathfrak{p}(\alpha) \neq 0$, where $v_\mathfrak{p}$ is the $\mathfrak{p}$-adic valuation. We then denote by $\mathrm{ind}_\mathfrak{p}(\alpha)$ the index of $(\alpha \bmod \mathfrak{p})$ in $k_\mathfrak{p}^\times$, and similarly define $\mathrm{ind}_\mathfrak{p}(G)$.

By density we always mean Dirichlet density. If clear from the context, we don't specify the dependency of densities on the field K. We call $\mathrm{dens}(\alpha)$ the classical Artin density of the primes $\mathfrak{p}$ of K for which $\mathrm{ind}_\mathfrak{p}(\alpha) = 1$, and for $t \geqslant 1$ we write $\mathrm{dens}(\alpha, t)$ if we require $\mathrm{ind}_\mathfrak{p}(\alpha) = t$. We write $\mathrm{dens}(\alpha, a \bmod d)$ if we require $\mathrm{ind}_\mathfrak{p}(\alpha) = 1$ and restrict to the primes whose norm lies in the arithmetic progression $a \bmod d$. Notice that such densities are known to exist conditionally under GRH, see [JP23, JPS25], because the condition on the norm is equivalent to the Frobenius automorphism at $\mathfrak{p}$ for the abelian field extension $K(\zeta_d)/K$ satisfying $\zeta_d \mapsto \zeta_d^a$.

Finally, we denote by $K(\zeta_n, G^{1/n})$ the cyclotomic-Kummer extension obtained by adjoining all n-th roots of the elements of G (equivalently, we may add the n-th roots of a set of generators for G). The above density notations will be used also for G instead of α.

3.2 Dirichlet Density in Extensions

Let Γ_K denote the set of primes of K of degree 1. The following result generalizes [Per18, Proposition 1] and it is unconditional:

Lemma 3.1 *Let L/K be an extension of number fields with Galois closure $\tilde{L}/K$, and let $\Gamma_{L,K} \subseteq \Gamma_K$ consist of the primes that split completely in L. Consider $S \subseteq \Gamma_{L,K}$ and let S_L be the primes of L lying over the primes of S. Then S has a Dirichlet density if and only if the same holds for S_L. In this case, we have*

$$\mathrm{dens}_L(S_L) = [\tilde{L} : K]\,\mathrm{dens}_K(S)\,.$$

Proof Call $\mathcal{P}_K$ the set of primes of K. Since $\mathrm{dens}_L(\Gamma_L) = 1$, by the definition of Dirichlet density we have

$$\mathrm{dens}_K(S) := \lim_{s \to 1^+} \frac{\sum_{\mathfrak{p} \in S} \mathrm{N}(\mathfrak{p})^{-s}}{\sum_{\mathfrak{p} \in \mathcal{P}_K} \mathrm{N}(\mathfrak{p})^{-s}}, \quad \mathrm{dens}_L(S_L) := \lim_{s \to 1^+} \frac{\sum_{\mathfrak{q} \in S_L} \mathrm{N}(\mathfrak{q})^{-s}}{\sum_{\mathfrak{q} \in \Gamma_L} \mathrm{N}(\mathfrak{q})^{-s}}\,.$$

We have

$$\sum_{\mathfrak{q}\in S_L} \mathrm{N}(\mathfrak{q})^{-s} = [L:K]\sum_{\mathfrak{p}\in S} \mathrm{N}(\mathfrak{p})^{-s}$$

and we may analogously relate the two sums over $\mathfrak{q} \in \Gamma_L$ and $\mathfrak{p} \in \Gamma_{L,K}$. We may conclude because we have

$$\frac{1}{[\tilde{L}:K]} = \lim_{s\to 1^+} \frac{\sum_{\mathfrak{p}\in\Gamma_{L,K}} \mathrm{N}(\mathfrak{p})^{-s}}{\sum_{\mathfrak{p}\in\mathcal{P}_K} \mathrm{N}(\mathfrak{p})^{-s}}$$

by the Chebotarev density theorem applied to $\tilde{L}$ taking into account that the primes $\mathfrak{p}$ of K that split completely in L also split completely in $\tilde{L}$. Indeed, the reductions of all roots of the minimal polynomial of a primitive element of L belong to the residue field at $\mathfrak{p}$ because the polynomial splits into linear factors modulo $\mathfrak{p}$. □

4 Primes in Arithmetic Progression

In this section we assume GRH. We consider primes of K whose norm lies in a given arithmetic progression and such that the index of the reduction of G belongs to a given set of positive integers. We also require a Frobenius condition with respect to a given finite Galois extension of K.

Theorem 4.1 *Let F/K be a finite Galois extension of number fields, and let C be a conjugacy-stable subset of* $\mathrm{Gal}(F/K)$. *Let G be a finitely generated and torsion-free subgroup of $K^\times$ of positive rank r. Let S be a non-empty set of positive integers, let $1 \leqslant a < d$ be coprime integers. Consider the set of primes of K given by*

$$\mathcal{P}_S(a,d) := \left\{\mathfrak{p} : \mathrm{N}\,\mathfrak{p} \leqslant x,\ \mathrm{N}\,\mathfrak{p} \equiv a \bmod d,\ \mathrm{ind}_{\mathfrak{p}}(G) \in S,\ \left(\frac{\mathfrak{p}}{F/K}\right) \subseteq C\right\},$$

where $\mathfrak{p}$ varies through the primes of K unramified in F and such that $v_{\mathfrak{p}}(g) = 0$ for all $g \in G$. Then we have

$$\pi_{\mathcal{P}_S(a,d)}(x) = \frac{x}{\log x}\sum_{t\in S}\sum_{n=1}^{\infty} \frac{\mu(n)c(nt)}{[F(\zeta_{[d,nt]}, G^{1/nt}):K]} + O\left(\frac{x}{(\log x)^{2-\frac{1}{r+1}}}\right),$$

where

$$c(m) := |\{\sigma \in \mathrm{Gal}(F(\zeta_{[d,m]}, G^{1/m})/K) : \sigma|_{K(\zeta_m, G^{1/m})} = \mathrm{id},\ \sigma(\zeta_d) = \zeta_d^a,\ \sigma|_F \in C\}|.$$

For $S = \{t\}$ the double sum is just one sum and when $x \geqslant t^3$ we have the better error term

$$O\left(\frac{x}{\log^2 x} + \frac{x \log\log x}{\varphi(t) \log^2 x}\right),$$

where φ is Euler's totient function

Notice that for $F = K$ the coefficient $c(m)$ is 1 if the automorphism induced by $\zeta_d \mapsto \zeta_d^a$ is the identity on the intersection $K(\zeta_m, G^{1/m}) \cap \mathbb{Q}(\zeta_d)$, and it is 0 otherwise. Theorem 4.1 is a combination of two standard conditions which have been studied extensively, namely, the index having prescribed value and the primes being in an arithmetic progression. For $K = \mathbb{Q}$ and G of rank 1 and $S = \{t\}$, Theorem 4.1 with $t = 1$ is stated in Moree's survey [Mor12, Theorem 1], and for general t in [HKMS19, Theorem 3.1].

Proof of Theorem 4.1 We may suppose that the primes $\mathfrak{p}$ that are being counted are of degree 1, as this restriction excludes only $O(\sqrt{x}/\log x)$ primes. The conditions $\mathrm{N}\,\mathfrak{p} \equiv a \bmod d$ and $(\mathfrak{p}, F/K) \subseteq C$ are equivalent to $(\mathfrak{p}, F(\zeta_d)/K) \subseteq C_{a,d}$, where

$$C_{a,d} := \{\sigma \in \mathrm{Gal}(F(\zeta_d)/K) : \sigma(\zeta_d) = \zeta_d^a, \sigma|_F \subseteq C\}.$$

Then for the second assertion it suffices to apply [PS19, Proposition 5.1] with $F' = F(\zeta_d)$ and $C' = C_{a,d}$. For the first assertion we apply [Mps24, Theorem 1] in a similar way. □

Lemma 4.2 *If m is an odd positive integer, then we have*

$$\mathrm{dens}(G^m) = \mathrm{dens}(G) - \mathrm{dens}(G, S_1 \bmod m)$$

where S_1 is the set of integers that are congruent to 1 modulo some prime divisor of m. Consequently, we have $\mathrm{dens}(G^m) = \mathrm{dens}(G^{\mathrm{sqf}(m)})$, where $\mathrm{sqf}(m)$ denotes the squarefree part of m.

Proof Let $\mathfrak{p}$ be a prime of K. The reduction ($G^m \bmod \mathfrak{p}$) has index 1 if and only if ($G \bmod \mathfrak{p}$) has index 1 and for every prime $q \mid m$ we have $\mathrm{N}\,\mathfrak{p} \not\equiv 1 \bmod q$. □

Remark 4.3 By Lemma 4.2 we have $\mathrm{dens}(G) = \mathrm{dens}(G^m)$ if and only if $\mathrm{dens}(G) = \mathrm{dens}(G^{\mathrm{sqf}(m)})$. Consequently, the condition $\mathrm{dens}(G) = \mathrm{dens}(G^m)$ implies that $\mathrm{dens}(G) = \mathrm{dens}(G^{m^n})$ holds for every $n \geqslant 1$.

Example 4.4 We have $\mathrm{dens}(G^3) = \mathrm{dens}(G) - \mathrm{dens}(G, 1 \bmod 3) = \mathrm{dens}(G, 2 \bmod 3)$.

Proposition 4.5 *If m is a squarefree integer, then we have*

$$\mathrm{dens}(G, 1 \bmod m) = \sum_{n \mid m} \mu(n)\, \mathrm{dens}(G^n).$$

Proof Let $\mathfrak{p}$ be a prime of K such that (G mod $\mathfrak{p}$) has index 1. The condition $\mathrm{N}\,\mathfrak{p} \equiv 1 \bmod m$ is equivalent to the assertion that for all primes $q \mid m$ the index of (G^q mod $\mathfrak{p}$) is not 1. Given two distinct prime numbers q, q', we have that both (G^q mod $\mathfrak{p}$) and ($G^{q'}$ mod $\mathfrak{p}$) have index 1 if and only if ($G^{qq'}$ mod $\mathfrak{p}$) has index 1, and the same holds for finitely many prime numbers. Hence, applying the inclusion–exclusion principle over the prime factors of m yields the formula in the statement. □

Let L/K be a Galois extension of number fields. Consider a prime $\mathfrak{p}$ of K such that the index of (G mod $\mathfrak{p}$) is well defined. If $\mathfrak{P}$ is a prime of L over $\mathfrak{p}$, then the index of (G mod $\mathfrak{P}$) is well defined and it is a multiple of the index of (G mod $\mathfrak{p}$). The two indices are the same if and only if $\mathfrak{p}$ splits completely in L.

Theorem 4.6 *For every set S of positive integers and for every positive integer m, the following holds:*

$$\mathrm{dens}_K(\mathfrak{p} : \mathrm{ind}_{\mathfrak{p}}(G) \in S(\mathfrak{p}, L/K) = \mathrm{id}) = \frac{1}{[L : K]} \cdot \mathrm{dens}_L(\mathfrak{P} : \mathrm{ind}_{\mathfrak{P}}(G) \in S);$$

$$\mathrm{dens}_K(\mathfrak{p} : \mathrm{ind}_{\mathfrak{p}}(G) \in S, \mathrm{N}\,\mathfrak{p} \equiv 1 \bmod m) = \frac{1}{[K(\zeta_m) : K]} \mathrm{dens}_{K(\zeta_m)}(\mathfrak{P} : \mathrm{ind}_{\mathfrak{P}}(G) \in S);$$

$$\mathrm{dens}_K(G, 1 \bmod m) = \frac{1}{[K(\zeta_m) : K]} \cdot \mathrm{dens}_{K(\zeta_m)}(G)\,.$$

Proof The first equality is a consequence of Lemma 3.1, the second equality is a special case of the first (taking $L = K(\zeta_m)$), and the third equality is a special case of the second (taking $S = \{1\}$). □

Remark 4.7 For every positive even integer m, G^m consists of squares and hence $\mathrm{dens}(G^m) = 0$. If m is an odd positive integer, then we always have

$$\mathrm{dens}(G^m) \leqslant \mathrm{dens}(G)$$

because G^m is a subgroup of G.

The two densities $\mathrm{dens}(G^m)$ and $\mathrm{dens}(G)$ may be the same though, even if m is prime number and $G \neq W^m$ holds for any subgroup W of $K^\times$:

Example 4.8 For $K = \mathbb{Q}$ and $\alpha = -3$ we have $\mathrm{dens}(\alpha) = \mathrm{dens}(\alpha^3)$. Indeed, α is a square in $K(\zeta_3)$, hence $\mathrm{dens}_{K(\zeta_3)}(\alpha) = 0$. By Theorem 4.6 we deduce that $\mathrm{dens}(\alpha, 1 \bmod 3) = 0$, and we may conclude by Lemma 4.2.

Lemma 4.9 *For all positive integers t, m, M with $m \mid M$, the following holds:*

(i) We have $\mathrm{dens}(G^M) = \mathrm{dens}(G^m)$ *if and only if for every prime $\ell \mid M$ we have* $\mathrm{dens}(G^m, 1 \bmod \ell) = 0$ *(this clearly holds if $\ell \mid m$).*
(ii) We have $\mathrm{dens}(G^M, t) = \mathrm{dens}(G^m, t)$ *if and only if for every prime $\ell \mid \frac{M}{m}$ we have* $\mathrm{dens}(G^m, t, 1 \bmod t\ell) = 0$.

Proof We may restrict to the primes $\mathfrak{p}$ of K of degree 1 (as the remaining primes have density 0) and we write $p = \mathrm{N}(\mathfrak{p})$ for the underlying rational prime.

(i) Suppose that ($G^m \bmod \mathfrak{p}$) has index 1. Then ($G^M \bmod \mathfrak{p}$) also has index 1 if and only if for every prime divisor ℓ of M/m the integer $\mathrm{N}(\mathfrak{p}) - 1$ is not a multiple of ℓ, which amounts to saying that $p \not\equiv 1 \bmod \ell$.
(ii) Suppose that ($G^m \bmod \mathfrak{p}$) has index t or, equivalently, order $(p-1)/t$. Then ($G^M \bmod \mathfrak{p}$) has index t or, equivalently, order $(p-1)/t$ if and only if each prime divisor ℓ of M/m does not divide $(p-1)/t$, which means $p \not\equiv 1 \bmod t\ell$. □

Theorem 4.10 *Let m, t be positive integers, with m odd and squarefree. Then the following are equivalent:*

(i) We have $\mathrm{dens}(G, t) = \mathrm{dens}(G^m, t)$.
(ii) For every prime $\ell \mid m$, we have $\mathrm{dens}(G, t) = \mathrm{dens}(G^\ell, t)$.
(iii) For every prime $\ell \mid m$, we have $\mathrm{dens}(G, t, 1 \bmod t\ell) = 0$.

Proof The equivalence of (ii) and (iii) can be obtained by fixing ℓ, hence it is a consequence of Lemma 4.9(ii) applied to G and G^ℓ. Lemma 4.9(ii), applied to G and G^m, gives the equivalence of (i) and (iii). □

5 Comparing General Artin Densities for Powers

In this section we assume GRH. We write ℓ to denote prime numbers. We consider a set of positive integers H and study the condition $\mathrm{ind}_\mathfrak{p}(G) \in H$ using definitions and results from [JPS25]. For a squarefree integer $Q > 1$ we consider the Q-adic valuation $v_Q : \mathbb{Z}_{>0} \to \prod_{\ell \mid Q} \mathbb{Z}_{\geqslant 0}$ mapping m to the tuple $(v_\ell(m))_{\ell \mid Q}$, and we denote by H_Q the preimage under v_Q of $v_Q(H)$. We say that H is *cut by valuations* if $H = \bigcap_\ell H_\ell$, and we set $V_\ell := v_\ell(H)$. Then, for $r \geqslant 1$, we define the following constants:

$$F_{v,r}(\ell) = \begin{cases} 1 - \frac{1}{\ell^r(\ell-1)} & \text{for } v = 0 \\ \frac{1}{\ell^{v(r+1)}} \cdot \frac{\ell}{(\ell-1)}(1 - \frac{1}{\ell^{r+1}}) & \text{for } v > 0, \end{cases}$$

and

$$A_{V_\ell,r}(\ell) := \sum_{v \in V_\ell} F_{v,r}(\ell), \qquad A_{V,r} := \prod_\ell A_{V_\ell,r}(\ell)\,.$$

We denote by μ_{Haar} the normalized Haar measure, and by r the rank of G. By [JPS25, Theorem 24] there is a (squarefree) integer $B \geqslant 1$ which depends only on K and G such that

$$\mathrm{dens}(\mathfrak{p} : \mathrm{ind}_\mathfrak{p}(G) \in H) = A_{V,r} \cdot \mu_{\mathrm{Haar}}(C_{H,B}(G)) \cdot \prod_{\ell \mid B} A_{V_\ell,r}(\ell)^{-1}, \tag{5.1}$$

where

$$C_{H,B}(G) := \bigcup_{h \in H_B} C_h(G) = \bigsqcup_{\substack{h \in H_B \\ h \mid B^\infty}} C_h(G) \subseteq \mathrm{Gal}(K_{B^\infty, B^\infty}/K) \tag{5.2}$$

and $C_h(G)$ is the conjugacy-stable set of those K-automorphisms which fix $K(\zeta_h, G^{1/h})$, and for all primes $q \mid B$ do not fix the field $K(\zeta_{qh}, G^{1/qh})$. The extension K_{B^∞,B^∞} denotes the union of all extensions $K(\zeta_{B^e}, G^{1/B^e})$ with $e \geqslant 1$. Notice that these definitions do not change if B is not squarefree.

We study how the above density changes if we replace G with G^m for some integer $m \geqslant 1$.

Remark 5.1 It follows from [JPS25, Pst21] that B depends on G only through its parameters of ℓ-divisibility and its adelic failure (i.e., whether roots of its elements lie in a cyclotomic extension of K). Therefore, considering G^m instead of G only affects its divisibility property, and not the entanglement of the radical extensions with the cyclotomic extensions. In practice, since B is divisible by all primes ℓ for which the parameters of ℓ-divisibility are not all zero, for G^m it suffices to replace B with Bm. Thus, we have

$$\frac{\mathrm{dens}(\mathfrak{p} : \mathrm{ind}_\mathfrak{p}(G) \in H)}{\mathrm{dens}(\mathfrak{p} : \mathrm{ind}_\mathfrak{p}(G^m) \in H)} = \frac{\mu_{\mathrm{Haar}}(C_{H,B}(G))}{\mu_{\mathrm{Haar}}(C_{H,Bm}(G^m))} \cdot \prod_{\ell \mid m, \ell \nmid B} A_{V_\ell,r}(\ell).$$

Theorem 5.2 *Suppose that m and B are coprime. We then have*

$$\mathrm{dens}(\mathfrak{p} : \mathrm{ind}_\mathfrak{p}(G^m) \in H) = A_{V,r} \cdot \mu_{\mathrm{Haar}}(C_{H,B}(G)) \cdot \prod_{\ell \mid m} A^{(m)}_{V_\ell,r}(\ell) \cdot \prod_{\ell \mid Bm} A_{V_\ell,r}(\ell)^{-1}, \tag{5.3}$$

where for $\ell \mid m$ we set

$$A^{(m)}_{V_\ell,r}(\ell) := \chi_{V_\ell}(0) \cdot \frac{\ell-2}{\ell-1} + \sum_{\substack{v \in V_\ell \\ 0 < v < v_\ell(m)}} \frac{1}{\ell^v} + \ell^{v_\ell(m) r} \cdot \sum_{\substack{v \in V_\ell \\ v \geqslant v_\ell(m)}} F_{v,r}(\ell),$$

where $\chi_{V_\ell}(0)$ is 1 if $0 \in V_\ell$, and it is 0 otherwise.

Proof We start with the formula given in (5.1) with G and B replaced by G^m and Bm, respectively. By [JP23, Proposition 3.1(iii)] for every integer t and any prime $\ell \mid m$ coprime to t (by assumption ℓ is then coprime to B), we have

$$K(\zeta_{\ell^{v_\ell(m)}}, G^{m/\ell^{v_\ell(m)}}) \cap K(\zeta_t, G^{m/t}) = K,$$

as the former field on the left-hand side equals $K(\zeta_{\ell^{v_\ell(m)}})$ and the latter is contained in $K(\zeta_t, G^{1/t})$. We deduce that the conditions on the K-automorphisms defining

$C_{H,B}(G^m)$ are independent from those defining $C_{H,\ell}(G^m)$ for $\ell \mid m$. Moreover, since $\gcd(B, m) = 1$, for every $v \geqslant 1$ we have $K(\zeta_{B^v}, G^{m/B^v}) = K(\zeta_{B^v}, G^{1/B^v})$. Hence, we obtain

$$C_{H,Bm}(G^m) = C_{H,B}(G) \times \bigoplus_{\ell \mid m} C_{H,\ell}(G^m).$$

This decomposition yields the second factor in (5.3), and it remains to justify the third factor in the formula.

By definition, for $\ell \mid m$, the set $C_{H,\ell}(G^m)$ is the disjoint union of the conjugacy classes $C_{\ell^v}(G^m)$ for $v \geqslant 0$ with $\ell^v \in H_\ell$. Recall that $C_{\ell^v}(G^m)$ consists of the K-automorphisms which fix $K(\zeta_{\ell^v}, G^{m/\ell^v})$ and do not fix $K(\zeta_{\ell^{v+1}}, G^{m/\ell^{v+1}})$. Hence, $C_{\ell^v}(G^m)$ has Haar measure equal to

$$\begin{cases} \frac{\ell-2}{\ell-1} & \text{if } v = 0; \\ \frac{1}{\ell^v} & \text{if } 0 < v < v_\ell(m); \\ \ell^{v_\ell(m) r} F_{v,r}(\ell) & \text{if } v \geqslant v_\ell(m). \end{cases}$$

Taking the Haar measure of the direct sum of the sets $C_{H,\ell}(G^m)$ for $\ell \mid m$ completes the proof. □

Corollary 5.3 *For m coprime with B we have*

$$\frac{\operatorname{dens}(\mathfrak{p} : \operatorname{ind}_{\mathfrak{p}}(G) \in H)}{\operatorname{dens}(\mathfrak{p} : \operatorname{ind}_{\mathfrak{p}}(G^m) \in H)} = \prod_{\ell \mid m} A_{V_\ell,r}(\ell) / A^{(m)}_{V_\ell,r}(\ell).$$

Proof This is an immediate consequence of Theorem 5.2. □

For the special case $\operatorname{ind}_{\mathfrak{p}}(G) = t$, we consider the r-rank Artin constant

$$A_r := \prod_\ell \left(1 - \frac{1}{\ell^r(\ell - 1)}\right).$$

Theorem 5.4 *The density* $\operatorname{dens}(G, t)$ *equals*

$$\frac{A_r}{t^{r+1}} \cdot \mu_{\mathrm{Haar}}(C_{(t,B^\infty)}(G)) \cdot \prod_{\ell \mid t} \frac{\ell^{r+1} - 1}{\ell^{r+1} - \ell^r - 1} \prod_{\ell \mid (t,B)} \frac{\ell^{(r+1)v_\ell(t)+r}(\ell - 1)}{\ell^{r+1} - 1} \prod_{\ell \mid B, \ell \nmid t} \frac{\ell^r(\ell - 1)}{\ell^{r+1} - \ell^r - 1},$$

where $(a, b) = \gcd(a, b)$.

Proof It suffices to notice that the set $H = \{t\}$ is trivially cut by valuations (by taking $V_\ell = \{v_\ell(t)\}$ for every ℓ) and to apply (5.1). In particular, we obtain

$$A_{V,r} = \frac{A_r}{t^{r+1}} \cdot \prod_{\ell \mid t} \frac{\ell^{r+1} - 1}{\ell^{r+1} - \ell^r - 1},$$

and $C_{H,B}(G) = C_{(t,B^\infty)}(G)$ by the definition in (5.2). □

Corollary 5.5 *For every odd positive integer m we have*

$$\frac{\mathrm{dens}(G^m,t)}{\mathrm{dens}(G,t)}=\frac{\mu_{\mathrm{Haar}}(C_{(t,(Bm)^\infty)}(G^m))}{\mu_{\mathrm{Haar}}(C_{(t,B^\infty)}(G))}\cdot\prod_{\ell\mid(t,m),\ell\nmid B}\frac{\ell^{(r+1)v_\ell(t)+r}(\ell-1)}{\ell^{r+1}-1}\prod_{\ell\mid m,\ell\nmid Bt}\frac{\ell^r(\ell-1)}{\ell^{r+1}-\ell^r-1}.$$

Proof This is an immediate consequence of Theorem 5.4 and Remark 5.1. □

Corollary 5.6 *For t coprime with B we have*

$$\mathrm{dens}(G,t)=\frac{A_r}{t^{r+1}}\cdot\mu_{\mathrm{Haar}}(C_1(G))\cdot\prod_{\ell\mid t}\frac{\ell^{r+1}-1}{\ell^{r+1}-\ell^r-1}\prod_{\ell\mid B}\frac{\ell^r(\ell-1)}{\ell^{r+1}-\ell^r-1}.$$

Proof The formula of Theorem 5.4 simplifies straightforwardly in the given special case. □

Remark 5.7 Suppose that m and B are coprime, and consider the ratio $\mathrm{dens}(G^m,t)/\mathrm{dens}(G,t)$, which we know how to express by Corollary 5.5. Following the arguments in the proof of Theorem 5.2 we have

$$\frac{\mu_{\mathrm{Haar}}(C_{(t,(Bm)^\infty)}(G^m))}{\mu_{\mathrm{Haar}}(C_{(t,B^\infty)}(G))}=\prod_{\substack{\ell\mid m\\ \ell\nmid t}}\frac{\ell-2}{\ell-1}\prod_{\substack{\ell\mid(m,t)\\ v_\ell(t/m)<0}}\frac{1}{\ell^{v_\ell(t)}}\prod_{\substack{\ell\mid(m,t)\\ v_\ell(t/m)\geqslant 0}}F_{v_\ell(t),r}(\ell).$$

References

[Hoo67] C. Hooley, *Artin's conjecture for primitive roots*, J. Reine Angew. Math. **225** (1967), 209–220.

[HKMS19] S. Hu, M.-S. Kim, P. Moree and M. Sha, *Irregular primes with respect to Genocchi numbers and Artin's primitive root conjecture*, J. Number Theory **205** (2019), 59–80.

[JP23] O. Järviniemi and A. Perucca, *Unified treatment of Artin-type problems*, Res. Number Theory **9** (2023), no. 10.

[JPS25] O. Järviniemi, A. Perucca and P. Sgobba, *Unified treatment of Artin-type problems II*, Res. Number Theory **11**, 42 (2025).

[Len77] H.W. Jr. Lenstra, *On Artin's conjecture and Euclid's algorithm in global fields*, Invent. Math. **42** (1977), 201–224.

[LMS14] H.W. Jr. Lenstra, P. Moree and P. Stevenhagen, *Character sums for primitive root densities*, Math. Proc. Cambridge Philos. Soc. **157** (2014), no. 3, 489–511.

[Mor99] P. Moree, *On primes in arithmetic progression having a prescribed primitive root*, J. Number Theory **78** (1999), 85–98.

[Mor08] P. Moree, *On primes in arithmetic progression having a prescribed primitive root. II*, Funct. Approx. Comment. Math. **39** (2008), 133–144.

[Mor12] P. Moree, *Artin's primitive root conjecture – a survey*, Integers **12** (2012), no. 6, 1305–1416.

[Mps24] P. Moree, A. Perucca and P. Sgobba. *The distribution of the multiplicative index of algebraic numbers over residue classes*, Abh. Math. Semin. Univ. Hambg. **94** (2024), 1–17.

[Per18] A. Perucca, *Reductions of algebraic integers II*, in Women in numbers Europe II, 19–33, Assoc. Women Math. Ser., 11, Springer, Cham, 2018.

[PS19] A. Perucca and P. Sgobba, *Kummer theory for number fields and the reductions of algebraic numbers*, Int. J. Number Theory, **15** (2019), no. 8, 1617–1633.

[Pst21] A. Perucca, P. Sgobba and S. Tronto, *The degree of Kummer extensions of number fields*, Int. J. Number Theory, **15** (2021), no. 5, 1091–1110.

[Zie06] V. Ziegler, *On the distribution of the order of number field elements modulo prime ideals*, Unif. Distrib. Theory **1** (2006), no. 1, 65–85.